Science and Technology
in the Industrial Revolution

A E Musson & Eric Robinson

Science and Technology
in the Industrial Revolution

Manchester University Press

Published by the University of Manchester
at the University Press
316–324 Oxford Road, Manchester 13

GB SBN 7190 0370 9

C

Printed in Great Britain by
Butler & Tanner Ltd Frome and London

Contents

Abbreviations

Ann. Chim.   *Annales de Chimie*
A.O.L.B.   Assay Office Library, Birmingham
B.M.   British Museum
B.R.L.   Birmingham Reference Library
D.N.B.   *Dictionary of National Biography*
Dol.   Doldowlod Papers
L.C.R.O.   Lancashire County Record Office
Phil. Trans. *Philosophical Transactions, Royal Society*
P.P.   Parliamentary Papers

Preface

The Industrial Revolution in Great Britain has probably attracted more attention than any other historical period or phenomenon. Students of economic growth, of business history, and of social revolution have long delved in this rich and fascinating field. Yet new veins continue to be discovered, as new areas are surveyed and new drillings undertaken. This book is the outcome of such research activities intermittently spread over more than ten years. Several samples of our work have, indeed, already been displayed in previously published articles, in a variety of academic journals; a few are reproduced here with comparatively little alteration, but others have been greatly amended and extended, sometimes to several times their original length. At the same time, we have included half a dozen entirely new chapters. The whole work is largely based on researches into hitherto unused or little-used sources, and the results, we hope, will provide new insights into the Industrial Revolution.

There are two main questions which we have always had in front of us and which we have sought to answer. What were the connections between the Scientific and the Industrial Revolutions? And how was technological knowledge developed and diffused? Our findings necessitate, in our opinion, considerable modification of the traditional view of the Industrial Revolution as being almost entirely a product of uneducated empiricism, and expand the suggestion thrown out twenty years ago by Professor Ashton that developments in science and in technology during the eighteenth century were not unrelated and that the Industrial Revolution was also an intellectual movement. At the same time, we have also found plentiful evidence of the continued importance of practical craftsmanship, so that we certainly would not wish to present the Industrial Revolution as simply a product of the Scientific Revolution. The situation was, in fact, immensely complicated and it is often difficult for modern researchers to distinguish, in the surviving evidence, between intelligent empiricism and applied science—if, indeed such a distinction would be valid in that period.

This theme of the interrelationship between science and technology runs through all the chapters of our book. At the same time we have also been greatly interested in the ways by which technological developments were diffused, especially in the engineering and chemicals industries. The field is, of course, far too vast for us to survey completely in a single volume, and we have therefore had to produce a collection of studies, rather than a unified monograph, but we hope that readers will find a basic unity in the whole.

We wish to express our thanks for permission to utilize articles previously published in the *Economic History Review, Journal of Economic History, Annals of Science*, and *Business History*. We also wish to acknowledge our debts to many libraries which have given us invaluable assistance in our researches. In all cases we hope that we have made proper acknowledgments, but we wish here particularly to express our appreciation of the unstinted help provided by the Municipal Reference Libraries of Manchester and Birmingham, the Assay Office Library in Birmingham, Chetham's Library in Manchester, and our own University Library. Various County Record Offices, especially that at Preston, have also helped us, and we have, of course, drawn on the vast resources of the British Museum in London. Many individuals, too, have aided us with references and suggestions, as will be apparent in our footnotes. Above all, we are grateful to Major David Gibson-Watt, M.C., M.P., for allowing us access to his family papers at Doldowlod. We also wish to express our warm thanks to Miss Pauline Higgins, History Department secretary in the University of Manchester, and also to Mrs. Anne Musson, for all their labours in typing our often difficult manuscripts.

University of Manchester   A. E. MUSSON
November 1968   ERIC ROBINSON

Introduction

The danger that historical research may be obscured by wrangles in semantics has increased in the last quarter-century to a point where the search for the right word to describe what has happened begins to take precedence over the search for the past itself. This has been specially true of the debate over the word 'science' when used by historians to describe 'The Scientific Revolution' or the place of 'science' in the development of the Industrial Revolution. Are the manifold interests of men of business and practical craftsmen of the eighteenth century to be called 'scientific', or are they better described as 'practical, un-regulated empiricism', or by some similar phrase which is taken to represent an alternative, or even an antithesis, to 'science' and to 'the scientific method'? We have already been criticized for using the word 'science' in reference to the industrial changes of the late eighteenth century.[1] Professor R. E. Schofield's recent work on *The Lunar Society of Birmingham* (1963) has been subjected to similar attacks,[2] and the Clows have not gone unscathed for their use of the word 'scientific' in their classic work, *The Chemical Revolution*. First of all, a difficulty arises from the difference between eighteenth- and twentieth-century terminology. The word 'science' was not that generally used in the eighteenth century, when the common term was 'natural philosophy'. This in turn leads to confusion over the words 'philosophy' and 'philosopher' where we, in the twentieth century, would normally speak of 'science' and 'scientist'. Similarly, though the word 'chymistry' was widely used, 'physics' was generally subsumed in the term 'natural philosophy'. It is in fact more usual to encounter in eighteenth-century

---

[1] By Dr. D. W. F. Hardie, for example, at the annual conference of the Economic History Society in Liverpool, April 1960.

[2] See Dr. Hardie's review of Schofield in *Business History*, Vol. VIII, no. 1 (Jan. 1966).

contexts the names of departments or sub-divisions of physics such as 'mechanics', 'hydraulics', 'pneumatics', 'optics', etc., rather than the inclusive words 'philosophy' or 'natural philosophy', probably because this was a period when men were aware that specialization was advancing and that the classification of the sciences was proceeding apace. The modern historian will thus tend to use the words 'science' or 'scientific' to describe situations in which the 'philosophers' of the eighteenth century would have required him to be more limited and exact.

If, however, a general equivalence can be accepted between the terms 'natural philosophy' and 'science', were the natural philosophers of the eighteenth century mistaken in the view that they took of their own activities? Were they dabbling in 'non-science' while believing that they were concerned with 'science'? In our opinion the historian is not entitled to dismiss out of hand the notions that people of the past had about themselves. Human activities are affected by the way in which human beings contemplate their own behaviour. If I think of myself as working within the structure of a rational and organized body of knowledge, I shall behave differently from the situation in which I conceive of myself as operating practically and empirically, even though I may be mistaken about the degree of coherence and rationality of the system in which I think of myself, in the former instance, as operating. Thus even 'alchemy' will result in an activity which can properly be described as 'alchemical', and will impose a certain order and regularity of terminology on operations which, in the view of modern scientists, may appear to be haphazard and empirical. In the same way, by the acceptance of a common terminology, the 'natural philosophers' of the eighteenth century made it possible for debate to take place about the experiments that they carried out and the phenomena that they observed. It is true, of course, that they also entered into debate about the scientific language they themselves employed and, in chemistry for example, most of them came to accept to a greater or lesser extent the new Lavoisierian nomenclature. But it is also true that the same process is continually taking place among modern scientists. Language is always metaphorical, and the scientist changes his language, his metaphors, as he alters his hypotheses. A term, or metaphor, such as 'phlogiston' may be replaced in the course of time, but during its useful life may make possible the advancement of scientific knowledge concerning the constitution and properties of newly discovered gases.[1] It is not helpful to dismiss as 'merely empirical'

[1] See S. E. Toulmin, 'Crucial Experiments: Priestley and Lavoisier', *Journal of the History of Ideas*, Vol. XVIII, no. 2 (April 1957), pp. 205-20; A. R. Hall, *The Scientific Revolution, 1500-1800* (1954), pp. 326-31.

the outworn metaphors of scientific debate in earlier historical periods than our own.

It may be further argued, however, that there is an important difference between the work of acknowledged 'scientists' of the eighteenth century such as Priestley or Cavendish and the operations of a Watt, a Wedgwood, or a Boulton. If so, one can only say that it is a distinction that Priestley and Cavendish themselves did not make, or at least did not make in such a way as to invalidate as 'natural philosophy' many of the experiments carried out by their industrialist friends who were also, it should be remembered, their colleagues in the Royal Society and other scientific clubs. Watt's experiments on the quantification of latent heat were judged by Joseph Black to be of sufficient importance to qualify Watt to be considered as Black's scientific collaborator; Wedgwood's work on pyrometry was generally considered by eighteenth-century natural philosophers to be of prime importance; Boulton's investigations into electricity obtained recognition by Benjamin Franklin, John Michell and others as early as the 1750s. Priestley and Cavendish would certainly not have thought of these men as 'non-scientists': all of them, virtuosi and industrialists alike, were, in their own views, partners and collaborators in the task of extending the frontiers of 'natural philosophy'. Similarly, though it is now realized that the chemistry of such processes as brewing, dyeing, and glazing pottery were not adequately understood by the industrial experimenters of the eighteenth century, yet it is undeniable that later theories would not have been possible without the body of detailed observations made by those actively concerned with those processes at an earlier date.

Science and empiricism are often regarded as opposites, but what foundation is there for this view? Modern science is a combination of theory and experiment: it has been developed, not in ivory towers, but in laboratories. It has left behind it a trail of exploded hypotheses: theories, found wanting by the test of practical experiment, have been successively replaced by revised ones. If by empiricism we mean procedure by trial and error, then the progress of science has been empirical.[1] As the great eighteenth-century Swedish chemist, Törbern

[1] Sir George Clark has pointed out that 'the experimental method'—the basis of modern science—appears to have been taken over from the mechanical arts in the sixteenth and seventeenth centuries. Bacon and Hooke, for instance, derived much knowledge from contacts with skilled craftsmen. Trial and error thus became the foundation of scientific as of industrial advance. G. N. Clark, *Science and Social Welfare in the Age of Newton* (2nd edn., 1949), pp. 76–7. See also J. Bronowski and B. Mazlish, *The Western Intellectual Tradition* (Penguin edn., 1963), p. 54, where it is stated that 'the essential point about modern scientific method [is] its combined use of empiricism and rationalism'.

Bergman, put it, 'We have no knowledge of bodies *a priori*: Every intelligence about them, must be acquired by proper observations and Experiments.'[1] Scientific empiricism, of course, has not been blind or irrational: practical experimentation has been directed by theory, while theory has been shaped by experiment. As soon as men begin to use guiding principles or hypotheses—which may have been arrived at inductively from systematic observation, rather than worked out deductively from so-called first principles—they are beginning to be 'scientific'. And the test of a hypothesis is not so much whether or not it is true, but whether or not it is fruitful, whether or not it is a help to understanding, and (especially in applied science and technology) whether or not it provides a generally reliable working rule.

Indeed many eminent scientists of today are discarding as unfruitful the nineteenth-century distinction between pure and applied science, recognizing that advances in science are most rapidly made where the phenomena of natural life are continuously under the most careful scrutiny. Moreover, the relationship between improved technology and scientific discovery is likely to be most fruitful where practical objectives are aimed at, and this may help to explain the impetus often given to scientific discovery by war. Dr. W. Grey Walter in his book, *The Living Brain*, gives one interesting example of the new scientific applications that may arise from the solution of a practical problem:

The first steam engine, left to itself, was unstable—pressure went down when power was used and boiler blew up when it was not used. Watt introduced the safety valve and automatic governor which stabilize by themselves both boiler pressure and engine speed. These two important devices were taken rather as a matter of course by engineers, but the great Clerk Maxwell devoted a paper to the analysis of Watt's governor. Maxwell was perhaps the first to realize the significance of this key process of feedback. Later, physiologists pointed out that Watt had incidentally constructed in principle the first working model of a reflex circuit similar to what they describe in the organization of sense organs, nerves, and muscles.'[2]

What is described here is the transference of a metaphor used in the solution of a practical engineering problem to the solution of a problem in physiology.

In eras of creative science, practical experimentation and theory have always cross-fertilized in this way. The fact that many erroneous theories prevailed two or three centuries ago, and that laboratory

[1] T. Bergman, Introduction to the *Chemical Observations and Experiments on Air and Fire* by C. W. Scheele (English trans., 1780), p. xvii.
[2] W. Grey Walter, *The Living Brain* (Penguin, 1961), p. 105.

equipment and procedures were rudimentary by modern standards, does not justify rejection of eighteenth-century natural philosophy as being 'unscientific'. At that time those theories were a help to the understanding of natural processes, and a basis for the eventual elaboration of more satisfactory explanations. Moreover, the fact that much of that science was not 'pure'; that it owed, indeed, as much to industry as did industry to science; that activities in laboratory and workshop were often hardly distinguishable; that men of learning, as Bischoff wrote in his *Essay on the History of the Art of Dyeing* (1780), 'no longer shunning the dirt and foul smell of the workshops, or despising the workman for his simplicity, nowadays . . . visit him at his work and discuss with him how and in what ways it may be improved'; that industrialists, on the other hand, were also actively interested in science, and in some notable cases were as able in applied science as the philosophers themselves, with whom they rubbed shoulders and minds in the Royal and other philosophical societies—these things do not render the technological advances of the time non-scientific: they are, indeed, the very essence of successfully applied science.[1]

But many, especially philosophers of science and technology, argue that, though the experimental *methods* of science and technology may be similar their *aims* or *purposes* are different: that science is concerned with 'knowing', with the search for truth or understanding, without regard to practical applications, whereas technology is concerned with 'doing', with utilitarian objects.[2] Undoubtedly there is such a distinction, but it should not be pressed too far. Many scientists and technologists have been interested in both knowing and doing: scientific knowledge has improved technology, while technological researches have led to advances in science. Innumerable examples of such fruitful interactions can be cited from the eighteenth century onwards.

With some of these we shall be concerned in the following chapters, but we wish to produce here a particularly striking example. Few people would question that Joseph Black was one of the most outstanding *scientists* of the eighteenth century, and he himself emphasized that 'Chemistry is not an Art [or Craft] but a Science. The Artist is

[1] Since writing the above paragraphs, we have read Sir Peter Medawar's book *The Art of the Soluble* (1967), which contains two chapters—one on 'Two Conceptions of Science' and the other on 'Hypothesis and Imagination'—expressing views very similar to our own regarding the artificial distinction between 'pure' and 'applied' science. This distinction, Professor Medawar points out, is 'terribly English' and is probably due to 'the conscious and deliberate perpetuation by our public schools of the Platonic conception of activities that did or did not become a gentleman' (*op. cit.*, p. 13).

[2] See, for example, the articles by M. Bunge, J. Agassi, J. O. Wisdom, H. Skolimowski, and C. Jarvie in *Technology and Culture*, Vol. VII, no. 3 (1967).

he who puts in practice what the phylosopher Conceives ... But he who studys deeply and thinks on rational Methods... is a Phylosopher ... The Study of Mechanical Arts has no doubt done a good dale of Service to this Science but these Arts are more Obliged to Chemistry than it to them.'[1] Yet Black also stated that the objects of Chemistry were utilitarian as well as philosophical: 'the improvements of Arts & Natural knowledge'.[2] Moreover, John Robison, his friend and former student and colleague, who published Black's lectures posthumously, made some remarkably interesting comments about Black whilst preparing them:

It will certainly be the clearest, simplest, and most intelligible Collection of Chemical knowledge in our Language; such that any sensible dyer or Blacksmith or druggist will understand completely, and by far the fittest for the elementary instruction of a Gentleman or any person entering.

[Dr. Black] made it his great Aim to be completely understood by the most illiterate of his hearers—and for this reason he avoids on all occasions all refined or abstruse reasonings, all subtile philosophical disquisition ... he never attempted to please their fancies with a fine theory which promised the explanation of everything. . . .[3]

In a later letter Robison emphasized these aspects of Black's teaching even more strongly:

dr Black never attempted to give a *System* of Chemistry. He was unfriendly to all systems of an experimental Science, and with respect to chemistry, he thought a system was an Absurdity, because altho' we think ourselves very wise, and have made many discoveries in it, the science is still but beginning, in its Infancy, and therefore it is nonsense to pretend to teach a system of it. dr Black pretended no such theory—but to make his hearers good Chemists—not able to talk about theories, but able to examine the Chemical properties of bodies and to apply their knowledge to the various purposes of Life which depend on Chemistry. His Lectures therefore *profess no system*, and, for this reason, may chance to be little thought of by the refined Theorists. But they will contain complete Instructions for those who are not acquainted with Chemistry, and wish to learn it. The philosophers, who want only Refinements and new discoveries, will perhaps be disappointed—the System mongers will throw the book aside—But the public will be instructed ... The simplicity which runs thro' the whole, and which makes it look more like the work of an operative chemist than that of a great philosopher, is a high recommendation of the performance.[4]

---

[1] Thomas Cochrane's *Notes from Doctor Black's Lectures on Chemistry 1767/8* (ed. D. McKie, 1966), pp. 2–4.
[2] *Ibid.*, p. 3.
[3] J. Robison to G. Black, jun., 1 Aug. 1800, Edinburgh University Library.
[4] Ditto, 18 Oct. 1800.

Yet Black, as Robison pointed out in the first letter, 'was considered as the first of Chemical philosophers', and he certainly did produce some extraordinarily perceptive hypotheses—in pneumatic chemistry and heat, for instance—but he combined philosophy very closely with experiment and applied chemistry to innumerable utilitarian purposes: he did not inhabit an intellectual ivory tower, remote from the realities of life, but saw its usefulness to artisans[1] as well as its intellectual interest to philosophers, and frequently acted as an industrial consultant.[2]

In the following pages we hope to show how fruitful was this collaboration between science and technology during the Industrial Revolution. Contrary to long-accepted ideas, the latter was not simply a product of illiterate practical craftsmen, devoid of scientific training. In the development of steam power, in the growth of the chemical industry, in bleaching, dyeing and calico-printing, in pottery, soap, and glass manufacture, and in various other industries, scientists made important contributions and industrialists with scientifically trained minds also utilized applied science in their manufactures. Whilst referring to developments in many fields, we shall concentrate our attention particularly upon the engineering and chemical industries, basically the most important and all-pervasive.

We are not, of course, the first to explore this territory. Wolf, Clark, Singer, Hall, and many others have examined the impact of the 'Scientific Revolution'.[3] Dr. and Mrs. Clow have revealed the ramifications of the 'chemical revolution', and Dr. Hans has written on the related changes in education.[4] But this is a vast terrain, the surface of which has as yet only been scratched: much more research is required on particular industries, industrialists, and scientists, and upon the agencies whereby scientific and technical knowledge was diffused. We have been able to tap many sources which have hitherto been inadequately if at all utilized: the unplumbed depths, for instance, of the Boulton and Watt papers, and other business records, the proceedings of philosophical societies, and the increasing plethora of contemporary books, journals, and newspapers.

Such sources, we think, reveal many new and fascinating aspects of the relationship between science and technology. At the same time, however, we do not wish to exaggerate the extent to which natural philosophy contributed to 'arts and manufactures' in the Industrial

---

[1] See below, p. 182.
[2] Dr. and Mrs. Clow, *op. cit.*, give a number of examples, to which more could be added. See below, chap. X.   [3] See below, chap. I.
[4] A. and N. L. Clow, *The Chemical Revolution* (1952); N. Hans, *New Trends in Eighteenth Century Education* (1951).

Revolution. This period was, of course, a transitional one, and traditional handicrafts, with their rule-of-thumb procedures, proved remarkably long-lasting in many industries, while even in these industries which were being most rapidly 'revolutionized' it is clear that practical empiricism was largely responsible for technical advance. Several of our studies, in engineering and dyeing, for example, will demonstrate its continued importance.

It is also true that scientific and technological factors were by no means entirely responsible for the Industrial Revolution, and that they were closely related to the economic and social changes of the time. But we do not propose in this book to deal explicitly with these complex interrelationships. An enormous amount has already been written on the causes and consequences of the economic changes of this period, including several recent and penetrating contributions.[1] We, in these studies, shall constantly be referring to the economic and social background, but we feel that it would be superfluous and superficial to attempt anything more in this book, which has a different, though related, central theme.

We do, however, wish to record one firm impression left upon us by study of the contemporary evidence. Although economic and social factors were undoubtedly of immense importance in motivating scientific and technological changes—growth of population, expansion of trade, development of transport, availability of capital and credit, social mobility, and the profit motive all impelled or encouraged men to develop new industrial techniques—we have been impressed by the fact that many of the leading scientists and scientifically-minded industrialists were motivated also to a considerable extent by innate curiosity, by a desire to discover more about how industrial processes worked, by an urge to make improvements, and to be esteemed by their fellows, not merely for the money they made, but for their contributions to scientific and technological advance.[2] This impression tends to suggest, indeed, that psychology and sociology may have as much as economics to teach us in such matters.

Our studies, like those of Professor Jewkes and his collaborators in the modern period, have necessarily been 'qualitative, selective, and

---

[1] See, for example, the impressive analysis by D. S. Landes in the *Cambridge Economic History of Europe*, Vol. VI (1965), Part I, chap. V, and M. W. Flinn, *Origins of the Industrial Revolution* (1966).

[2] A similar conclusion has been reached from a study of modern inventors: that 'the intuition, will and obstinacy of individuals spurred on by the desire for knowledge, renown or personal gain [remain] the great driving forces in technical progress'. J. Jewkes, D. Sawers, and R. Stillerman, *The Sources of Invention* (1958), p. 223.

impressionistic'.[1] At a time when quantitative aggregate studies of industrial production and national income are becoming fashionable in economic history, we consider that the real sources of economic growth are to be found rather in 'disaggregated' studies of particular industries, firms, and individuals.[2] Some economists, indeed, are beginning to recognize that aggregate statistics of, say, capital investment are inadequate to explain economic growth, and that detailed research is needed into the causes and effects of scientific progress and technological innovation,[3] instead of regarding these as 'exogenous' variables. Samuel Smiles is coming back into favour. There are, however, vast areas of industrial history to be surveyed—much more than can be comprised in a single book. We have not, therefore, attempted a comprehensive study, but have made a series of probes in particular areas, which, we hope, will shed some new light upon the way in which Great Britain established her technological leadership in the Industrial Revolution.

[1] *Ibid.*, p. 25.
[2] Professor Postan recently emphasized the need for such detailed studies in a paper on 'The Historic and Economic Problems of Technological Change', at the Annual Conference of the Economic History Society in Manchester, April 1966. It must, however, be noted that Professor Postan has recently expressed the view, very categorically, that there was little or no connection between science and industry in the Industrial Revolution, and that 'it was not until the turn of the [nineteenth] century that a regular flow of ideas between laboratories and firms set in': M. M. Postan, *An Economic History of Western Europe 1945–1964* (1967), p. 154. The evidence which we present in this volume necessitates, we think, some revision of this opinion.
[3] In addition to the work of Professor Jewkes and his associates, see also, for example, C. F. Carter and B. R. Williams, *Industry and Technical Progress* (1957), *Investment in Innovation* (1958), and *Science in Industry* (1959). These two authors, however, while producing interesting evidence on modern relationships between science and industry, have repeated uncritically the traditional view of the first Industrial Revolution as almost entirely the product of uneducated empiricism, though they make an exception of James Watt and the steam engine.

I Scientific Prelude to the Industrial Revolution[1]

The debate in which we are concerned—whether the Industrial Revolution in the eighteenth century saw a fusion between science and technology which rapidly advanced man's control in these islands over material forces—is not confined to that period. Other scholars, whose interests are primarily in the sixteenth and seventeenth centuries, are also fiercely engaged with each other, duelling about the precise character of the Scientific Revolution of the seventeenth century, whether the great names of Galileo, Kepler, and Newton are to be forced to hob-nob with scarcely known shipwrights and surveyors, with mechanics and almanac-makers, and whether there is any significant relationship between Puritanism and Science. Is the Scientific Revolution to be regarded as a take-over bid by the artisans, was London the true home of learning rather than Oxford or Cambridge, was the lofty science of astronomy promoted basically by the economic demands of a seafaring nation or by the disinterested curiosity of great minds remote from the hurly-burly of the wharf-side and the tumult of tarpaulins: these are the questions that are being asked.[2] But they are not new questions and many times one can see the men of the sixteenth and seventeenth centuries asking themselves the very same questions. Thus Gabriel Harvey wrote in 1593:

He that remembereth Humphry Cole, a mathematical mechanician, Matthew Baker, a shipwright, John Shute, an architect, Robert Norman, a navigator,

---

[1] This is a previously unpublished chapter, produced jointly by the two authors.
[2] See C. Hill, *Intellectual Origins of the English Revolution* (Oxford, 1965), and A. R. Hall's review of that book in *History*, Vol. L (1965), pp. 332–7; also C. Hill, *op. cit.*, appendix, 'A Note on the Universities', and M. H. Curtis, *Oxford and Cambridge in Transition, 1558–1642* (1959); R. K. Merton, 'Science, Technology and Society in Seventeenth-century England', *Osiris*, Vol. IV (1938), pp. 360–632, and A. R. Hall, 'Merton Revisited, or Science and Society in the Seventeenth Century', *History of Science*, Vol. II (1963), pp. 1–16; H. F. Kearney,

William Bourne, a gunner, John Hester, a chemist, or any like cunning and subtle empiric, is a proud man, if he contemn expert artisans, or any sensible industrious practitioners, however unlectured in schools or unlettered in books.[1]

While a century later, in 1694, John Flamsteed, the Astronomer Royal, could write to Pepys: 'All our great attainments in science and in the mechanic part also of Navigation have come out of the Chambers and from the fire-sides of thinking men within doors that were schollers and mechanics, and not from Tarpaulins, tho' of never so great experience.'[2]

The men of the sixteenth and seventeenth centuries, however, were well aware of the fruitful collaboration between scholars and artisans, and it may be that there is artificiality in the modern conflict between those who point to Newton on their side and those who find in the lives of many humble artisans, merchants, seamen, and the like the broad base of mathematical and scientific advance. Perhaps our confusion is the result of a false system of categories, which distorts the fact that when the scholar and the instrument-maker co-operated in the sixteenth and seventeenth centuries, as they often did, they were both acting in the character of proto-scientists.

A generation ago, the 'Scientific Revolution' itself had only recently become a subject of intensive historical study and its possible links with technology were generally regarded as tenuous, though the late Professor Wolf had done much to open up this field.[3] Sir George Clark, indeed, though observing many interesting examples of such interactions in the early years of the Royal Society, had come to the conclusion that 'technology as a whole was far less suitable for systematic study than theoretical science', and that natural philosophers of the seventeenth century had done little more for it than to amass information and throw out a few practical hints.[4] In Newton's day, 'the greater part of the scientific labour ... brought no practical return until long afterwards; the growth of science ... touched the needs of human life only here and there'.[5] These views have more

---

Origins of the Scientific Revolution (1964). See also the articles by Kearney, Hill, and Rabb, and the various other works referred to, in *Past and Present*, Vols. 28, 29, 31, and 32 (1964–5).

[1] Quoted by E. G. R. Taylor, *Tudor Geography, 1485–1583* (1930), p. 161.

[2] Quoted by E. G. R. Taylor, *The Mathematical Practitioners of Tudor and Stuart England* (Cambridge, 1954), pp. 5–6.

[3] A. Wolf, *A History of Science, Technology and Philosophy in the Sixteenth and Seventeenth Centuries* (1935); *A History of Science, Technology and Philosophy in the Eighteenth Century* (1938).

[4] G. N. Clark, *Science and Social Welfare in the Age of Newton* (2nd edn., 1949), pp. 17 and 28.   [5] *Ibid.*, pp. 89 and 90–1.

recently been echoed by Professor Nef: 'The revolutionary scientific discoveries of Gilbert, Harvey, Galileo and Kepler, like the new mathematics of Descartes, Desargues, Fermat and Pascal, were of no immedate practical importance.'[1]

Evidence is accumulating, however, which suggests that these distinctions between 'pure' and 'applied' science in the sixteenth, seventeenth, and eighteenth centuries are to a large extent artificial, although there are still strong differences of opinion on this subject. There was, of course, a considerable gulf between, say, a humble instrument-maker and a Kepler or a Newton. But there were also important links between them, and what was happening among the mechanics of England cannot be divorced from the 'Scientific Revolution' in Europe. Though the Middle Ages, as Dr. Crombie has shown, were not 'Dark Ages', devoid of scientific and technological progress,[2] they were less fertile than the sixteenth and seventeenth centuries, precisely because in medieval times prestige was conferred by excellence in theology rather than by distinction in *techné*. However much we revise our knowledge of the medieval period and point to advances in navigation, in astronomy, in ballistics, and even in mathematics, there was undoubtedly a 'Scientific Revolution' in the sixteenth and seventeenth centuries, which served to emphasize the comparative slowness of technical and scientific advance in the preceding centuries. The chief characteristics of the Scientific Revolution were an increasing concentration on 'physical rather than metaphysical problems ... accurate observation of the kind of things there are in the natural world', and 'systematic use of the experimental method'. These were associated with greater interest in the utilitarian possibilities of applied science. Expanding trade and industry caused philosophers to develop 'a vigorous interest in the study of the technical processes of manufacture, and this helped to unite the mind of the philosopher with the manual skill of the craftsman'.[3]

The area in which one would expect this union to be most clearly observable is naturally that of engineering. Although the link between technical activity and scientific preoccupations was never entirely lost,[4] it was with the Italians of the Renaissance city-states of the late

---

[1] J. U. Nef, *Cultural Foundations of Industrial Civilization* (1958), p. 64.

[2] A. C. Crombie, *Augustine to Galileo* (2 vols., Mercury Books, 1961).

[3] *Ibid.*, Vol. II, pp. 121–2. For more detailed treatment of these developments, see H. Butterfield, *The Origins of Modern Science, 1300–1800* (1949), A. R. Hall, *The Scientific Revolution, 1500–1800* (1954), R. Taton (ed.), *The Beginnings of Modern Science from 1500–1800* (trans. by A. J. Pomerans, 1964). Wolf's older works, previously cited, are still useful.

[4] G. Beaujouan, *L'Interdépendence entre la science scolastique et les techniques utilitaires* (Paris, 1957).

fifteenth century and the first half of the sixteenth century that it developed most strongly, in fortification, architecture, hydraulics, and other branches of engineering.[1]

This was the period ... in which technical literature, with its specialization, its continuous research, and its sometimes spectacular discoveries, was making its appearance ... even before the end of the fifteenth century there were important technical works. . . . In the first half of the sixteenth century there was a veritable avalanche of these treatises, in nearly all fields.[2]

Many of these and later works, mostly Italian and German, are famous—Leonardo da Vinci's technical drawings, Biringuccio's *De la pirotechnia* (1540), Agricola's *De re metallica* (1556), and Ercker's *Beschreibung* (1574)—but there are many others, demonstrating the interaction between science and technical developments,[3] in which, according to the French scholar just quoted, it was the precise concrete questions posed by technology which induced scientists to elucidate them and thus 'set scientists on new paths which led them to important discoveries'.[4] It was particularly the application of mathematics in architecture, ballistics, and clockwork which produced this efflorescence in Italy and Germany.

These developments were before long paralleled in England, where, again, the practical applications of science were strongly evident. The study of mathematics was especially important, as A. N. Whitehead has stressed: 'Apart from this progress of mathematics, the seventeenth-century developments of science would have been impossible.'[5] But mathematics was used not only to plumb the depths of the universe, but also less loftily, in many utilitarian applications. Certainly there is evidence that mathematics was more sought after as a practical tool in the early seventeenth century by artisans than it was as an intellectual discipline in the universities. Dr. John Wallis remarked, in 1697, about his early life:

Mathematicks at that time, with us, were scarce looked upon as Academical Studies, but rather Mechanical; as the business of Traders, Merchants, Seamen, Carpenters, Surveyors of Lands or the like, and perhaps some Almanack Makers in London. . . . For the Study of Mathematicks was at that time more cultivated in London than in the Universities.[6]

[1] B. Gille, *The Renaissance Engineers* (English trans., 1966).
[2] *Ibid.*, p. 192.   [3] See Hall, *op. cit.*, p. 70.   [4] Gille, *op. cit.*, p. 238.
[5] A. N. Whitehead, *Science and the Modern World* (1925) (Mentor Books, 1949), p. 31. See also Crombie, *op. cit.*, Vol. II, pp. 121–2, 125 *et seq.*, and Hall, *op. cit.*, pp. 69 and 224–34. For more detailed recent treatment, see the works of the late Professor E. G. R. Taylor, previously cited.
[6] Taylor, *Mathematical Practitioners of Tudor and Stuart England*, p. 4.

The possibilities for self-education in mathematics were quite extensive at this time and the 'Printing Revolution' had brought many cheap books of instruction in the subject on to the market.[1] These text books were themselves often written by men who had held no academic post and had never attended a university, but were in touch with the practical part of the population—gunners, surveyors, instrument-makers, and the like—who needed to acquire a better grasp of mathematics to earn their daily bread.

As the demand developed, particularly in London, educational institutions evolved to satisfy that demand, and Gresham's College in particular boasted several leading mathematicians in the seventeenth century, whose instructions were 'happily extended to mechanics, and even the meanest artificers: to trade and all the manufactures of the nation'.[2] Henry Briggs (*c.* 1560–1631), the first professor of geometry (1597–1620), not only visited Napier and discussed logarithms with him, but also published tables of logarithms and navigational tables. He was in fact the popularizer of Napier, who was himself interested in improving industrial processes, designing, for example, a hydraulic screw for pumping water out of mines. Briggs numbered among his friends not only William Gilbert, author of *De Magnete*, but also Edward Wright (1558–1615), who was eminent for his publications on mathematical instruments, navigation, and magnetism, and whose work 'effected a revolution in the science of navigation'; Thomas Blundeville (ob. 1602), who wrote about maps, globes, and navigational instruments ; William Barlowe (1544–1625), Mark Ridley (1560–1624), and Sir Thomas Challener (1561–1615), who were all interested in the same kind of work ; John Wells (ob. 1635) Keeper of the Naval Stores at Deptford, who was a practical mathematician ; and the famous mathematician, William Oughtred (1576–1660), who links the first professors at Gresham College with John Wallis, who was his pupil, and others who took part in those later meetings at the college in 1648 which eventually led to the formation of the Royal Society. Oughtred also numbered among his pupils Seth Ward, Lawrence Rooke (1622–62), professor of astronomy 1652–7, and of geometry 1657–62, and Christopher Wren (astronomy, 1657–61), while other Gresham professors were

---

[1] L. B. Wright, *Middle-Class Culture in Elizabethan England* (Chapel Hill, 1935), chap. XV; F. R. Johnson, *Astronomical Thought in Renaissance England* (Baltimore, 1937), pp. 3–13; and the works by E. G. R. Taylor already cited.

[2] John Woodward, Gresham Professor of Physic, 1693–1728, writing probably in the late 1720s, quoted by Taylor, *Tudor Geography*, p. 61.

Isaac Barrow (geometry, 1662-4) and Robert Hooke (geometry, 1665-1703).[1]

But to return to the sixteenth century and the earlier years of the seventeenth, the mathematicians and men of science at Gresham's College had their predecessors in such men as Robert Recorde (1510-1558), who wrote and published four text-books in English between 1542 and 1557 providing a course in arithmetic and elementary mathematics for those who did not understand Latin ; John Dee (1527-1608), whose fame in mathematics was European-wide, and who for thirty years gave advice to gunners, cartographers, pilots, navigators, and astronomers, as well as assembling a fine mathematical and scientific library, and a collection of mathematical instruments; and Leonard Digges (1510-58), who undertook to make what had so far been 'locked up in strange tongues' available for the English craftsman. From such men descended spiritually not only Briggs, but Edmund Gunter (1581-1626), who was professor of astronomy at Gresham's (1619-26) and whose name is associated with a number of instruments—Gunter's Line of Numbers, Gunter's Scale, Gunter's quadrant and Gunter's chain—all of which were important in the development of navigation and surveying. Gunter was succeeded in 1627 by Henry Gellibrand (1597-1637), who wrote on navigation, trigonometry, and magnetic declination. All these men and their work have been described in detail by the late Professor Eva Taylor, who first accurately charted these seas.

The close co-operation between professors at Gresham College and naval officers, naval administrators and ship-builders, is only one instance of the collaboration between men of the study and men of the workshop or dockyard. The last years of the sixteenth century saw an expansion in the engraving and clock- and watch-making trades, which continued in the seventeenth century in response to the demands for estate-survey, chart and map-making, and generally greater precision in astronomical and navigational study. Land-surveying in particular might lead to an interest in drainage, mining, water-supplies, and a host of other industrial activities. For example, Edward Wright (1538-1615), who co-operated with Briggs and Gilbert and produced the Wright–Mercator projection which was a fundamental advance in marine cartography, was a skilled instrument-maker and

[1] D. McKie, 'The Origins and Foundation of the Royal Society of London', *Notes and Records of the Royal Society of London*, Vol. 15, Tercentenary Number, July 1960; F. R. Johnson, 'Gresham College: Precursor of the Royal Society', *Journal of the History of Ideas*, Vol. I (1940), pp. 413-38; C. Hill, *op. cit.*, chap. ii; E. G. R. Taylor, *Mathematical Practitioners of Tudor and Stuart England, passim*.

surveyor employed by Sir Hugh Myddelton in his New River Scheme for bringing water to London.[1]

These developments in England, however, required a unifying philosophy and a voice to express it before they could have their greatest impact: Francis Bacon was that voice, especially in his treatise on *The Usefulness of Experimental Natural Philosophy*, which demonstrated the ways in which applied science might improve methods of manufacture.[2] As Christopher Hill has said: 'Men like Recorde, Dee, Digges, Hood, Gilbert, Briggs had practised new methods and glimpsed some of their possibilities. But Bacon gave men a noble and all-embracing programme of co-operative action, in which the humblest craftsman had a part to play.'[3] Bacon's influence can be perceived everywhere among men of science in the seventeenth and eighteenth centuries, constantly encouraging them to comprehend workshop practices. His spirit permeated the members of the informal society which met at Gresham College and Oxford between 1648 and 1660. The best account of these meetings is given by John Wallis, Savilian Professor of Geometry at Oxford from 1649 until his death. The subjects discussed included: 'Physick, Anatomy, Geometry, Astronomy, Navigation, Staticks, Magneticks, Chymicks, Meckanicks, and Natural Experiments'. The members also considered such matters as the improvement of telescopes and the grinding of optical glasses (engaging an operator for that purpose), 'the Possibility or Impossibility of Vacuities', 'the Toricellian Experiment in Quicksilver', etc. The persons who met included Wallis himself, John Wilkins, Dr. Jonathan Goddard, Dr. (later Sir George) Ent, Dr Glisson, Dr. (later Sir Charles) Scarbrough, and Dr. Merrett. About 1648, some of these people moved to Oxford and their meetings there included Dr. Ralph Bathurst, William Petty, Thomas Willis, Robert Boyle, and Dr. Ward, but later, just before the Restoration, meetings were resumed at Gresham College in London. Out of these meetings grew the Royal Society of London.[4] Most of the members of this group had strong practical interests and it is typical of them that William Petty's large fortune

---

[1] D. Chilton, 'Land Measurement in the Sixteenth Century', *Transactions of the Newcomen Society*, Vol. XXXI, p. 127, quoted by C. Hill, *op. cit.*, pp. 39–40; J. W. Gough, *Sir Hugh Myddelton, Entrepreneur and Engineer* (1964), pp. 38–9 and 49–50.

[2] B. Farrington, *Francis Bacon, Philosopher of Industrial Science* (1951), *Francis Bacon: Pioneer of Planned Science* (1963), *The Philosophy of Francis Bacon* (1964); see also C. Hill, *op. cit.*, chap. iii. J. U. Nef, *Industry and Government in France and England, 1540–1640* (Cornell, 1964), p. 147, points out, however, that Bacon failed to observe the connection between freedom from government interference and the material improvement sought for by the 'new philosophy'.

[3] *Op. cit.*, p. 88.   [4] D. McKie, *op. cit.*

and the Shelburne dynasty were established on the basis of the 'mechanical' activity of the land-surveyor, when he made a profit of about £10,000 from his geometrical survey of Ireland. Trades now became a proper subject for 'philosophers'. Samuel Hartlib proposed to Boyle in 1647 that Petty should be assisted to compile a history of trades, and Petty's first printed work was 'The Advice of W[illiam] P[etty] to Mr. Samuel Hartlib for the advancement of some particular parts of learning', in which he advocated the establishment of a new college where not only would the pupils, even though from the highest ranks of society, be taught manual skills, but the central part would be a *gymnasium mechanicum* or a college of tradesmen.[1]

Not only did the virtuosi begin to look more closely at trades, but the tradesmen were increasing in number, becoming more diversified in techniques and beginning to publish their findings. It is in this half-century, or sometimes just a little earlier, that we first discover 'the incipient engineers (surveyors, millwrights, military engineers, smiths and clockmakers); industrial chemists (metal-smelters, assayers, distillers, and pharmacists) and instrument-makers (opticians, rule-makers, gaugers)'.[2] Though there are only 'dim images of the highly technical crafts', and though few craftsmen wrote on anything much removed from their own skills, a corpus of knowledge on technical

---

[1] W. Letwin, *The Origins of Scientific Economics* (1963), pp. 126–7. Since the above paragraph was written, Miss Margery Purver's book on *The Royal Society: Concept and Creation* (1967) has appeared. Miss Purver differentiates sharply between previous manifestations of the 'New Philosophy'—such as earlier societies in Europe, the meetings in London in the 1640s described by John Wallis, the activities in Gresham College, Boyle's 'Invisible College', and the ideas of Hartlib and Comenius—and the Royal Society, originating (as she considers) solely in the meetings at Oxford from 1648 onwards. She emphasizes, in fact, 'the unique character of the Royal Society', based on Baconian philosophy, and rejects the alleged influence of the 'Puritan ethic'.

We note, however, that Miss Purver admits (*op. cit.*, pp. 127 and 128) that 'earlier societies with an interest in "the New Philosophy" had helped to create the climate in which the Oxford club originated and developed'. Some of the same people and ideas were involved, Bacon's publications had appeared many years previously, and it seems doubtful whether they can be entirely separated from the earlier origins of the 'Scientific Revolution' in Europe.

There is, however, no difference between Miss Purver and ourselves in regard to the effects of the 'New Philosophy' in the Oxford club and Royal Society, with their strong emphasis on experimental research, combining science and empiricism —or 'pure' and 'applied' science—and bringing forth fruit in 'useful' works. This spirit may be summed up in Seth Ward's words, rejecting Aristotelian philosophy in 1654, that, 'instead of verball Exercises, we should set upon experiments and observations, that we should lay aside our Disputations, Declamations, and Public Lectures, and betake ourselves to Agriculture, Mechanicks, Chymistry, and the like' (*op. cit.*, p. 66).

[2] A. R. Hall, *From Galileo to Newton, 1630–1720* (1963), pp. 30–3. The remainder of this paragraph is based on this valuable work.

matters was beginning to emerge to which craftsmen began to make their contribution. Naturally it was the scientific instrument-makers who were in closest touch with the virtuosi, and from the 1650s onwards their dialogue became ever closer. The first important studies made with the microscope date from this period; so do the first attempts to obtain vacua by pumps for experimental purposes, and in the second half of the seventeenth century a number of new instruments were introduced, such as the pendulum clock, telescopic sights, the bubble-level, and the screw-micrometer. Professor Hall's conclusion is that: 'The standards [in experimental science] prevailing about the time of Newton's death (1727) were utterly different from those of the age of Galileo and Kepler. Although the drive behind this change came from the scientists it was largely made possible by the success of the instrument-makers who served their needs.'[1]

Bacon's ideas exercised a profound influence upon the Royal Society, established in 1662, which was actively concerned with the practical applications of natural philosophy.[2] As its first historian, Bishop Thomas Sprat, emphasized, its members aimed especially at improving the 'mechanick arts', such as building, smith's work, 'chymical' operations, shipbuilding, and agriculture.[3]

They have recommended the advancing of the *Manufacture of Tapestry*: the improving of *Silk-making*: the propagating of Saffron: the melting of Lead-Oar with Pit-coal; the making of Iron with Sea-coal: the using of the Dust of black Lead instead of Oil in Clocks: the making *Trials* on English Earths, to see if they will yield so fine a substance as *China*, for the perfecting of the *Potter's Art*.

They have *propounded* and *undertaken* the comparing of several *Soils* and *Clays*, for the better making of *Bricks* and *Tiles* . . .[4]

Among the Fellows of the Royal Society, Robert Boyle was the chief follower of Bacon in forwarding 'the new philosophy', arguing strongly *That the Goods of Mankind may be much increased by the Naturalist's Insight into Trades*, and combining his philosophical interests with practical knowledge of mining, assaying, and other industrial processes.[5] Robert Hooke, another brilliant Baconian and professional experimentalist of the Royal Society, was described by John Aubrey

[1] *From Galileo to Newton, 1630–1720* (1963), p. 33.

[2] See especially G. N. Clark, *Science and Social Welfare in the Age of Newton*, (2nd edn., 1949).

[3] T. Sprat, *The History of the Royal Society of London, for the Improving of Natural Knowledge* (4th edn., 1734), pp. 149–50.   [4] *Ibid.*, pp. 190–2.

[5] M. Boas, *Robert Boyle and Seventeenth-Century Chemistry* (Cambridge, 1958). It should be remembered that Boyle and other scientists of the Royal Society had the assistance of Denys Papin in their pneumatic experiments. A link is thus established between Boyle and the steam-engineers of the later seventeenth century.

as 'certainly the greatest mechanick this day in the world';[1] an able scientist and mathematician, he was also inventive in clock- and instrument-making, and had wide practical interests.[2] Hooke's diary shows him to have been in regular touch with instrument-makers, including the famous clock-maker, Thomas Tompion. One entry describing a visit to Tompion reads: '[Saturday 2nd May 1764] Told him the way of making an engine for finishing wheels, and a way how to make a dividing plate; about the forme of an arch; about another way of Teeth work; about pocket watches and many other things.'[3] We also find him 'discoursing' at Garaway's, Man's, and other London coffee-houses with many of those who were in the van of manufacturing and commercial progress. Thus he discoursed with Andrew Yarranton, the river engineer and 'improver', about hammering and rolling mills and about 'cast iron pillars for bridges, hard[en]ing iron into steel quite through, pressing of cloth &c.'[4] Later he discussed Yarranton's proposal 'about [an] iron rowl for staining Printed stuffs', i.e. for roller-printing,[5] and 'had much discourse with Yarington about Registers, steel wire work, [and] his plates'.[6] Similarly he talked with Joseph Moxon, author of *Mechanick Exercises* (1677–83), 'about mixing mettalls';[7] with Denys Papin 'about his Engines';[8] and with many other craftsmen and with fellow-philosophers such as Wren, Hoskins, Ward, etc., about a wide variety of manufacturing subjects —about 'engines' and mills of all kinds, glass-making, steel-making, coal-mining, cloth-fulling, metal-working, 'chymical' operations, etc. An interesting link with the earlier 'Scientific Revolution' on the continent is provided by references in Hooke's diary to Biringuccio's *Pirotechnia*,[9] which, it has been pointed out, displays a similar interest in industrial (especially metallurgical) operations and is said to have been widely used by practical men in workshops and foundries.[10]

Other members of the Royal Society shared these practical interests, which were strongly encouraged by their royal patron, with a view

---

[1] J. Aubrey, *Brief Lives and Other Selected Writings* (ed. A. Powell, 1949), p. 128.   [2] M. Espinasse, *Robert Hooke* (1956).

[3] H. W. Robinson and W. Adams (eds.), *The Diary of Robert Hooke, 1672–80* (1935), p. 100.

[4] *Ibid.*, pp. 76–7, 26 Dec. 1673. He spells Yarranton's name as 'Yarrington' and 'Yarington'.

[5] *Ibid.*, p. 79, 6 Jan. 1673/4.   [6] *Ibid.*, p. 204, 27 Dec. 1675.

[7] *Ibid.*, p. 90, 7 March 1673/4. See below, p. 22, for Joseph Moxon.

[8] *Ibid.*, p. 366, 14 July 1678. See also p. 391, 5 Jan, 1678/9.

[9] E.g. 12 Nov. 1675, regarding Biringuccio's method of making steel, and 1 March 1677/8, regarding purchase of a copy of *Pirotechnia*.

[10] J. R. Zietz, ' "The Pirotechnia" of Vannoccio Biringuccio', *Journal of Chemical Education*, Vol. 29 (1952), pp. 507–10. See also J. Read, 'Biringuccio in English', *Nature*, Vol. 151 (1943), pp. 169–70.

to the development of English industry and trade. One need only mention Sir Christopher Wren's architectural achievements, Sir William Petty's works on statistics and navigation, and on weaving and dyeing, and John Evelyn's book on trees and afforestation, and his interests in chemistry,[1] to illustrate the range of their activities. Even the Olympian Newton was interested in scientific experiments performed by craftsmen, as, for instance, when he wrote Hans Sloane on 14 September 1705: 'Dear Sir, I beg the favour of you to get Mr. Hawksbee to bring his air-pump to my house, and then I can get some philosophical persons to see his experiments . . .'[2]

Apart from the academic man of science and the scientifically-minded artisan, the virtuoso also came into his own. Sir Francis North, Baron Guildford, seems to exemplify the new type of aristocratic amateur of science. As a judge on circuit he was in the habit of studying the trades of the area in true Baconian fashion: 'His lordship's entertainment at Newcastle was very agreeable, because it went most upon the trades of the place, as coal-mines, salt-works, and the like . . .'[3] On another occasion he visited the coal mines of Sir Roger Bradshaw,[4] took notes of wooden railways and studied engines for mine-drainage. Since his brother-in-law was Robert Foley, the ironmaster, these interests may well have been strengthened. The Lord Keeper, who eventually became President of the Royal Society,

was an early virtuoso; for after his first loose from the university, where the new philosophy was then but just entering, by his perpetual inquisitiveness, and such books as he could procure, he became no ordinary connoisseur in the sciences so far as the invention of the then latter critics had advanced them. And the same course he pursued, more or less, all the rest of his life; whereby all discoveries at home and from abroad, came to his notice, and he would have been loth to have let any escape him.[5]

Among his scientifically-minded acquaintance he numbered John Aubrey, John Evelyn, 'Mr. Weld, a rich philosopher, who lived in Bloomsbury', Sir John Werden, a commissioner of the customs, who 'was very far gone in the mystery of algebra and mathematics', Sir Jonas Moor, 'a capital mathematician' who had been employed by the commissioners for dividing the fens, Sir Samuel Moreland, with whom

---

[1] See F. Sherwood Taylor, 'The Chemical Studies of John Evelyn', *Annals of Science*, Vol. VIII (1952), pp. 285–92.

[2] J. Nichols, *Illustrations of the Literary History of the Eighteenth Century*, Vol. IV (1822), p. 59. For Francis Hauksbee (or Hawksbee), the London instrument-maker, who became operator to the Royal Society, see below, pp. 37, 40–1.

[3] Sir Dudley North, *The Lives of the Norths* (3 vols., 1826), Vol. I, p. 280. We are indebted to our colleague, Dr Brian Manning, for this reference.

[4] *Ibid.*, pp. 294–5.   [5] *Ibid.*, Vol. II, pp. 176–7.

he discussed such matters as mercurial barometers, Flamsteed the Astronomer Royal, Jones the clock-maker in the Inner Temple, and Winn
the famous instrument-maker in Chancery Lane.[1] North himself
wrote papers on the bladders of fishes and the doctrine of pulses in
music. Another well-known virtuoso of the period was 'Sir Kenelm
Digby, Courtier, Chemist', as he is described when proposed as a
founder-member of the Royal Society. Digby is another amateur of
science linking Gresham College with the Royal Society, since it was
at Gresham College that he carried out in the 1640s experiments on
magnetism, the circulatory system, the growth of the embryo, refrac-
tion and reflection, and other matters, including the absorption of
gases by plants.[2] He was well known to Christiaan Huygens, Petty,
Wren, Goddard, and others.

Though he was not a member of the Royal Society group, space
should also be found for the Marquis of Worcester, author of *A Century
of Inventions* (written 1655; published 1663). He employed Gaspar
Kaltoff as his private engineer for forty years and claimed to have spent
over £9,000 'in building of a house called Fauxhall, for an operatory
for engineers and artists to work public works in', and a further
£50,000 on experiments.[3] His own inventions were diverse but unsuccessful; the best known is his 'water-commanding engine', steam-
driven and erected as a public entertainment at Vauxhall. It seems
certain that by the 1650s it had become fashionable for men of leisure
to be interested in science and experimentation, and particularly to
take note of recent knowledge of the applications of mathematics and
science. The tradition is strengthened in the eighteenth century.

The gentleman virtuoso, however, had his counterpart in the lower
ranks of society; among teachers and academics; and among the new
type of professional civil-servant, represented at its highest distinction
by Samuel Pepys, but also at a lower level by such a man as John
Collins (1625–83).[4] Collins held several minor administrative posts
in his day and was valuable largely for his mathematical knowledge.
He kept up an extensive correspondence with mathematicians like
Newton, Wallis, Pell, Vernon, Gregory, Beale, Strode, Baber, and
Leibnitz. His publications included *Sector on a Quadrant* (1658),

---

[1] *Ibid.*, Vol. II, pp. 179–203.
[2] R. T. Petersson, *Sir Kenelm Digby* (1956), p. 8. He also carried out experiments
in the early manufacture of vitriol (sulphuric acid) by the 'bell' process, in which
he was assisted by George Hartman, his 'Chymist and Steward', who published
Digby's *Chymical Secrets and Rare Experiments in Physick and Philosophy* in 1683,
after Digby's death. J. Mactear, Glasgow Phil. Soc. *Proceedings*, Vol. XIII (1880–2),
p. 417.
[3] H Dircks, *The Life, Times and Scientific Labours of the 2nd Marquis of Worcester*
(1865), pp. 286–7.   [4] Letwin, *op. cit.*, pp. 99–109.

Geometrical Dyalling (1659), *The Mariner's Plain Scale new plain'd* (1659), and the *Doctrine of Decimal Arithmetick* (1685), and he was a great disseminator of mathematical knowledge. His friend, Michael Dary, a gauger of wine-casks, was a similar character. The fact that some men like Petty and Pepys were successful and rose quickly in society does not differentiate them in their mathematical and scientific interests from Collins and Dary who were less fortunate.

Joseph Moxon (1627–1700) is another interesting figure.[1] Well known in the later seventeenth century as a hydrographer, mathematician, and instrument-maker, through his popularizing works on maps, geography, navigation, astronomy, mathematics, architecture, and 'mechanick exercises' generally (especially printing), he had, according to his own account,[2] 'for many years been conversant in . . . Smithing, Founding, Drawing, Joynery, Turning, Engraving, Printing Books and Pictures, Globe and Map-making, Mathematical Instruments, &c.', all of which, he pointed out, 'work upon Geometrical Principles'. As we have seen, he was among those with whom Boyle 'discoursed',[3] he was elected F.R.S. in 1678, and his best-known work, *Mechanick Exercises: or the Doctrine of Handy-Works* (1677–83), was based on Baconian principles, as is shown by the advertisement of it which appeared in the *London Gazette*, of which Sir Joseph Williamson, President of the Royal Society, was then editor:

Forasmuch as all Natural Knowledge was Originally produced (and still eminently depends) upon Experiments, and all or most Experiments are couched among the Handy-crafts; and also that Handy-works themselves may be improved: There was begun (by *Joseph Moxon*, Hydrographer to the King's Most Excellent Majesty), *Jan.* 1, 1677, Monthly Exercises, upon the Mechanicks, and hath been since continued in Six Monthly Exercises, till *July* 1, 1678.[4]

Moxon's works were read by Boyle, Evelyn, and other members of the Royal Society, but they were in language understandable by literate craftsmen. They provide, in fact, a good example of how the scientific spirit was percolating down to lower levels of society.[5]

At both the higher and lower levels, mathematics continued to be

---

[1] See *D.N.B.* and the Introduction to the modern reprint of the second volume of his *Mechanick Exercises*, that on the Art of Printing (ed. H. Davis and H. Carter, 1958).   [2] J. Moxon, *Mechanick Exercises* (1677), Preface.
[3] See above, p. 19.   [4] Davis and Carter, *op. cit.*, p. xlv.
[5] C. H. Timperley, *A Dictionary of Printers and Printing* (1839), p. 567, pointed out that 'the pursuits of Mr. Moxon were those of general science . . . Moxon was the first of English letter-founders who reduced to rule the art which before him had been practised but by guess . . .; by nice and accurate divisions, he adjusted the size, situation, and form of the several parts and members of letter, and the proportion which every part bore to the whole.'

important. The basis of the new scientific method, it was used not merely to calculate the movement of heavenly bodies, but also in practical arts such as navigation, cartography, ballistics, mining, and surveying, and these gave rise to the craft of instrument-making: the manufacture of telescopes, microscopes, barometers, chronometers, micrometers, dividing and gear-cutting engines, etc. The application of mathematics to gear-cutting provides an interesting example of this inter-relationship. It derives from two different starting points in the sixteenth century—on the one hand, 'an approach through the mathematical analysis of curves and the geometry of motion', and on the other 'an empirical approach through the efforts of clock-makers to cut gears for their wheel work'.[1] The question arises as to when these two lines came to converge. In most cases the answer must be not until well into the nineteenth century,[2] but it is also clear that some clock-makers drew upon their mathematical knowledge even as early as the seventeenth century. Professor Woodbury, however, is self-contradictory on this question. In some places he says that in the eighteenth century 'only a few scientists, such as Réaumur, and even fewer mathematicians, were interested in the problems of the engineer';[3] that the theory of epicyloidal and involute gearing was 'familiar only to mathematicians';[4] that craftsmen such as clock-makers and instrument-makers arrived at 'purely empirical solutions for the form of their gear teeth'[5]; and that it was only 'after 1800' that science was applied to 'production' engineering.[6] But elsewhere he states that 'by 1700 the attention of geometers was attracted to the theory of the form of gear teeth . . . by the engineers' practical interest in mills and water power';[7] that 'the interest of the mathematicians, such as Desargues, de La Hire, Euler and Camus, seems to have arisen from a desire to increase the efficiency and reduce the wear in mills of various types';[8] that 'the French mathematician Desargues (1593–1661) . . . had interests in architecture and engineering', and that while 'building some machinery near Paris he designed and constructed the first gears having epicycloidal teeth', in about 1650;[9] that Phillipe de la Hire (1640–1718), who in 1694 was 'the first to treat gear teeth mathematically and systematically', is also 'said to have applied his discoveries to the design of a

---

[1] A. P. Usher, Introduction, p. ii, to R. S. Woodbury, *History of the Gear-Cutting Machine* (Cambridge, Mass., 1958).
[2] See, for example, J. W. Roe, *English and American Tool Builders* (1916), chap. VI, especially pp. 65–6, referring to the general lack of applied theory in gear-cutting among leading British engineering firms in the 1830s.
[3] Woodbury, *op. cit.*, p. 9. [4] *Ibid.*, p. 17. [5] *Ibid.* ,p. 4.
[6] *Ibid.*, pp. 5 and 65 ff. [7] *Ibid.*, p. 3.
[8] *Ibid.*, p. 4. [9] *Ibid.*, p. 10.

large waterworks';[1] that Robert Hooke invented a wheel-cutting engine, later described by Le Roy, the French clockmaker;[2] that the works of these and later mathematicians such as Euler, Camus, and Kaestner, in the eighteenth century, were familiar not only to French clock-makers, but also to leading British clock-makers such as Henry Hindley (1710–71), of York, Thomas Reid (1710–96), of Edinburgh, and Samuel Rehé (?–1806), of London, who produced 'wheel-cutting engines', etc., which eventually evolved into heavy gear-cutting machines;[3] that instrument-makers such as Graham and Ramsden applied similar mathematical precision in their 'dividing engines', etc.;[4] and that the popular scientific writings of such men as Ferguson, Imison, and Buchanan, and translations such as John Hawkins's English edition of Camus, helped to spread such knowledge more widely in the second half of the eighteenth and early nineteenth centuries.[5] As we shall see, engineers such as John Smeaton (originally an instrument-maker) and John Rennie (an Edinburgh graduate), who made important developments in iron gearing, were also familiar with engineering science. (Professor Woodbury himself points out that the teeth of Rennie's iron gears in the Albion Mills 'had their teeth carefully chipped and filed into epicycloids'.[6]) While, therefore, we would certainly accept Professor Woodbury's view that mathematical theory was not generally applied to gear-cutting by early millwrights and engineers, and that empirical methods were widespread, we consider that there is plenty of evidence of links between science and practice among clock-makers and instrument-makers, and that these craftsmen were making important contributions to mechanical engineering some considerable time before 1800.[7]

At a lower level, scientific knowledge was also becoming more widely diffused among millwrights and engineers, who learned to measure, plan, calculate velocities, etc. and who therefore studied mathematics and, later on, other subjects such as mechanics, hydrostatics, and hydraulics.[8] The diffusion of mathematical knowledge among such artisans in the eighteenth century is illustrated by Francis Walkinghame's text-book, *The Tutor's Assistant*, which was first issued

[1] Woodbury, *op. cit.*, p. 11.
[2] *Ibid.*, p. 46.
[3] *Ibid.*, pp. 50–6.
[4] *Ibid.*, pp. 56–9.
[5] *Ibid.*, pp. 18–22 and 31.
[6] *Ibid.*, p. 5, n. 5.
[7] See below, especially chap. XIII.
[8] See below, pp. 43–51, 73–8, chap. III, chap. XI, and pp. 429, 480–1, 487–8, 490, 500, 508–9. See also G. Doorman, *Patents for Inventions in the Netherlands during the 16th, 17th and 18th Centuries* (abridged English version trans. by John Meijer, The Hague, 1942), which shows the surprising number of patents taken out by Englishmen in the Netherlands during the seventeenth century for dredging and excavating machines, windmills, systems of navigation, drainage mills, etc.

in 1751 and became the most popular arithmetic text-book for over a century afterward.[1] Among the original subscribers, apart from some gentlemen and clergy, there were 'a large number of writing masters and accomptants', together with

carpenters, bricklayers, plumbers, builders, a brewer, a painter, a timber merchant and an instrument-maker. This brief list illustrates the kind of craftsmen who were to buy Walkinghame's and other mathematical books, formed so many mathematical clubs and read the mathematical periodicals of the time. Without the old, classical education, these men were much more interested in mathematics and science and helped to carry through the Industrial Revolution which was to transform England.[2]

Walkinghame's text-book was for many years a 'best seller'. Before the author died in 1783 there had been eighteen different editions, each comprising 5,000–10,000 copies, and successive editions continued to be published far into the nineteenth century. Walkinghame, as Mr. Wallis has pointed out, was in a continuous mathematical tradition, stretching back to the sixteenth century, and there were many other authors of similarly popular text-books on mathematics during the eighteenth century.[3]

In other branches of science, there is similar evidence of continuity in the Baconian tradition. Chemistry, developing from medieval alchemy, was applied in many manufacturing processes.

Useful chemistry was not any longer medical, but rather industrial, and many members of the Royal Society brought in accounts of everything from the mining of minerals to soap-making and dyeing; though this did not influence industry to any great extent, it did help spread the latest technical innovations, and indicates a kind of chemical climate to be found in the Royal Society circle.[4]

John Houghton, in his *Letters* and *Collection*, represents the type of man who was concerned at the end of the seventeenth century with the applications of chemistry to agriculture, manufactures, and trade. He frequently quoted the works of Hooke, Evelyn, and others, and also drew attention to some of the first public lectures on chemistry,

[1] P. J. Wallis, 'An Early Best Seller. Francis Walkinghame's "The Tutor's Assistant" ', *Mathematical Gazette*, Vol. XLVII, no. 361, Oct. 1963, pp. 199–208.

[2] *Ibid.*, p. 201.

[3] See E. G. R. Taylor, *The Mathematical Practitioners of Hanoverian England 1714–1840* (1966), and also P. J. Wallis, 'British Mathematical Biobibliography', *Journal of the Institute of Navigation*, Vol. 20, no. 2 (April 1967), pp. 200–5. There were close links between these mathematical practitioners and crafts or professions such as instrument-making, gauging, surveying, and architecture.

[4] M. Boas, *op. cit.*, p. 73. See also Hall, *op. cit.*, pp. 220–4 and chap. XI; C. Singer and others (eds.), *History of Technology*, Vol. III, chap. XXV.

by George Wilson (1631–1711), who was active in this field until his death and who has been described as 'probably the most noted writer in English on practical chemistry in the period from Robert Boyle to Peter Shaw' (1649–1763).[1] Nor were the British unaware of developments on the continent. The work of J. R. Glauber[2] (1604–68), for example, who had practical interests in chemistry extending outside medicine to agricultural fertilizers and metallurgy, was well known, and Glauber provided a link with the earlier German writers, Agricola and Ercker, whom we have mentioned.[3] By the early eighteenth century, 'the point had been reached where science, and particularly chemistry, could begin to give a lead to several manufactures'.[4]

England was also about to enter upon the era of effective steam pumping engines, with the emergence first of Savery and then of Newcomen. The development of the steam engine took place, as Sir George Clark has pointed out, 'by almost regularly alternate steps of experimental research and practical application . . . the classic example of science in alliance with practice'.[5] This view has received support from Professor A. P. Usher, referring not only to the development of the suction-pump and steam engine, but also to the manufacture of scientific instruments, clocks and watches: 'The sixteenth and seventeenth centuries mark the transition from complete empiricism to engineering techniques fully grounded in mathematics and applied science.'[6]

For these varied scientific-technological interests there grew up a supporting literature on 'mining and mineralogy, smelting and casting, the extraction of saltpetre and the manufacture of gunpowder, the making of glass and mineral acids, the purification of mercury and the precious metals . . . on machinery for lifting, pumping, sawing, textile manufacture, etc. . . .'[7] The widening range of scientific and technical knowledge is also illustrated by the early encyclopaedias or 'dictionaries of arts and sciences', such as John Harris's *Lexicon Technicum* (1704; 2nd edn., 2 vols., 1708–10) and Ephraim Chambers's *Cyclopaedia* (2 vols., 1728), which had run to five editions by 1746.

There is still, however, considerable debate as to the extent to which 'science' was applied to industry in the sixteenth and seventeenth

---

[1] F. W. Gibbs, *Annals of Science*, Vol. 8 (1952), pp. 125 and 274, giving the references: *Collection*, Vol. IV, p. 89 and subsequent issues in April, May, and June 1694. Wilson's *Complete Course of Chemistry* was first published in 1691, running to several later editions in the eighteenth century.

[2] Boas, *op. cit.*, p. 54.   [3] See above, p. 13.

[4] F. W. Gibbs in Singer, *op. cit.*, Vol. III, p. 706.

[5] Clark, *op. cit.*, p. 21. See also the Introduction by A. E. Musson to the 2nd edn. of H. W. Dickinson, *Short History of the Steam Engine* (1963). See also below, pp. 28, 39, 47–9, 73, 79–81.   [6] Singer, *op. cit.*, Vol. III, p. 344.

[7] Hall, *op. cit.*, p. 70. See also Clark, *op. cit.*, pp. 30–2.

centuries. Much of this debate, as we have suggested, arises from differing interpretations of 'science', which, in our view, has always contained a substantial applied or empirical element: 'pure' theories have rarely, if ever, been concocted out of thin air, without practical observation and experiment—indeed the science that evolved in the 'Scientific Revolution' was essentially 'experimental science', often linked with utilitarian applications. There is no doubt, of course, that the farther back into history one goes, the more the empirical element predominates in industry, and even during and after the 'Scientific Revolution' of the sixteenth and seventeenth centuries links between science and technology were often tenuous and took long to develop. The slow penetration of science into engineering practices, for instance, has been emphasized by Professor Finch, of Columbia University.[1] By contrast with the views more recently expressed by Professor Gille,[2] for example, he considers that, despite the appearance of the first printed engineering books in the Renaissance, there was 'little or no interlinking of theory and practice' at that time, and that 'the courtship of science and engineering occupied at least two centuries'.[3] Nevertheless, by the eighteenth century, in such technological activities as bridge-building, construction, and hydraulics, engineering science was developing, especially in France, aided by techniques of testing and measurement.[4]

In the same number of *Technology and Culture* Professor Hall expresses similar views,[5] but more strongly. In the sixteenth and seventeenth centuries, he maintains, 'theoretically-derived rules were employed in no branch of engineering . . . the engineer, like every other technologist, was only rarely educated or mathematically literate . . . there was no engineering revolution alongside the scientific revolution, barely indeed a minor disturbance.'[6] These statements, it seems to us, are exaggerated. Whilst we agree that most engineering was empirical and that there was not an 'engineering revolution', there were some significant developments. Professor Hall has to admit, for instance, that 'the notebooks of Leonardo da Vinci undoubtedly demonstrate a true scientific outlook and the sound, original design ability of a true engineer'. But 'Leonardo . . . is an enigma', and 'nothing is known' of other Renaissance engineers.[7] A very different impression is obtained from Professor Gille's book. Moreover, Professor Hall's statement as to the almost total mathematical ignorance

[1] J. K. Finch, 'Engineering and Science: A Historical Review and Appraisal' *Technology and Culture*, Vol. II (1961), pp. 318–32.
  [2] See above, pp. 12–13.   [3] *Ibid.*, p. 323.   [4] *Ibid.*, pp. 323–5.
  [5] 'Engineering and the Scientific Revolution', *ibid.*, pp. 333–41.
  [6] *Ibid.*, pp. 333, 335, and 337.   [7] *Ibid.*, pp. 338–9.

of engineers and, indeed, all craftsmen in this period is hardly compatible with the evidence produced by the late Professor Eva Taylor and others. Moreover, Professor Hall himself suggests the possibility that engineers may have absorbed something of the 'scientific spirit', particularly 'the experimental method', though it might also be said that science adopted the latter from engineering and other crafts.[1] He admits that in the key trades of instrument-making and clock-making, and also in survey and levelling, as well as in the development of air-pumps and other pumping engines, leading up to the steam engine, there were important links between science and technology: 'the scientific notion of exactitude began to affect engineers'.[2] Thus,

the clockmaker's gear-cutter and the instrument-maker's dividing-engine were the ancestors of many later machine tools. Long before large-scale engineering could attempt such tasks these craftsmen were making fairly accurate gears, turning exact steel shafts, cutting good screws, measuring to an accuracy of one part in a quarter of a million or better, and so on, for their scientific employers and even for open market. From their ranks some great engineers like Smeaton and Watt were recruited, and the total effect of their craftsmanship on productive mechanical engineering may well have been considerable. For besides their craftsmanship the leading men were scientifically educated (in England a few of them attained to the Fellowship of the Royal Society) and well able to look beyond the bounds of manual dexterity . . .

At the same time, philosophers were dependent upon these craftsmen in performing scientific experiments: 'as Robert Boyle was fond of saying, scientists could learn from craftsmen . . . Experimental science had to master certain techniques: distillation, lense-grinding, glass-blowing, even turning and metal working and the art of assay . . . To that extent a minimum level of technological and engineering competence was necessary before serious experimental science could begin. . . .'[3]

With all these statements we warmly agree. It is a pity, therefore, that Professor Hall has to mar his article with the exaggerations previously mentioned. In any case, is he not tilting at windmills? Does anyone really believe there was an 'engineering revolution' in the sixteenth and seventeenth centuries? Would not the great majority of scholars agree that developments in these centuries were but a *prelude* to the Industrial Revolution?

Professor Hall tends to dismiss the 'scientific propaganda' of the seventeenth century which 'stressed the potential benefits flowing to medicine and technology from scientific inquiry'; he considers that

[1] *Technology and Culture*, Vol. II (1961), p. 338.
[2] *Ibid.*, pp. 337-8.   [3] *Ibid.*, p. 340.

'there is far bulkier evidence that in their work scientists investigated only the questions that took their fancy . . . science was apt to be concerned at any moment not with urgent problems of engineering and agriculture but with matters that scientists considered significant and interesting'.[1] Other scholars, however, as we have shown, have stressed that the essence of the Scientific Revolution was the change from metaphysical to experimental science, and that this change was stimulated by technological, economic, and social factors, and there can hardly be doubt of the increasing extent to which scientists became interested in technological problems during this period; this is not to say, of course, that such problems became their sole or even their main concern, or that they actually applied their developing theories to any great extent.

This view is supported by other authors writing in the same number of *Technology and Culture*.[2] Professor Leicester, for example, considers that it was 'the union of theory with technology which gave birth to the science of chemistry'.

After the fourteenth century theory and practice in Chemistry were never again separated in western Europe. There was a continuous development in which physicians, apothecaries, and technologists shared, and nearly every great man in the history of chemical thought from the fifteenth to the seventeenth century was also noted for practical discovery. Thus Paracelsus and Van Helmont proposed significant theories and did equally important technical work. Libavius, Glauber, Agricola, Biringuccio, and Ercker were chiefly interested in technology, but they helped to advance the quantitative approach to chemistry; by their descriptions and discoveries they systematized alchemy and converted it into a true science of chemistry.[3]

This coherence of philosophers and technologists led to the rapid progress of chemistry in eighteenth-century France, as shown by the illustrious examples of Lavoisier, Berthollet, and others.

Professor Smith has made similar investigations into the interaction of science and practice in the history of metallurgy, but he uses the words 'science' and 'empiricism' inconsistently. He suggests for example, that rational empiricism may be regarded as scientific:

[1] *Ibid.*, p. 339.
[2] H. M. Leicester, 'Chemistry, Chemical Technology, and Scientific Progress', *ibid.*, pp. 352–6; C. S. Smith, 'The Interaction of Science and Practice in the History of Metallurgy', *ibid.*, pp. 357–67. See also J. K. Feibleman, 'Pure Science, Applied Science, Technology, Engineering: An Attempt at Definitions', *ibid.*, pp. 305–12, and P. F. Drucker, 'The Technological Revolution: Notes on the Relationship of Technology, Science, and Culture', *ibid.*, pp. 342–51. The last two articles are mainly concerned with the modern period, from the Industrial Revolution onward, but they are interesting on the relationships referred to.
[3] *Ibid.*, p. 355.

'... although he was motivated only by the desire to make something useful or beautiful, or merely profitable, the artisan who experimented with different materials to select the best for his purpose was as much a contributor to the development of science, was as much a scientist, as the observational astronomer or the classifying mineralogist or naturalist.'[1] The sixteenth and seventeenth centuries witnessed, with the publication of Biringuccio's *Pirotechnia*, Ercker's *Treatise on Ores and Assaying*, and later works, a great deal of descriptive and classifying work in metallurgy, with some attempts at theoretical explanation. Intelligent factual observation and experimentation of this kind, Professor Smith points out, are an essential basis for scientific theory. But he then goes on, abandoning the use of the term scientific as applied to craft developments in metallurgy, to emphasize that the latter owed little or nothing to science—though science was considerably indebted to metallurgy—until fairly recent times. A change did begin, he points out, with the work of Réaumur in the early eighteenth century, continued by other chemists, especially in France, which contributed substantially to the revolution in chemistry; but the great British iron and steel inventions of the eighteenth and nineteenth centuries 'owed little or nothing to the direct influence of science'. Theory, in fact, followed in the wake of 'empirical experiment'. Even in present-day metallurgical operations, 'science cannot yet eliminate trial'.[2] But is this not true of nearly all modern industrial developments? And is it not also true, even of the purest science, that theories have to be tested by practical experiment? Where, then, does 'empiricism' end and 'science' begin, especially in earlier centuries, when 'natural philosophy' was far less developed and more nearly 'empirical'?

The relationship between science and industry has been studied much less closely in the first half of the eighteenth century than in the early years of the Royal Society in the seventeenth century, or than in the latter part of the eighteenth century. The explanation is probably that the Scientific Revolution and the Industrial Revolution have both tended to draw attention to themselves and away from the intervening period. There was, however, a good deal of interest in the applications

---

[1] *Technology and Culture*, Vol. II (1961), p. 357. See also Professor Smith's *History of Metallurgy* (Chicago, 1960), which is dedicated 'To those craftsmen whose intuitive understanding of materials provided the seed from which metallurgical science grew'. In his introduction Professor Smith expresses his 'belief that scientific metallurgy provides a good illustration of the complexity of the development of human knowledge, for it depends on the interaction of theory and empirical knowledge in a way that is not so evident when tracing the growth of the pure sciences'.   [2] *Ibid.*, p. 365.

of science in the first half of the eighteenth century. The Royal Society, which continued to embrace a considerable number of 'virtuosi, and perhaps dilletanti', whose interests ranged over a wide field, was still, to some extent, under Baconian influence and continued its interest in applied science.[1] This is illustrated by William Watson's words referring to William Brownrigg's book on *The Art of Making Common Salt* (1748),[2] which was presented to the Royal Society: 'The making and refining Salt must certainly be considered as one of the mechanic Arts, the History of which, as we are taught by the noble Verulam, is a necessary Part of that Knowledge, that true Science of Nature, which is not taken up in vain and fruitless Speculations, but effectually labours to relieve the Necessities of human Life.'[3] Dr. Trengrove also refers to papers on potash, glass-making, soap-boiling, sal-ammoniac manufacture, tin-plating, tanning, porcelain manufacture, agriculture, etc., but considers that, for the most part, these did not contribute a great deal to industrial progress. In the second half of the eighteenth century, indeed, 'fewer papers were read on industries, though men like John Roebuck, James Watt, Josiah Wedgwood, and James Keir were members of the Royal Society. The Society of Arts was established in 1754 and it was recognized that this society was concerned with manufactures in a way which the Royal Society was not.'[4] In 1771, for example, Sir John Pringle, President of the Society, replying to a communication from David Macbride of Dublin regarding an improved process of tanning, referred this to the Dublin Society and also 'to the Societies which are established in London and Edinburgh, for the purpose of encouraging trade and manufactures; as judging it will be more in their way than in the Royal Society's to extend the utility of this invention'.[5] Already, it appears, a distinction was developing between 'pure' and 'applied' science, as exemplified in the Royal Society and the Society of Arts.[6]

The universities of Oxford and Cambridge continued to be closely associated with the Royal Society.[7] New science chairs had been

[1] L. Trengrove, 'Chemistry at the Royal Society of London in the Eighteenth Century—I', *Annals of Science*, Vol. XIX (1963, published 1965), pp. 183–237.

[2] William Brownrigg (1711–1800), M.D., F.R.S., a doctor of Whitehaven, also carried out other industrial-chemical investigations, e.g., into coal-mine gases and mineral waters. See the articles on Brownrigg by J. Russell-Wood in *Annals of Science*, Vol. VI (1950), pp. 186–96, 436–47, and Vol. VII (1951), pp. 77–94, 199–206.   [3] *Phil. Trans.*, Vol. XLV (1748), p. 372.

[4] Trengrove, *op. cit.*, p. 192.   [5] *Phil. Trans.*, Vol. LXVIII (1778), p. 111.

[6] See D. Hudson and K. W. Luckhurst, *The Royal Society of Arts, 1754–1954* (1954), p. 58.

[7] For general accounts of early science teaching in Oxford and Cambridge, see C. Wordsworth, *Scholae Academicae: Some Account of the Studies at the English*

established in the seventeenth century: at Oxford in Geometry and Astronomy (Savilian) and in Natural Philosophy (Sedleian), 1619–21; but at Cambridge only in Mathematics (Lucasian), 1663. Professors such as Seth Ward, Sir Christopher Wren, John Wallis, and Sir Thomas Millington were among the original members of the Royal Society, while Boyle and Petty had been closely associated with Oxford, and the great Sir Isaac Newton, second holder of the Lucasian Chair of Mathematics at Cambridge, exercised immense influence both there and in the Royal Society. It is doubtful, however, whether the natural philosophy, mathematics, and astronomy taught in the universities had much practical relationship with technology in this period, apart from the links with instrument-making, clock-making, surveying, etc., to which we have previously referred. Chemistry, however, was a more 'practical' science. Peter Stahl (or Sthael), the first public teacher of chemistry at Oxford,[1] was brought over from the continent in 1659 through the influence of Hartlib and Boyle, primarily because of his knowledge of mining, mineralogy, and metallurgy—including secret manufacturing processes—as well as of medicine, and his pupils included such notable names as Christopher Wren, John Wallis, Thomas Millington, and John Locke. Chemical studies were also furthered in Oxford by the establishment of a chemical laboratory in the Ashmolean Museum, erected in the late seventeenth century.

The trend towards scientific studies continued into the early eighteenth century. Oxford established a Readership in Chemistry in 1704 and the post was first held by Dr. John Freind, or Friend (1675–1728), 'one of the most eminent physicians of the century',[2] who lectured to large audiences in the Ashmolean Museum. He emphasized the importance of accurate experimentation and was described by George Wilson as 'Well-skill'd in Speculative and Practical Chymistry'.[3]

Universities in the Eighteenth Century (Cambridge, 1877); R. T. Gunther, *Early Science in Oxford* (Oxford, 1922) and *Early Science in Cambridge* (Oxford, 1937); Hans, *op. cit.*, pp. 41–54; D. M. Turner, *History of Science Teaching in England* (1927), pp. 37–8 and 45–50; D. A. Winstanley, *Unreformed Cambridge* (Cambridge, 1935). H. M. Sinclair and A. H. T. Robb, *A Short History of Anatomical Teaching in Oxford* (Oxford, 1950), provides information on the close links between medical and chemical teaching.

[1] G. H. Turnbull, 'Peter Stahl, The First Public Teacher of Chemistry at Oxford', *Annals of Science*, Vol. IX (1953), pp. 265–70.

[2] Wordsworth, *op. cit.*, pp. 175–6. See also Gunther, *Early Science in Oxford*, Vol. I, pp. 52–63.

[3] Gunther, *op. cit.*, p. 53. Freind's lectures for 1704, delivered in Latin, were published in 1709 and translated into English in 1712, as *Chymical Lectures in which almost all the Operations of Chymistry are reduced to their True Principles and Laws of Nature.*

Another popular lecturer was Nathan Alcock (1707-79), F.R.S., a pupil of Boerhaave at Leyden, who came to Oxford in 1738 to lecture on chemistry and anatomy.[1] The links between medicine and chemistry were, of course, very close and many of the leading chemists of the eighteenth century were medical men, both inside and outside the universities. Interest in natural philosophy, including mechanics, hydrostatics, optics, etc., as well as astronomy, also increased at Oxford. John Keill (1671-1721), F.R.S., who came there from Edinburgh in the wake of David Gregory (Savilian Professor of Astronomy, 1691-1708), is said to have been 'the first who publickly taught Natural Philosophy by Experiments in a mathematical Manner', in about 1704-5;[2] he lectured on Newtonian philosophy at Hart Hall, deputizing for Sir Thomas Millington, the Sedleian Professor of Natural Philosophy.[3] Keill was appointed to the Savilian Chair of Astronomy in 1712, but his lectures in 'experimental philosophy' had been taken over in 1710 by John T. Desaguliers (1683-1744), F.R.S.,[4] of French Huguenot extraction, who continued them for about three years before moving to London, where, as we shall see, he gave similar public lectures.[5] James Bradley, F.R.S., Keill's successor as Savilian Professor in 1721 and as Whiteside Lecturer in Experimental Philosophy in 1729, lectured regularly to substantial audiences till his resignation in 1760, though his interests were mainly in astronomy.[6]

The story of science teaching in Cambridge in the eighteenth century demonstrates a stronger interest in mathematics and natural philosophy, while the study of chemistry was also vigorously developing. Keill's counterpart in Cambridge was Roger Cotes (1682-1716), F.R.S., first holder, 1705-16, of the newly established Plumian Chair of Astronomy and Natural (or Experimental) Philosophy. In 1707, together with William Whiston (1667-1752), who had succeeded Newton as Professor of Mathematics, he started a course of lectures on hydrostatics and pneumatics, illustrated by experiments, in the

[1] T. Alcock, *Some Memoirs of the Life of Dr. Nathan Alcock* (1780); Wordsworth, *op. cit.*, p. 185; Hans, *op. cit.*, pp. 48-9.

[2] J. T. Desaguliers, *A Course of Experimental Philosophy*, Vol. I (1734), Preface; D. Brewster, *Ferguson's Lectures* (2nd edn., 1806), pp. xxi-xxii.

[3] His lectures, published in Latin in 1701 and translated into English in 1720, influenced all subsequent lecturers in natural philosophy during the eighteenth century. His brother, James Keill, who had studied medicine at Edinburgh, Paris, and Leyden and then lectured on anatomy at both Oxford and Cambridge, translated Lemery's *Course of Chemistry*, which became the standard 'text-book of chemistry in the early part of the eighteenth century, and gave the first account in English of current theories of acids and alkalis'. Sinclair and Robb-Smith, *op. cit.*, pp. 19-20; Wordsworth, *op. cit.*, pp. 182 and 187.

[4] Desaguliers, *op. cit.*, Vol. I, Preface.   [5] See below, pp. 37-40.

[6] Wordsworth, *op. cit.*, p. 247; Hans, *op. cit.*, p. 48.

observatory at Trinity College.[1] These were public lectures, delivered to 'large assemblies'.[2] Whiston was dismissed for Arian heresy in 1710, when he departed to London and began to lecture there publicly, like Desaguliers, [3] but Cotes carried on with the lectures at Cambridge, and after his death in 1716 they were continued by his successor, Robert Smith, until his resignation in 1760. Nicholas Sanderson (1682–1739), the remarkable blind mathematician who followed Whiston in the Lucasian Chair, also delivered lectures on Newtonian philosophy and mathematics to 'numerous classes'.[4]

Others at Cambridge who contributed to the development of this subject and to its widespread dissemination were John Rowning, Fellow of Magdalene College, who produced *A Compendious System of Natural Philosophy* (Cambridge, 1738), and Dr. Thomas Rutherford, Professor of Divinity, 1756–61, who had previously published several scientific works, including *A System of Natural Philosophy* (1748). Rowning, described as 'an ingenious mechanic, mathematician, and philosopher',[5] was an active member of the Gentlemen's Society at Spalding, the earliest provincial literary and philosophical society;[6] he had a brother who was 'a great mechanic and famous watch-maker, at Newmarket'.[7]

Meanwhile, chemical studies were also being developed at Cambridge. John Francis Vigani (c. 1650–1713), a native of Verona, who appears to have started teaching chemistry there privately in 1683, continued until his retirement in 1708.[8] His lectures were probably public, including 'local physicians, pharmacists and amateurs', as well as members of the university; Vigani himself dabbled in pharmacy. In February 1703 he was granted the honorary title of Professor of Chemistry, and in 1707 a laboratory was fitted up for him in Trinity College; he was on intimate terms with Newton, who enjoyed discussing chemistry with him—until Vigani told him an improper

---

[1] *Memoirs of the Life and Writings of Mr. William Whiston . . . Written by Himself* (1753), Part I, p. 118; Wordsworth, *op. cit.*, p. 247; Hans, *op. cit.*, pp. 49–50 and 137.

[2] According to Robert Smith, Cotes' successor, in the preface to Cotes' *Hydrostatical and Pneumatical Lectures*, published in 1738.

[3] Whiston's *Memoirs*, Part I, pp. 118, 150, and 201. See below, pp. 40–1.

[4] Wordsworth, *op. cit.*, pp. 68–70; Hans, *op. cit.*, p. 50.

[5] J. Nichols, *Literary Anecdotes of the Eighteenth Century*, Vol. VI, pt. i (1812), p. 109. Joseph Priestley studied his *Compendious System of Natural Philosophy* at Daventry. R. E. Schofield, 'Joseph Priestley, Natural Philosopher', *Ambix*, Vol. XIV, no. 1, Feb. 1967, pp. 1–15.

[6] See below, p. 139.   [7] Nichols, *loc. cit.*

[8] L. J. M. Coleby, 'John Francis Vigani, First Professor of Chemistry in the University of Cambridge', *Annals of Science*, Vol. VIII (1952), pp. 46–60; Wordsworth, *op. cit.*, p. 188.

story about a nun! Vigani, so Dr. Coleby tells us, was 'essentially a practical chemist', skilled in experimental work. His *Medulla Chemiae* (apparently first published in Dantzig in 1682, with later enlarged editions in London and elsewhere) and surviving notes of his lectures are composed mainly of 'clear straightforward instructions for the preparation of chemical compounds and pharmaceutical recipes', often with practical hints regarding apparatus, etc., and with very little theorizing.

Some impression of the experimental science at Cambridge in Vigani's day is provided by the memoirs of Dr. William Stukeley.[1] 'We saw . . . many Philosophical Experiments in Pneumatic Hydrostatic Engines & instruments performed at that time by Mr. [John] Waller[2] . . . & the doctrine of Optics & Telescopes & Microscopes, & some Chymical Experiments with Mr. Stephen Hales then Fellow of the College, now of the Royal Society.'[3] Stukeley himself 'made air pumps & 20 inventions to try mechanical & philosophical experiments'.[4] He was particularly interested by Vigani's 'Chymical Lectures'[5] and managed to obtain a room in college in which to 'practise Chymical Experiments', as well as animal dissections. 'I had sand furnaces, Calots, Glasses, & all sorts of Chymical Implements.'[6] He was also friendly with Stephen Gray, assistant first to Cotes and later to Desaguliers, and with his nephew, John Gray, with whom he often tried 'Various Experiments in Philosophy'.[7]

John Waller, second Professor of Chemistry at Cambridge, 1713–18, is a rather obscure figure, but we know a good deal about the third holder, John Mickleburgh (or Mickleborough), 1718–56, and the fourth, John Hadley, 1756–60, thanks to Dr. Coleby.[8] Like Vigani,

[1] *The Family Memoirs of the Rev. William Stukeley, M.D.*, Surtees Society Publications, Vol. LXXIII (1882, for 1880).   [2] Later Professor of Chemistry.

[3] Stukeley *Memoirs*, p. 21. Hales is another interesting figure. Primarily a 'pure' scientist, he did, however, have some strong utilitarian interests, as evidenced by his collaboration with Desaguliers in producing a sea-sounding device (see below, p. 39) and with Thomas Yeoman, the engineer, in the development of ventilators for use in large buildings and ships (see below, pp. 39, 378–9, 384–7). He also analysed mineral waters. See A. E. Clark-Kennedy, *Stephen Hales, D.D., F.R.S.* (Cambridge, 1929). He was a founder-member of the Society for the Encouragement of Arts, Manufactures, and Commerce. Hudson and Luckhurst, *op. cit.*, chap. I.   [4] *Ibid.*, p. 32. See also pp. 142–3.

[5] *Ibid.*, pp. 28, 33, 39. He refers to the establishment of the 'New Chymical Laboratory' in Trinity College and to Vigani's Professorship on p. 40.

[6] *Ibid.*, pp. 32–3.

[7] *Ibid.*, p. 41. Stukeley refers to the establishment of the astronomical observatory in Trinity College, describing Cotes as 'a very ingenious Man, well versed in Philosophy, Astronomy, Optics, Mechanics &c.'

[8] L. J. M. Coleby, 'John Mickleburgh, Professor of Chemistry in the University of Cambridge, 1718–56', *Annals of Science*, Vol. VIII (1952), pp. 165–74; 'John Hadley, Fourth Professor of Chemistry in the University of Cambridge', *loc. cit.*, pp. 293–301. See also Wordsworth, *op. cit.*, pp. 188–9.

Mickleburgh combined his professorial duties with a commercial side-line as a dispensing chemist. He also followed Vigani's practice of delivering public lectures, which attracted a few medical men, but attendances were not very high, and he apparently gave no lectures during his last fifteen years' tenure of the Chair. Like Vigani, too, he illustrated his lectures with experiments, dealing with such matters as calcination, distillation, fermentation, pharmaceutical preparations, metals and minerals, acids and alkalis, etc. But he was much more interested in theory than Vigani, especially in Newton's attempt to find a physical explanation of chemical affinity and reaction. John Hadley, F.R.S., was similarly interested in theory—following the German and French chemists such as Stahl and Macquer and adopting the phlogiston theory—but his lectures, like those of his predecessors, were illustrated by a great many experiments and also showed a noticeable interest in industrial-chemical processes, such as the manufacture of alum at Whitby, of green vitriol at Deptford, and Joshua Ward's method of making sulphuric acid in his works at Twickenham. Here was no cloistered academic.

In addition to the chairs of astronomy, experimental philosophy, and chemistry, new ones were also established at Cambridge in other scientific subjects such as botany, geology, and geometry. Clearly, the two ancient universities were showing considerable flexibility and making increasing provision for study of the physical sciences.[1] Their scientific influence spread far beyond their walls: the lectures and publications in 'experimental philosophy', for example, were widely utilized and popularized in London and the provinces during the eighteenth century, whilst their *alumni* were active in many walks of life. As Dr. Hans has shown, they produced a high proportion of British scientists in the eighteenth century: about a third of the total, according to his estimate.[2] They were clearly not the benighted institutions so frequently depicted. On the other hand, however, 'the number of scientists who received no secondary and University education . . . increased with every generation both absolutely and relatively', and 'the percentage of Oxford and Cambridge graduates among the scientists decreased steadily from 67 per cent in the seventeenth century to 20 per cent at the end of the eighteenth'.[3] There are signs of stagnation at Oxford and Cambridge in the middle decades of the eighteenth century.[4] In many ways, and especially in applied science, it is evident

---

[1] See also the various contemporary lists of scientific courses and books in Wordsworth, *op. cit., passim*; Turner, *op. cit.*, pp. 47–9; Hans, *op. cit.*, p. 51.
[2] Hans, *op. cit.*, p. 53.   [3] *Ibid.*, p. 34.
[4] See below, pp. 166–78, however, for revival in the later eighteenth century.

that, even in the first half of the century, the most significant developments were taking place outside the universities and the Royal Society, as Dr. Hans and the late Dr. F. W. Gibbs have shown.

In London, as we have previously noticed, George Wilson was giving public lectures on chemistry in the later seventeenth and early eighteenth centuries.[1] Dr. John Harris (1666?–1719), F.R.S., scholar of Trinity College, Cambridge, author of *Lexicon Technicum: or an Universal English Dictionary of Arts and Sciences* (1704; 2nd edn., 2 vols., 1708–10),[2] began about 1698 to read free public lectures on mathematics at the Marine Coffee House in Birchin Lane; these were instituted 'for the public good' by Mr. (later Sir) Charles Cox, M.P.[3] Harris was still lecturing in 1704. In 1702–15 he ran a private school in his house at Amen Corner, teaching mathematics and other 'modern' subjects.[4] Francis Hauksbee, the instrument-maker, also lectured publicly on 'experimental philosophy' in London from about 1704–5.[5] Desaguliers was not, therefore, as Dr. Hans asserts, the first to give such public lectures there.[6] He was certainly, however, an early pioneer and an outstanding figure. After leaving Oxford,[7] he began lecturing on 'mechanical and experimental philosophy' in London in about 1713,[8] at his house in Channel Row, Westminster, assisted by William Vream, instrument-maker. The lectures were intended for those 'but little vers'd in Mathematical Sciences' and were made attractive by experiments;[9] his public audiences included 'merchants, craftsmen and clerks',[10] while he also lectured privately to gentry and nobility, and even to King George I and his court. In 1714 he was elected a Fellow of the Royal Society, which invited him to become its demonstrator and curator; he was held in high esteem by Sir Isaac Newton, who was then President. He continued to lecture with great success until his death in 1744; from 1738 in the Bedford Coffee House, Covent Garden. His lectures were published in various forms[11] and provided models for

[1] See above, pp. 25–6. [2] See above, p. 26.
[3] *D.N.B.;* Hans. *op. cit.*, p. 153. [4] Hans, *op. cit.*, pp. 114 and 153.
[5] See below, p. 40. [6] Hans, *op. cit.*, p. 137. [7] See above, p. 33.
[8] Desaguliers, *A Course of Experimental Philosophy*, Vol. I (1734), Preface. Charles Hutton, *Mathematical and Philosophical Dictionary* (1795), states that Desaguliers began lecturing as soon as he moved to London, in 1712; Wordsworth, *op. cit.*, p. 246, states 1713; A. De Morgan, *A Budget of Paradoxes* (1872), p. 93, says that 'Desaguliers removed to London soon after 1712, and commenced his lectures soon after that.'
[9] In addition to Desaguliers' own lectures, see *D.N.B.*, Wordsworth, *op. cit.*, pp. 246–7, and Hans, *op. cit.*, pp. 137–41. [10] Hans, *op. cit.*, p. 141.
[11] *Physico-Mechanical Lectures* (1717); *Lectures of Experimental Philosophy* (1719, edited by Paul Dawson, but repudiated by Desaguliers); *A Course of Mechanical and Experimental Philosophy* (1724); *A Course of Experimental Philosophy*, Vol. I (1734), Vol. II (1744).

successive lecturers later in the century. He read several papers to the Royal Society on light, colours, the barometer, etc., and produced a dissertation on electricity in 1742. He also translated a number of foreign works on natural philosophy.[1]

'Practical' engineering historians, such as the late Dr. H. W. Dickinson and Mr. L. T. C. Rolt, have sneered at Desaguliers' attempts to apply theory to engineering and have resented his criticisms of the smiths, millwrights, etc. who were engaged in it. But their view of Desaguliers is largely distorted. He was not purely a theorist, devoid of engineering experience, but constantly emphasized the need for union of theory with practice. It is true that he made some considerable errors, as Smeaton and Watt discovered, and perhaps he was too critical of 'the Plumbers and Mill-Wrights now set up for Engineers', many of whom he regarded as incompetent and even fraudulent projectors;[2] but he fully appreciated the abilities of outstanding contemporaries such as Hadley and Sorocold, who had 'a strong natural Genius for Mechanicks' and, though 'without any previous Knowledge of Mathematicks and Philosophy', had yet 'performed great Works, and built such Rules upon Facts and Observation as have stood them in good stead'.[3] Desaguliers hoped that engine-makers, 'generally quite ignorant of mathematics', would, by familiarizing themselves with science, be better able to design mills, engines, etc., and not attempt impossibilities.[4] At the same time, he hoped that philosophers 'wou'd not think it below them to direct Workmen, and consider Engines a little more than they do, which wou'd render their Speculations more useful to Mankind'. He stressed the need for practical experiment and pointed out that 'the incomparable Sir Isaac Newton, whom They with all other Philosophers admire, has made as many as (if not more Experiments than) any Man living'.[5] Not merely did practical engineers often waste their efforts through lack of scientific knowledge, but philosophers often failed in designing water-mills, etc. through lack of practical experience: the workmen's distrust and contempt of 'Men of Theory' was often justified.[6] Desaguliers emphasized not only that practical engineers should acquaint themselves with mathematics, hydrostatics, and hydraulics, but also that philosophers attempting utilitarian

---

[1] For example, Marriotte's on *The Motion of Water and other Fluids. Being a Treatise of Hydrostaticks* (1718), and Gravesande's on the *Mathematical Elements of Natural Philosophy* (1720).

[2] See, for example, *A Course of Experimental Philosophy*, Vol. I, p. 69, Vol. II, p. 415.   [3] *Ibid.*, Vol. II, p. 414.

[4] *Ibid.*, Vol. I, p. 69. But see below, pp. 43–51, for the mathematical and scientific knowledge of some of these early engineers, including Hadley and Sorocold.   [5] *Ibid.*   [6] *Ibid.*, Vol. II, p. 415.

objectives must have a thorough knowledge of practical operations:[1]

> to have a compleat Theory, the Undertaker must understand Bricklayer's Work, Mason's Work, Mill-Wright's Work, Smith's Work, and Carpenter's Work; the Strength, Duration, and Coherence of Bodies; and must be able to draw not only a general Scheme of the whole Machine, but of every particular Part; and small Parts must be drawn by a larger Scale, in order to be fully examined before any thing is begun.

Desaguliers' lectures were packed with information about industrial practices, water-wheels, steam engines, etc. He drew not only upon scientific theory (including innumerable references to the works of French scientists such as Belidor, Parent, Camus, etc.), with mathematical calculations, but also upon wide-ranging practical knowledge. He had made a particularly close study of water-wheels 'for many Years'[2] and described a large number of the best industrial examples in corn-milling, mine-drainage, water-supply, etc. He also had a detailed knowledge of steam engines, in the design of which as well as of water-wheels he had not merely suggested theoretical improvements, but had actually engaged in practical construction, with the aid of mill-wrights and other craftsmen.[3] Nor were these his only utilitarian interests. He also demonstrated his practical abilities in instrument-making, in the development of 'a new and improved engine for raising water',[4] as the inventor (with Dr. Stephen Hales) of a 'Sea-Gage' or sounding device,[5] and also (again with Hales) of engines for ventilating coal mines, large buildings, and ships, and for drying malt, etc.,[6] and as consultant on the building of Westminster Bridge in 1738–9.[7] Not surprisingly, he was awarded the Royal Society's Copley gold medal

---

[1] *Ibid.*   [2] *Ibid.*, Vol II, Preface, p. iv.

[3] See, for example, his improved water-mill, which he demonstrated to the Royal Society (*ibid.*, pp. 459–61) and his improved Savery 'fire engine' (*ibid.*, pp. 484–90). The latter was an outcome of discussion with Gravesande, the later famous Dutch philosopher, while he was attending Desaguliers' lectures in 1716. See also J. Farey, *A Treatise on the Steam Engine* (1827), pp. 110–16.

[4] *D.N.B.*, quoting *Daily Post*, 6 Jan. 1721. His claims to originality were disputed.

[5] *A Course of Experimental Philosophy*, Vol. II, pp. 223–4 and 241–7.

[6] *Ibid.*, pp. 556–68. Both the sounding and ventilating 'engines' were demonstrated to the Royal Society.

[7] *D.N.B.* Desaguliers referred to the works engineer, Charles Labelye, as 'formerly my Disciple and my Assistant' (*A Course of Experimental Philosophy*, Vol. II, p. 506). He also gave a description of a horse-powered pile-driving engine, invented by Vauloué, a watch-maker, which was used in building this bridge (*ibid.*, pp. 417–18). Both Desaguliers and Labelye (see below, pp. 44–5) were of French–Huguenot extraction, and so, too, perhaps, was Vauloué. This combination of French science and engineering is particularly interesting.

in 1741. He always emphasized that 'the business of . . . Science' is not merely 'to contemplate the Works of God, [and] to discover Causes from their Effects', but also to 'make Art and Nature subservient to the Necessities of Life, by a Skill joining proper Causes to produce the most useful Effects'.[1] His activities, it may be noted, extended outside London, to membership of the Gentlemen's Society at Spalding, the earliest provincial philosophical society,[2] and to lecturing in Holland in 1730, where he interested scientists such as Boerhaave and Huygens.

Several years before Desaguliers began lecturing in London, Francis Hauksbee, F.R.S., a celebrated metropolitan instrument-maker, had given public lectures there, illustrated by experiments. According to Desaguliers, these were started at about the same time as Keill's first Oxford lectures, in 1704/5, dealing with pneumatics, hydrostatics, optics, and electricity, but were not grounded on mathematics like those of Keill, 'tho' perhaps perform'd more dexterously and with a finer Apparatus: They were Courses of Experiments, and his [Keill's] a Course of Experimental Philosophy'.[3] Francis Hauksbee, and also his nephew and namesake, were both advertising such courses in 1712, at their respective London establishments.[4] Hauksbee senior (d. 1713) was instrument-maker and operator to the Royal Society,[5] published *Physico-Mechanical Experiments* in 1709, and contributed to the Society's *Transactions*. The younger Hauksbee (1687–1763) followed in his uncle's footsteps and in 1723 became the Royal Society's clerk or librarian and keeper of the museum in Crane Court. These two men typify the skilled London craftsmen who, in addition to providing instruments for the scientists and often assisting in their experiments, were themselves no mean philosophers. They collaborated with William Whiston, who, after dismissal from the Cambridge chair of mathematics in 1710,[6] came to London and started lecturing on astronomy and experimental philosophy.[7] Surviving syllabuses of their joint courses show that Whiston did the lecturing while Hauksbee

[1] *A Course of Experimental Philosophy*, in the dedication at the beginning of Vol. I.

[2] Nichols, *op. cit.*, Vol. VI, pt. i (1812), p. 81, See below, p. 139.

[3] Desaguliers, *A Course of Experimental Philosophy*, Vol. I, Preface.

[4] *The Spectator*, 7 to 19 Jan. 1712. (These references were kindly supplied by Mr. G. L'E. Turner, of the Oxford Museum of the History of Science.) Hauksbee junior was assisted by Humphry Ditton (1675–1715), a master in the new mathematical school at Christ's Hospital. For Ditton, see Hans, *op. cit.*, pp. 38, 58, 59, 217. Ditton also collaborated with Whiston (Whiston *Memoirs*, Part I, p. 202).

[5] See above, p. 20.   [6] See above, p. 34.

[7] Whiston's *Memoirs*, Part I, pp. 118, 150, and 201; Wordsworth, *op. cit.*, p. 242; Hans, *op. cit.*, p. 142; *D.N.B.*

carried out the experiments.[1] Whiston may possibly have begun lecturing in London just before Desaguliers,[2] and he continued this activity till late in life (d. 1752).

Other courses in 'experimental philosophy' were also provided in London during this period, notably by men connected with the Little Tower Street Academy.[3] In 1722, for example, the lectures of Thomas Watts and Benjamin Worster were published in *A Compendious and Methodical Account of the Principles of Natural Philosophy: As they are Explained and Illustrated in the Course of Experiments performed at the Academy in Little Tower Street*. This course, which included mechanics, hydrostatics, pneumatics, optics, etc., was intended 'for qualifying young gentlemen for business'.[4] Watts was primarily a teacher of mathematics, while Worster, a Cambridge graduate, appears to have been mainly responsible for the lectures in natural philosophy, which were delivered publicly as well as in the Academy. After Worster's death, James Stirling (1692–1770),[5] a mathematician of international repute, continued these lectures. Stirling had been educated at Glasgow University and Balliol College, Oxford, but was expelled for Jacobite sympathies in 1715; he then went to Venice, where he studied mathematics and became friendly with Nicolas Bernoulli, the great Italian mathematician. He also maintained a correspondence with Newton and Desaguliers, who assisted him on his return to London in 1725, when he began to lecture at the Little Tower Street Academy and also in the Bedford Coffee House.[6] He was elected F.R.S. next year and continued lecturing until 1735, when he was appointed manager to the

[1] (i) *A Course of Mechanical, Optical, Hydrostatical, and Pneumatical Experiments, to be performed by Francis Hauksbee and the Explanatory Lectures read by William Whiston, M.A.* (ii) *An Experimental Course of Astronomy proposed by Mr. Whiston and Mr. Hauksbee*. Neither of these is dated, but Whiston himself gives the date of publication of the first as 1714, so his collaborator must have been Hauksbee junior. The latter also assisted Desaguliers, together with William Vream, another instrument-maker, in developing his ventilators (Desaguliers. *op. cit.*, Vol. II, pp. 556 ff.).　　　　[2] De Morgan, *op. cit.*, p. 93.

[3] Turner, *op. cit.*, p. 44; Hans, *op. cit.*, pp. 82–7; Taylor, *Mathematical Practitioners of Hanoverian England*, p. 7. The Soho Academy, founded by Martin Clare, M.A., F.R.S., a friend of Desaguliers, was a similar institution, providing courses in mathematics, natural philosophy, etc., John Barrow, author of a *New Dictionary of Arts and Science* (1753), appears later to have become a master there. Hans, *op. cit.*, pp. 87–92.

[4] In addition to natural philosophy, other scientific and commercial subjects were taught, such as mathematics, book-keeping, and foreign languages.

[5] In addition to the works previously cited, see C. Tweedie, *James Stirling* (1922) and the *D.N.B.*

[6] *A Course of Mechanical and Experimental Philosophy*, by Stirling and others. was published in 1727. Stirling was assisted not only by Watts, but also by William Vream, the instrument-maker, who, as we have seen, had previously assisted Desaguliers and also taught at the Academy.

D

Scots Mining Company at Leadhills in Lanarkshire, where he proved himself a successful engineer and administrator, a good example of science combined with industry. Later he surveyed the Clyde with a view to rendering it navigable by a series of locks.

Another who did much to spread natural philosophy in London and later in the provinces was Benjamin Martin (1704–82).[1] Beginning life as a ploughboy in Surrey, studying mathematics and other scientific subjects in his spare time, becoming a schoolmaster and maker of optical instruments, and finally an optician in London, he also gave lectures in natural philosophy and published a variety of works on the subject, after acting as assistant to Desaguliers[2] in the early 1740s. His *Bibliotheca Technologica* (1737) was 'a kind of "Vade-mecum" for technicians'.[3] It was followed by *A Course of Lectures in Natural and Experimental Philosophy, Geography and Astronomy* (Reading, 1743), *Philosophia Britannica* (2 vols., 1747), and a host of other publications; his best-known work was the *General Magazine of Arts and Sciences* (1755). He continued the encyclopaedic line started by Harris and Chambers and his books had a wide sale, since he popularized the work of the great natural philosophers and provided many practical illustrations. He wrote many elementary text-books on mathematics and philosophy, and on mathematical and scientific instruments, including their uses in surveying, distilling, etc.; his utilitarian outlook is also illustrated by his book on *The Principles of Pump Work* (1767). His books were still being read in the second half of the nineteenth century.[4] As another instance of the way in which scientific knowledge was transmitted, it may be mentioned that John Robison, the later famous Edinburgh professor of natural philosophy and friend of James Watt, when he came, as a young man, to London, attended Martin's lectures.[5] Martin's example was followed by many others, especially instrument-makers and watch-makers, who similarly lectured and published works on natural philosophy, both in London and the provinces.[6] Some of these works were specifically intended for artisans, e.g. Batty Langley's publication of *The Builder's Compleat Assistant, or a Library of Arts and Sciences, absolutely necessary to be understood by builders and workmen in general* (1738), which included arithmetic, geometry, architecture, mensuration, trigonometry, land-surveying, mechanics, and hydrostatics, in two large volumes.[7]

[1] *D.N.B.;* Hans, *op. cit.,* pp. 145 and 152–3.   [2] De Morgan, *op. cit.,* p. 91.
[3] Hans, *op. cit.,* p. 152.   [4] De Morgan, *loc. cit.*
[5] J. Robison to J. Watt, 15 Dec. 1798, Dol.
[6] Hans, *op. cit.,* pp. 144–55. See also below, pp. 109–10.
[7] Langley was a notable land-surveyor, builder and architect in London. He had previously published *Practical Geometry applied to . . . Building, Surveying, Garden-*

Thus there is no doubt of the increasing provision, especially in London, of courses in natural philosophy, knowledge of which, as we shall see, was later spread in the provinces by itinerant lecturers, academies, schools, books, and periodicals, and was to prove of utilitarian value to engineers and craftsmen of many sorts. Even in this earlier period, there is evidence to suggest that Desaguliers exaggerated the unscientific empirical practices among engineers of his day. John Hadley and George Sorocold, for example, whom Desaguliers referred to as simply practical engineers, appear to have had some knowledge of mathematics, surveying, etc., which they utilized in their waterworks and river navigation schemes.[1] Hadley, notable as the inventor of a mechanism for raising and lowering water-wheels according to the level of the stream (patented 1693), and for his waterworks schemes in Worcester, Chester, London, etc., and also as engineer for the Aire and Calder navigation, was described by Sir Godfrey Copley, F.R.S., of Copley Medal fame, as 'a man of Mathemat. & Bookish';[2] Ralph Thoresby, F.R.S., referred to him as 'Mr. Hadley, the hydrographer';[3] and somewhat later William Maitland, in his *History of London* (1756), called him 'that great Master of Hydraulicks'.[4] George Sorocold, who collaborated with Hadley in Marchant's London Waterworks (1696) and the more famous waterworks at London Bridge (1701–2),[5] and who also engineered similar schemes in Derby, Leeds,

---

ing and *Mensuration* (1726), as well as many other works, and had established a school or academy of architectural drawing. See *D.N.B.*

[1] In the first edition of *D.N.B.*, Hadley the engineer is confused with the famous John Hadley (1682–1744), instrument-maker, inventor of the sextant, and Vice-President of the Royal Society; but this error is corrected in the 1908 edition. Sorocold does not appear in the *D.N.B.*, though he was the leading engineer of his day. For the work of Hadley and Sorocold on river improvements, see A. W. Skempton, 'The Engineers of the English River Navigations, 1620–1760', *Newcomen Soc. Trans.*, Vol. XXIX (1953–5), pp. 25–54. For more detailed accounts of these engineers and their other activities, see Rhys Jenkins, 'George Sorocold: A Chapter in the History of Public Water Supply', *The Engineer*, Vol. 126 (1918), pp. 333–4; F. Williamson, 'George Sorocold of Derby', *Journ. Derbyshire Arch. & Nat. Hist. Soc.*, Vol. 57 (1936, pub. 1937), pp. 43–93; F. W. Williamson and W. B. Crump, 'Sorocold's Waterworks at Leeds, 1694', *Thoresby Soc. Pub.*, Vol. XXXVII (1945; originally published 1941); W. H. G. Armytage, 'George Sorocold and Sir Godfrey Copley', *ibid.*, Vol. LXXII (1953), pp. 105–7. See also the article by Peet on Thomas Steers, cited below, p. 46, n. 6, which has an appendix on Sorocold.

[2] Sir Godfrey Copley to Thomas Kirke (also F.R.S.), 4 June 1696, cited by Skempton, *op. cit.*, p. 41, and Armytage, *op. cit.*, p. 106.

[3] Williamson and Crump, *op. cit.*, p. 175.   [4] Skempton, *op. cit.*, p. 41.

[5] It is interesting to note that the former waterworks was described in S. Switzer, *Hydrostaticks and Hydraulicks* (1729) and the latter by Henry Beighton, the engineer, in the Royal Society's *Philosophical Transactions*, Vol. XXXVII (1731), pp. 5–12. Particularly interesting is Sorocold's use of cast-iron for some of the parts.

Norwich, Yarmouth, Portsmouth, Bristol, Sheffield, and many more towns in the 1690s and early 1700s, is also famous for his engineering work on the waterpowered silk-throwing mills of Thomas Cotchet (1702) and the Lombe brothers (1717–21) at Derby;[1] he was also consulting engineer for the first great wet dock in England, at Rotherhithe (1696–1700), and for the first Liverpool dock (1708), in addition to which he was engineer for the River Derwent navigation scheme, and for draining coal mines at Alloa and perhaps also at Newcastle; he also patented (1704) a horse- or water-powered sawing machine, and he similarly bored water-pipes. He, like Hadley, was well known to and greatly admired by such notable scientific virtuosi as Sir Godfrey Copley and Ralph Thoresby, and was referred to in 1717 by the Mayor of Derby as the 'ingenious . . . mathematician, Mr. Sorocold';[2] his surviving plans and maps show an obvious ability in surveying.

Professor Skempton has pointed out that some of these early professional engineers did, indeed, supplement their income with surveying, or, like John Grundy, by teaching 'mathematicks'.[3] We shall deal later with Grundy, a leading fen-drainage and river-improvement engineer,[4] and also with Thomas Yeoman, similarly notable for river navigation schemes as well as for his mechanical ventilators and for his engineering work with Thomas Wyatt at the Northampton cotton-spinning mill in the early 1740s.[5] These men, too, as we shall see, were certainly not without knowledge of mathematics and natural philosophy, and had close contacts with men of science. Another outstanding figure was Charles Labelye (1705–81?),[6] born in Switzerland of French-Huguenot extraction, who came to England in about 1725 and later became famous as the engineer of the first Westminster Bridge and especially for his use of the caisson method of building underwater foundations. He also may be considered an early scientific engineer.[7] He was on close and friendly terms with Desaguliers, who, as we have noticed, referred to him as 'formerly my Disciple and my Assistant'.[8] We find him writing to Desaguliers in 1735 regarding the laws of motion, of which he displayed considerable

---

[1] It has generally been overlooked by economic historians that the Lombes' famous silk mill was preceded by that of Cotchet, but Williamson, *op. cit.*, pp. 55–64, has made this perfectly clear. Sorocold was responsible for the waterwheels and millwork in both these mills.

[2] Williamson, *op. cit.*, p. 51; Skempton, *op. cit.*, p. 43.

[3] Skempton, *op. cit.*, p. 37.

[4] See below, pp. 140–1.   [5] See below, chap. XI.

[6] See *D.N.B.* There are only scattered references to him in Smiles, *Lives of the Engineers*, Vol. II.

[7] As agreed by Professor Skempton and Mr. L. E. Harris, in Skempton, *op. cit.*, p. 51.   [8] See above, p. 39, n.7.

philosophical and mathematical knowledge;[1] he also provided Desaguliers with a technical description and scale drawings of Newsham's
'fire-engine'.[2] His surviving plans and calculations for Westminster
Bridge also demonstrate his application of mathematics to bridge-
building, on which subject he appears to have been in touch with
scientifically-minded French engineers such as Belidor.[3]

Desaguliers also had contacts with other leading engineers including
John Grundy,[4] Henry Beighton,[5] and more obscure improvers of the
time, who were similarly knowledgeable in natural philosophy,
among them 'the learned and ingenious Dr. [Robert] Barker', whose
improved breast-wheel Desaguliers described and 'whose skill in
Mechanicks as well as all Parts of Mathematicks and [Natural] Philosophy is well known';[6] Desaguliers also described 'a new invented Mill'—an
early water turbine—introduced by Dr. Barker and developed by himself, which was based on a scientific proposition put forward by Parent.[7]

There is no doubt that knowledge of mathematics and natural
philosophy, including the writings and experiments of philosophers
such as Desaguliers, Belidor, Parent, etc., as well as earlier Italian
works—some of these made available in translation—was much more
widespread among leading engineers and millwrights of that time than
has hitherto been appreciated. The emphasis upon hydrostatics and
hydraulics, with many practical examples, was particularly important
in a period of rapidly developing water power and water transport.
Professor Willan considers that 'it cannot be said that the river improvements of this period were inspired by any great advance in
technical knowledge',[8] but he does provide some evidence of the
influence of science on civil engineering as it developed during these
years. He admits, for instance, that 'the mathematician merged into the
surveyor and probably the surveyor into the engineer'.[9] He also

---

[1] Desaguliers, *A Course of Experimental Philosophy*, Vol. II, pp. 77–91.

[2] *Ibid.*, pp. 505–18.

[3] Belidor's *Architecture Hydraulique* (Paris, 1753), Vol. II, p. 198, describes
Labelye's method of laying the foundations of Westminster Bridge.

[4] They were both members of the Gentlemen's Society at Spalding. See below,
pp. 139–40.   [5] See below, pp. 47–8.

[6] Desaguliers, *op. cit.*, Vol. II, pp. 453–4.

[7] *Ibid.*, pp. 459–61. Desaguliers had demonstrated it to the Royal Society.
He referred on several occasions to the works of 'that excellent Mechanick
Mons. Parent, of the Royal Academy of Sciences' (see, for example, *ibid.*, pp.
422–7). 'Barker's Mill' frequently featured in later courses of experimental
philosophy.

[8] T. S. Willan, *River Navigation in England 1600–1750* (1964), p. 79.

[9] *Ibid.* He gives as an example Sir Jonas Moore, who 'did everything from
writing mathematical treatises to founding the Royal Observatory, surveying the
Great Level of the Fens, and studying the connexion of the Cam and the Thames'.

recognizes that 'there was ... some interest shown in the more theoretical side of hydrodynamics, chiefly through Italian influence', and mentions the works of Castelli and Guglielmini.[1] As we shall see, engineers such as John Grundy certainly utilized these works;[2] Professor Willan does, indeed, refer to Grundy's *Philosophical and Mathematical Reasons*, as an example of the 'engineer pamphleteers who were prepared to give their opinion on the technical problems of their day and occupation'.[3] It may be, as Professor Willan observes, that they were wrong on some points, and perhaps some of their schemes suffered thereby, but this does not mean that their scientific-technical knowledge was of no practical significance. If so, why should an engineer such as Grundy, a man of great practical experience, have so strongly emphasized the importance of mathematics and natural philosophy? Is it not significant that it was from Italy, where hydrodynamics was early studied, that the pound-lock originated? This invention, which Professor Willan considers was 'almost as important for economic development as the discovery of steam power',[4] was 'perhaps part of the scientific side of the Renaissance spirit of inventiveness'.[5] All these observations hardly square with Professor Willan's opening assertion.

Professor Skempton and Mr. L. E. Harris, by contrast, have no doubts that engineers such as Grundy and Labelye 'viewed their work as well as a science as an art', and Professor Skempton considers that earlier men, especially Hadley and Steers,[6] 'were also [as] scientifically-minded as the age in which they were working permitted'.[7] Indeed, even earlier, in the first half of the seventeenth century, there is some evidence of applied science in this field; for instance, 'much of the technical work, and certainly the surveying', for the New River scheme of Sir Hugh Myddelton was the responsibility of Edward Wright, the mathematician.[8]

---

[1] Willan, *op. cit.*, p. 79.
[2] See below, p. 140. [3] *Ibid.*, p. 81.
[4] *Ibid.*, p. 94. [5] *Ibid.*, p. 88.
[6] For Thomas Steers (1672–1750), see *ibid.*, pp. 44–5, and especially H. Peet, 'Thomas Steers, the Engineer of Liverpool's First Dock', *Trans. Hist. Soc. Lancs. & Chesh.*, Vol. 82 (1930), pp. 163–242 (published in book form 1932). In addition to his work on Liverpool docks, Steers was also notable for his surveys and engineering of the Mersey–Irwell Navigation and the Newry Canal in Northern Ireland.
[7] Skempton, *op. cit.*, p. 51. At the same time, however, he points out that it was not till 1729 that Belidor's *La Science des Ingénieurs*—'one of the first text-books containing scientific matter of any real engineering value'—was published, and that 'the theoretical principles of hydraulics, soil mechanics and structures were not correctly formulated until the latter half of the century by men such as Coulomb, Chezy and Euler'.
[8] Skempton, *op. cit.*, p. 36, n. 35. See above, pp. 15–16.

Water power, as we shall later emphasize, was the basis of the early Industrial Revolution,[1] and occupied a considerable part of the developing literature of hydrostatics, hydraulics, and mechanics during the period we have been surveying. In the early development of the steam engine, too, science played some part. Desaguliers, it is true, referred to Newcomen and Calley as being neither mathematicians nor philosophers, but simply practical men, working by trial and error, aided by various skilled craftsmen;[2] but whilst this is almost certainly true as far as the construction and improvement of the engine was concerned, the principles upon which the Savery and Newcomen engines were based had been earlier demonstrated in the scientific researches of Boyle and Huygens and the scientific-technological experiments of Papin, and there is a strong probability that Savery and Newcomen may have acquired knowledge of them through the Royal Society.[3] Later, when these engines were coming into widespread use, detailed descriptions of them, and of various improvements, together with theoretical explanations, appeared in the works of Leupold, Belidor, Desaguliers, etc., which were studied by James Watt before he made his great discoveries,[4] and doubtless by other engineers.

Henry Beighton (1636–1743), F.R.S., has been referred to by the late Dr. H. W. Dickinson as 'the first scientific man to study the performance of the [Newcomen] engine'.[5] He was, in fact, one of the outstanding land-surveyors, mathematicians, and engineers of his day.[6] His survey of Warwickshire in the late 1720s is particularly notable. In 1713 he became editor of the *Ladies' Diary*, a position which he held until his death, and he was largely responsible for making this most unlikely sounding journal one of the main channels for the diffusion of mathematical science in the eighteenth century.[7] In addition to his surveying and mathematical work, he was also an expert in waterwheel construction and on the Newcomen steam engine. His engineering knowledge is amply demonstrated in the pages of Desaguliers' *Course of Experimental Philosophy*. Most of the technical descriptions,

---

[1] See below, pp. 67–72.   [2] *Op. cit.*, Vol. II, pp. 474 and 533.
[3] See the introduction by A. E. Musson to the second edition of H. W. Dickinson, *Short History of the Steam Engine* (1963).
[4] See below, p 80.   [5] *Op. cit.*, p. 43.   [6] *D.N.B.*
[7] Hans, *op. cit.*, p. 156. Beighton was succeeded as editor in the second half of the eighteenth century by equally outstanding mathematicians, Thomas Simpson and Charles Hutton, both F.R.S. (see below, pp. 117 and 160–1), while many other leading mathematicians were contributors. *The Edinburgh Review*, Vol. XI (1808), p. 282, commented: ' . . . a certain degree of mathematical science, and indeed no inconsiderable degree, is perhaps more widely diffused in England, than in any other country of the world. The Ladies' Diary, with several other periodical and popular publications of the same kind, are the best proofs of this assertion.'

drawings, and calculations in the second volume, dealing with water-wheels and steam engines, were, as Desaguliers acknowledged, supplied by his 'ingenious and very good friend Mr. Henry Beighton',[1] who based his accounts on close, practical investigation of the machinery in question and obviously had very wide knowledge and experience. He was elected F.R.S. in 1720 and made numerous contributions to the *Philosophical Transactions*; his description of the famous London Bridge waterworks, built by Sorocold and Hadley, is an outstanding example, complete with technical descriptions, calculations, drawings, and suggested improvements.[2] Moreover, his combination of science and technology in water-wheel design and construction was paralleled by a similar union of theory with practical experience in steam engines. Residing at Griff, near Coventry, he witnessed the introduction of Newcomen's first mine-drainage engines and soon became vigorously engaged himself in the construction and improvement of such engines, especially on the Northumberland and Durham coalfield.[3] He was particularly notable for the self-acting valve gear (1718) and other improvements which he made (some apparently suggested by Desaguliers), and also for the table which he produced on the performance of these engines, 'A Physico-Mechanical Calculation of the Power of an Engine', first brought out in 1717 and based on careful, scientifically controlled experiments.[4] This table, as eventually developed, provided clear directions as to the quantities of water that could be pumped per stroke, per minute, and per hour, from various depths, according to the diameter of the engine cylinder, strokes per minute, and bore of pump. Even before Beighton produced his table, Newcomen himself had developed similar though less refined rules for calculating engine power, according to the diameter of the cylinder, including allowance for variations in barometric pressure and also for friction;[5] he was

---

[1] See especially the innumerable references in Desaguliers, *op. cit.*, Vol. II, pp. 431–535 and pp. 539 ff. Desaguliers most certainly did not, as Dickinson (*op. cit.*, p. 44) asserts, retail Beighton's information 'as if it were his own', but gave repeated acknowledgements.

[2] *Phil. Trans.*, Vol. XXXVII (1731), pp. 5–12.

[3] See Desaguliers, *op. cit.*, Vol. II, pp. 464–90 and 533–5; Robison, *Encyclopaedia Britannica* (3rd edn., 1797), article on 'Steam'; Farey, *op. cit.*, pp. 133, 149, 150, 157, 231; R. H. Thurston, *A History of the Growth of the Steam Engine* (1878), pp. 61–3 and 67; Dickinson, *op. cit.*, pp. 41 and 43; E. Hughes, 'The First Steam Engines in Durham Coalfield', Society of Antiquaries of Newcastle-on-Tyne, *Archaeologia Aeliana*, 4th ser., Vol. XXVII (1949).

[4] It was published in the *Ladies' Diary*, 1721, p. 22 (reproduced in Dickinson, *op. cit.*, p. 44); it appears in a more developed form in Desaguliers, *op. cit.*, Vol. II, p. 535.

[5] Desaguliers, *op. cit.*, Vol. II, p. 482; Farey, *op. cit.*, p. 156; Dickinson, *op. cit.*, p. 45.

obviously not a simple empiricist, nor mathematically ignorant, and in endeavouring to establish rational rules for calculating engine power was following a rudimentary scientific method. Beighton adopted a more sophisticated procedure, including experiments 'to know what Quantity of Steam a Cubical Inch of Water produces',[1] in which he observed, though he could not explain, the phenomenon later termed latent heat, which Dr. Joseph Black and James Watt were to investigate much more thoroughly. He strongly emphasized the importance of natural philosophy and mathematics to the correct design and construction of mine-drainage engines:[2]

It were much to be wish'd, they who write on the Mechanical Part of the Subject, would take some little Pains to make themselves Masters of the Philosophical, and Mechanical Laws of (Motion or) Nature; without which it is Morally impossible, to proportion them [i.e. the Machines for draining of Mines] so as to perform the desired End of such Engines. We generally see, those who pretend to be *Engineers*, have only guess'd, and the Chance is, they sometimes succeed; else they have made them like others that have done pretty well. But he who has skill enough in *Geometry*, to reduce the *Physico-Mechanical* Part to Numbers, when the Quantity of Weight or Motion is given, and the Force designed to move it, can bring forth all the Proportions, in a Numerical Calculation, so as it may be almost impossible to Err.

This statement, of course, like the similar one by Desaguliers,[3] indicates the widespread empirical practices among most 'Engineers' of the early eighteenth century, and perhaps it exaggerates the possibilities of applied science at that time, but it does mark the development of more scientific methods and Beighton's own researches were of considerable practical importance; his table of steam-engine powers and proportions 'came into general use' and was still quoted respectfully by Farey in the early nineteenth century.[4] His mechanical improvements 'brought the machine [the atmospheric engine] into the form in which it has continued, to the present day, without any material change',[5] and his experimental scientific methods foreshadowed those of Smeaton and Watt.[6]

In the early development of the steam engine, as Desaguliers pointed out, the skills and tools of traditional metal-working crafts were of considerable importance. We know very little about the obscure craftsmen who helped Savery and Newcomen to construct their

[1] Desaguliers, *op. cit.*, Vol. II, pp. 533–4.
[2] *Ladies' Diary*, 1721, p. 21, as quoted in Dickinson, *op. cit.*, p. 44.
[3] See above, p. 38.   [4] Farey, *op. cit.*, p. 157.
[5] *Ibid.*, pp. 128 and 133, repeating a statement by Robison, *op. cit.*
[6] See below, pp. 73 and 79–81.

engines, but there is little doubt that instrument-makers and clock-makers made important contributions in later developments. Not only did they provide the tools and techniques for scientists in whose experimental investigations they shared; they also made vital contributions to advancing technology. Smeaton and Watt, of course, both instrument-makers by training, provide the most illustrious examples, but there were other, earlier men who played more obscure yet significant roles. Henry Hindley (1701–71), of York, 'a celebrated clock-maker and mechanist in his time',[1] was one of these. He was 'a very able maker of watches, clocks and turret clocks', including several large ones, such as that for York Minster in 1750.[2] Moreover, he was an outstanding tool-maker, particularly notable for his improved 'wheel-cutting engine' and circular 'dividing engine', which he used for cutting toothed wheels for clocks and which contributed to the development of heavy industrial gear-cutting machines.[3] These early machine-tools he showed to John Smeaton in 1741,[4] and like John Wyke, another leading clock-maker and tool-maker, on the other side of the Pennines, he may well have supplied these and other 'engines' such as lathes to early machine-making or engineering firms.[5] Hindley himself, in fact, entered the field of engineering in his later life. Farey has described an interesting, though unsuccessful, attempt by Hindley to develop a direct-acting steam engine, which would utilize the expansive force of

[1] Farey, *op. cit.*, pp. 257 and 426.
[2] G. H. Baillie, *Watchmakers and Clockmakers of the World* (1963); F. J. Britten, *Old Clocks and Watches and their Makers* (7th edn., 1956), entries under Hindley.
[3] Woodbury, *op. cit.*, pp. 51–2 and 58; Singer *et al.* (eds.), *History of Technology*, Vol. IV, p. 391. See above, p. 24.
[4] See Smeaton's 'Observations on the Graduations of Astronomical Instruments; with an Explanation of the Method invented by the late Mr. Henry Hindley, of York, Clock-maker, to divide Circles into any given Number of Parts', *Phil. Trans.*, Vol. LXXVI (1786), pp. 1–47. 'In the autumn of the year 1741, I was first introduced to the acquaintance of that then eminent artist, Mr. Henry Hindley, of York, clockmaker;—he immediately entered with me into the greatest freedom of communication, which founded a friendship that lasted to his death, which did not happen till the year 1771. . . . On the first interview, he shewed me not only his general set of tools, but his *engine*, at that time furnished with a dividing plate, with a great variety of numbers for cutting the teeth of clock wheels, &c.' Hindley also showed Smeaton 'how, by the single screw of his lathe, he could cut, by means of wheel-work, screws of every necessary degree of fineness' (*op. cit.*, pp. 19–21). According to Smiles, *Lives of the Engineers* (1874), Vol. II, p. 91, quoting from Holmes, *Short Narrative of the Genius, Life and Works of the late Mr. John Smeaton, C.E., F.R.S.* (1793), 'Mr. Hindley was a man of the most communicative disposition, a great lover of mechanics, and of the most fertile genius. Mr. Smeaton . . . spent many a night at Mr. Hindley's house till daylight, conversing on these subjects.'
[5] See below, p. 437. Joseph Hindley, his son, visited Matthew Boulton at Birmingham in 1760: J. Hindley to M. Boulton, 14 Nov. 1760. In this letter he also mentioned dining with Smeaton, after returning to York.

steam as well as atmospheric pressure, a foreshadowing of Watt's direct-action engine; he died before it was completed, but his engine was used subsequently, in a modified form.[1] Hindley, as we shall see, was to be followed by many other clock-makers and instrument-makers who were similarly to turn their mechanical knowledge, tools, and techniques to producing machinery for the Industrial Revolution.[2]

Equally important in this period was the spread of industrial chemistry. Chemistry was already coming to be regarded as distinct from natural philosophy (mechanics, hydrostatics, etc.), but the distinction was not clear-cut. The younger Hauksbee, for instance, who collaborated with Whiston and Desaguliers in 'experimental philosophy', was also associated with Dr. Peter Shaw (1694–1763), F.R.S., in chemical lectures.[3] About Shaw's early life and education little is known, except that his father was a Cambridge graduate and Master of Lichfield Grammar School; he is sometimes said to have been a pupil of Boerhaave at Leyden, but this is uncertain.[4] There is no doubt, however, about his extensive knowledge of science and medicine, as well as of the classics. More than anyone else in the first half of the eighteenth century, he emphasized the importance of chemistry to the improvement of arts and manufactures.[5] This belief he appears to have derived from the works of Boyle and Bacon, of which he brought out scholarly abridged versions in 1725 and 1733 respectively. He also disseminated the teachings of great foreign chemists such as Boerhaave and Stahl, by collecting copies of lecture-notes taken by students and publishing them in English, much to the annoyance of their authors.[6] At the same time, and in addition to his medical practice, he published several pharmaceutical and medical works.[7] He also began to turn his attention to industrial chemistry, proposing to write a set of essays 'design'd for the farther application and advancement of Genuine Chemistry in

---

[1] Farey, *loc. cit.*   [2] See below, chap. XIII.

[3] F. W. Gibbs, 'Peter Shaw and the Revival of Chemistry', *Annals of Science*, Vol. VII (1951), pp. 211–37, and 'Essay Review: Prelude to Chemistry in Industry', *ibid.*, Vol. VIII (1951), pp. 271–81; Hans, *op. cit.*, pp. 143–4; *D.N.B.*

[4] He acquired a Cambridge M.D. by royal *mandamus* in 1751, at fifty-seven years of age, but was styling himself M.D. from the early 'twenties.

[5] He was not, however, as Hans (*op. cit.*, p. 144) states, 'the first to deliver public lectures on Chemistry'. See above, p. 26, for George Wilson. ,

[6] Together with Ephraim Chambers, he published an edition of Boerhaave's chemical lectures in 1727; he also published in 1730 a digest of Stahl's chemical courses at Jena, and in 1741 a new English translation of Boerhaave's *Elementa Chemiae*.

[7] For example, John Quincy's lectures on pharmacy in 1723; *New Practice of Physic* in 1726, based mainly on the teachings of Sydenham and Boerhaave; the Edinburgh *Pharmacopoeia* in 1727, and various lesser works.

England; with regard to Science, Arts, Trades, and Commerce'.[1] Three of the promised essays appeared in 1731, dealing with practical applications of chemistry in, for example, distillation of alcoholic spirits and fermentation of wines, and emphasizing its contribution towards useful discoveries.[2] These were followed by an essay on a portable laboratory, equipped with furnace and other apparatus and also with chemical substances, together with instructions for carrying out chemical experiments; these portable laboratories were supplied by Hauksbee.[3] At the same time Shaw and Hauksbee announced their *Proposals for a Course of Chemical Experiments with a view to Practical Philosophy, Arts, Trades and Business* (1731); Hauksbee would provide the apparatus and experimental demonstrations, while Shaw read the lectures. These lectures, delivered in 1731–3 and published in 1734,[4] included practical examples of the use of chemistry in improving arts and manufactures, such as the preservation and curation of vegetables, the manufacture of wines and spirits, the arts depending on distillation, the extraction of vegetable oils and their uses, improvements in colours, dyes, and stains, pharmacy, mineralogy and mining, metallurgy and pyrotechny. Improvements were also suggested in the manufacture of sal ammoniac, varnishing and japanning, glass-making, etc.

Shaw and Hauksbee thought of themselves as pioneers, distinguishing themselves from those who were interested in chemistry only for philosophical reasons or for entertainment. 'We seem . . . to have been travelling in a new Road, too little frequented either by Philosophers, Chemists, or the Men of Business. Our endeavour has been to improve the useful Arts, by means of a more Philosophical Chemistry; and at the same time to shew the Method of conducting Enquiries, so as that they may terminate in useful Discoveries.'[5] They were pursuing, in fact, 'the Verulamian Method', which, after the enthusiasm of the early years of the Royal Society, appears to have cooled off considerably and needed reviving.[6] In his later years, however, Shaw devoted himself mainly to his medical pursuits, in which he was very successful,[7] but he freely gave his expert advice to the Society of Arts, established in 1754, suggesting ways of improving industrial-chemical processes and advising on awards of premiums.

[1] *Philosophical Principles of Universal Chemistry* (1730), Preface.
[2] *Three Essays in Artificial Philosophy, or Universal Chemistry* (1731).
[3] *Essay for Introducing a Portable Laboratory* (1731).
[4] *Chemical Lectures . . .* (1734).   [5] *Ibid.*, p. 438.
[6] Gibbs, *Annals of Science*, Vol. VII (1951), p. 149, and Vol. VIII (1952), pp. 274–6.
[7] He became Physician-Extraordinary to George II in 1753 and Physician-in-ordinary to George III in 1760.

Shaw was not alone in giving public lectures on chemistry during this period. Henry Pemberton, for example, a product of Leyden, gave a course of such lectures at Gresham College in 1730–1.[1] These reflected a similar interest in the development of manufactures, which he was in the habit of studying on tours of inspection.

Another lecturer in London in the 1730s and 1740s was Dr. William Lewis (1708–81), F.R.S.,[2] whose chemical writings on a variety of subjects were still regarded as authoritative by Brewster's *Edinburgh Encyclopaedia* in 1830. His place in a line of influence is established by his edition of George Wilson's *Complete Course of Chemistry* in 1746, to which Lewis had added much new matter from French, Dutch, and German chemists who had written since Wilson's day. This was a book used by William Cullen of Edinburgh, although Cullen had criticisms to make of it. Lewis's own lectures, for 'the Improvement of Pharmacy, Trades, and the Art [of Chemistry] itself', were published as *A Course of Practical Chemistry* in 1746. His access to the work of foreign scientists through his patrons and friends was an important part of his success, and he was greatly helped by his laboratory assistant, Alexander Chisholm, a graduate of Marischal College, Aberdeen, later Wedgwood's assistant, who did much of the translation for him, as well as chemical experiments. Besides publishing several works on pharmacy,[3] Lewis first projected and then completed his *Commercium Philosophico-Technicum: or the Philosophical Commerce of Arts* (1753, Proposals; Part I, 1763; Parts II and III, 1765). Lewis's first intention was fully in the Baconian tradition: to 'lay the foundation of a philosophical and experimental history of arts'. He also wanted, by laboratory studies, to seek out new materials for industry, to find an outlet for by-products then regarded as waste, and to replace expensive imported materials by domestic ones. Though he later revised his intentions and claimed rather to write a work describing the properties of materials in wide use, his interests were still predominantly technological. His work marks the arrival of the professional chemist, applying scientific knowledge systematically to industrial problems. His four Royal Society papers on platina earned him the Copley Medal for 1754; his researches on the physics and chemistry of gold were of immediate use

[1] Gibbs, *Annals of Science*, Vol. VIII (1952), p. 277. Pemberton's lectures were not published until 1771 (*A Course of Chemistry*, ed. James Wilson).

[2] F. W. Gibbs, 'William Lewis, M.B., F.R.S. (1708–1781)' *Annals of Science*, Vol. VIII (1952), pp. 122–51, and 'A Notebook of William Lewis and Alexander Chisholm', *loc. cit.* pp. 202–20. Lewis studied at both Oxford and Cambridge, taking the M.B. degree at the latter.

[3] F. W. Gibbs lists Lewis's publications in 'William Lewis . . .', *op. cit.* A further list is to be found in N. Swin, 'William Lewis (1708–1781) as a chemist', *Chymia*, Vol. 8 (1962), pp. 63–88.

to gilders; his investigations into black materials, complementary to his work on gold, covered all kinds of black paints and dyes, the manufacture and varnishing of papier-mâché goods, sheep-marking, preservatives for ships, inks, and many other matters; while his work for the Society of Arts on methods of testing potash (for which he was awarded their gold medal) resulted in a report which was the standard guide to analysis of potash until the nineteenth century. Although many of his papers have been lost, some still remain to confirm that he regularly visited industrial premises such as glass-works, dye-works, breweries, etc.; of particular interest are his visits to iron foundries in 1768 to investigate processes for producing bar-iron in coke-fired furnaces.[1] He was frequently consulted on these matters, carrying out analyses, etc. He was an active member of the Society of Arts, regularly undertaking assay work for it and advising on prizes for industrial discoveries. Of particular interest is the link with Wedgwood, into whose service Alexander Chisholm entered at Lewis's death.[2]

Robert Dossie (1717–77) provides further evidence of London activities in applied science, though he originated from Sheffield where he was apprenticed to an apothecary and where, according to Dr. Gibbs, one of his schoolfellows had been John Roebuck, the pioneer of the manufacture of sulphuric acid in lead chambers.[3] Dossie in his youth attended scientific lectures and studied chemistry, botany, physics, and machine-making. Later he appears to have earned a living as a consulting chemist, but nothing much is known of him before 1757 when he settled in London and quickly established his reputation by his publications, *An Elaboratory Laid Open* (1758), *Handmaid to the Arts* (1758), and *Institutes of Experimental Chemistry* (1759), the first of a series of books and articles, some of the latter appearing in the *Museum Rusticum et Commerciale* before its demise. Like Lewis, he soon came to use the Society of Arts as a centre for the collection of information on applied science and for the diffusion of it among industrialists and craftsmen. Though he was unsuccessful in his application for the secretaryship of the Society of Arts, despite the friendship of such persons as Dr. Samuel Johnson, he continued to serve on many of the Society's committees, and frequently alternated in the chairmanship

---

[1] F. W. Gibbs, 'A Notebook . . .', *op. cit.* He visited ironworks at Staveley, John Cockshutt's works near Rotherham, and Richard Reynolds's at Ketley and Coalbrookdale, where the Cranage brothers' process was being used. Lewis, it may be mentioned, was on friendly terms with John Cockshutt junior, who was the London representative of the firm and a supporter of the Society of Arts.   [2] See below, pp. 78 and 143.

[3] F. W. Gibbs, 'Robert Dossie (1717–1777) and the Society of Arts', *Annals of Science*, Vol. VII (1951), pp. 149–72.

with Dr. Benjamin Franklin. He was associated with Lewis in the assays of potash, and his memoir on the methods of cultivating kali and manufacturing barilla, written at the request of the Society (which awarded him its gold medal), became the standard work on the subject. The work of Dossie and Lewis contributed some twenty years later to the researches of Richard Kirwan on the use of alkalis in bleaching.

Dossie's industrial interests also included the purification of train oil, varnishing, japanning, gilding, etc., the manufacture of sal ammoniac, bleaching, agriculture, sheathing of ships' bottoms, silk production, cordage, insecticides for fruit-trees, and flax cultivation. He edited for the Society of Arts, *Memoirs of Agriculture and other Oeconomical Arts* (1st vol., 1768; 2nd vol., 1771; and the 3rd vol., published after his death in 1782). Like Lewis's work, Dossie's also was still regarded as authoritative in Brewster's *Edinburgh Encyclopaedia* (1830). Dossie deserves particular recognition with Lewis as a contributor to the Birmingham and Black Country 'toy' industry of the eighteenth century, for his discoveries in the relevant metallurgy and finishing processes, and also for his recognition of the importance of design in capturing these fashion-markets from the French. Dossie also had connections with Manchester through his brother-in-law, Robert Staniforth, a Manchester dyer, to whom he referred the Society of Arts for some tests on cudbear.

The industrial activities of the provinces were both studied and stimulated by the London chemists, such as Shaw, Lewis, and Dossie, so that there was no actual divorce between the metropolis and the nation as a whole in matters of applied science. In the same way there was much exchange between Britain and continental countries, not only by the subterranean routes of industrial espionage, but also openly by the exchange of publications. An interesting example of the latter is the publication in 1728 in the *Philosophical Transactions* of a paper by Réaumur, previously published in 1725 in the *Memoires de l'Académie Royale des Sciences*. This paper dealt with Réaumur's visit to a tinplate manufactory at Beaumont-la-Ferrière in Nivernais, and contains a full account of what he saw. It was probably the main channel for disseminating information about the use of sal ammoniac in descaling iron sheets before the tinning process. Réaumur's work on tinplating, oils and varnishes, iron and steel manufacture, porcelain, thermometry, and other subjects was widely known in Britain in the first half of the eighteenth century, paralleling similar exchanges of information between continental and British scientists and industrialists in the second half of the century. Native publications remained important, however, and the first intelligible account of tin-plating in English was published in 1724 in W. Derham's edition of the *Philosophical Experiments of . . . Robert Hooke, 1724.*

The studies of the late Dr. Gibbs on Shaw, Lewis, Dossie, and others[1] have, as he claimed, made possible 'a new approach to the history of the alliance of chemistry and the manufactures in England at the middle of the century'.[2] 'With the advent of the private laboratory and the chemical consultant, in substantially increasing numbers during the second quarter of the eighteenth century, chemistry was already playing a part in the manufactures commensurate with its own development.' In these ways, and through chemical lectures and literature, 'already there was a free flow of ideas and information between chemists and the leading manufacturers in many parts of the country'.

This opinion must not be pushed too far. William Cullen, for instance, the famous Edinburgh professor of chemistry, though himself a keen advocate of applied chemistry in industry and agriculture, considered that Shaw had exaggerated its immediate practical possibilities:[3] 'As to the Arts, Manufactures, Trade, Commerce &c. in Life to which Chemistry may be subservient, the Drs. Boerhaave & Shaw may be consulted, but with Caution, lest a Person shou'd Acquire too high an Opinion of it . . .' This view, as we shall see, was reiterated by various other well-informed people later in the century, who emphasized the empirical aspects of industrial-chemical advances.[4] Nevertheless, it is clear that applied chemistry was beginning to make significant contributions.

As Dr. Gibbs has emphasized, there is need for much more research into the activities of particular firms in this early period. Such early business records, however, have mostly been destroyed. It is therefore particularly interesting to have some information about the history of probably the earliest firm of industrial chemists in this country— Godfrey & Cooke as it later came to be known[5]—which provides

[1] See also Dr. Gibbs's articles on the tinplate industry and the japanning trade in *Annals of Science*, Vol. VII, VIII, and IX (1951, 1952, and 1953).

[2] F. W. Gibbs, 'Essay Review: Prelude to Chemistry in Industry', *ibid.*, Vol. VIII (1952), pp. 271–81.

[3] London Medical Society, MS. 79A, f.7. For Cullen's own industrial-chemical interests, see Clow, *op. cit.*, especially pp. 588–9.   [4] See below, pp. 82–3.

[5] Godfrey and Cooke, acquired by Savory and Moore Ltd., in 1916–17; founded by Ambrose Gottfried Hanckwitz (1660–1714), F.R.S.; continued by Ambrose Godfrey [Hanckwitz] II (d. 1756) and his younger brother John Godfrey (d. 1766); then by Ambrose Godfrey III (1730–97), Ambrose Towers Godfrey IV (1769–1807), Charles Gomand Cooke (1765–1842); finally by Cooke's daughter and her husband, Samuel Platt, William Ince (d. 1853), Joseph Ince (d. 1907), and the Greenish family. See R. E. W. Maddison, 'Studies in the Life of Robert Boyle, F.R.S. Part V. Boyle's Operator: Ambrose Godfrey Hanckwitz, F.R.S.', *Notes and Records of the Royal Society*, Vol. XI, no. 2 (March 1955), pp. 159–88.

further evidence of the relationship between science and industry in the late seventeenth and eighteenth centuries. The founder of the firm, Ambrose Gottfried Hanckwitz (1660–1741), F.R.S., was an assistant to Robert Boyle and while working for him learnt a successful method for manufacturing phosphorus from human urine and excrement. By 1710 he was firmly established in business and attracted the attention of foreign visitors to this country who wished to learn his methods. Besides working as a manufacturing chemist, Hanckwitz also acted as a consultant analyst for the Royal Society and other persons. His chemical laboratory, erected behind his house in Southampton Street, London, was famous and was visited by travellers who compared it favourably with the chemical laboratory attached to the Ashmolean Museum, Oxford. His eldest son, Boyle Godfrey, who was a spendthrift and was obliged to travel abroad, gave courses of lectures on chemistry in Paris in the 1730s which were well attended by aristocrats and by students.[1] Ambrose Godfrey III was a subscriber to Bryan Higgins's Society for Philosophical Experiments and Conversations instituted in January 1794,[2] and was elected into a dining-club called the Friendly Medical Society in December 1786.[3] The family's history demonstrates the importance of apothecaries to the development of the chemical industry in the eighteenth century.

All this evidence demonstrates not only the importance of developments outside the Royal Society and the universities, but also the continued predominance of London in scientific activities. What was 'in general true [of] the seventeenth century [that] science moved from the universities into the capital cities; from colleges to clubs and academies',[4] is also largely true of the first half of the eighteenth century. Until the fast developing provincial towns like Birmingham, Manchester, Leeds, and Sheffield really came into their own in the second half of the eighteenth century, London remained in a dominant position for scientific culture. It is true, as we shall show, that scientific clubs sprang up in many cities and market-towns all over England in the course of the eighteenth century,[5] but London still acted as a magnet for men with scientific and technical interests. The coffee-houses of London, which proliferated after the Restoration, were the

---

[1] R. E. W. Maddison, 'Notes on some members of the Hanckwitz Family in England', *Annals of Science*, Vol. XI, no. 1 (March 1955), pp. 64–73.

[2] See below, p. 122.

[3] Maddison, 'Notes on . . . the Hanckwitz Family', p. 70. The club was founded *c.* 1725, and was limited to twenty-six members of the Company of Apothecaries and the clerk of the Company for the time being.

[4] A. R. Hall's review of C. Hill, in *History*, Vol. L (1965).

[5] See below, Chap. III.

E

centres of several scientific groups. From the 1660s and 1670s, when Robert Hooke discoursed at Garaway's or Man's or other coffee-houses,[1] to the Philosophical Society that met at the Chapter Coffee House in the 1780s there is an unbroken succession.[2] Dr. John Harris, for example, lectured on mathematics from 1698 to 1704 at the Marine Coffee House in Birchin Lane;[3] William Whiston lectured at Button's Coffee House; James Stirling and later J. T. Desaguliers both lectured on experimental philosophy at the Bedford Coffee House;[4] at the Rainbow Coffee House in 1735 a society was established with the general aim of promoting the arts and sciences; Rawthmell's was the birthplace in 1754 of the Society of Arts;[5] and the Chapter Coffee House was the meeting-place of a most impressive society, including John Whitehurst, Edward Nairne, Richard Kirwan, Boulton, Watt, Wedgwood, Keir, Dr. Hutton, J. H. de Magellan, and many other distinguished scientists.[6] Here, and sometimes at the Baptist's Head Coffee House or in the room of Adam Walker, the itinerant lecturer in science, they discussed not only electricity and the composition of air, but such matters as Watt's blowing engines at Wilkinson's iron-works, Cort's iron-founding methods, and the recovery of silver from copper and other base materials. The Royal Society itself carried out many of its functions at informal meetings of the dining-club in coffee-houses, which have been well described as 'the Penny Universities'.[7] Manufacturers like Boulton, Watt, and Wedgwood when they visited London often spent their leisure time among 'philosophical' friends in a coffee-house. The minute-book of the Chapter Coffee House Society also shows us how by correspondence the network of international science and technology was strengthened. In his chapter 'Academic Honey-combs', Professor Armytage has described the great number of academies and societies all over the continent of Europe which collected and diffused the useful knowledge of the world through their members, their correspondents and their publications: 'Technology had found its clerisy in the academies—220 of them by 1790.'[8] In England, the societies concerned with useful knowledge covered a wide social range; at the one extreme is the club frequented by John Collins, 'consisting of divers ingenious mechanics, gaugers, carpenters, ship-wrights, some seamen, lightermen, etc., whose whole discourse is

---

[1] See above, p. 19.
[2] W. H. G. Armytage, *The Rise of the Technocrats* (1965), n. 48, pp. 371–2.
[3] See above, p. 37.   [4] See above, pp. 37 and 41.
[5] Hudson and Luckhurst, *op. cit.*, chap. I. See also below, p 127.
[6] MS. Gunther 4, Oxford Museum of the History of Science.
[7] Aytoun Ellis, *The Penny Universities: A History of the Coffee Houses* (1956).
[8] Armytage, *op. cit.*, p. 37.

about equations',[1] an association which suggests self-education at a fairly low level, and at the other is the Society for the Encouragement of Arts, Commerce and Manufactures, or the Royal Society itself, boasting among their members the most distinguished men of the day from professional or aristocratic families. In the pursuit of useful knowledge, in the study of the ways in which science could be applied to industry, and industry provide the material for science, class-divisions in Great Britain seem to have been largely overlooked. Royal princes could sit at the feet of self-educated lecturers in science like Adam Walker and James Ferguson; great landowners like the Shelburnes and the Egertons discussed engineering and workshop practice with business-men and semi-literate artisans; the instrument-maker and the technical author could meet, in the philosophical club, men of much greater wealth and social standing and be listened to. When Adam Smith referred to technological improvements not only by practical craftsmen but also by 'philosophers, or men of speculation', and remarked how with the growth of trade and specialization 'the quantity of Science is considerably increased',[2] he was not, as has been suggested, 'ahead of his time',[3] but was simply observing the developments of his own day.

[1] Quoted by Letwin, *op. cit.*, p. 109. Another such was the 'Mathematical Society' established in Spitalfields, London, in 1717, which survived till the mid-nineteenth century. Its members were 'chiefly tradesmen and artisans', but included such notable men as Dollond, the instrument-maker, and Thomas Simpson, the mathematician. The Society loaned instruments—air-pumps, telescopes, microscopes, electrical machines, surveying instruments, etc.—and also books from a library of eventually nearly 3,000 volumes. J. Timbs, *Clubs and Club Life in London* (new edn., 1886), p. 403; H. H. Cawthorne, 'The Spitalfields Mathematical Society (1717-1845)', *Journal of Adult Education*, Vol. III, no. 2 (April 1929).
[2] *Wealth of Nations*, Book I, chap. I.
[3] C. F. Carter and B. R. Williams, *Investment in Innovation* (1959), pp. 149-50.

II The Diffusion of Technology in Great Britain during the Industrial Revolution[1]

Great Britain was the first country to experience an Industrial Revolution—the transformation of a predominantly rural, agricultural, and handicraft society into a predominantly urban, industrial, and mechanized society—and the modern world undoubtedly owes a great debt to British pioneers of the eighteenth and nineteenth centuries, such as Abraham Darby and John Wilkinson, Richard Arkwright and Samuel Crompton, Thomas Newcomen and James Watt, Richard Trevithick and George Stephenson. Dr. Henderson has shown in his book on *Britain and Industrial Europe 1770–1870* how British skill and capital helped to spread this technological revolution on the continent. But we in Great Britain are inclined to take too much credit for ourselves and to exaggerate our insular achievements. The scientific and technological traffic was not all one way: Great Britain owed much in many fields to Frenchmen, Germans, Dutchmen and others.

The tendency nowadays is to push the origins of the Industrial Revolution farther back into history, and there is no doubt that in these earlier developments—in what Professor Nef has called 'the industrial revolution of the sixteenth century'—Britain drew heavily upon continental technology.[2] In the smelting and refining of metallic ores, for example, by the Mines Royal and the Mineral and Battery Works Companies, German skill and capital, applied by Höchstetter and others, was extremely important. It was from the continent, moreover, that the blast-furnace and iron casting were introduced into England in the late fifteenth century, and it seems fairly certain that English tool-makers, instrument-makers, and clock-makers—important

---

[1] This chapter, by A. E. Musson, is based on a previously unpublished paper read at the Third International Economic History Conference in Munich, Aug. 1965, but has been considerably extended.

[2] Samuel Smiles was well aware of this in the nineteenth century. See *Lives of the Engineers*, Vol. I (1874), pp. iv–viii.

in the early development of mechanical engineering—owed a great deal to continental influences. From the continent, too, in this period came other technological developments, such as the invention of printing, the paper mill, wire-drawing machines, the 'new draperies' in the woollen industry, the Saxony spinning wheel, and the Dutch swivel (or inkle or ribbon) loom. Foreign influences can also be traced in many other industries, such as silk, pottery, glass-making, sugar refining, tinplate manufacture, brewing, etc. Alien craftsmen were attracted here in considerable numbers to develop these new industrial processes.[1] Flemish and Dutch engineers were also brought over to construct water-mills, harbour-works, and land-drainage systems.[2]

These continental influences were not confined to the sixteenth and seventeenth centuries, but continued during the Industrial Revolution of the eighteenth and nineteenth centuries, though they were perhaps of less importance. In textiles, for example, the Lombe brothers acquired from Italy the plans for the silk-throwing machines installed in 1717–21 in their famous water-powered factory. In the cotton industry, Lewis Paul, inventor of roller spinning, was apparently of French Huguenot extraction, and in the bleaching, dyeing, and calico-printing trades French achievements were immensely important. It was from France that chlorine bleaching, discovered by Berthollet, was brought to Britain; chemists from a firm set up near Liverpool by a Frenchman, Bourboulon de Boneuil, in the late 1780s, played an important role in the development of chlorine bleaching in Lancashire.[3] It is also clear that French influences were very important in dyeing and calico-printing: these are clearly evidenced, for example, in the researches and papers of Thomas Henry, of Manchester, while Frenchmen such as Borelle were important in the early development of Turkey-red dyeing in the Manchester area.[4] Similarly, in the closely related alkali trade, Leblanc's process is famous; he was followed in the later nineteenth century by the Belgian, Solvay, whose ammonia-soda process was introduced into England by the German Ludwig Mond. In allied trades, too, one finds important continental contributions: in soap-making, for example, Chevreul's researches were important, and in glass-making Professor Barker has clearly shown French influences.[5]

---

[1] See for example, E. Taube, 'German Craftsmen in England during the Tudor Period', *Economic History*, Feb. 1939; W. C. Scoville, 'Minority Migrations and the Diffusion of Technology', *Journal of Economic History*, Vol. II (1951), and 'The Huguenots and the Diffusion of Technology', *Journal of Political Economy* Vol. 60 (1952).

[2] Smiles, *op. cit.*, pp. v–vi; L. E. Harris, *Vermuyden and the Fens* (1953).

[3] See below, chap. VIII.   [4] See below, chap. IX.

[5] T. C. Barker, *Pilkington Brothers and the Glass Industry* (1960), chaps. 2 and 5.

Dutch technical knowledge and experience similarly contributed to
fen drainage, river improvements, and canal building, while the
French also made important contributions to early British civil
engineering.¹ In the early development of the steam engine, likewise,
it seems clear that this country owes more to the experiments of
continental scientists such as Huygens and Papin than has hitherto been
acknowledged.² In mechanical engineering, it is clear that Maudslay
owed a good deal to earlier continental influences³ and to the great
engineer, Marc Isambard Brunel, a French refugee.⁴ Other foreign
engineers were also important, including the extraordinary Swiss,
J. G. Bodmer,⁵ and the American, J. C. Dyer,⁶ both of whom
established works in Manchester; another American, Jacob Perkins,
pioneer in the development of high-pressure steam engines and as versatile as Bodmer, had a workshop in London,⁷ where other names such

---

¹ S. B. Hamilton, 'Continental Influences on British Civil Engineering to
1800', *Archives Internationales d'Histoire des Sciences*, Vol. 11 (1958), pp. 347–55.
² See the introduction by A. E. Musson to the new edition of H. W. Dickinson's *Short History of the Steam Engine* (1963).
³ C. Singer and others (eds.), *A History of Technology*, Vol. IV (1963), chaps.
13 and 14.
⁴ R. Beamish, *Memoir of the Life of Marc Isambard Brunel* (2nd edn., 1862);
S. Smiles, *Industrial Biography* (1863), pp. 215–22; Roe, *op. cit.*, chaps. III and
IV; C. B. Noble, *The Brunels, Father and Son* (1938).
⁵ Roe, *op. cit.*, pp. 75–80; D. Brownlie, 'John George Bodmer, His Life and
Work', *Newcomen Soc. Trans.*, Vol. VI (1925–6). See also other bibliographical
references in W. O. Henderson, *J. C. Fischer and his Diary of Industrial England
1814–1851* (1966), p. 75, n. 37.
⁶ *Select Committee on Patent Laws*, P.P. 1829, Vol. III, p. 134 (evidence of
John Farey); *Select Committee on Exportation of Machinery*, P.P. 1841, Vol. VII,
First Report, Q. 1531 ff. (evidence of Matthew Curtis); *Newcomen Soc. Trans.*,
Vol. XIII (1932), pp. 21–2, 45–6 (Joshua Field's Diary of a tour through the
provinces, 1821); *D.N.B.* Dyer was originally a merchant, who saw the possibilities of patenting and developing American inventions in England, and who
made himself into an engineer after settling here. He introduced a number of
machines, several of which he patented, in 1810: a nail-making machine; a card-
making machine (originally invented in this country by Sharp and Whittemore,
in the late 1790s, but not commercially successful, so taken to America, and then
brought back here by Dyer, who perfected it); Perkins's engraving process for
calico-printing; a reed-making machine; and a shearing machine. Subsequently
he took out further patents for improvements in these machines, and also began
to manufacture others, including his patent tube roving-frame (another American
invention) and Smith & Orr's patent self-acting mule. He retired in the late
1830s, handing the business over to former employees, Parr, Curtis, and Madeley
(textile machine-makers) and Curtis, Parr, and Walton (wire-card makers), who
by the mid-nineteenth century were the largest manufacturers of cotton-spinning
machinery in the country (see below, p. 477).
⁷ *Minutes of Proceedings, Institution of Civil Engineers*, Vol. XXV (1865–6),
pp. 516–19; G. and D. Bathe, *Jacob Perkins* (Historical Society of Pennsylvania,
1943), and 'The Contribution of Jacob Perkins to Science and Engineering',
Newcomen Soc. Trans., Vol. XXIV (1943–5); Henderson, *op. cit.*, pp. 31–3. For

as Holtzappfel[1] show continental influences. In steel-making, one need only refer to Bessemer and Siemens as outstanding examples of French and German genius respectively.

Further examples can be drawn from other industries. Thus the first paper-making machine was produced by Frenchmen, Louis Robert and the Fourdrinier brothers.[2] The German, Friedrick Koenig, invented the first steam-powered cylinder printing press, and later on the American Hoe machines were to be very important.[3] The earliest attempts at mechanical typesetting, by such men as Church, Delcambre, and Rosenberg, were mostly foreign, and the American linotype and monotype machines were eventually to solve the problems involved; the names of Senefelder (lithography) and Kronheim (stereotyping) show the importance of continental innovations in other branches of the printing trade. It seems very likely that further examples could be drawn from other industries. In agriculture, too, foreign examples were followed, as in Flemish or Dutch crop rotations; Jethro Tull's famous seed-drill was apparently of continental origin.

The importance of foreign inventions was clearly apparent to intelligent observers in the early nineteenth century. John Farey, an eminent engineering consultant of London, for instance, told the Select Committee on Patent Laws in 1829 that

we have derived almost as many good inventions from foreigners, as we have originated among ourselves. The prevailing talent of the English and Scotch people is to apply new ideas to use, and to bring such applications to perfection, but they do not imagine so much as foreigners; clocks and watches, the coining press, the windmill for draining land, the diving bell, the cylinder paper machine, the stocking frame, figure weaving loom, silk throwsting mill, canal-lock and turning bridge, the machine for dredging and deepening rivers, the

---

Perkins's engraving process in calico-printing, see E. Baines, *History of the Cotton Manufacture in Great Britain* (1835), p. 267.

[1] John Jacob Holtzappfel, a German who settled in London as a tool-maker about 1787, established a famous engineering concern, which was carried on by his son, Charles (1806–47). See *D.N.B.*; *Annual Register* (1847), Appendix, p. 223; *Gentleman's Magazine* (1847), pt. ii, p. 213; *Proceedings of the Institution of Civil Engineers* (1848), pp. 14–15; Roe, *op. cit.*, pp. 74–5. See also below, p. 78. Some Holtzappfel business records have recently come to light, showing an annual turnover of more than £10,000 by 1811. Included in that year's sales were some fifty lathes, ranging from small wood-turning ones costing around £12 to what must have been very big machines costing about £300; two 'rose engines' were sold for well over £400 each. I am indebted to Professor S. B. Saul for this information.

[2] D. C. Coleman, *The British Paper Industry, 1495–1860* (1958), chap. VII.

[3] See A. E. Musson, *The Typographical Association* (1954), chaps. II and VI, and 'Newspaper Printing in the Industrial Revolution', *Econ. Hist. Rev.*, Vol. X, no. 3 (1958), pp. 411–26, for references to foreign innovations in printing.

manufacture of alum, glass, the art of dyeing, printing, and the earliest notions of the steam engine, were all of foreign origin; the modern paper-making machine, block machinery, printing machine, and steam boats, the same; there are a multitude of others . . .[1]

Matthew Curtis, the Manchester machine-maker, expressed a similar view to the Select Committee on Exportation of Machinery in 1841:

I should say that the greatest portion of new inventions lately introduced into this country have come from abroad; but I would have it to be understood that by that I mean not improvements in machines, but rather entirely new inventions. There are certainly more improvements carried out in this country; but I apprehend that the chief part, or a majority, at all events, of the really new inventions, that is, of new ideas altogether, in the carrying out of a certain process by new machinery, or in a new mode, have originated abroad, especially in America.[2]

These are only a few brief examples, but it seems probable that a book could be written on foreign influences in the Industrial Revolution in Britain, to set beside that of Dr. Henderson. Some of the continental immigrants, as is well known, came to this country as religious or political refugees, bringing their industrial skills with them. But there were other factors attracting them here to develop their inventions: the British patent system; greater availability of capital; greater freedom for private enterprise. Both Farey and Curtis, in their evidence referred to above, emphasized the greater opportunities of developing or improving new processes in this country. Farey stated that foreigners came here 'because the means of executing and applying inventions abroad are so very inferior to ours'.[3] And Curtis attributed British superiority in this respect to the larger scale of industry, greater 'sub-division of labour' and specialization, producing greater knowledge of particular processes and machines and of possible improvements;[4] he also emphasized the importance of British machine-tools, in improving and cheapening machinery.[5]

It would be wrong, however, to attribute too much to foreign innovations. Despite all this evidence, the Industrial Revolution still remains primarily a British achievement. That, after all, is why foreigners came here. The question then arises as to the causes of

---

[1] P.P. 1829, Vol. III, p. 153.

[2] P.P. 1841, Vol. VII, *First Report*, Q. 1544. Curtis referred particularly in his evidence to the new inventions brought in by Dyer and Perkins, but he also mentioned several others. On the other hand, of course, there were many, probably more, examples of British technological skills and machinery exported to America. See H. J. Habakkuk, *American and British Technology in the Nineteenth Century* (1962). [3] *Op. cit.*, p. 153.

[4] *Op. cit.*, Q. 1556 and Q. 1608. [5] *Ibid.*, Q. 1600.

British technological leadership. Was it, as Professor Landes has said, achieved through the efforts of 'practical tinkerers',[1] or did it have some scientific basis? There is no doubt that native empiricism was of immense and probably predominant importance. It was strongly evident, for instance, in the development of engineering and industrial motive power—the bases of modern mechanized mass-production. Mr. Robinson and I have investigated the origins of engineering in Lancashire, where the early factory system had its most striking development.[2] We have traced how early engineering workers were recruited from a wide range of traditional skilled craftsmen: smiths, wheelwrights, millwrights, carpenters, turners, clock-makers, etc., in fact from all kinds of workers in metal and wood. The millwright— the jack-of-all-trades described by Fairbairn[3]—was especially important. So, too, as we have demonstrated, was the clock-maker, whose skills and tools were turned to cutting gear-wheels, etc. for the 'clockwork' of early textile machinery. A similar transition is uniquely illustrated in Pyne's *Microcosm* of 1808,[4] where we see how the making of wooden cart wheels metamorphosed into the construction of wooden 'wheel machinery' (spur wheels, crown wheels, pinions, etc.) for the early mills. Brindley was such a wheelwright and millwright. Similarly, craftsmen from other trades—notably Smeaton (instrument-maker) and Rennie (millwright)—developed the use of cast-iron gears and other mechanical improvements; Rennie, for example, appears to have brought the centrifugal governor for steam engines—the invention of which is often attributed to James Watt—from his millwright practice in windmills.[5] In the same way, turners in wood changed easily into turners in metal.[6] By the late eighteenth century, it is clear that a mechanical engineering industry was emerging, and in Lancashire, and perhaps elsewhere, it was beginning to show signs of specialization—

[1] D. S. Landes, 'Entrepreneurship in Advanced Industrial Countries: The Anglo-German Rivalry', *Entrepreneurship and Economic Growth*, papers presented at a conference at Cambridge, Massachusetts, Nov. 1954, p. 8. He has somewhat revised this view, however, in the *Cambridge Economic History of Europe*, Vol. VI (1965), pt. I, pp. 293–6, in the light of the evidence we have produced.

[2] A. E. Musson and E. Robinson, 'The Origins of Engineering in Lancashire', *Journal of Economic History*, June 1959. See below, chap. XIII.

[3] W. Fairbairn, *Treatise on Mills and Millwork* (2 vols., 1861–3), Vol. I, pp. v–vi. See below, pp. 73, 429, 481.

[4] W. H. Pyne, *Microcosm*, Vol. II (1808), plate 69, reproduced in W. H. Chaloner and A. E. Musson, *Industry and Technology* (1963), plate 64.

[5] C. T. G. Boucher, *John Rennie 1761–1821* (1963), pp. 11 and 81–4. Mr. Robinson, however, considers that Watt developed the use of this device independently, though also from windmills.

[6] In addition to our own researches in this field, see R. Willis, 'Machines and Tools for Working in Metal, Wood, and other Materials', *Lectures on the Results of the Great Exhibition of 1851* (1852).

e.g. in textile machine-making, and even in the specialized production of particular parts, such as spindles and rollers—to cater for the rapidly growing market.

Ironfounders, of course, were especially important in this development of engineering. The achievements of such great eighteenth-century pioneers as the Darbys, Wilkinson, and others do not need elaborating here.[1] It should, however, be emphasized that some mechanical engineering firms—variously described in trade directories as ironfounders, millwrights, machine-makers, etc.—had reached a much greater size by the early nineteenth century than has hitherto been appreciated. Bateman & Sherratt of Salford, for example, were a considerable firm by the late eighteenth century, using power-operated lathes and boring engines in the manufacture of textile machinery, steam engines, and a wide range of other products.[2] Peel, Williams & Co., of Manchester, were another large engineering firm, probably employing several hundred workers by the early 1820s; for some years they had been mass-producing engine parts, such as gear-wheels, which they advertised by printed catalogues.[3] Richard Roberts was using templets for standardized production of parts for textile machines and railway locomotives in the 1820s and 1830s.[4] James Nasmyth in the 1830s proceeded further in this direction, by developing the manufacture of standardized machine-tools and assembly-line production.[5] At the same time, Joseph Whitworth was also emphasizing the necessity of standardization for mass-production, and was similarly developing the manufacture of machine-tools.[6]

Manchester engineers such as Roberts, Fairbairn, Nasmyth, and Whitworth owed a great deal to the famous Henry Maudslay, of London,[7] whose works were a training-ground for many of these and other engineers. London, in fact, was much more important as a centre

[1] See T. S. Ashton, *Iron and Steel in the Industrial Revolution* (3rd edn., 1963), and A. Raistrick, *Dynasty of Ironfounders. The Darbys and Coalbrookdale* (1953).

[2] Musson and Robinson, *op. cit.*, pp. 226–7, and 'The Early Growth of Steam Power', *Econ. Hist. Rev.*, 2nd ser., Vol. XI (1958–9), pp. 425–6. See below, chaps. XII and XIII.

[3] A. E. Musson, 'An Early Engineering Firm: Peel, Williams & Co., of Manchester', *Business History*, Vol. III, no. 1 (Dec. 1960). See below, chap. XIV.

[4] See below, p. 478.

[5] A. E. Musson, 'James Nasmyth and the Early Growth of Mechanical Engineering', *Econ. Hist. Rev.*, 2nd ser., Vol. X (1957–8). See below, pp. 494–500.

[6] A. E. Musson, 'Joseph Whitworth, Toolmaker and Manufacturer', *Chartered Mechanical Engineer* (March 1963), reprinted in *Engineering Heritage* (Inst. Mech. Eng., 1963); a revised version of this article, 'The Life and Engineering Achievements of Sir Joseph Whitworth, 1803–87', appears in the *Whitworth Exhibition* catalogue, July–Aug. 1966 (Inst. Mech. Eng.).

[7] See J. F. Petree, *Henry Maudslay, 1771–1831, and Maudslay, Sons and Field Ltd.* (Maudslay Society, 1949).

of engineering than has been generally recognized. To Maudslay's illustrious name, and that of his erstwhile employer, Joseph Bramah, we can add many others such as Rennie, Clement, Galloway, Donkin, etc. These engineers played a vital role in the diffusion of advancing technology in the early nineteenth century.

Traditional empirical skills were undoubtedly important in this early development of engineering. It should also be emphasized that traditional forms of power remained far more important in the Industrial Revolution than is generally realized. Water power was especially important. We know, of course, that water-wheels had long been used for grinding corn and fulling cloth, and increasingly, from the sixteenth century onwards, for working mine-drainage pumps and winding engines, for operating furnace-bellows, hammers, and rollers in ironworks, for driving saw-mills and paper mills, and for many other industrial purposes. We know, too, that early textile factories, such as Lombe's silk mill and Arkwright's cotton mills, were powered by water-wheels. But the enormous number and wide variety of uses of such wheels have never been fully appreciated, nor have the improvements in their design and their increasing size during the Industrial Revolution, though Mr. P. N. Wilson and Mr. A. Stowers have done much in the last decade or so to emphasize their importance.[1]

Over 5,000 water-mills were recorded in Domesday Book. One can hardly imagine how many there were in the whole country by 1800, but research into their uses in particular areas has produced some very interesting results. Within a few miles of Kendal, in Westmorland, for example, 112 mills have been traced.[2] These were used in corn-milling, fulling, iron manufactures, textile mills, paper mills, snuff mills, and a wide range of other trades. In the Sheffield area, similarly, water power was the basis of industrial development, mainly for working tilt hammers and grinding wheels in the iron and steel trades, making cutlery, files, etc., but also in many other manufactures.[3] An anonymous

[1] P. N. Wilson, 'Water Power and the Industrial Revolution', *Water Power* (Aug. 1954); *Water Mills: an introduction* (1956), a pamphlet issued by the Society for the Protection of Ancient Buildings; 'Water-driven Prime Movers', *The Chartered Mechanical Engineer* (Jan. 1961). A. Stowers, 'Observations on the History of Water Power', *Newcomen Soc. Trans.*, Vol. XXX (1955-7), and his chapter on water-mills in C. Singer and others (eds.), *A History of Technology*, Vol. IV (1958), pp. 199-213. See also C. P. Skilton, *British Windmills and Watermills* (1947).
[2] J. Somervell, 'Water Power and Industries in Westmorland', *Newcomen Soc. Trans.*, Vol. XVIII (1937-8). See also his *Water Power Mills of South Westmorland* (Kendal, 1930).
[3] A. Allison, 'Water Power as the Foundation of Sheffield's Industries', *Newcomen Soc. Trans.*, Vol. XXVII (1949-51); W. T. Miller, *The Water-mills of Sheffield* (4th edn., 1949).

pamphlet of 1794 lists 111 water-powered cutler's 'wheels' in that area, but only five powered by steam, and water-wheels remained numerous and important far into the nineteenth century, a few even surviving today. Altogether over 150 such wheels have been recorded in this area.

In the midlands metal trades, Dr. R. A. Pelham has produced similar evidence. 'In Birmingham', he concludes, 'the Industrial Revolution of the nineteenth century, in so far as it was a revolution at all, was based almost entirely on the water mill.'[1] Here water-wheels were used mainly—apart from their use in corn-mills—in blade mills (in the cutlery and edge-tool trades) and in rolling and slitting mills. Boulton and Watt erected only two of their rotative engines in the Birmingham area (outside their own works) in the period up to 1800, and water-powered mills remained important throughout the nineteenth century. In Cornwall, similarly, we are told that 'there were forges or hammer mills, driven by water-wheels, scattered over the county from one end to the other'.[2]

Similar evidence comes from the mainly rural county of Lincolnshire, where, it is said, water-mills 'must [once] have been numbered in their hundreds'.[3] They were used mainly for corn-milling, but also for fulling, paper-making, iron manufacture, furniture-making, and grinding bones for manure; a number of 'farm wheels' were also used for grinding, chaff cutting, etc. By 1828, however, according to Bryant's one-inch Lincolnshire map, working water-mills were reduced to 67, and by 1905 there were only 46.

Even in the textile industries, the *locus classicus* of the steam-powered factory system, water-mills were very slowly displaced. They were frequently encountered by the Factory Commissioners of 1833, and by the factory inspectors in later years. In 1839 there were 3,051 steam engines and 2,230 water-wheels in the textile factories of the United Kingdom, though the total horse-power of the former (74,094) considerably exceeded that of the latter (27,983).[4]

An example of the proliferation of water-wheels in Scotland is provided by a surviving token halfpenny, issued by John Ferriar of Perth in 1797, which records the fact that there were then '46 Water-Mills

---

[1] R. A. Pelham, 'The Water-Power Crisis in Birmingham in the Eighteenth Century', *University of Birmingham Historical Journal*, Vol. IX, no. 1 (1963).
[2] R. Jenkins, 'Hammer Mills in Cornwall', *Newcomen Soc. Trans.*, Vol. XVIII (1937–8).
[3] W. E. R. Hallgarth, 'The Water Mills of Lincolnshire', *Lincolnshire Life*, Vol. 4, no. 2 (April–May 1964).
[4] A. Ure, *The Philosophy of Manufactures* (1861), p. 428, based on the factory inspectors' returns.

for Bleaching, Printing, Cotton-Works, Corn, &c. within 4 miles of Perth'.[1] How many, one wonders, were there in the whole country?

In the second half of the eighteenth century, engineers such as Smeaton and Rennie erected water-wheels in many parts of Great Britain, for a wide variety of industrial purposes.[2] In the first half of the nineteenth century their work was continued by Hewes, Fairbairn, and others. The article on 'Water' in Rees's *Cyclopaedia* (1819) and Joseph Glyn's book *On the Power of Water to Turn Mills* (1853), which had run into five editions by 1875, show the continued importance of water power in this later period. Early water-wheels were small, made of wood, and mostly undershot; many probably developed well under 1 h.p. But during the Industrial Revolution their construction was greatly improved and they were made much more powerful and efficient. Smeaton, for example, demonstrated the superiority of over-shot and breast-wheels over the undershot type, and he also introduced iron axles and gears.[3] John Farey, writing in the early nineteenth century on the development of the steam engine, considered that this would have been more rapid had it not been for Smeaton's improvements in water-wheel design and construction,

whereby the performance of almost all the mills then existing, could be doubled with the same supplies of water. He [Smeaton] had an extensive practice in directing the construction of such mills during 30 years; and the numerous specimens which he left in all parts of the kingdom, were copied by others in their respective neighbourhoods, till the improved methods of constructing water-mills became general, and was [sic] attended with great benefit to trade.[4]

Thus, Farey pointed out, 'the necessity of employing steam power for mills was not felt until some years afterwards'. Moreover, Smeaton and others then used steam engines to pump water for water-wheels.[5]

Other engineers also contributed improvements. Rennie invented the depressing sluice, whereby the water always flowed over the top and the maximum fall was obtained; he also extended the use of iron parts. Hewes and Fairbairn developed far bigger wheels of all-iron construction. Fairbairn has left accounts of his improvements, such as ventilated buckets, and of the many wheels which he erected in both

[1] This token coin is now in the possession of Dr. W. H. Chaloner, of Manchester University, to whom I am indebted for this piece of evidence.
[2] P. N. Wilson, 'The Waterwheels of John Smeaton', *Newcomen Soc. Trans.*, Vol. XXX (1960); C. T. G. Boucher, *John Rennie 1761–1821* (1963), chap. 6.
[3] See below, p. 73, for reference to Smeaton's experimental researches in this field.
[4] J. Farey, *A Treatise on the Steam Engine* (1827), p. 296.
[5] See below, p. 398

the United Kingdom and Europe;[1] but one would like to know more about the activities of T. C. Hewes, one of the most important engineers in Manchester in the first quarter of the nineteenth century, who was said to have built water-wheels all over the country, including some for the Arkwright–Strutt mill at Belper and for Greg's mill at Styal, and who is especially notable for his invention of the 'suspension wheel', driving machinery by means of gearing fitted round its inner circumference, instead of via the shaft or axle, thus enabling much more power to be developed. The *Manchester Guardian*, in its obituary on Hewes, declared that

his great improvements in water wheels, and especially the introduction of the *suspension* wheel, which he brought to its present state of perfection (and of which he erected a great number in different parts of the United Kingdom,) will hand down the names of 'Hewes' to a distant period . . . In short, as the name of 'Watt' is associated with the steam engine, so with similar excellence is the name of 'Hewes' associated with the water wheel.[2]

Like Watt, too, he was 'a man of science', which he applied in his engineering improvements.[3] Unfortunately, however, his name has tended to be forgotten and comparatively little is known about his mill-building and machine-making activities in different parts of the country.[4] William Fairbairn testified to Hewes's achievements in water-wheel design and construction:[5]

It was reserved for Mr. T. C. Hewes, of Manchester, to introduce an entirely new system in the construction of water-wheels, in which the wheels, attached to the axis by light wrought-iron rods, are supported simply by suspension . . . [This principle] is due originally, it is said, to the suggestion of Mr. William Strutt and was carried out fifty years ago by Mr. Hewes . . .

There is in Adam Rees's *Cyclopaedia* (1819) a most interesting description of two of his suspension wheels, erected in the Strutts' mill at

[1] See his *Treatise on Mills and Millwork* (1861–3) and W. Pole's *Life of Sir William Fairbairn* (1877). See also below, pp. 482 and 486.

[2] *Manchester Guardian*, 11 Feb. 1832. See also *ibid.*, 19 May 1827, for a reported address by Benjamin Heywood to the Mechanics' Institution, in which, after referring to 'four enormous water-wheels now constructing by Messrs. Fairbairn and Lillie, 50 feet in diameter, and 12 feet wide, and each of 96 horse-power', he went on to explain 'the superiority of this description of water-wheel, invented by Mr. T. C. Hewes, of this town, over the wheels previously in use'. The wheels referred to were almost certainly those constructed by Fairbairn and Lillie for Buchanan and Finlay's famous cotton mills at Catrine, Ayrshire (see below, p. 482). [3] See below, p. 98. [4] See, however, below, pp. 98, 425, 436, 445–7, 481.

[5] Fairbairn, *Mills and Millwork* (2nd edn., 1864–5), Part I, p. 120, and also 'The Rise and Progress of Manufactures and Commerce, and Civil and Mechanical Engineering in these Districts', in T. Baines, *Lancashire and Cheshire, Past and Present* (1867), pp. cxliv–cxlv.

Belper.[1] These were breast-wheels, each $21\frac{1}{2}$ feet in diameter and 15 feet broad, made of cast-iron, with wrought-iron 'arms' or spokes— 'small round rods, which are very light'—and instead of the mill-drive being via the axle, the wheel was merely 'suspended' on the latter, while power was transmitted 'by a ring of cogs screwed to the circular rim of the wheel [on the inside edge] and working in a pinion which conveys the motion to the mill'. To these wheels Hewes fitted what was apparently the earliest water-wheel 'regulator' or governor, so as to maintain a regular drive, so important in cotton-spinning.[2]

These much larger and more efficient wheels continued to be built far into the nineteenth century, long after steam engines began to be used for mill-drives. A fine example was that installed by Hewes and Wren in a Bakewell (Derbyshire) cotton mill in 1827, measuring 25 feet in diameter by 18 feet wide and developing about 200 h.p.[3] Other engineers followed Hewes's example. The Mersey Iron Works, of Liverpool, for instance, built the famous 'Big Wheel' at Laxey, Isle of Man, in 1854, for draining lead mines; measuring $72\frac{1}{2}$ feet in diameter and 6 feet in breadth and weighing about 100 tons, it also developed about 200 h.p.[4] These and other wheels, such as those built by Fairbairn, might well justify our calling the early Industrial Revolution 'the age of water power'.

These advances in water power deserve stressing because they have been unduly overshadowed by the early development of steam engines. In the long run, of course, steam power was to be of far greater importance, but it was a longer run than is generally realized. Even though, as we have shown elsewhere, steam engines were built in greater numbers than would appear from earlier estimates by such as Lord, based almost exclusively on the Boulton and Watt papers—even though other steam engines, including both Savery and Newcomen types, as well as 'pirated' versions of Watt's engine, were being built in considerable numbers during the late eighteenth century[5]—the early pro-

---

[1] Article on 'Water'. Rees does not state that Hewes built these water-wheels, but we know that he did from D. P. Davies, *A New Historical and Descriptive View of Derbyshire* (Belper, 1811), pp. 346–7. See also R. S. Fitton and A. P. Wadsworth, *The Strutts and the Arkwrights 1758–1830* (Manchester, 1958), p. 207, n. 1. A surviving example of a Hewes and Wren water-wheel, of the 'suspension' type, is to be seen in the Department of Technology of the Royal Scottish Museum, Edinburgh.

[2] Rees, *op. cit.*, article on 'Mill-Work'.

[3] See Chaloner and Musson, *op. cit.*, plate 12. It remained in operation until the 1950s. [4] *Ibid.*, plate 13.

[5] A. E. Musson and E. Robinson, 'The Early Growth of Steam Power', *Econ. Hist. Rev.*, 2nd ser., Vol. XI (1958–9). See below, chap. XII. See also J. F. Rowe, *Cornwall in the Age of the Industrial Revolution* (1953), p. 101.

gress of steam power should not be exaggerated out of proportion to that of water power, or to the later development of steam power itself. All the engines (about 500) built by Boulton and Watt up to 1800 totalled only about 7,500 h.p., and the really massive growth of steam power occurred in the later nineteenth century, rising from 500,000 h.p. in 1850 to 9,650,000 h.p. in 1907.[1] Moreover, the early steam engines were of comparatively low power. In the early 1820s, for example, a 60-h.p. cotton-mill engine was described as a large one,[2] but by the early twentieth century steam engines of 3,000–4,000 h.p. were to be found in big cotton mills.

The early steam-engine makers came from the ranks of ironfounders, millwrights, and other traditional trades, as previously mentioned. Watt himself was originally an instrument-maker, like Smeaton. They were able to apply their practical skills to the solution of new engineering problems. Probably most of the achievements of the early Industrial Revolution were similarly products of practical empiricism. But during the past decade or so an increasing amount of evidence has been accumulating to show that scientific technology was also developing at this time and assisting in these achievements. Dr. and Mrs. Clow, for example, have shown how the Scottish universities contributed to advancements in chemical technology in a wide range of industries, such as the alkali, soap, glass, bleaching, dyeing, and other trades.[3] Many similar examples can be culled from the pages of *Isis*, the *Annals of Science*, and the growing number of similar journals. Mr. Robinson and I have also tried to show the industrial significance of philosophical societies in Birmingham, Manchester, and other towns, of dissenting academies, embryo technical colleges and schools, of books, encyclopaedias, periodicals, and libraries, and of itinerant lecturers—all of which helped to spread scientific-technological knowledge and interests more widely.[4] Dr. Schofield has more recently demonstrated in copious detail the links between science and industry in the Birmingham Lunar Society.[5]

Professor Jewkes and his collaborators have suggested that 'the disposition of modern writers to regard nineteenth [and eighteenth] century inventors as uneducated and empirical in their methods is a direct outcome of the difficulty which academically educated people

---

[1] M. G. Mulhall, *Dictionary of Statistics* (4th edn., 1903), pp. 545 and 807; *First Census of Production, Final Report* (1907), p. 17.
[2] By Joshua Field, the London engineer, *Newcomen Soc. Trans.*, Vol. XIII (1932–3), p. 23.   [3] A. N. and N. Clow, *The Chemical Revolution* (1952).
[4] Musson and Robinson, 'Science and Industry in the Late Eighteenth Century', *Econ. Hist. Rev.*, Vol. XIII, no. 2 (Dec. 1960). See below, chap. III.
[5] R. E. Schofield, *The Lunar Society of Birmingham* (1963).

often have in understanding the possibilities of self-education'.[1] As they briefly observe, facilities for self-education certainly existed, and some of the most notable scientists and industrialists were self-educated.[2] Many of the early engineers, for example, combined scientific knowledge with practical experience. Even the ordinary eighteenth-century millwright, according to Fairbairn, was generally 'a fair arithmetician, knew something of geometry, levelling, and mensuration, and in some cases possessed a very competent knowledge of practical mechanics. He could calculate the velocities, strength and power of machines; could draw in plan and in section . . .': indeed they were commonly regarded as men of 'superior attainments and intellectual power'.[3] The truth of this statement is clearly evident, not only in Fairbairn's own works, but in those of earlier engineers such as Smeaton, Telford, and Rennie. Smeaton's high abilities in surveying, designing, mechanics, etc. are revealed in the four volumes of his *Reports*, published by the Smeatonian Society of Engineers in 1812, and also in his drawings,[4] while his capacity for scientific-technological experiment is brilliantly demonstrated in 'An Experimental Enquiry concerning the natural Powers of Water and Wind to turn Mills, and other Machines',[5] and by his similar investigations into the atmospheric steam engine, all scientifically controlled, with results tabulated mathematically.[6] Hence his election as a Fellow of the Royal Society, to which he read numerous other papers on mechanics,[7] scientific instruments,[8] and astronomy.

[1] Jewkes, *op. cit.*, pp. 63–4.

[2] Faraday, Davy, Sturgeon, and Wheatstone are among the 'self-educated' scientists mentioned by Jewkes; Young, Hodgkinson, and others could be added. We shall be referring to numerous industrialists who similarly acquired scientific knowledge.

[3] W. Fairbairn, *Treatise on Mills and Millwork* (1861–3), Vol. I, pp. v–vi.

[4] See *A Catalogue of the Civil and Mechanical Engineering Designs, 1741–92, of John Smeaton, F.R.S.*, preserved in the Library of the Royal Society, Newcomen Society Extra Publication no. 5, 1950.

[5] Smeaton made his experiments in 1752–3. The results formed the subject of papers read to the Royal Society in May and June 1759 (*Phil. Trans.*, Vol. LI, 1759, pp. 100–74), for which he was awarded the Society's Copley Medal. These papers were republished in 1794. These and his other papers on mechanics were frequently referred to by later engineers.

[6] These experiments, made in the early 1770s, are explained and analysed in detail by J. Farey, *A Treatise on the Steam Engine* (1827), pp. 134, 158, and 166 ff. The experiments, of which he recorded over 130, were spread over four years.

[7] Notably 'An Experimental Examination of the Quantity and Proportion of Mechanic Power necessary to be employed in giving different Degrees of Velocity to Heavy Bodies from a State of Rest', read 25 April 1776, *Phil. Trans.*, Vol. LXVI (1776), pp. 450–75, and 'New Fundamental Experiments upon the Collision of Bodies', read 18 April 1782, *ibid.*, Vol. LXXII (1782), pp. 337–54. Smeaton refuted the theoretical opinions of earlier philosophers such as Belidor, Parent, Desaguliers, Maclaurin, etc. on these matters, especially in regard to the operation of water-wheels. [8] See, for example, above, p. 50, n. 4.

During his visits to London, 'it was a source of great pleasure to him to attend the meetings of the Royal Society, as well as to cultivate a friendship with the distinguished members of the Royal Society Club'.[1] He was one of the leading figures in the earliest society of engineers, founded in 1771, which, after his death, became the Smeatonian Society in 1793 and was the forerunner of the Institution of Civil Engineers, founded in 1818.[2] The members, who included most of the leading engineers of that day, discussed engineering theory as well as practice. One of their meetings in 1778, for example, was spent 'canallically, hydraulically, mathematically, philosophically, mechanically, naturally, and socially'.[3]

Smeaton, so Smiles tells us, deliberately 'limited his professional employment, that he might be enabled to devote a certain portion of his time to self-improvement and scientific investigation'; he was frequently engaged in study and experiment in the specially erected tower, combining workshop, study, and observatory, at his home at Austhorpe, near Leeds.[4] He was also a member of the scientific coterie in Leeds, including Joseph Priestley and others.[5] George Stephenson later referred to Smeaton as

the greatest philosopher in our profession that this country has yet produced. He was indeed a great man, possessing a truly Baconian mind, for he was an incessant experimenter. The principles of mechanics were never so clearly exhibited as in his writings, more especially with respect to resistance, gravity, and the power of water and wind to turn mills. His mind was as clear as crystal, and his demonstrations will be found mathematically conclusive. To this day there are no writings so valuable as his in the highest walks of scientific engineering . . .[6]

Telford, originally a stonemason, is often regarded, like Brindley before him, as a purely practical engineer, devoid of, and even hostile to, scientific theory.[7] There is no doubt, as Sir Alexander Gibb has shown,[8] that Telford emphasized above all the value of practical

---

[1] Smiles, *Lives of the Engineers* (1874), Vol. II, p. 169.
[2] S. B. Donkin, 'The Society of Civil Engineers (Smeatonians)', *Newcomen Soc. Trans.*, Vol. XVII (1936–7).   [3] *Ibid.*, p. 58.
[4] Smiles, *op. cit.*, chap. VI.   [5] See below, pp. 157–8.
[6] Quoted by Smiles, *op. cit.*, p. 177. Smiles rightly emphasized, however, Smeaton's distrust of pure theory unsupported by practical experiment.
[7] This view appears to have been based on the statement by Sir David Brewster in the *Edinburgh Review*, Oct. 1839, that 'Telford had a singular distaste for mathematical studies, and never even made himself acquainted with the elements of geometry'; he is even said to have considered that mathematical acquirements unfitted a man for practical engineering. Dr. Boucher has recently repeated this view uncritically, in an unfavourable comparison with Rennie (*op. cit.*, p. 30.)
[8] In *The Story of Telford* (1935).

experience, but he was 'no enemy to either mathematics or theory'. Indeed, he acquired a considerable fund of theoretical knowledge. His early letters reveal him reading very widely on architecture, mathematics, chemistry, mechanics, hydrostatics, etc., on which he made copious notes. Thus on 1 February 1796 he wrote to his friend, David Little:[1]

Knowledge is my most ardent pursuit . . . I am now deep in Chemistry—the manner of making Mortar led me to inquire into the Nature of Lime and in pursuit of this, having look'd into some books on Chemistry I perceived the field was boundless—and that to assign reasons for many Mechanical processes it required a general knowledge of that Science. I have therefore had the loan of a MSS Copy of Dr. Black's Lectures. I have bought his Experiments on Magnesia and Quick Lime and likewise Fourcroy's Lectures translated from the French by a Mr. Elliott. And I am determined to study with unwearied attention until I attain some general knowledge of Chemistry as it is of Universal use in the [practical] Arts as in Medicine.

A year later, he was still 'Chemistry mad' and specially interested in 'Calcareous matters', with the practical objective of securing the best type of cement for building bridges, etc., both above and below water. But his interests were 'not . . . confined to that alone', and in addition to the works of Black and Fourcroy, previously mentioned, he had read Scheele's *Essays*, Macquer's *Dictionary*, Watson's *Essays*, and writings by his friend Dr. Irving [Irvine?]. From these he had taken notes in a pocket-book, 'which I always carry with me', and into which he had also 'cramm'd Mechanics, Hydrostatics, Pneumatics, and all manner of Stuff, and to which I keep continually adding'.[2]

Telford became friendly with various professors at Edinburgh University, such as Professors Stewart, Gregory, Playfair, and Robison, who secured his election to the Royal Society of Edinburgh. Towards the end of his life, in 1827, he was elected F.R.S., in recognition of his experimental research as well as of his civil engineering achievements. His researches generally had a strong practical bias, such as those on the strength of iron chains and rods for bridge-building, or on the movement of canal boats through water, but, as Sir Alexander Gibb has remarked, they were 'always carried out scrupulously, in logical order and with real scientific care'. Moreover, though he himself lacked a thorough scientific education, he could draw on the scientific help of others, as in 1814 when he carried out experiments on the strength of

---

[1] Quoted in S. Smiles, *The Life of Thomas Telford* (1867), pp. 128–9, and Gibb, *op. cit.*, p. 10.

[2] Letter to David Little, 27 Jan. 1797, quoted in Smiles, *op. cit.*, p. 134, and Gibb. *op. cit.*, pp. 266–7. This practice of noting down information, gained from reading and observation, in a pocket memorandum book, 'a sort of engineer's vade mecum', was continued by Telford throughout his life.

iron with the assistance of Peter Barlow, Professor of Mathematics at Woolwich Military Academy, who published the results in his book on *The Strength of Materials*.[1] Telford himself contributed long articles on architecture, bridge-building, and canals to the *Edinburgh Encyclopaedia*. He also wrote a treatise *On Mills* (1798).[2] He accumulated a large collection of scientific and technological books, many of which he donated to the Institution of Civil Engineers when he was elected its first President in 1820. But he always stressed the practical side of engineering, and was scornful of engineering theories untested by experience.

John Rennie's career, as Dr. Boucher has emphasized, provides 'a happy blend of theory and practice'.[3] He was educated at Dunbar High School, where he displayed remarkable ability in mathematics and in natural and experimental philosophy, before proceeding to Edinburgh University, where he developed a lifelong friendship with John Robison, the famous Professor of Natural Philosophy. In his youth he is said to have read 'the treatises of Belidor, Parent, and Lambert; Emmerson's *Fluxions and Mechanics*, Switzer's *Hydraulics*, Smeaton's *Experimental Enquiry*, and others'. At the same time he had a thorough practical training under the celebrated millwright, Andrew Meikle, inventor of the threshing machine and the spring sail for windmills. In his subsequent civil and mechanical engineering structures, 'he applied scientific theory, as well as practical experience . . . and used his university training to seek out the root causes of problems which he encountered'. In regard to his water-wheel constructions, for example, Dr. Boucher remarks:

It should not be thought that Rennie's application of water power was based on a mere rule of thumb and tradition. He was a practical student of the science of hydraulics, as has been shown by reference to the textbooks he studied. Among his possessions is a handbook on hydraulics in his own writing compiled by himself. Along with the principal formulae are tables incorporating their application.

Rennie also helped John Robison with articles published by the latter in 1797 under the title *Outlines of a Course of Lectures in Mechanical Philosophy*, which then appeared in the third edition of the *Encyclopaedia*

---

[1] For Barlow, see *D.N.B.* and also E. G. R. Taylor, *Mathematical Practitioners of Hanoverian England* (1966), pp. 330–1. Fairbairn later carried out similar experiments with the scientific aid of Eaton Hodgkinson. See below, pp. 77, 117–18, 481, 487. See also below, p. 79, for the earlier experiments of Charles Bage.

[2] E. L. Burne (ed.), 'On Mills', by Thomas Telford, *Newcomen Soc. Trans.*, Vol. XVII (1936–7).

[3] In addition to Dr. Boucher's recent book on Rennie, previously cited, see also the older life by Smiles, *Lives of the Engineers* (1874), Vol. II, in which Rennie's successful combination of theory and practice is similarly emphasized.

Britannica, and were afterwards published separately in four volumes on *Mechanical Philosophy*. It is evident that Rennie had a very good knowledge of structural theory, and Dr. Boucher has demonstrated in detail how he applied this to bridge-building, etc. Rennie was one of the founders of the London Literary and Philosophical Institution, which he attended whenever possible, and he was also a member of the Smeatonian Society. Like Telford, he built up a library rich in works on engineering and other subjects.

Other engineers of this period, such as Thomas Hewes, Peter Ewart, and George Lee in Manchester, were similarly knowledgeable in science and technology.[1] Henry Maudslay, although he started work at twelve years of age, was another remarkable example of 'self-education'. 'He was not merely a mechanical genius wholly ignorant of science; astronomy and the manufacture of telescopes for his own use was his chief hobby. Faraday was his close friend and a frequent visitor to his works. Like many inventors of this period, scientific discoveries were of as much interest to Maudslay as his practical achievements were to the scientist.'[2]

William Fairbairn, another great self-educated engineer, has left his own account of how he spent his youthful evenings studying arithmetic, algebra, geometry, trigonometry, etc.[3] After starting his own engineering business in Manchester, he became a member of the Literary and Philosophical Society, carried out important scientific experiments with Eaton Hodgkinson, the mathematician, on the strength of iron beams and pillars,[4] was eventually elected a Fellow of the Royal Society, and wrote a considerable number of books on engineering and industrial development.[5] Other leading engineers acquired scientific knowledge in much the same way, but they were not all 'self-educated'; Nasmyth, for example, like Rennie, had the benefit

[1] See below, pp. 98–101.
[2] Jewkes, *op. cit.*, pp. 46–7. Similar examples of the relationships between scientists such as Faraday, metallurgists, and engineers in the early nineteenth century are provided in the diaries of the Swiss industrialist, J. C. Fischer. See W. O. Henderson, *J. C. Fischer and his Diary of Industrial England 1814–51* (1966), pp. 5, 12, 13, 33, 36–7, 65, 106, 117, 160.
[3] W. Pole (ed.), *The Life of Sir William Fairbairn . . . Partly written by himself* (1877), pp. 72–80, 101, 103. See below, pp. 480–1.
[4] See *Manchester Lit & Phil. Soc. Memoirs*, 3rd ser., Vol. V (1831); *British Association, Third Report* (1833), *Fourth Report* (1835), *Fifth Report* (1835), *Seventh Report* (1837); *Royal Society Transactions*, 14 May 1840. See also below, pp. 117–18, 481, and 487.
[5] See, for example, his *Useful Information for Engineers* (1856), *The Rise and Progress of Civil and Mechanical Engineering* (1859), *Iron: Its History, Properties and Process of Manufacture* (1861), and *Treatise on Mills and Millwork* (2 vols., 1861–3).

of a good academic education.[1] Charles Holtzappfel also appears to have been well educated and 'devoted himself assiduously to the acquirement of scientific and practical knowledge'.[2]

It was not only in engineering that science was applied. The famous potter, Josiah Wedgwood, was greatly interested in experimental scientific research into clays, glazes, colouring, and temperature control.[3] Together with Watt, Boulton, and other industrialists, he associated with scientists such as Darwin, Priestley, and Withering, and was elected a Fellow of the Royal Society. Moreover, in his factory at Etruria, he introduced men who could apply scientific knowledge to pottery manufacture:

Scientific men were engaged, at liberal salaries, in the various departments of the business—in chemistry, in design, in modelling, in painting, &c. The ingenious Mr. Alexander Chisholme, who had been employed in experimental chemistry by Dr. William Lewis, the celebrated author of the 'Commercium Philosophico-Technicum', was taken into Wedgwood and Bentley's service in 1781, and for many years . . . up to the period of his death, enjoyed the bounty of Mr. Wedgwood.[4]

John Marshall, the great Leeds flax-spinner, also had an enthusiasm for natural philosophy, attending scientific lectures and putting his knowledge to practical use:

Before the age of forty his intellectual interests had revolved around the mills and new scientific techniques. He studied science because it helped him: a knowledge of chemistry was useful in bleaching, the theory of machines in providing power, the properties of materials in construction. Beyond this, he took lecture-notes on optics, electricity, and astronomy. He studied and talked about geology . . . In short he wanted to keep up with the rush of modern knowledge because it could be useful.[5]

He was a member of the Literary and Philosophical Society of Leeds, and participated in the founding of a Lancastrian school and Mechanics' Institute in that town; later he proposed the establishment of a university

[1] See below, pp. 489–90.

[2] *Proceedings of the Institution of Civil Engineers*, 1848, p. 14. See also the references given above, p. 63, n. 1. Holtzappfel's book on *Turning and Mechanical Manipulation* (3 vols., 1843) 'displays a masterly knowledge of technical art and of the scientific principles underlying it'. *D.N.B.*

[3] R. E. Schofield, 'Josiah Wedgwood and a Proposed Eighteenth-Century Industrial Research Organization', *Isis*, Vol. 47 (1956), pp. 16–19; 'Josiah Wedgwood, Industrial Chemist', *Chymia*, Vol. 5 (1959), pp. 180–92.

[4] J. Ward [and S. Shaw], *The Borough of Stoke-upon-Trent* (1843), pp. 433–4. See also S. Parkes, *Chemical Essays* (1815), Vol. I, pp. 48–9. For Lewis and Chisholm, see above, pp. 53–4.

[5] W. G. Rimmer, *Marshall's of Leeds, Flax Spinners 1788–1886* (1960), p. 103. See also below, pp. 153–5 and 329–32.

there, and subscribed towards the new university in London, becoming a member of its council.

Professor Rimmer has also shown how Charles Bage, who was associated in business with Marshall, made similar use of applied science. Thus in designing the Castle Foregate mill at Shrewsbury, 'Bage calculated the strength of his mill beams and proved them by full-scale tests. In this, and in other fields like bleaching, he was abreast if not ahead of current developments in science and engineering.'[1]

Bleaching provides a particularly good example of applied science. The discoveries of the scientists Scheele and Berthollet have already been mentioned. We have found abundant evidence of applied chemistry in the correspondence of James Watt, Thomas Henry (of Manchester), and others, who utilized their chemical knowledge in bleaching, dyeing, and calico-printing experiments.[2] As early as the mideighteenth century one comes across a reference to chemical analysis of water-supplies preparatory to location of a bleachworks,[3] while later on, in the early nineteenth century, one finds the scientist John Dalton providing similar services to industry.[4] A great deal of evidence has also come to light regarding James Watt's collaboration with Dr. Joseph Black in the development of synthetic soda manufacture.[5]

In the development of the steam engine, science also played a considerable role.[6] Papin, Savery, and Newcomen owed much to the earlier scientific investigations of such men as Boyle and Huygens. Watt's relations with Black—discoverer of the principle of latent heat—at the

[1] Rimmer, *op. cit.*, p. 59. Bage combined mathematical theory with experimental testing in ascertaining the strength of iron beams and columns for mill construction. T. C. Bannister, 'The First Iron-framed Buildings', *Architectural Review*, Vol. 108 (April 1950), pp. 231–46; A. W. Skempton, 'The Origin of Iron Beams', *Actes VIII Int. Conf. History of Science* (Florence, 1956), Vol. 3, pp. 1029–39; H. R. Johnson and A. W. Skempton, 'William Strutt's Cotton Mills, 1793–1812', *Newcomen Soc. Trans.*, Vol. XXX (1955–7), pp. 179–205.

[2] See below, chaps. VII, VIII, and IX.

[3] F. Home, *Experiments on Bleaching* (1756), pp. 281–8; S. Parkes, *Chemical Essays* (1815), Vol. IV, pp. 211–12.

[4] Sir A. J. Sykes, *Concerning the Bleaching Industry* (1925), p. 91. Sykes, of Edgeley, Stockport, 'employed John Dalton, the famous scientist, as consultant on the quality of their water supply'. Parkes, *op. cit.*, Vol. IV, pp. 181 ff., strongly emphasized the necessity of chemical analysis of water supplies for bleaching, dyeing, and calico-printing.

[5] See below, chap X. Mr. Robinson and the late Professor McKie are bringing out an edited version of the Black–Watt correspondence. This will show the importance of Watt's scientific-experimental abilities.

[6] See M. Kerker, 'Science and the Steam Engine', *Technology and Culture*, Vol. 2 (1961), pp. 381–90; the introduction by A. E. Musson to the new edition of H. W. Dickinson, *Short History of the Steam Engine* (1963); D. S. L. Cardwell, *Steam Power in the Eighteenth Century* (1963).

time of his development of the steam engine have often been referred
to, but Watt's own scientific abilities have been inadequately appreciated. It is abundantly clear, however, that, in the words of his
intimate friend, Professor John Robison, he 'was a person of truly
philosophical mind, eminently conversant in all branches of natural
knowledge'.[1] He was on very friendly terms, not only with Robison,
but with Black, Anderson, and other professors at Glasgow and
Edinburgh universities,[2] and later with Priestley, Darwin, and most
other eminent philosophers in England, and also with many French,
German, and other continental scientists. He was not merely a brilliant
mechanic, but a truly scientific engineer and chemist, well versed in
contemporary scientific knowledge and constantly engaged in scientific
experiments. He inherited the abilities of his grandfather, a teacher of
mathematics, and of his father, a shipwright; educated at Greenock
Academy and trained as an instrument-maker, he early exhibited
strong interests in mathematics, mechanics, and chemistry—before he
was fifteen he had read Gravesande's *Mathematical Elements of Natural
Philosophy*, translated by Desaguliers. Later, when appointed instrument-
maker to Glasgow University, he started learning German so that he
might read Leupold's *Theatrum Machinarum*, and Italian for a similar
purpose; to construct an organ, he read the *Harmonics* of Dr. Robert
Smith, of Cambridge.[3] Before his famous improvements on the steam
engine, he had read Desaguliers and Belidor on this subject, and these
improvements were the outcome of careful experiments on steam and
on models of Savery and Newcomen engines, in which he consulted
with Professor Black. His famous tea-kettle, usually dismissed nowadays as mythical, was in fact, as his diary reveals, used as a miniature
boiler in a series of laboratory experiments, and he produced tables on
the thermal efficiency of steam engines, based on mathematical theory
as well as practical experiment. He also appears to have rediscovered
the principle of latent heat, independently of Black.[4] Watt himself, of
course, like Boulton, Wedgwood, Keir, and other industrialists, was
made a Fellow of the Royal Society.

That science could make useful contributions to early steam engineering is also shown by other evidence. The scientist Davies Gilbert (or

[1] Article on 'Steam Engine', *Encyclopaedia Britannica* (3rd edn., 1797). Robison's
articles on steam and steam engines were later reprinted, with notes and additions
by Watt (Edinburgh, 1818). See also F. Arago, 'James Watt', in *Biographies of
Distinguished Scientific Men* (English trans., 1857); J. P. Muirhead, *The Life of
James Watt* (2nd edn., 1859); G. Williamson, *Memorials of . . . James Watt* (1856).
[2] See below, pp. 179–80, for Watt's relations with Anderson.
[3] For Smith, see above, p. 34.
[4] On this latter point, see D. Fleming, 'Latent Heat and the Invention of the
Watt Engine', *Isis*, Vol. 43 (1952), pp. 3–5.

Giddy), for example, provided Jonathan Hornblower, the practical engineer, with a great deal of advice in regard to the latter's attempts to develop his compound and 'rotary' (turbine) engines.[1] Similarly he rendered valuable assistance to Richard Trevithick, inventor of the high-pressure, non-condensing steam engine.[2]

In many ways, then, applied science was helping to bring about the Industrial Revolution. On the other hand, one must not fall into the error of supposing that the latter was simply a product of the Scientific Revolution. France was ahead of any other country in scientific achievements during the eighteenth century, yet the Industrial Revolution did not occur there. Instead it occurred in Great Britain, where there were richer resources of coal and iron, where foreign trade was greatly expanding, where capital could more easily be mobilized, where enterprise was freer, where the social and political climate was more favourable to economic development, and where there were therefore greater opportunities of *applying* scientific knowledge.

In the early mechanization of the cotton industry, applied science appears to have played a very minor role.[3] Inventors such as Kay, Paul, Wyatt, Hargreaves, Arkwright, Crompton, and Cartwright appear to have had little or no scientific training, though they often utilized the knowledge and skills of clock- and instrument-makers. They have fairly recently been described as 'mostly...men of some social standing and good education',[4] but this is questionable. Kay, Hargreaves, and Arkwright, for instance, appear to have had a very limited and rudimentary schooling before being put to a trade, and Wyatt, though educated at Lichfield Grammar School, became a carpenter. Cartwright, it is true, was a Fellow of Magdalen College, Oxford, but he was a Doctor of Divinity and was apparently unacquainted with mechanical matters prior to his invention of the power-loom. Crompton, however, is an interesting case. He had a good schooling in writing, arithmetic, book-keeping, geometry, mensuration, and

---

[1] A. C. Todd, 'Davies Gilbert—Patron of Engineers (1767–1839) and Jonathan Hornblower (1753–1815)', *Newcomen Soc. Trans.*, Vol. XXXII (1959–60), pp. 1–13.

[2] H. W. Dickinson and A. Titley, *Richard Trevithick, the Engineer and the Man* (1934), p. 36 *et passim*.

[3] As was pointed out by Arkwright's Counsel during the great patent case of 1785, in oft-quoted words: 'It is well known that the most useful discoveries that have been made in every branch of art and manufacture have not been made by speculative philosophers in their closets but by ingenious mechanics, practically acquainted with the subject matter of their discoveries.'

This statement may be contrasted, however, with that of defending Counsel in the Tennant bleaching patent case of 1802, that 'most of these discoveries arise from scientific men engaging in them, for the purpose of science and speculation in their closet'. [4] Jewkes, *op. cit.*, pp. 44–5.

mathematics, under an outstanding master, William Barlow, of Little Bolton, and during his later 'teens he continued his education by attending evening school, improving his knowledge of algebra, trigonometry, and mathematics generally.[1] According to French, his biographer, his invention of the mule 'appears to have been the result of pure inductive philosophy, followed out step by step with a mathematical precision for which his mind had been duly prepared by previous education'.[2]

Sometimes one comes across contemporary scientists and industrialists deploring the general lack of applied science in industry. Theophilus L. Rupp, for instance, a German who settled in Manchester as a cotton manufacturer in the late eighteenth century, wrote as follows.[3]

The arts [manufactures], which supply the luxuries, conveniences, and necessaries of life, have derived but little advantage from philosophers . . . In mechanics, for instance, we find that the most important inventions and improvements have been made, not through the reasonings of philosophers, but through the ingenuity of artists [craftsmen], and not unfrequently by common workmen. The chemist, in particular, if we except the pharmaceutical laboratory, has but little claim on the arts: on the contrary, he is indebted to them for the greatest discoveries and a prodigious number of facts, which form the basis of his science. In the discovery of the art of making bread, of the vinous and acetous fermentations, of tanning, of working ores and metals, of making glass and soap, of the action and applications of manures, and in numberless other discoveries of the highest importance, though they are all chemical processes, the chemist has no share. . . . The art of dyeing has attained a high degree of perfection without the aid of the chemist, who is totally ignorant of the rationale of many of its processes, and the little he knows of this subject is of late date.

It is significant, however, that Rupp went on to refer warmly to the recent experimental and theoretical work of Thomas Henry of Manchester, in dyeing, and of the French chemist, Berthollet, in bleaching. And it is also significant that Rupp himself was described by his fellow-countryman, P. A. Nemnich as

a German, who possesses great knowledge of chemistry, and who at the same time, as a manufacturer, can carry out most splendid experimental applications for the benefit of scientific knowledge . . . Mr. Rupp at the time of my visit to Manchester was very busy with the building of a big spinning mill to be

[1] G. J. French, *The Life and Times of Samuel Crompton* (1859), pp. 30 and 32.
[2] *Ibid.*, p. 53.
[3] T. L. Rupp, 'On the Process of Bleaching with the oxygenated muriatic Acid', *Manchester Lit. & Phil. Soc. Memoirs*, Vol. V, pt. i (1798).

erected according to his own ideas, and many improvements are expected from him.[1]

Rupp was a member of the Manchester Literary and Philosophical Society, to which he read several papers, including a notable one on chlorine bleaching and another making an anti-phlogiston attack on Priestley and defending 'the new chemical theory', with numerous references to the works of foreign chemists such as Lavoisier, Bergman, etc.[2] William Henry, second only to John Dalton as a scientific chemist in Manchester in this period,[3] referred to some of his experiments having been witnessed by 'my friend Mr. Rupp . . . who is much conversant in the observation of chemical facts'.[4]

Like Rupp, however, William Nicholson also referred to the many matters on which chemists were ignorant, or which they could not adequately explain.[5] The great controversy over phlogiston and the new chemical theory, he said, 'amounts to a confession of ignorance in our theoretical explanation'. But he pointed out that continuous experimental investigation was leading to the discovery of the laws of nature, and that one must expect the 'successive emendation' of scientific theories, and 'the rejection of principles formerly held to be essential to the science' of chemistry. Nicholson himself did much to spread a knowledge of chemical theories and facts, by his own works and by his translations of those of foreign chemists.

In engineering, too, one comes across similar remarks emphasizing the role of empiricism. George Atwood, for instance, Fellow of Trinity College, Cambridge, and lecturer in natural philosophy,[6] pointed out inconsistencies in the theory of motion and doubted whether it could provide any assistance to the practical mechanic in constructing power-driven machinery. 'Machines of this sort owe their origin and improvement to other sources: it is from long experience of repeated trials, errors, deliberations, [and] corrections, continued through the lives of individuals, and by successive generations

---

[1] P. A. Nemnich, *Beschreibung einer in Sommer 1799 von Hamburg nach und durch England geschehenen Reise* (1800), p. 324.

[2] *Manchest Lit. & Phil. Soc. Memoirs*, Vol. V, pt. i (1798). See below, pp. 317-19, for his article on bleaching. [3] See below, pp. 246-50.

[4] *Phil. Trans.* (1797), p. 410. Manchester Central Reference Library possesses a copy of Henry's *General View of the Nature and Objects of Chemistry* (1799) inscribed to 'Mr. Rupp from the Author'.

[5] See, for example, his remarks in his *Dictionary of Chemistry* (1795), pp. iii–v and 1–2. He had previously published *An Introduction to Natural Philosophy* (1782) and *The First Principles of Chemistry* (1790), both of which ran into several editions. Later he started *Nicholson's Journal of Natural Philosophy, Chemistry and Arts*, an important scientific periodical. See *D.N.B.* and Hans, *op. cit.*, pp. 115 and 158. [6] See below, p. 171.

of them, that sciences, strictly called practical, derive their gradual advancement . . .'[1]

Peter Ewart, however, an outstanding Manchester engineer of that day, well versed in engineering science,[2] though regretting 'that theory should appear to be at variance with practice', thought that Atwood had 'pressed his argument too far'.[3] There was no doubt 'that ingenious men, of rare natural endowments, have, without any scientific aid, accomplished wonders in the invention and improvement of machinery', but there were also some notable examples to the contrary in 'the history of useful discoveries in mechanics'.

If Huygens and Hooke had not been scientific as well as ingenious men, we might possibly have been still ignorant of the properties of the balance regulated by springs. If Smeaton had not availed himself of just theory, as well as experiment, we might still have had to learn the principles by which we must be guided in applying water to the best advantage as a moving power. If a clear and strong understanding, and a mind richly stored with scientific attainments, had not been combined with wonderful fertility of invention in the justly celebrated improver of the steam engine [Watt]; incalculable labour might still have been wasted in performing operations which are now accomplished with as much ease and regularity as the gentle motions of a time-piece.

Similarly, in regard to the application of chemical science, one frequently comes across contemporary statements of its importance. Thus we find William Henry putting forward *A General View of the Nature and Objects of Chemistry, and of its Application to Arts and Manufactures* (Manchester, 1799), in which, after quoting Bacon in support of 'the union of theory with practice', he referred to the illustrious examples of Watt and Wedgwood, 'both of whom have been not less benefactors of philosophy, than eminent for practical skill', and went on to demonstrate the utility of chemistry in a wide range of industries —in metallurgy, in the production of alkalis, acids, etc., in the glass and pottery manufactures, in brewing, in bleaching, dyeing, and calico-printing.

Some parts of the natural philosophy of the eighteenth century, as of the seventeenth, had no immediate practical application. Though in some cases they were ultimately to have revolutionary industrial consequences, they were originally developed mainly out of intellectual curiosity, out of a desire to unravel the mysteries of nature. Electricity and magnetism, for instance, which were very popular subjects in

---

[1] G. Atwood, *A Treatise on the Rectilinear Motion and Rotation of Bodies* (1784), p. 381, quoted by P. Ewart, 'On the Measure of Moving Force', *Manchester Lit. & Phil. Soc. Memoirs*, 2nd ser., Vol. II (1813), p. 111.
[2] See below, p. 99.   [3] Ewart, *op. cit.*, pp. 112–14.

contemporary lecture courses, played little or no part in the early Industrial Revolution. 'Electricity', said the German traveller, C. P. Moritz, 'is the plaything of the English.'[1] But other subjects, such as chemistry, mathematics, mechanics, hydraulics, and hydrostatics, as we shall see, certainly were studied for utilitarian as well as scientific reasons. Even the most purely philosophical investigations sometimes had very practical connections and consequences. Priestley, for example, says that he began his researches into 'airs' or gases as a result of living in Leeds next to a brewery, where he noticed that 'fixed air' (carbon dioxide) was evolved in the fermentation vats;[2] his investigations led immediately to the manufacture of artificial mineral waters,[3] while their long-term consequences were of far wider importance.

Thus there is abundant evidence that applied science contributed substantially to the Industrial Revolution in this country. Why, then, it may be asked, were there so many criticisms in the later nineteenth century of the lack of scientific training and technology in British industry? Only a few suggestions can be offered here. It seems not unlikely that these criticisms have been overdone, that British industry was not so scientifically and technologically backward by comparison with other countries as is so often, almost masochistically, reiterated. Moreover, there were many other, non-scientific, causes of the slowing-down in Britain's rate of industrial growth. Nevertheless, there are signs that the early flowering of applied science tended to wither away, and that there was an 'educational setback' in the nineteenth century.[4] As early as 1830 Charles Babbage was deploring the decline of science in England.[5] The Dissenting Academies went into decline, and the Mechanics' Institutes fizzled out. Sons and grandsons of successful industrialists tended to go to public schools and to Oxford and Cambridge, which were still gripped by the dead hand of the classical tradition. The amateur-scientific agencies of the early Industrial Revolution were inadequate for the later, more advanced stages of industrialization: 'self-help' no longer sufficed. But in this country the

[1] C. P. Moritz, *Journeys of a German in England in 1782* (trans. and ed. by R. Nettel, London, 1965), p. 70.
[2] J. Priestley, *Experiments and Observations on Different Kinds of Air*, Vol. II (1775), pp. 269–70; F. W. Gibbs, *Joseph Priestley* (1965), pp. 57–60.
[3] See below, pp. 234–9.
[4] N. A. Hans, *New Trends in Eighteenth Century Education* (1951), pp. 209–12.
[5] C. Babbage, *Reflections on the Decline of Science in England, and on some of its causes* (1830). He was mainly concerned about the state of the Royal Society, but referred also to the domination of the classics in English public schools and universities, the almost complete ignorance of science among the ruling classes, the lack of government support, and inadequate material rewards for scientists. See also *Quarterly Review*, Vol. XLIII (1830), pp. 305–42.

State made far less provision for the teaching of science and technology than in France, Germany, and the U.S.A. Much more research however, will have to be carried out before one can speak confidently on this problem.

III Science and Industry in the Late Eighteenth Century[1]

It is a deeply entrenched tradition that the technical achievements of the early Industrial Revolution were the products of uneducated empiricism.[2] It is gradually becoming evident, however, that the Industrial Revolution was not unrelated to the Scientific Revolution, which became so marked in Restoration England, with the founding of the Royal Society and the great discoveries of Newton and Boyle, and which was developed in the later eighteenth century by such men as Black, Priestley, Cavendish, and Herschel.[3] James Keir, F.R.S., the

[1] Based on a jointly-produced article in the *Econ. Hist. Rev.*, 2nd ser., Vol. XIII, no. 2 (Dec. 1960), pp. 222–44. But very substantial additions have been made, trebling the length of the original article.

[2] This view has been reiterated fairly recently in R. Mousnier, *Progrès Scientifique et Technique au XVIIIe Siècle* (Paris, 1957), a work based almost entirely on secondary sources, with little new evidence. For a very critical review of this book, see *Annales Historiques de la Révolution Française* (Oct.–Dec. 1959), pp. 376–82. M. Daumas, 'Le mythe de la révolution technique', *Revue d'Histoire des Sciences*, Tome XVI (Paris, 1963), also argues against any connection between science and technology in the Industrial Revolution; but he, too, neglects the wealth of contemporary evidence, and discusses applied science very much in the modern sense. These views may be compared with the evidence provided by M. Crosland, *The Society of Arcueil* (1967), of the interplay between science and technology in Revolutionary and Napoleonic France.

Professor A. F. Burstall, in his *History of Mechanical Engineering* (1963), pp. 201 and 203, has similarly repeated the traditional view that British engineers in the Industrial Revolution 'were for the most part uneducated in the scientific knowledge of their time', that they 'had little understanding or appreciation of science'. On the other hand, however, in an earlier chapter (p. 144), he states, contradictorily, that 'the sixteenth and seventeenth centuries mark the birth of engineering science. Until then, engineering had been an art based on empirical rules . . .' But as a result of work by scientists such as Galileo, Newton, Napier, Boyle, Hooke, and 'many others', engineering was then placed on a 'scientific foundation'.

[3] See, for example, G. N. Clark, *Science and Social Welfare in the Age of Newton* (1938); A. Wolf, *A History of Science, Technology and Philosophy in the Eighteenth Century* (1938); A. Ferguson (ed.), *Natural Philosophy through the 18th Century* (Commemoration number of the *Philosophical Magazine*, 1948); J. D. Bernal, *Science in History* (1954); A. R. Hall, *The Scientific Revolution, 1500–1800* (1954);

famous industrial chemist, declared in his *Dictionary of Chemistry* (1789): 'The diffusion of a general knowledge, and of a taste for science, over all classes of men, in every nation of Europe, or of European origin, seems to be the characteristic feature of the present age.'[1] And the *Philosophical Magazine* referred in 1798 to the 'well known' fact that 'the Arts and Manufactures of Great Britain . . . have been much improved by the great Progress that has lately been made in various Branches of the Philosophical Sciences'.[2]

Professor Ashton has already suggested that collaboration between men of science and of industry, in societies such as the Literary and Philosophical Society in Manchester and the Lunar Society in Birmingham, was an important factor in the Industrial Revolution.[3] It is clear that outstanding figures like Boulton, Watt, and Wedgwood were men of considerable intellectual and scientific as well as technical and commercial ability. Other important industrialists such as John Kennedy of Manchester, the Strutts of Derby, and Thomas Bentley of Liverpool were also members of their local philosophical societies and were not strangers to liberal conversation and intellectual pursuits.[4] In recent years historians of education have done much to illuminate this question and their findings have suggested that a knowledge of science was more widely diffused through industrial society than has hitherto been suspected.[5] In this chapter we shall first make a detailed examination of the relationship between science and industry in Manchester—centre of the most rapid and important technical advances in the later eighteenth century—and then make a general survey of the situation in other industrial centres.

I

When the Scientific Revolution began in the seventeenth century it was principally a metropolitan movement centred on the Royal Society.

A. and N. L. Clow, *The Chemical Revolution* (1952); R. E. Schofield, *The Lunar Society of Birmingham* (1963); S. Pollard, *The Genesis of Modern Management* (1965), chap. IV.

[1] *Op. cit.*, Preface, p. 111.   [2] *Op. cit.*, Vol. I (1798), Preface.

[3] T. S. Ashton, *The Industrial Revolution, 1760–1830* (1948), pp. 16 and 20–1.

[4] See R. Angus Smith, *A Centenary of Science in Manchester* (1883); R. B. [Richard Bentley], *Thomas Bentley, 1730–1780* (Guildford, 1927); and Eric Robinson, 'The Derby Philosophical Society', *Annals of Science*, Vol. IX (1953), pp. 359–67 (see below, chap. IV).

[5] N. Hans, *New Trends in Eighteenth Century Education* (1951); W. H. G. Armytage, *Civic Universities* (1955) and *The Rise of the Technocrats* (1965); T. Kelly, *George Birkbeck* (Liverpool, 1957); B. Simon, *Studies in the History of Education 1780–1870* (1960).

By the end of the eighteenth century it was largely a provincial movement, healthily expressed through the many philosophical societies in Norwich, Northampton, Exeter, Bristol, Bath, Plymouth, Birmingham, Derby, Manchester, Newcastle, and many other places. The preface to the first volume of the Manchester Literary and Philosophical Society's *Memoirs* pointed out the utilitarian as well as the scientific value of these learned societies. Those founded in different parts of Europe in the seventeenth and eighteenth centuries had 'been not only the means of diffusing knowledge more extensively, but have contributed to produce a greater number of important discoveries, than have been effected in any other equal space of time'. Their journals had publicized 'many very valuable discoveries, or improvements in arts, and much useful information in the various branches of science'. But although several French provincial societies had been instituted, 'in England, they have almost been confined to the Capital'. It was therefore urged that 'the promotion of arts and sciences' would be 'more widely extended by the forming of Societies, with similar views, in the principal towns of this kingdom'.[1]

The Manchester Society was founded in 1781, growing out of informal meetings, which had been held for some years previously, of 'a few Gentlemen, inhabitants of the town, who were inspired with a taste for Literature and Philosophy', and who had 'formed themselves into a kind of weekly club, for the purpose of conversing on subjects of that nature'.[2] These meetings had first been held in the house of Dr. Thomas Percival,[3] the leading figure in the Society's foundation and later its president for many years. Its subjects of conversation were to include 'Natural Philosophy, Theoretical and Experimental Chemistry, Polite Literature, Civil Law, General Politics, Commerce, and the Arts' (manufactures).[4] The medical profession was strongly represented in its membership and a few clergy also participated, but 'the great majority of the members were either engaged or interested in the extension of Science and Art to manufacturing purposes'.[5]

---

[1] *Memoirs of the Manchester Literary and Philosophical Society*, Vol. I (1785), pp. v–vi. Erasmus Darwin's *Address to the First Meeting of the Literary Society at Derby, July 18, 1784* (Derby Public Library) makes the same kind of case.
[2] *Manchester Lit. & Phil. Soc. Memoirs*, Vol. I (1785), p. vii. There is no modern history of the Society, but see Angus Smith, *op. cit.*, and F. Nicholson, 'The Literary and Philosophical Society, 1781–1851', *Manchester Lit. & Phil. Soc. Memoirs*, Vol. LXVIII (1923–4), pp. 97–148.
[3] E. Percival, *Memoirs of the Life and Writings of Thomas Percival, M.D.* (1807), p. lxvii.   [4] *Manchester Lit. & Phil. Soc. Memoirs*, Vol. I (1785), pp. xii–xiii.
[5] A. A. Mumford, *The Manchester Grammar School 1515–1915* (1919), p. 218. This statement is borne out by the membership lists in the *Memoirs*.

G

The Society was strongly influenced by Warrington Academy, several of its leading members, such as Thomas Percival and Thomas Barnes, being Warrington alumni.[1] The important role of the Dissenting Academies in introducing more liberal, scientific and utilitarian studies into English education is well known,[2] and Warrington provided one of the most illustrious examples. Joseph Priestley, of course, was the most eminent of the Academy's tutors (1761–7), and in his *Essay on a Course of Liberal Education* (1765) he argued that natural philosophy and chemistry should form part of the general educational curriculum, though he himself taught languages and 'belles lettres' at Warrington and did not become very actively interested in chemical research until a few years later. Already, in 1760, however, the Academy's Trustees had granted £100 for the purchase of a philosophical apparatus, acquired from Kendal Academy,[3] and announced their intention of adding 'the more important Processes in Chemistry, especially that Part of it which has connection with our Manufactures and Commerce'.[4] Such chemical lectures were subsequently provided in the 1760s and 1770s by Matthew Turner[5] and John Aikin. Lectures in natural philosophy were also given, the content of which is revealed by the *Institutes of Natural Philosophy, Theoretical and Experimental* (1785), by Dr. William Enfield, tutor there from 1770 to 1786, who wrote this as a text-book for his classes. It deals with the properties of matter, mechanics, hydrostatics, pneumatics, optics, astronomy, magnetism and electricity. It is mathematical in emphasis, requiring 'a previous knowledge of the elements of Geometry, Trigonometry, the Conic Sections, and Algebra', but many parts are illustrated with experiments. And just as the Academy's experimental apparatus was inherited, so, too, was the content of the scientific teaching, for, as

[1] H. McLachlan, *Warrington Academy: Its History and Influence* (Manchester, 1943), pp. 114–28. See also W. Turner, *The Warrington Academy* (reprinted from the *Monthly Repository*, 1813–15), edited with an introduction by G. Carter (Warrington, 1957).

[2] Irene Parker, *Dissenting Academies in England* (1914); H. McLachlan, *English Education under the Test Acts* (Manchester, 1931).

[3] Dr. Caleb Rotheram (1694–1752), of Kendal Academy, had himself acquired it from John Horsley (1685–1732), M.A., F.R.S., Dissenting minister and lecturer in natural philosophy in Morpeth and Newcastle in the early eighteenth century. For Rotheram and Horsley, see below, pp. 101, 103–4 and 160–1. See also McLachlan, *English Education*, pp. 190 and 228, and Hans, *op. cit.*, p. 59.

[4] F. W. Gibbs, 'Itinerant Lecturers in Natural Philosophy', *Ambix*, Vol. VI (1960), p. 112.

[5] Dr. Matthew Turner, a physician of Liverpool, was a close friend of Priestley, Josiah Wedgwood and his partner, Thomas Bentley. It was Turner who introduced Wedgwood to Bentley, and who carried on a chemical correspondence with both of them. He also advised Matthew Boulton in 1776–7 about varnishes used in japanning.

Enfield pointed out, he 'made use of the works of NEWTON, *Keil, Whiston, Gravesande, Cotes, Smith, Helsham, Rowning,* and lastly *Rutherforth'*.[1] Undoubtedly the origins of the developments we shall be surveying go back to the Scientific Revolution of the seventeenth century, and the Dissenting academies played an important role in preserving and disseminating scientific knowledge.

The attempt to build up a culture based on liberal studies and science was continued in Manchester in a similar spirit of combined utility and idealism. Thomas Henry pointed out that 'several branches of Natural Philosophy seem peculiarly adapted to fill up the vacant hours in which the tradesman can withdraw from his employments. A general knowledge of all will tend to open and enlarge his understanding . . . While the study of some, *in particular* (such as mechanics, hydrostatics, hydraulics, and chemistry), may . . . supply him with a kind of information which he may turn to good account . . .'[2] In a similar mood the Rev. Thomas Barnes expressed the hope that 'the happy art might be learned, of CONNECTING TOGETHER, LIBERAL SCIENCE and COMMERCIAL INDUSTRY.'[3] It is clear, however, from remarks by both Henry and Barnes that there was considerable antipathy among tradesmen in Manchester to liberal education, but this antipathy was being combated. It is true that Thomas Henry was an apothecary and that Barnes was a religious minister, but a large number of their fellow-members were men of business. John Barrow, John Drinkwater, the Philips, George Walker, and John Wilson as well as several others were engaged in trade.

The dividing line between professional man and manufacturer was not so marked then as it is today. Thomas Henry, F.R.S., an apothecary, developed into a manufacturing chemist, and though he trained his sons as doctors, brought them into business alongside him.[4] James Dinwiddie, A.M., lecturer in natural philosophy, appears to have entered the calico-printing business, though not, it would seem, with

---

[1] For the works of these earlier natural philosophers and mathematicians, see above, chap. I.

[2] Thomas Henry, 'On the Advantages of Literature and Philosophy in General, and especially on the consistency of Literary and Philosophical with Commercial Pursuits', read 3 Oct. 1781, *Manchester Lit. & Phil. Soc. Memoirs*, Vol. I (1785), pp. 7–29.

[3] Thomas Barnes, 'A Plan for the Improvement and Extension of Liberal Education in Manchester', read 9 April, 1783, *Manchester Lit. & Phil. Soc. Memoirs*, Vol. II (1785), pp. 16–29. See also his paper, 'On the Affinity subsisting between the Arts, with a Plan for promoting and extending Manufactures by Encouraging those Arts, on which Manufactures principally depend', read 9 Jan. 1782, *Manchester Lit. & Phil. Soc. Memoirs*, Vol. I (1785), pp. 72–89.

[4] See below, chap. VII.

any great success.[1] Thomas Cooper, who failed to become a F.R.S. though his certificate was signed by Priestley and Watt, contributed his chemical knowledge to the bleaching firm of Baker and Cooper. Charles Taylor, the Manchester dyer, became Secretary to the Society for the Encouragement of Arts, Commerce and Manufactures.[2] Men of chemical knowledge quickly put it to practical test in the bleaching and dyeing trades.[3] The result was that in the Manchester Literary and Philosophical Society there was from the beginning an easy exchange of opinion, and in its rooms a manufacturer could consult with half a dozen Fellows of the Royal Society among the Manchester members, without taking the Corresponding Members into account. In the very first published list of members the names of Charles White, Thomas Henry, Thomas Butterworth Bayley, and Thomas Percival appear, all of whom were Fellows of the Royal Society,[4] and there were more to come in a few years.

As a result of the papers read to the Manchester Literary and Philosophical Society by Thomas Henry and the Rev. Thomas Barnes, it was decided to establish a 'College of Arts and Sciences' in Manchester in 1783, apparently the first institution of its kind in England.[5] Barnes was the chief architect of this scheme. He stressed the great need for an institution of higher education, after elementary tuition at private or grammar schools. Here young tradesmen would be able to study '*Natural Philosophy*, the *Belles Lettres*, and *Mathematics*; together with some attention to *History*, *Law*, *Commerce*, and *Ethics*'. The course on Natural Philosophy 'should pay a very particular attention to CHEMISTRY and MECHANICS because of their intimate connection with our manufactures'.[6] This scheme was supported by the Literary and Philosophical Society.[7] It was 'principally accommodated to young

---

[1] *Manchester Mercury*, 22 Nov. and 6 Dec. 1796. (Advt. re bankruptcy of Messrs. Dinwiddie and Bewicke.)

[2] Sir Henry Trueman Wood, *A History of the Royal Society of Arts* (1913), pp. 334–5.   [3] See below, chaps. VII, VIII, and IX.

[4] *Manchester Lit. & Phil. Soc. Memoirs*, Vol. I (1783), pp. xvi–xvii.

[5] In addition to the papers of Henry and Barnes, previously cited, see the first report of the College, 'College of Arts and Science, Instituted at Manchester, June VI, MDCCLXXXIII' (Manchester Central Reference Library, S. & A. 94/42. Reprinted in *Manchester Lit. & Phil. Soc. Memoirs*, Vol. II (1785), pp. 42–6); *Manchester Mercury*, 29 July, 30 Sept. 1783; E. Percival, *op. cit.*, pp. lxx–lxxi; W. C. Henry, 'A Tribute to the Memory of the Late President [Thomas Henry] of the Literary and Philosophical Society of Manchester', *Manchester Lit. & Phil. Soc. Memoirs*, 2nd ser., Vol. III (1819), pp. 224–6. There are brief accounts of the College in R. Angus Smith, *op. cit.*, pp. 104 ff.; J. Thompson, *The Owens College* (1886), pp. 7–10; W. Bowden, *Industrial Society in England towards the end of the Eighteenth Century* (New York, 1925), pp. 45–6, H. McLachlan, *op. cit.*, p. 123.

[6] *Manchester Lit. & Phil. Soc. Memoirs*, Vol. II (1785), p. 22.   [7] *Ibid.*, pp. 29–41.

men designed for a respectable line of *trade*', but would also provide a general education for those entering the professions. Lectures would be given in the evenings, 'so as not to interfere with the regular hours of business'. Those on natural philosophy would be illustrated by experiments on fire-engines, air-pumps, microscopes, etc., while the chemical lectures would have 'a reference to the arts of *Dyeing, Bleaching*, &c. which, depending upon chemical principles, might probably, by the knowledge of those principles, be very greatly extended and improved'.

The 'College of Arts and Sciences' was instituted at Manchester on 6 June 1783.[1] Its president, Thomas Percival, and eight governors were all members of the Literary and Philosophical Society, and so were all the lecturers: Henry Clarke on Practical Mathematics, Natural and Experimental Philosophy, and Geography; Thomas Henry on Chemistry with reference to Arts and Manufactures; George Bew on the Theory and History of the Fine Arts; and the Rev. Thomas Barnes on the Origins, History, and Progress of Arts, Manufactures and Commerce, Commercial Law, &c. But the College—started with such high hopes—had a brief existence, lasting only until 1787-8.[2] The most successful of the science lectures, delivered to large audiences, were those of Thomas Henry on chemistry, which included 'a course on the Arts of Bleaching, Dyeing, and Calico-Printing', which was made available to skilled operatives in those trades.[3] John Banks also lectured at the college on natural philosophy.[4] Charles White, F.R.S., and his son Thomas White, M.D., delivered lectures on anatomy.[5] The causes of the College's demise are not clear, but are said to have included 'a superstitious dread of the tendency of science to unfit young men for the ordinary details of business'.[6]

The College of Arts and Sciences was a non-denominational institution which has sometimes been confused with the Manchester Academy, or 'New College', forerunner of the modern Manchester College, Oxford, founded by local Dissenters on 22 February and opened on

[1] *Manchester Mercury*, 29 July 1783, and the College's first annual report, cited above.

[2] *Ibid.*, 28 Aug., 25 Sept., 2 Oct. 1787. James Watt junior attended lectures by Charles White as late as December 1788. J. Watt, jun., to J. Watt, 7 Dec. 1788, Dol. Cf. Percival, *op. cit.*, p. lxxi.

[3] W. C. Henry, *op. cit.*, p. 226. Thomas Henry continued his lectures 'long after' the College's collapse. See below, p. 243, n.7

[4] See below, p. 108.

[5] *Manchester Mercury*, 10 and 31 Oct. 1786. See also 'A Discourse delivered at the Commencement of the Manchester Academy, September XIV, MDCCLXXXVI. By Thomas Barnes, D. D.', p. 25.

[6] W. C. Henry, *op. cit.*, p. 225.

14 September 1786.[1] The Manchester Academy was formed in order to save something from the wreckage of the Warrington Academy.[2] Hackney College, London, received the scientific equipment from Warrington, while the library, of about 3,000 volumes, was inherited by Manchester Academy. The most important figures in the promotion of the Manchester Academy were Thomas Percival, president of the Literary and Philosophical Society, Dr. Thomas Barnes, co-secretary of that society, and the Rev. Ralph Harrison, another member. These three, however, involved the Society too far in the projected academy, especially by their suggestion in the trustees' first report that the Literary and Philosophical Society supported the foundation of the new college. On 15 March 1786, in a statement signed by Thomas Barnes and Thomas Henry as secretaries, the Society, *as such*, declared its independence of the Manchester Academy on the grounds that it had always maintained its independence of any religious opinion or sect.[3] Nevertheless individual members of the Literary and Philosophical Society continued to be prominent in the Manchester Academy. Barnes and Harrison were appointed professors of theological and classical studies and were to be responsible for choosing a professor of mathematics and natural philosophy. Most of the officers and committee were members of the Literary and Philosophical Society. Many

---

[1] Examples of this confusion are in Hans, *op. cit.*, p. 159, and E. Halévy, *History of the English People in the Nineteenth Century* (revd. edn., 1949), Vol. I, p. 561.

[2] See the trustees' report on its institution, 22 Feb. 1786 (Manchester Central Reference Library, S. & A. 94/48), which is reprinted, together with later reports, in a booklet devoted mainly to two sermons on the Academy by Dr. Thomas Barnes and the Rev. Ralph Harrison, ministers at Cross Street Chapel, Manchester. (Volume entitled *Manchester Academy* in Manchester Cen. Ref. Lib., ref. no. 378.42/H2). These sermons were 'A Sermon Preached at the Dissenting Chapel in Cross Street, Manchester, March XXVI, MDCCLXXXVI, on the Occasion of the Establishment of an Academy in that Town. By Ralph Harrison', and 'A Discourse delivered at the Commencement of the Manchester Academy, September XIV, MDCCLXXXVI. By Thomas Barnes, D.D.' The latter was reprinted in 1879 under the ambiguous title *Free Teaching and Free Learning in 1786* (Manchester, 1879). See also *Manchester Mercury*, 21 March, 18 July, 12 Sept. 1786, 3 April 1787, 16 Dec. 1788, 6 July 1790; *The Monthly Magazine*, Vol. IV (July–Dec. 1797), pp. 105–7, containing a report of 'the Academical Institution, or New College, at Manchester', reprinted in J. Harland, *Collectanea Relating to Manchester and Its Neighbourhood* (Manchester, 1867); E. Percival, *op. cit.*, pp. lxxvii–lxxxi; F. Nicholson, *op. cit.*, p. 140; R. Wade, *The Rise of Nonconformity in Manchester* (Manchester, 1880), pp. 148–9; McLachlan, *Warrington Academy*, pp. 119–20, 128–37, and *English Education*, pp. 255–62. There is a good modern history of the Academy by V. D. Davis, *A History of Manchester College* (1932), but this shows little appreciation of the Academy's early significance as a training-ground for industry and trade.

[3] *Manchester Mercury*, 18 July 1786; Nicholson, *op. cit.*, p. 140; McLachlan, *Warrington Academy*, pp. 119–20.

of the founder-members were also former pupils of Warrington Academy or other dissenting academies.

The main purpose of the Academy was to train candidates for the Dissenting ministry, but it would also prepare students who wished to enter the medical and legal professions and give a general education to those 'designed for Civil and Commercial life'.[1] In regard to the third category, Barnes proceeded to 'imagine . . . a system of education, for a commercial man, which shall contain all the parts of science proper for him to know, as much as possible in a practical form'. The first tutor in mathematics and natural philosophy at the Academy was Thomas Davies, appointed 1787, followed in 1789 by Francis Nicholls.[2] He was succeeded in 1793 by the famous scientist, John Dalton, who had previously been running a school at Kendal.[3] In 1794, when Dalton became a member of the Literary and Philosophical Society, he is said to have had twenty-four pupils for mathematics, mechanics, geometry, algebra, book-keeping, natural philosophy, and chemistry.[4] He left the Academy in 1800, to engage in private teaching. In that year he was elected secretary of the Literary and Philosophical Society, by which time he had begun to publish important scientific papers.[5]

When Barnes resigned as principal of the Academy in 1798, he was succeeded by George Walker, F.R.S.,[6] who had been a student at Kendal Academy under Caleb Rotheram,[7] then at Edinburgh and Glasgow Universities. He had been a tutor in mathematics and natural philosophy at Warrington Academy in 1772–4, after which he became minister of High Pavement Chapel, Nottingham. A distinguished mathematician as well as theologian, he was elected president of the Manchester Literary and Philosophical Society on Percival's death in 1804.

Up to Barnes's retirement in 1798, the total number of students who

---

[1] Barnes's 'Discourse'.

[2] We also find Nicholls taking 'a few private Pupils in Mathematics and the Principles of Natural Philosophy' (*Manchester Mercury*, 25 Oct. 1791).

[3] On Dalton see W. C. Henry, *Memoirs of the Life and Scientific Researches of John Dalton* (1854); R. Angus Smith, 'Memoir of John Dalton, and History of the Atomic Theory', *Manchester Lit. & Phil. Soc. Memoirs*, 2nd ser., Vol. XIII (1856); H. Lonsdale, *John Dalton* (1874); H. E. Roscoe, *John Dalton* (1901); J. P. Millington, *John Dalton* (1906); H. McLachlan, 'John Dalton and Manchester (1793–1844)', *Manchester Lit. & Phil. Soc. Memoirs*, Vol. LXXXVI (1943–5), pp. 165–77; F. Greenaway, 'The Biographical Approach to John Dalton', *ibid.* (1958–9), pp. 1–98.   [4] R. Angus Smith, 'Memoir of John Dalton', p. 18.

[5] For Dalton's analyses of water for bleachers, see above, p. 79. Dalton also carried out scientific researches on 'oxymuriatic acid' (chlorine), used in bleaching (see below, p. 336).

[6] Davis, *op. cit.*, pp. 67–9.

[7] On Kendal Academy and the Rotherams, see below, p. 103.

had entered the Academy was 137,[1] of whom 89 were destined for commerce and industry. Even before Barnes's resignation, however, the Academy was in difficulties, which finally resulted in its transference to York in 1804.

II

Some impression of the intellectual life available to a young business-man in Manchester in the late eighteenth century can be seen from the correspondence of James Watt junior, though it must be admitted that he was a rather exceptional individual. He was apprenticed to the firm of Maxwell and Taylor, fustian manufacturers and dyers; the Taylor was Charles Taylor to whom we have already referred.[2] The introduction to this firm in 1788 seems to have been made in the first place by Thomas Henry, who had been corresponding with Watt about bleaching.[3] In a short while young Watt struck up friendships with Thomas Henry's sons, with Joseph Priestley junior, who was also working in Manchester, with John Ferriar, an assistant of Thomas Percival, with Thomas Cooper, the bleacher, and with the merchants, Thomas and Richard Walker. By November 1788 he was attending meetings of the Literary and Philosophical Society, and read extracts from his translation of a German treatise by Professor Winterl,[4] even before he was made a member, in February 1789.[5] He was also encouraged by James Keir in his translation of a work by Christopher Meiners, Professor of Philosophy at Gottingen,[6] and he was visited by several foreign scientists while in Manchester. At this time he helped Keir with an article on bleaching for his *Dictionary of Chemistry* (1789), to which his master, Charles Taylor, also contributed articles on cotton dyeing and calico-printing.[7]

In 1789 Watt junior was appointed joint secretary of the Literary and Philosophical Society with John Ferriar. Part of the Society's programme was to establish relations with foreign scientists and scientific societies. Presumably with this aim in view, Watt wrote a letter of congratulation to the famous German scientist and editor, Lorenz Crell, and at the same time sent him copies of two papers he

[1] *Roll of Students entered at the Manchester Academy* (Manchester, 1868); Percival, *op. cit.*, p. lxxxi; McLachlan, *Warrington Academy*, pp. 136–7. Davis, *op. cit.*, p. 62, followed by McLachlan, 'John Dalton and Manchester', p. 171, gives 135, but appears to have overlooked two students at the end of the official *Roll*.
[2] See above, p. 92.   [3] See below, pp. 244–5, 299–303, 345.
[4] James Watt, jun., to James Watt, 27 Nov. 1788 and 7 Dec. 1788, Dol.
[5] James Watt, jun., to James Watt, 24 Feb. 1789, Dol.
[6] James Keir to James Watt, jun., 20 Nov. 1789, A.O.L.B.
[7] Ditto, 10 Feb. 1790, A.O.L.B. See below, p. 346.

had published in the *Memoirs* of his own society. Crell accepted the compliments and the papers for publication in his *Annalen* and asked for a regular scientific correspondence to be maintained between himself and the Manchester Society.[1] Watt was eventually instrumental in securing Crell's election as a Corresponding Member of the Society.

Apart from establishing connections with these foreign scientists, young Watt did something to bring foreign publications on science and engineering to the knowledge of his Manchester friends. He wrote to his father in 1789, saying:

I am glad to hear that Messrs Lavoisier, Bertholet & Morveau intend translating Crells Annals it will be doing an essential service to Chemistry & will, I have no doubt, amply repay their trouble. Mr [Charles] Tayler & Mr [Thomas] Henry wish to subscribe for one copy each & Mr [Thomas] Cooper for two, one for himself and one for his partner, Mr [Joseph] Baker. These are all our chemical men here, Dr Percival not professing any partiality to that science . . .[2]

On another occasion, George Lee, who had been manager of Drinkwater's mill and later became a partner in the cotton-spinning firm of Philips and Lee, wrote to James Watt junior in Paris: 'if in the course of y^r travels you meet with any good or curious Books upon mechanics and it does not incommode you materially I wish you would purchase them upon my Account'.[3] In the same letter he delivered a message from Peter Ewart, the Manchester millwright, who wanted to buy a copy of '*Architecture Hydraulique par Monsr Porny*'.[4] Even before he left Manchester, Watt had been buying German scientific books for his father. His connections with Manchester ended for a time in February 1792, when he left for Paris, together with Thomas Cooper, as a delegate of the Manchester Constitutional Society to the Club des Jacobins.

It must not be thought that young men like Watt and Joseph Priestley junior were the only men in Manchester who had contacts with French and German scientists. Charles Taylor described to James Watt, senior, a day he spent in Berthollet's company,[5] and when he applied for the post of secretary to the Society for the Encouragement of Arts, Commerce and Manufactures in 1799 he stated to the committee that he was known to almost all the chemists in Europe.[6] Thomas Henry, in a paper on dyeing, refers to similar publications

[1] Lorenz Crell to James Watt, jun., 26 Jan. 1790, Dol.

[2] James Watt, jun., to James Watt, 1 Feb. 1789, Dol. Taylor, Henry, Cooper, and Baker played leading roles in the application of chemistry to chlorine bleaching, dyeing, and calico-printing in Lancashire (see below, chaps. VII, VIII, and IX).

[3] George Lee to James Watt, jun., 18 April 1792, M. IV, B.R.L. See below, pp. 99–100, for George Lee.

[4] The author's name should read 'Prony'. For Peter Ewart, see below, pp. 99 and 441–2.   [5] Charles Taylor to James Watt, 21 June 1789, Dol.

[6] Sir H. T. Wood, *A History of the Royal Society of Arts* (1913), p. 335.

by foreign chemists such as d'Apligny, Macquer, Hellot, Berthollet, and many others.[1] In 1776 he translated some of Lavoisier's works, *Essays, Physical and Chemical*, and in 1783 a further selection of the same writer's *Essays on the effects produced by various processes on atmospheric air.* Adam Walker, the Manchester schoolmaster and lecturer, had several friends among French scientists, and Henry Clarke was of international repute.[2] In 1794 John Dalton at the Manchester Academy is said to have been teaching science to his pupils from Lavoisier's *Elements of Chemistry* and the similar work of Chaptal.[3] The *Memoirs* of the Literary and Philosophical Society teem with references to the works of foreign chemists, while the Manchester *Memoirs* were themselves translated into French and German.

It may be felt, however, that our argument has not yet been sufficiently concerned with practical men of business of the sort who must have formed the majority in Manchester's industrial life at this time. Let us examine four such men.[4] The obituary of Thomas C. Hewes (1768–1832), 'a machine maker and general millwright', who settled in Manchester in 1792, describes him as 'a man of science' with 'eminent talents in the various branches of the mechanical art'. He is said to have had 'a valuable knowledge of mechanics in general' and 'a particular acquaintance with hydrostatics' and to have 'studied with most satisfactory results the effects of currents of wind as a means of power'. Hewes planned and erected mills, made mill-gearing and spinning machines, and constructed water-wheels all over the country. He was famous for his improvements to water-wheel machinery, especially for the introduction of the suspension wheel.[5] Such a man evidently

---

[1] T. Henry, 'Considerations relative to the Nature of Wool, Silk, and Cotton, as Objects of the Art of Dyeing', *Manchester Lit. & Phil. Soc. Memoirs*, Vol. III (1790). See below, pp. 242–3. A translation of the French works mentioned was advertised in the *Manchester Mercury*, 30 June 1789.

[2] For Walker and Clarke, see below, pp. 104–6 and 114–15.

[3] R. Angus Smith, 'Memoir of John Dalton', p. 18. Dr. William Henry, of Manchester, also recommended these and other French works in his *Epitome of Chemistry* (1801).

[4] For Henry, Taylor, Cooper, and others, see below, chaps. VII, VIII, and IX. See also above, pp. 81–2 (Crompton) and 82–3 (Rupp), and below, pp. 480–1, 487–8 (Fairbairn) and 490, 508–9 (Nasmyth).

[5] *Manchester Guardian*, 11 Feb. 1832. Hewes had one of the biggest engineering works in the country in 1824, when he gave evidence before the Select Committee on Artizans and Machinery (*Fourth Report*, pp. 340–50). See J. H. Clapham, *An Economic History of Modern Britain* (1939), Vol. I, p. 355, where, however, his name is misspelt 'Herves'. See above, pp. 69–71, regarding Hewes' water-wheels, and below, pp. 445–7, for his millwork and machine-making activities. He also built bridges, e.g. a suspension bridge over the River Aire at Armley Mill. W. O. Henderson, *J. C. Fischer and his Diary of Industrial England 1814–1851* (1966), pp. 63–4.

had a sound knowledge of the mechanical principles underlying the machines he built.

A rather more celebrated millwright than Hewes was Peter Ewart (1767–1842), a pupil of John Rennie and later an employee of Boulton and Watt, who established himself in Manchester in the 1790s first of all as an engineer and then as a cotton manufacturer. He was specially interested in theoretical mechanics and was 'sufficiently conversant with mathematics to read with facility most works, which were then the standard authorities in mechanics'. His extensive familiarity with such works, by British and foreign scientists, is demonstrated by his lengthy and learned paper 'On the Measure of Moving Force', in which he sought to achieve a closer union of mechanical theory and practice.[1] He was 'profoundly learned in all that has reference to steam', both theoretical and practical. He had a close friendship of nearly half a century with John Dalton and published a paper in the *Annals of Philosophy*, Vol. VI, on Dalton's 'Theory of Chemical Composition'. He was also a friend of Thomas Henry and one of the executors of his will. He was elected a member of the Manchester Literary and Philosophical Society in 1798 and was a vice-president from 1812 to 1835 when he left the town. He read several more papers to the Society, chiefly on mechanical theory.[2]

George A. Lee (1761–1826), of the great Manchester cotton-spinning firm of Philips and Lee, is said to have been 'imbued with a love of the sciences, and was . . . remarkable for the extent and precision of his acquirements in them'. It was for this reason that he was so perceptive of 'the advantages to be derived from applying to useful purposes the great inventions that distinguished the era in which he lived'.[3] The spinning machinery in his mills, constructed under his supervision, was remarkable for its 'correct and excellent workmanship'. He was on friendly terms with James Watt and Matthew Boulton and also

[1] *Manchester Lit. & Phil. Soc. Memoirs*, 2nd ser., Vol. II (1813), pp. 105–258.

[2] Obituary notice in *Manchester Lit. & Phil. Soc. Memoirs*, 2nd ser., Vol. VII (1846).

[3] Obituary notice in *Manchester Guardian*, 12 Aug. 1826, and *Gentleman's Magazine*, n.s., Vol. XCVI, pt. ii (July–Dec. 1826), pp. 281–2. Robert Owen described Lee as 'one of the most scientific men of his day'; he had been 'highly educated', and was 'a finished mathematician'. He 'possessed great talent as a scientific machinist and engineer'. He had been manager of Drinkwater's mill, the first Manchester mill (1789) to instal one of Watt's rotary engines. R. Owen, *The Life of Robert Owen, Written by Himself* (1857), Vol. I, pp. 26–8 and 32. The Swiss steel manufacturer, J. C. Fischer, said that 'Mr. Lee is an expert in many branches of knowledge and he is certainly one of the most outstanding Englishmen of today'. Henderson, *op. cit.*, p. 143. See above, p. 97, for his request to James Watt junior in Paris to acquire for him 'any good or curious Books upon Mechanics'.

with Peter Ewart, and was 'fully sensible of the advantages of the steam engine'. He soon made himself 'master of the abstrusest parts of its theory', with the result that his firm's engines were 'the finest specimens of perfect mechanism'. He was the first to improve on the fireproof mills of his friend William Strutt, by the employment of cast-iron beams,[1] and he was among the first to employ steam for warming mills. He was friendly with Boulton and Watt's foreman, William Murdoch, whose invention of coal-gas lighting he immediately utilized, Philips and Lee's mill being the first (1805) in Manchester to do so;[2] and he gave Dr. William Henry assistance in his chemical researches to improve the process.[3]

John Kennedy (1769-1855), another of the leading Manchester cotton spinners, trained originally as a 'machine-maker' or engineer and later a partner in the great firm of McConnel and Kennedy, was similarly interested in applied science.[4] In the course of his Scottish village schooling, his attention was fortunately directed by his teacher, Alexander Robb, 'to the elementary principles of mechanics and mechanical movements',[5] and later, when he travelled down to Chowbent in Lancashire, these scientific interests were stimulated by a lecture on natural philosophy by John Banks, which 'laid the foundation of his future tastes and desires in these pursuits'.[6] After he had established himself in Manchester, 'every discovery in mechanical science received his cordial support. He was a friend and admirer of Watt, and there were few distinguished men in the scientific world with whom he was not acquainted, and on terms of friendly inter-course. Round his table were at all times to be found men who were noted for intellectual attainments.' He was a member of the Literary and Philosophical Society, and throughout his life 'remained the friend of Dalton, Henry, and other men eminent for their discoveries and writings in science'. 'Mr. Kennedy never pursued business for the sake of money, but for the love of improvements in his favourite mechanical pursuits. To these he devoted nearly the whole of his time,

---

[1] The Salford mill of Philips and Lee (1799-1801) was one of the earliest iron-framed buildings. See T. C. Bannister, 'The First Iron-framed Buildings', *Architectural Review*, Vol. 108 (April 1950), pp. 231-46; H. R. Johnson and A. W. Skempton, 'William Strutt's Cotton Mills 1793-1812', *Newcomen Soc. Trans.*, Vol. XXX (1956), pp. 179-205.   [2] *Phil. Trans.*, Feb. 1808.
[3] W. Buckley and A. McCulloch, 'A Record of Some Early Experiments on the Carbonisation of Oil and Coal, by Dr. W. Henry, F.R.S. (1808-1824)', *Manchester Lit. & Phil. Soc. Memoirs*, Vol. LXXIV (1929-30), p. 69. See also below, pp. 248-9.
[4] According to William Fairbairn, 'A Brief Memoir of the late John Kennedy, Esq.', *Manchester Lit. & Phil. Soc. Memoirs*, 3rd ser., Vol. I (1862), pp. 147-57. See below, p. 440, for Kennedy's early career in engineering.
[5] *Ibid.*, p. 147.   [6] *Ibid.*, p. 149. See below, p. 108.

and there was scarcely any discovery in the arts [manufactures] that he did not make himself acquainted with. He was fond of mechanical discussion . . .'[1]

These men—Hewes, Ewart, Lee, and Kennedy—were among the leading engineers and cotton spinners in Manchester. But even a common engine erector like Isaac Perrins was not an absolutely un-lettered man, as is evident from the style in which Boulton and Watt wrote to him: 'From long experience & knowledge of Steam Engines we doubt not but you are fully competent to discriminate between a variation of form & a variation of principle in the construction of an Engine . . .'[2] Even the ordinary millwright in the eighteenth century, as we have seen, usually had a fairly good knowledge of mathematics, mechanics, and machine-drawing.[3]

III

One is naturally prompted to ask how these men acquired such scientific and technical knowledge. They were undoubtedly influenced to some extent by the itinerant lecturers in natural philosophy who regularly visited the principal towns in the later eighteenth century.[4] In Manchester, for example, men like Caleb Rotheram, master of Kendal Dissenting Academy, and his son, John Rotheram, James Ferguson, John Arden, Adam Walker and his son, of the same name, John Warltire, Henry Moyes, John Banks, Thomas Garnett, Gustavus Katterfelto,[5] and many others gave lectures on science and mechanics,

---

[1] *Ibid.*, p. 155.

[2] Boulton and Watt to Isaac Perrins, 7 Jan. 1795 (should be 1796), Boulton and Watt Letter Book (Office), Sept. 1795 to Aug. 1796, B.R.L.

[3] See above, p. 73.

[4] The influence of these men has been seriously underestimated by economic historians, e.g. W. H. B. Court, *The Rise of the Midland Industries, 1600–1838* (1938), p. 225: 'A sign of this side of the town's [Birmingham's] life was the occasional giving of scientific lectures by visitors, whose qualifications, however, are not always apparent . . . The circle of people affected by such lectures would be exceedingly small, the effects on invention probably negligible . . .'

The best general account of the lecturers is given by N. Hans, *op. cit.*, chap. VII. There are also important articles on individual lecturers such as: Douglas McKie, 'Mr. Warltire, a good chymist', *Endeavour*, Vol. X (1951), pp. 46–9, and J. A. Harrison 'Blind Henry Moyes, "An Excellent Lecturer in Philosophy" ', *Annals of Science*, Vol. XIII (1957), pp. 109–25. See also D. Layton, 'Newton's Effect on English Thought', *History Today*, Vol. VII (1957), pp. 388–95. At the same time as our article was originally published, the late Dr. F. W. Gibbs also produced some interesting evidence on 'Itinerant Lecturers in Natural Philosophy', *Ambix*, Vol. VI (1960), pp. 111–17.

[5] Katterfelto, a German, appears to have been less of a philosopher and more of an entertainer than the others. The German traveller, C. P. Moritz, wrote that he 'understands, besides electricity and a few other tricks of physics, a little of the

besides those lecturers we have already mentioned in connection with the College of Arts and Sciences and the Manchester Academy.[1] Most of these lecturers were itinerant. Some of them were nationally famous in their day; others were rather more local figures, but their importance in the general pattern of scientific education has not been generally recognized. Several of them took with them on their tours quite large collections of working models and scientific apparatus. In addition, they usually published their lectures in book form, or if they did not publish the complete lectures, they printed and sold a synopsis of them.

Such public lectures were not, as we have seen, a novelty of the late eighteenth century, but a continuation of those started in the first half of that century.[2] James Ferguson (1710–76), for example, after referring to the foundation of 'experimental philosophy' by Bacon, Boyle, and Newton, pointed out that

The method of teaching and laying the foundation of physics, by public courses of experiments, was first undertaken in this kingdom, I believe, by Dr. John Keill, and since improved and enlarged by Mr. Hauksbee, Dr. Desaguliers, Mr. Whiston, Mr. Cotes, Mr. Whiteside, Dr. Bradley, our late Regius and Savilian Professor of Astronomy, and Dr. Bliss his successor. Nor has the same been neglected [in Scotland] by Dr. James, Dr. David Gregory, Sir Robert Stewart, and after him Mr. Maclaurin—Dr. Helsham in Ireland . . .[3]

Ferguson and other itinerant lecturers in the eighteenth century did not themselves contribute much to scientific discovery, but they were often in close touch with scientists such as Priestley,[4] and, as Dr. Brewster

---

art of conjuring. By this . . . he bewilders the whole public . . . Every intelligent man regards Katterfelto as a windbag.' C. P. Moritz, *Journeys of a German in England in 1782* (trans. and ed. by R. Nettel, 1965), p. 70.

[1] We have compiled a long list from the *Manchester Mercury*, which shows that almost every year for the last forty years of the eighteenth century Manchester was visited by one and sometimes by several of these lecturers, each usually giving a course, occasionally two courses, of lectures, numbering sometimes as many as thirty, spread over several weeks or even months.

[2] See above, pp. 34–42 and 51–9.

[3] D. Brewster, *Ferguson's Lectures* (2nd edn., 1806), Vol. I, pp. xxi–xxii, reproducing Ferguson's preface to the original edition of 1760. (By 1793 the seventh edition had appeared.) The lectures were on mechanics, hydrostatics, hydraulics, pneumatics, optics, geography, astronomy, and dialling. Ferguson, the son of a Scottish day labourer in Banffshire, educated himself and developed a strong interest in astronomy and natural philosophy. He came to London in 1740, where he combined the teaching of mathematics, etc. with instrument-making. In the later 'forties he began his popular scientific publications and lectures. His books had a very wide circulation, while his lectures were delivered in towns all over England. He was, in fact, one of the first and most famous itinerant lecturers in natural philosophy. See Brewster's account of Ferguson's life, and also *D.N.B.*, Hans, *op. cit.*, pp. 145–6, and Taylor, *Math. Pract. of Hanov. England*, p. 176.   [4] Gibbs, *op. cit.*, stresses this connection.

pointed out, had outstanding abilities in popularizing science.[1] They were able 'to give a familiar view of the various branches of physical science, and to render them accessible to those who are not accustomed to mathematical investigation', although many craftsmen did, in fact, acquire mathematical knowledge. It was largely to their labours that

we must attribute that general diffusion of scientific knowledge among the practical mechanics of this country, which has, in a great measure, banished those antiquated prejudices, and erroneous maxims of construction, that perpetually mislead the unlettered artist [artisan].

[Ferguson's book has been] widely circulated among all ranks of the community. We perceive it in the workshop of every mechanic. We find it diffused into the different Encyclopaedias which this country has produced; and we may easily trace it in those popular systems of philosophy which have lately appeared.

Such works, on the 'application of science to the practical purposes of life', as Brewster emphasized, were of great importance 'in a commercial country like ours, which depends so much on the improvement of its manufactures, and the progress of the useful arts'.

Some of the earliest lectures in Manchester on 'experimental philosophy', those delivered in 1743 by Caleb Rotheram, of Kendal Academy, and his son John, are still preserved in manuscript in Chetham's Library.[2] Natural philosophy, Caleb Rotheram emphasized, was of both intellectual and utilitarian interest, being 'one of the most usefull and entertaining branches of Learning, a high and refined part of Speculative knowledge, and of great importance in the common affairs of human life'. His lectures, and those of his son, covered many subjects: the nature of matter, the laws of motion, mechanics (including not only simple 'machines' such as the balance, lever, pulley, wheel, inclined plane, wedge, and screw, but also 'compound engines'), projectiles and pendulums, hydrostatics, hydraulics, pneumatics (including the air-pump, barometer, etc.), heat (including the pyrometer), optics, light and colours, and astronomy. Rotheram illustrated his lectures with an extensive apparatus, which had previously belonged to John Horsley, Dissenting minister and lecturer in natural philosophy at Morpeth and Newcastle.[3]

Richard Kaye, a doctor of Baldingstone, Bury, attended these

---

[1] *Op. cit.*, Vol. I, pp. v–x.

[2] Chetham's Library, Mun. A. 2.68, and Mun. A. 2.81.

[3] This, in turn, as we have seen, was later acquired by Warrington Academy and then by Hackney College. See above, pp. 90 and 94. For Horsley, see below, p. 160.

lectures, which he referred to in his diary.[1] But these were not the first he had heard. In February 1738/9, for example, he went to Bury 'to see some Mechanical and Mathematical Observations', including lectures on mechanics, hydrostatics, optics, and pneumatics. In June 1741 he attended geographical lectures in Bury by a Mr. Hamer, mathematician, followed in February 1741/2 by a lecture from Hamer on trigonometry and logarithms, and by anatomical lectures in March and June; later in June 1744/5 he attended more philosophical lectures by Hamer.

It is worth recalling that one of the greatest ironfounders of the age, John Wilkinson, was educated at Kendal Academy under Caleb Rotheram, and so were George Walker, principal of the Manchester Academy from 1798 and President of the Manchester Literary and Philosophical Society, 1804–7, and John Banks, one of Manchester's lecturers on science at the College of Arts and Sciences.

Another early lecturer in natural philosophy in Manchester was John Arden (1721–91/2), a 'teacher of experimental philosophy at Beverley', Yorkshire, who gave courses there in 1756, 1762, 1774, and 1776.[2] He also lectured in Birmingham and other towns, and eventually established himself in Bath, where he continued to teach and lecture for many years. James Ferguson also lectured in Manchester in the 1760s[3].

One of the most famous itinerant lecturers of the century, Adam Walker (1731–1821),[4] had close connections with Manchester, where he established a school in January 1762.[5] He is said to have 'conceived

---

[1] Entries for June and July 1743. We have been very kindly allowed to see the galley proofs of a new edition of Kaye's diary which is being brought out by the Chetham Society. The diary was written by Richard Kaye (1716–51), of Bury, not Samuel Kaye (1708–82), of Manchester, and the lectures were not *given by* Kaye, as is wrongly stated in Turner, *History of Science Teaching in England*, p. 53, and followed by Sir Eric Ashby, *Technology and the Academics* (1963), p. 115.

[2] *Manchester Mercury*, 14 Sept. 1756, 9 Nov. 1762, 9 Aug. 1774, 23 Jan. 1776. See Hans, *op. cit.*, pp. 148–9, and Gibbs, *op. cit.*, pp. 113–14, for brief references to Arden's career. That he was John and not James Arden (as stated by Hans and Gibbs) is shown by the *Beverley Guardian*, 18 Oct. 1890 (obituary of Charles Arden; we are indebted to Mr. G. P. Brown, of Beverley Public Library, for this reference). See also K. N. Cameron, *Shelley and his Circle* (Harvard U.P., Cambridge, Mass., 1961), Vol. II, pp. 936–7. John Arden was apparently an Oxford graduate, and taught at Heath Academy, near Wakefield (see below, p. 156), before moving to Beverley in about 1750. His son, James, published an *Analysis of Mr. Arden's Course of Lectures on Natural and Experimental Philosophy*, in successive editions in the 1770s and 1780s.

[3] *Manchester Mercury*, 28 Sept. 1762, 29 May 1764.

[4] For accounts of Walker's life see *European Magazine, and London Review*, Vol. XXI (Jan.–June 1792), pp. 411–13; *Gentleman's Magazine*, Vol. XCI, pt. i (Jan.–June 1821), pp. 182–3; *D.N.B.;* Bowden, *Industrial Society in England*, pp. 20–1; Hans, *op. cit.*, pp. 146–8; Gibbs, *op. cit., passim.*

[5] *Manchester Mercury*, 26 Jan. 1762.

a system of education more adapted to a Town of Trade than the Monkish system still continued in our Public Schools'.[1] His ideas were widely approved and 'many of the first people in that town' are said to have benefited from his tuition during the five years (1762–6) which he spent in Manchester. He gave several public lectures there on subjects of natural philosophy, especially on astronomy. Then in 1766 he purchased 'the celebrated Philosophic Apparatus' of another lecturer, William Griffith,[2] and thereafter proceeded to travel throughout the north of England, southern Scotland, and Ireland, lecturing on 'Mechanics, Hydrostatics, Pneumatics, Chemistry; on Optics and Astronomy; and on Magnetism, Electricity, and the general Properties of Matter'.[3] The remarkable extent of his 'Philosophic Apparatus' is revealed by a newspaper advertisement of the early 'seventies.[4] It included not only astronomical equipment (orrery, globes, spheres, planetariums, etc.), air-pumps, telescopes, microscopes, pyrometer, electrical machine, magnets, etc., but also

All the mechanical Powers, with working Models of various Cranes, Pumps, Water-Mills, Pile-Drivers, Engines, the Centrifugal Machine, and a working Fire-Engine for draining Mines, of the latest Construction. Also another working Fire-Engine, which by Means of an inverted Piston, will do the same Work at about Two Thirds the Expence. Another working Model for the same Purpose, which saves the Expence of Fire by forming the Vacuum with cold Water; and Blakey's Patent Engine. A new-contrived Bucket-Engine, which works with a remarkable small Quantity of Water; and a Wind-Mill, which, by cloathing and uncloathing itself, moves with the same Velocity in high or low Winds. An horizontal Wind-Mill invented by Joseph Henry, Esq.; Working Models of the Duke of Bridgewater's curious Canal and Colliery, near Manchester; and a Machine that will raise Water to any Height to supply Gentlemen's Houses, Towns, or Villages . . .

Walker was obviously abreast of the latest developments in mechanical construction, and was later able to impress Matthew Boulton with his collection of steam-engine models, including one of a Boulton and Watt engine.[5]

[1] *European Magazine*, Vol. XXI, p. 411. The subjects taught in his school included English, writing, accounting, mathematics, book-keeping, drawing, and geography (*ibid.*, p. 412).

[2] *Manchester Mercury*, 21 Jan. 1766. See below, p. 164, for William Griffith (or Griffis).　　　　　　　　　　　　　[3] *European Magazine, loc. cit.*

[4] *York Courant*, 24 March 1772, advertising 'a Course of Lectures on Natural and Experimental Philosophy. By A. Walker.' The course, which included twelve lectures of two hours each, had been 'last read to upwards of 600 subscribers in Manchester and Liverpool'. Tickets for the course cost a guinea for gentlemen and half a guinea for ladies.

[5] M. Boulton to J. Watt, 30 July 1781, Box 20, B.R.L. Earlier he had informed Boulton & Watt that he had 'working models of most engines for Hydraulic

H

After several years' residence firstly in Ireland and then in York, Walker established himself as a lecturer in London. He was a friend of Joseph Priestley and other scientists, whose discoveries he publicized.[1] But he was not merely an academic lecturer, for he had a long string of practical inventions to his credit.[2] He was sufficiently well versed in engineering to be consulted by the Richmond Water Company when they were considering whether to instal a Boulton and Watt engine in 1780,[3] and three years earlier Boulton had even thought it worth while to sign an article of agreement with Walker for an option on one of Walker's inventions.[4] Unfortunately the invention is not described, but we know that Walker invented a steam engine, a drawing of which appears in his published lectures and another in the Boulton and Watt Collection at the Birmingham Reference Library.[5] When he lectured in Birmingham in 1781 he dined with Boulton, who expected that he would 'set people a talking about Engines'.[6] Both Adam Walker and his son continued to lecture from time to time in Manchester,[7] where they were clearly held in high esteem. Walker senior published his lectures, which had gone into eight editions by 1792.[8]

John Warltire was similarly active during these years. He gave

---

purposes—amongst others I have every kind of steam Engine that has been used, except the very ingenious & powerful one for which you have a patent'. He therefore requested 'a small working Model', which, he pointed out, 'might extend the knowledge of its powers & utility, as I have the honour to collect all the genteel & scientific people in every place I visit'. A. Walker to Boulton & Watt, 23 Sept. 1777, A.O.L.B., quoted by Gibbs, *op. cit.*, p. 114, n. 17.

[1] This is well brought out by Gibbs, *op. cit.*

[2] *Ibid.* and *Gentleman's Magazine, loc. cit.*

[3] M. Boulton to J. Watt, 26 Dec. 1780, B.R.L.

[4] M. Boulton to J. Watt, 26 May 1777, B.R.L.

[5] Parcel E, Boulton & Watt Collection, B.R.L.

[6] M. Boulton to J. Watt, 2 July 1781, Matthew Boulton Letter Book, 1780–3, A.O.L.B.

[7] Walker's early lectures were mainly on geography and astronomy (*Manchester Mercury*, 23 March, 5 and 21 Sept. 1762, 22 March 1763), but he later widened them to 'Natural and Experimental Philosophy; Consisting of every new and useful Discovery, that has been made in Astronomy, Use of the Globes, Pneumatics, Hydrostatics, Hydraulics, Mechanics, Engineering, Fortification, Magnetism, Electricity, Optics, &c.' (*ibid.*, 3 Sept. 1771; see also 12, 19, 26 Nov. and 10, 24, 31 Dec. 1771, where advertisements show that he also lectured in Liverpool and Bolton and probably in other north-western towns). In 1778 he advertised a course of lectures on 'Mechanical and Chemical Philosophy' (*ibid.*, 11 March 1788). His son lectured mainly on astronomy (*ibid.*, 12, 19, 26 Aug. and 2, 9, 16 Sept. 1783, 13 and 20 June and 19 Dec. 1797).

[8] *European Magazine, loc. cit.* The second edition of his *Analysis of a course of lectures*, published at the request of his students in Manchester, appeared in 1770. The date of the first edition is not known. Hans, *op. cit.*, p. 147.

courses of lectures in Manchester, for example, in 1772/3 and 1779,[1] specializing in chemistry; like Walker, he was a friend of Priestley and publicized the latter's recent discoveries concerning 'different kinds of air'. He was followed by many others, such as James Booth[2] and more obscure figures such as Pitt,[3] Long,[4] and Burton.[5] Henry Moyes, the blind philosopher, lectured in Manchester in 1781,[6] Katterfelto was performing there in 1790,[7] and in the later 'nineties Dr. Thomas Garnett and Dr. William Henry gave lectures on natural philosophy, chemistry, and their industrial applications.[8]

A more frequent lecturer in Manchester and the north-west generally was John Banks, who stated in the preface to his *Treatise on Mills* (1795) that this work was based largely on the public lectures he had been delivering for twenty years. The book did not attempt to teach the mechanic how to perform the detailed engineering work, but examined the general principles of mechanics and hydrostatics. In justifying the publication of his book, Banks referred to the condition of mechanical knowledge in the country at that time, in terms reminiscent of Desaguliers: 'It is true, that we have in the kingdom many intelligent engineers, and excellent mechanics; and there are others who can execute better than they can design, otherwise there would not have been so much money expended in attempting what men of science know to be impossible.' The *Treatise on Mills*, together with his earlier *Epitome of a course of lectures* (1775) and his book, *On the Power of Machines* (Kendal, 1803), which describes many experiments on steam engines and on the strength of materials, did much to inform those who could 'execute better than they can design' by discussing engineering problems in a realistic manner.[9]

---

[1] *Manchester Mercury*, 15 Dec. 1772 and 30 March 1779. These were part of a general lecturing programme in Lancashire and Cheshire towns (Gibbs, *op. cit.*, p. 113).   [2] *Ibid.*, 8 and 15 Sept. 1778.

[3] *Ibid.*, 16, 23 Feb. and 16 March 1779, advertising lectures on experimental philosophy.   [4] *Ibid.*, 16 Sept. 1783, advertising lectures on electricity.

[5] *Ibid.*, 31 July and 7 Aug. 1787, advertising lectures on experimental philosophy; he was recommended by Priestley.

[6] *Ibid.*, 3 April 1781.   [7] *Ibid.*, 13, 20, 27 July and 3, 10 Aug. 1790.

[8] *Ibid.*, 26 Jan. 1796 (Garnett) and 18 Dec. 1798 (Henry). For Garnett, see below, pp. 146 and 181; for Henry, see above, p. 84, and below, pp. 246–8.

[9] It must be pointed out, however, that Banks, like Desaguliers, was perhaps too supercilious in his attitude towards practical engineers and that his theorizing was sometimes seriously at fault, as was shown by Robertson Buchanan in an article 'On the Velocity of Water Wheels', *Philosophical Magazine*, Vol. X (1801), pp. 278–81, based on a paper read to the Edinburgh Philosophical Society. Buchanan, however, referred to Banks with respect (and also to the earlier researches of Smeaton), and himself provides a fine example of a practising millwright-engineer who was also interested in experimental philosophy. His paper was based on experiments in cotton mills and he was soon to acquire fame with his *Practical Essays on Millwork* (1814).

There are several references in the *Manchester Mercury* to Banks's lectures in the last quarter of the eighteenth century,[1] and we know also that he lectured in Bolton, Birmingham, Doncaster, and Kendal. By 1785 his lectures on natural philosophy in Manchester were being given under the auspices of the College of Arts and Sciences, and when he lectured in Doncaster, he advertised himself as 'Lecturer in Philosophy to the College of Arts and Science in Manchester'.[2] John Kennedy, the great Manchester cotton spinner, refers to Banks in his 'Early Recollections', where he tells how, at the age of fourteen, he (Kennedy) was apprenticed with Adam Murray to the firm of Cannan and Smith, machine-makers, at Chowbent in Lancashire, and was brought down from Scotland by James Smith, one of his masters. On the journey, Smith had some calls to make:

amongst others, on Mr. Banks, a lecturer on mechanics and natural philosophy, whom he advised to come to Chowbent, where he would strive . . . to get him a sufficient audience to attend his lectures. I remember Adam Murray and I had a ticket between us, each paying half-a-crown, and we attended alternately. This was to us a wonderful insight into the laws of nature.[3]

The price, five shillings, for the course of lectures, was evidently not prohibitive even for two apprentices, so that the lecturers at least to some extent anticipated the work of the Mechanics' Institutes. Leaders of science and industry often recommended lecturers to each other so that an audience was guaranteed at the next port of call.[4]

Like Walker, Banks had an extensive experimental apparatus, consisting of 'various models and instruments for illustrating the principles of mechanics'. In addition to the usual philosophical and astronomical apparatus, including Smeaton's improved air-pump, microscopes, telescopes, orrery, electrical machine, etc., he also had working models of various sorts of pumps, 'water engines', and steam engines—so

---

[1] Beginning, apparently, in 1775 (*Manchester Mercury*, 9 Jan. 1776, advertising a second course of his lectures). He was a former pupil of Dr. Caleb Rotheram.

[2] *Yorkshire Journal and General Weekly Advertiser*, 10 and 24 March 1787. We are grateful to Mr. J. A. Harrison for this reference. There are numerous references to Banks's lectures for the College of Arts and Sciences in the *Manchester Mercury* between 1785 and 1787. He continued his public lectures in the town after the College's demise (*ibid.*, 12 Feb. and 4 March 1788; 26 Jan., 9 and 16 Feb. 1790; 22 Jan., 26 Feb., 7 May, 30 July and 6 Aug. 1793).

[3] J. Kennedy, *Miscellaneous Papers* (Manchester, 1849), p. 14. See above, p. 100.

[4] Joseph Priestley (Calne) to Matthew Boulton (Birmingham), 28 Sept. 1776, A.O.L.B., introducing John Warltire; John Roebuck (Bo'ness) to Matthew Boulton (Birmingham), 30 Oct. 1777, Dol., recommending Henry Moyes; and Thomas Walker (Manchester) to James Watt, jun. (Birmingham), 8 May 1796, Box 3.W, B.R.L., recommending Dr. Garnett.

many, in fact, that 'the whole require a waggon to convey them from place to place'.[1]

Banks's *Treatise on Mills* was subscribed to by a most impressive array of scientists, engineers, and professional men.[2] Such subscription lists for the works of itinerant lecturers show the wide interest in lectures on mechanics and science. A similar impression is formed from the study of the membership lists of the philosophical societies of the day—from the great societies of Birmingham and Manchester to those in market towns. The subscribers to the books and the members of the philosophical societies were largely the same people. Quite often the lecturers were themselves members of scientific societies.

The itinerant lecturers were varied in their educational backgrounds and social antecedents. Many of them, such as the two Rotherams, Warltire, and Moyes, had degrees from Scots universities, where scientific research was most alive at this period, and they may also have attended one of the Dissenting Academies; others, like Adam Walker and James Ferguson, are said to have been almost entirely self-educated. Nearly all of them had strong practical interests and several had one or two minor inventions to their credit; some had watch-making or mathematical instrument-making as their principal occupation. Peter Clare, the Manchester clock-maker, was one of the latter type. He lectured in Manchester from about 1772.[3] In 1778 he proposed to establish 'a Philosophical School in which young Persons will be instructed in the general principles of Mechanics, Hydrostatics, Pneumatics, Electricity, Chemistry, &c . . .',[4] but he also continued to give public lectures for many years.[5] Clare was a very close friend of John Dalton, in fact his closest companion and secretary in later years,

---

[1] *Syllabus of a Course of Twelve Lectures in Experimental Philosophy and Astronomy by John Banks* (Bolton, undated), Wedgwood MSS., 21674–114.

[2] The list of subscribers shows the wide variety of people who were interested in buying such a book. The names included Richard Bradley, engineer, Manchester; David Broad, architect, Manchester; Peter Clare, watch-maker, schoolmaster and lecturer, Manchester; John Dalton; William Hirst, engineer, Gomersal; William Kirk, watch-maker, Manchester; John Knowles, teacher of mathematics, Liverpool; Robert Owen; Joseph Priestley; John Raistrick, Low-moor furnace, Bradford; John Rennie; Thomas Rider, ironfounder, Salford; John Samuels, engineer, Manchester; Michael Satterthwaite, Ulverston, and Benjamin Satterthwaite, Lancaster; William Sharrett (Sherratt), engineer, Manchester; John Southern, Birmingham; Thomas and Richard Walker, merchants, Manchester; and John Wilkinson.

[3] *Manchester Mercury*, 26 May 1772 and 8 March 1774 (lectures on electricity).

[4] *Ibid.*, 6 Jan. 1778.

[5] *Ibid.*, 5, 12 May and 22 Dec. 1778; 9 March and 27 July 1779; 24 April 1798. He gradually expanded his lectures from electricity and magnetism to include pneumatics, mechanics, etc.

and was an executor of Dalton's will.[1] He also became one of the secretaries of the Literary and Philosophical Society. That he also had industrial interests outside clock-making, etc., is evidenced by an advertisement in 1789, which seems to indicate that he was involved in a 'new-erected' water-powered spinning mill at Hatton, near Warrington,[2] and by another of 1794 in which he advertised 'A New invented Pumping Machine, for raising Water', powered by a horse.[3]

Another Manchester clock- and watch-maker, as well as optician, was John Imison.[4] In 1785 he published *The School of Arts; or, An Introduction to Useful Knowledge*,[5] which dealt with such 'Branches of Practical Science' as mechanics, electricity, optics, astronomy, instrument-making, clock- and watch-making, and other crafts. The first section was separately published in 1787, as *A Treatise on the Mechanical Powers*,[6] to which Imison had added items on millwork, including the 'Construction of Water Mills', which he hoped would prove useful to 'those for whose information they are principally intended, viz. working mechanics'. This revised section on mechanical powers was included in a second edition of *The School of Arts*, published about 1794,[7] which contained additional sections on pneumatics, hydrostatics, and hydraulics, including pumps and steam engines. Imison also published other works, including *Elements of Science and Art*.[8] The instrument-makers and clock-makers like Imison and Clare thus played a very important role in diffusing scientific knowledge, by making apparatus for lecturers, sometimes by demonstrating the experiments or lecturing themselves, by running schools, and by publishing books on mechanical and scientific subjects.[9]

The subscription fee for a course of lectures might be anything from a few shillings to a guinea, according to the duration of the course and the popularity of the lecturer. No doubt professional men and some of

[1] H. McLachlan, 'John Dalton and Manchester', p. 177; R. Angus Smith, 'Memoir of John Dalton', p. 277.

[2] *Manchester Mercury*, 31 March 1789.

[3] *Ibid.*, 21 Jan. 1794.   [4] Hans, *op. cit.*, p. 151.

[5] There is a copy in the Manchester Cen. Ref. Lib. The work had been published in parts at an earlier date, under the title of *An Introduction to Useful Knowledge*. See *Manchester Mercury*, 30 Sept. 1783.   [6] Copy in the British Museum.

[7] Copy in Chetham's Library, Manchester. A second edition of *A Treatise on Mechanical Powers* was also published in 1794 (copy in Manchester Cen. Ref. Lib.).

[8] Revised edition (2 vols., 1803), in the British Museum and Manchester University Library.

[9] Hans, *op. cit.*, pp. 150–2. Clock-making, as we have shown elsewhere, was important in the development of mechanical engineering. A. E. Musson and E. Robinson, 'The Origins of Engineering in Lancashire', *Journal of Economic History*, June 1960, pp. 209–33. See below, chap. XIII.

the more prosperous manufacturers tended to make up the bulk of the audiences, but, as we have seen, craftsmen like John Kennedy and Adam Murray also attended, before they had risen into the manufacturers' ranks, and Thomas Henry encouraged artisans to come by charging them lower fees. Practical men must have been attracted to the lectures by the continual attempts to adapt them to local needs:

This Course is confessedly more adapted to every Branch of Business than any ever read in this Town . . .[1]

If a sufficient Number wish to attend, a Lecture will be delivered on the application of Water to Bucket Wheels, accompanied with many Experiments . . .[2]

[Chemistry's] Application to the common Purposes of Life, as well as to the useful Arts of Bleaching, Dyeing, Agriculture, &c.[3]

A good deal of the apparatus used also shows a sense of practicality in the lecturers, as the lists of equipment used by Walker and Banks so strikingly demonstrate.[4] A similar list advertised by Mr. Pitt in 1779 included optical instruments, cranes, pile-drivers, sawing engines, pumps, steam engines, and many other items.[5] These machines could be seen by the public, free of charge. Most itinerant lecturers had a similar collection.

No one has yet studied in detail the wealth of books published in the eighteenth century on science and technology.[6] The *Manchester Mercury* in the second half of the century printed many advertisements of such books as Benjamin Martin's *General Magazine of Arts and Sciences*;[7] *The New and Compleat Dictionary of Arts and Sciences*, by a Society of Gentlemen;[8] *The Circle of Sciences*, sold by A. Clark, bookseller, Manchester;[9] Charles Hutton's *Treatise on Mensuration*, 'adapted

[1] *Manchester Mercury*, 26 Nov. 1771. Adam Walker's lectures.
[2] *Ibid.*, 26 Jan. 1790. John Banks's lectures.
[3] *Ibid.*, 26 Jan. 1796. Dr. Thomas Garnett's lectures.
[4] See above, pp. 105 and 108–9.
[5] *Ibid.*, 16 March 1779.
[6] Since our article was originally published, this gap has been to some extent filled by D. A. Kronick, *A History of Scientific and Technical Periodicals* (New York, 1962), and by D. Layton, 'Diction and Dictionaries in the Diffusion of Scientific Knowledge: An Aspect of the History of the Popularization of Science in Great Britain', *The British Journal for the History of Science*, Vol. I (1962–3), pp. 265–79. There is still room, however, for much more research in this field.
[7] *Manchester Mercury*, 7 Jan. 1755. For Martin, see Hans, *op. cit.*, pp. 141, 145, 152, 206, and above, p. 42. Many of these books were advertised in the printed catalogue of the first large-scale booksellers of cheap books: *A Catalogue of Books, for the year 1804–5; Comprising upwards of 850,000 Volumes, in all Languages and Classes of Learning . . . by Lackington, Allen, & Co. Temple of the Muses, Finsbury Square, London.*
[8] *Manchester Mercury*, 21 Oct. 1755.   [9] *Ibid.*, 30 Dec. 1755.

particularly to the Uses of Schools, Mathematicians and Mechanics';[1] William Nicholson's *Introduction to Natural Philosophy*;[2] B.W. Emerson's *The Principles of Mechanics*, described as being 'extremely useful to all sorts of Artificers, particularly to Architects, Engineers, Shipwrights, Millwrights, Watch-makers, &c or any that Work in a Mechanical Way';[3] A. Fletcher's *The Universal Measurer and Mechanic*, 'a Work equally useful to the Gentleman, Tradesman and Mechanic';[4] William Casson's proposed book on fire engines which was to cover every side of the mechanical operation of steam engines;[5] John Imison's books which we have mentioned above;[6] John Curr's *The Coal Viewer and Engine Builder's Practical Companion*;[7] Adam Walker's lectures;[8] William Henry's *A General View of the Nature and Objects of Chemistry And of its Application to Arts and Manufactures*,[9] and many others. Some provided elementary mathematical knowledge, such as J. Bonnycastle's *Scholar's Guide to Arithmetic* and his *Introduction to Mensuration and Practical Geometry*, which was adapted 'as much as possible to practical Uses . . . Workmen and Artificers of all Denominations will find it a ready Assistant'.[10] To these must be added the many encyclopaedias and dictionaries which discussed machines and mechanical processes, often illustrating them with drawings. Books and dictionaries were sometimes published in parts so that they were within range of the artisan's pocket. Some were printed locally, others were printed in London for well-known publishers of scientific books such as Joseph Johnson, who was the publisher for at least three famous Manchester scientists at this period—Thomas Percival, Thomas Henry, and John Dalton. In addition to scientific books, there was a great increase in scientific periodi-

---

[1] *Manchester Mercury*, 21 May 1771. See below, pp. 160–1, for Hutton.

[2] *Ibid.*, 3 Dec. 1782. See above, p. 83, for Nicholson.

[3] *Ibid.*, 7 Dec. 1773. See below, p. 117, for Emerson.

[4] *Ibid.*, 7 Dec. 1773.   [5] *Ibid.*, 18 Jan. 1785.

[6] See above, p. 110.

[7] *Manchester Mercury*, 30 May 1797. John Curr, of Sheffield, was one of the most outstanding coal viewers of his day, notable especially for his introduction of wheeled corves, running on cast-iron rails, his invention of flat-rope winding with rotative steam engines, introduction of steam haulage underground, etc. T. S. Ashton and J. R. Sykes, *The Coal Industry of the Eighteenth Century* (2nd edn., Manchester, 1964), pp. 26, 61, 64–9. John Farey in his *Treatise on the Steam Engine* (1827), p. 204, referred very favourably to Curr's book, with its 'copious tables of dimensions for all parts, proportioned to different sizes of cylinder. It is a useful guide for those who require to construct such engines, being a complete manual for their instruction.'

[8] *Manchester Mercury*, 23 July 1799.

[9] *Ibid.*, 10 Dec. 1799. See below, pp. 246–50, for William Henry.

[10] *Ibid.*, 29 Jan. 1782. See *D.N.B.* and Hans, *op. cit.*, pp. 112, 183, 185, for Bonnycastle (1750–1827), a teacher of mathematics in London, who wrote many books on the subject.

cals from 1780 to 1800.[1] Journals such as Tilloch's *Philosophical Magazine*, Nicholson's *Journal of Natural Philosophy, Chemistry, and Arts*, and the *Repertory of Arts and Manufactures* were immensely important in disseminating knowledge of scientific papers, the proceedings of philosophical societies, new industrial processes, and patents of inventions.

Those who could not afford to buy these books and periodicals could read many of them in Chetham's Library, Manchester, which had been founded in 1653, the first free public library in the country. John Dalton found this a great asset: 'There is in this town', he wrote in 1794, 'a large library, furnished with the best books in every art, science and language, which is open to all, gratis . . .'[2] Examination of the library MS. accessions book and of its first printed catalogues[3] reveals that a considerable number of scientific works, both British and foreign, were acquired in the later eighteenth century. The Literary and Philosophical Society also began to build up its own library, which was particularly strong in works on natural philosophy. In addition, there was a subscription library established in the Exchange in 1757, which acquired many scientific works.[4] Another subscription library was founded in 1792, by the 'Manchester Reading Society', the first name on whose membership list was that of Joseph Priestley junior, who made a presentation of 85 volumes, including some of his father's works, which apparently formed the library's original nucleus.[5]

When all has been said, however, about philosophical societies, itinerant lecturers, and scientific books and periodicals, the strongest educational influence in the lives of many Manchester men must have been their schooling. There were at this period several academies and schools in and about Manchester which taught the principles of general science and provided a sound basis in mathematics. The most important of all, the Warrington Academy (1757–1786), we have previously mentioned.[6] Many leading Dissenters in Manchester were educated there in a liberal and scientific spirit, which, as we have seen, was

---

[1] Douglas McKie, 'The Scientific Periodical from 1665 to 1798', *Philosophical Magazine Commemoration Number*, 1948.

[2] Letter of Dalton dated 20 Feb. 1794, quoted in Nicholson, 'The Literary and Philosophical Society, 1781–1851', p. 114.

[3] *Bibliotheca Chethamensis*, Vols. I and II (Manchester, 1791) and Vol. III (Manchester, 1826). See also Mumford, *op. cit.*, pp. 222–3.

[4] J. Aston, *A Picture of Manchester* (1816), p. 175, gives 1757, but the correct date may be 1765. See *A Classed Catalogue of the Manchester Subscription Library, instituted 1765* (Manchester, 1831), and W. E. A. Axon, *Handbook of the Public Libraries of Manchester and Salford* (Manchester, 1877), p. 178.

[5] Axon, *op. cit.*, p. 61. *A Catalogue of the Manchester Subscription Library . . . instituted Aug. 29th, 1792* (Manchester, 1834). [6] See above, pp. 90–1.

transmitted to the short-lived College of Arts and Sciences and the Manchester Academy. Manchester was also fortunate in possessing a grammar school which preserved in the eighteenth century a strong scientific and mathematical tradition.[1] The High Master from 1764 to 1807 was Charles Lawson, a keen mathematician, who promoted the study of his subject in the school and recruited the services of two former pupils of the school, Henry Clarke and Jeremiah Ainsworth. Several eminent scientists were educated at Manchester Grammar School at this time,[2] and it is also interesting to observe that the largest proportion of pupils came from among the artisans, shopkeepers, and minor occupations.[3]

Henry Clarke (1743–1818) was an important figure in local scientific education.[4] After leaving Manchester Grammar School in 1756, he taught for some years in a school run by Robert Pulman, land-surveyor and writing master, at Leeds, and then travelled on the continent, visiting several important mathematicians, among them Lorgna of Verona, whose works he later translated. He established his 'Commercial and Mathematical School', or 'Salford Academy', in Chapel Street, Salford, as early as 1765. In 1772 he began also to provide evening courses in a variety of subjects,[5] including mathematics and natural philosophy, in which subjects he was appointed to lecture at the College of Arts and Sciences.[6] His school at Salford was still being advertised in 1789–90.[7] In 1792 he moved to Liverpool and there established the 'Literary, Commercial and Mathematical School', Mount Pleasant (Martindale Hill).[8] But on 25 March 1794 he removed his school back to Manchester.[9] He particularly stressed in his advertisement that he prepared youths for 'the Scientific Parts of Mechanic Professions in general', as well as for 'the Mercantile Line', the Army and Navy, and that a regular course of lectures would be instituted 'on the most interesting Parts of Science, Theoretical and Experimental, illustrated occasionally by an improved and extensive modern Appara-

[1] Mumford, *op. cit.*, p. 193; Hans, *op. cit.*, pp. 39–41.
[2] Hans, *op. cit.*, p. 40.   [3] Mumford, *op. cit.*, p. 170.
[4] See *D.N.B.* and Hans, *op. cit.*, pp. 95–7. We have, however, corrected these accounts on some points. See also Taylor, *Math. Pract. of Hanov. England*, pp. 79 and 257–8.
[5] *Manchester Mercury*, 29 Dec. 1772. Advertisements appeared frequently in the *Mercury* in the following years.
[6] See above, p. 93. He was a friend of Joseph Priestley.
[7] *Manchester Mercury*, 29 Dec. 1789 and 23 March 1790. It was then described as the 'Commercial, Mathematical, and Philosophical School'.
[8] Hans, *op. cit.*, p. 96. The syllabus included writing, languages, mathematics, book-keeping and commercial subjects, geography and astronomy, land-surveying, gauging, and navigation, and natural and experimental philosophy.
[9] *Manchester Mercury*, 25 Feb. 1794.

tus'. About 1797 he moved to Bristol,[1] and in 1802 he was appointed Professor of History, Geography and Experimental Philosophy at the Royal Military College, Marlow (later Sandhurst).[2] His publications on mathematics were many and varied, and he was altogether a most distinguished scholar and teacher.

We have already referred to the schools of Peter Clare and Adam Walker.[3] These were not the only schools in Manchester with a mathematical and scientific basis. Earlier, in 1754, Leonard and Thomas Burrow advertised their school 'down Fountain Court behind the Exchange', which provided instruction not only in writing, English, Latin, and Greek, but also in arithmetic—'with an application of it to all the useful purposes of life and branches of trade'—together with book-keeping, mensuration, etc.[4] Jeremiah Ainsworth (1743-84), educated at Manchester Grammar School, ran a Mathematical School in Long Millgate for many years and took some of the grammar school boys for mathematical lessons.[5] Holt and Son's 'Mercantile and Mathematical School', Oldham Street, was described in 1788 as having been going for several years, and catered for both day and evening students.[6] In the same year R. Hartley's 'Commercial and Mathematical School', Princes Street, was advertised.[7] In 1796 John Fern opened a 'Mathematical Academy' at 26 Spring Gardens.[8] We can conclude therefore that there were numerous schools in Manchester directed towards satisfying the demand for scientific, mathematical, and commercial studies.

The emphasis on mathematics in all these schools, academies, and lectures is of considerable interest. As the late Professor Eva Taylor has demonstrated, and as we have previously observed, mathematics had long been an essential tool of applied science.[9] Even at a humble level, some knowledge of mathematics was required by many skilled craftsmen, for whom, as we have seen, many of the elementary textbooks were intended. At a higher level, it was of fundamental importance—as a basis, for example, of engineering design and construction.

The study of mathematics is said to have become widespread in Lancashire during the eighteenth century, even among 'operatives of the humbler class, and these chiefly weavers', who were able to use the

---

[1] Rev. George Heath, *The History, Antiquities, Survey and Description of the City and Suburbs of Bristol* (Bristol, 1797), p. 81.

[2] In that same year he received the LL.D. degree from the University of Edinburgh.　　　　　[3] See above, pp. 109 and 104-5.

[4] Mumford, *op. cit.*, p. 173, and Hans, *op. cit.*, p. 98.

[5] Mumford, *op. cit.*, p. 197; Hans, *op. cit.*, p. 111.

[6] *Manchester Mercury*, 1 Jan. 1788.　　　　　[7] *Ibid.*, 18 March 1788.

[8] *Ibid.*, 29 March 1796.　　　　　[9] See above, chap. I and pp. 73-84.

text-books provided by Simpson, Lawson, Emerson, etc.[1] A Mathematical Society was established in Manchester in the early eighteenth century, for which the 'Ingenious Mathematician', John Jackson, provided the inaugural lectures in 1718,[2] in which he stressed the practical utility of mathematics. Geometry, for example was

useful in all sorts of Professions . . . There is no State of Life that does not require the use of all sorts of Measures, its the Soul of all Trades, all the Arts which tend to the supplying of Mankind with Food and Raiment, subsist only by the Measures of Lengths and Breadths, Circles and Squares. Those whose Lot is to drive a Trade in Provisions, must know the Measures of Dry, and Liquid Things and their Reductions. Husband-men understand Surveying . . . Brewers understand Gauging, *Masons* and *Carpenters* know how to Lay out, to Measure Angles, Squares, and all the Dimensions of a Piece of Building. They who Work in Wood, Iron or Stone, cannot be without Knowledge of those Measures.[3]

These practical applications of mathematics were strongly emphasized in the schools of Clarke, Ainsworth, Holt, and others in Manchester, and also in the text-books advertised in the press. A similar interest was aroused in other Lancashire towns, such as Oldham, where a Mathematical Society was established in about 1770,[4] and Bury, where an evening Mathematical School was opened in that year.[5]

Detailed evidence of the ways in which mathematics contributed to the industrial developments of the time is not easy to find owing to lack of records, but we can provide a striking example in Manchester

[1] *Notes and Queries*, Vol. II (May–Dec. 1850), pp. 8, 57–60, 436–8, and Vol. IV (July–Dec. 1851), p. 300. See also Thomas de Quincy, *Recollections of the Lakes and the Lake Poets, Coleridge, Southey and Wordsworth* (1862 edn.), pp. 113–14: 'Mathematics, it is well known, are extensively cultivated in the north of England. Sedburgh, for many years, was a sort of nursery or rural chapel-of-ease to Cambridge . . . and, indeed, so widely has mathematical knowledge extended itself throughout Northern England that, even amongst the poor Lancashire weavers, mechanic labourers for their daily bread, the cultivation of pure geometry in the most refined shape, has long prevailed. . . .'

[2] *Mathematical Lectures, Being the First and Second, That were Read to the Mathematical Society at Manchester. By the late Ingenious Mathematician John Jackson* (Manchester, 1719). These lectures, preserved in Chetham's Library, Manchester, were delivered on 12 and 19 Aug. 1718. It was stated in the Preface that the Mathematical Society had been 'lately set up'. According to W. E. A. Axon, *The Annals of Manchester* (1886), p. 77, these lectures formed the first book printed in Manchester. Jackson was later called 'the father of the Lancastrian School of Geometers' (Hans. *op. cit.*, p. 144). See above, p. 59, n. 1, for the Mathematical Society established in Spitalfields, London, about the same time, with a similarly practical appeal to artisans.

[3] *Mathematical Lectures*, p. 14. See also *ibid.*, pp. 16–17, for similar remarks on the utility of arithmetic, geometry, and trigonometry.

[4] See the references in *Notes and Queries* previously cited.

[5] Hans, *op. cit.*, p. 116.

during the early nineteenth century, in the collaboration between William Fairbairn, the engineer, and Eaton Hodgkinson, the mathematician. We have previously referred to Fairbairn's self-education and his abilities in applied engineering science.[1] He realized, however, like Telford,[2] that he did not have sufficient mathematical knowledge and skill for advanced engineering research, and therefore turned for help to Eaton Hodgkinson (1789–1861).[3] Hodgkinson's own mathematical training is of great interest. The son of a Cheshire farmer, he was sent to Northwich Grammar School, but revolted from the classical education and discipline and so left to attend the private school of a Mr. Shaw, a man of 'superior mathematical attainments' in that town. On leaving school, Hodgkinson was first put to farming, but then, in 1811, the family moved to Salford, where his mother ran a pawnbroking business.[4] Young Hodgkinson now read 'any standard works on science which he could procure', including those of the eighteenth-century mathematicians, Thomas Simpson and William Emerson. 'Many of the self-taught men of the last and beginning of the present century have expressed their obligations to Thomas Simpson and William Emerson. Their works . . . were the best standard works of the age . . . These humble but highly gifted men . . . wrote to instruct the mass of mankind.'[5]

Hodgkinson now 'became acquainted with the most gifted men then living in Manchester', including Dalton, Holme, Henry, Ewart, Sibson, Johns, and Fairbairn. He became one of Dalton's pupils, studying the mathematical works of Lagrange, Laplace, Euler, Bernoulli, etc. He developed an interest in the strength of materials, especially cast and wrought iron, and their industrial applications, e.g. in beams, pillars, bridges, etc. These interests brought him into touch with Fairbairn, who provided him with experimental research facilities at his works

---

[1] See above, p. 77. See also below, pp. 480–1 and 487–8.   [2] See above, pp. 75–6.
[3] R. Rawson, 'Memoir of the late Eaton Hodgkinson, F.R.S.', *Manchester Lit. & Phil. Soc. Memoirs*, 3rd ser., Vol. II (1865), pp. 145–204; and the section by W. Fairbairn in T. Baines, *Lancashire and Cheshire, Past and Present* (1867), Vol. II, pt. ii, pp. xv–xxviii.
[4] His father having died when Hodgkinson was only six years of age.
[5] Rawson, *op. cit.*, pp. 151–2. Thomas Simpson (1710–61), originally a hand-loom weaver of Market Bosworth, eventually became professor of mathematics at Woolwich Military Academy and a Fellow of the Royal Society. William Emerson (1701–82) was a somewhat eccentric village schoolmaster at Harworth, near Darlington. Both wrote voluminously on mathematics, surveying, navigation, etc. Their works were amongst those advertised in the *Manchester Mercury*, where their practical importance was emphasized (see above, p. 112). Engineers such as Rennie (see above, p. 76) were familiar with their works. See *D.N.B.* and Taylor, *Math. Pract. of Hanov. England*, pp. 191–2 (Simpson) and 34–5, 157 (Emerson).

and who himself participated enthusiastically in these investigations. The results were demonstrated in papers to the Literary and Philosophical Society, especially that on 'Theoretical and Experimental Researches to ascertain the Strength and Best Form of Iron Beams'.[1] His collaboration with Fairbairn was commented upon at some length in the *Manchester Guardian*, which also mentioned that the experiments were observed by Peter Ewart, 'whose extensive scientific attainments rendered his presence peculiarly valuable'. The results of these researches were immediately applied in building an iron bridge for the new Manchester–Liverpool Railway over Water Street and also in the construction of cotton factories at Macclesfield and elsewhere.[2]

Hodgkinson continued to develop these scientific-industrial interests in the following years and he read a series of papers on iron beams, pillars, etc. to the British Association,[3] and to the Royal Society,[4] of which he was elected a Fellow in 1841. In 1847 he was appointed Professor of Mechanical Engineering at University College, London. In his publications he referred to the earlier works on the strength of materials, etc. by Galileo, Newton, Leibniz, Bernoulli, Euler, Lagrange, etc. (Here we see the continuity of the 'Scientific Revolution'.) At the same time, whilst demonstrating his status in this distinguished scientific line, he also continued the Baconian tradition of linking science with industry. Not merely did his researches make important contributions to Fairbairn's engineering achievements; he was also consulted by George and Robert Stephenson in their railway engineering, especially in regard to the construction of the Menai tubular bridge.[5] 'The most eminent engineers of the age place unbounded confidence in the results of his experiments . . . As a confirmation of this, it may be stated that the engineers' pocket- and text-books of the present time are full of Hodgkinson's formulae for calculating the strength and deflection of pillars and beams.'[6]

---

[1] *Manchester Lit. & Phil. Soc. Memoirs*, 3rd ser., Vol. V (1831). For references to other (earlier and later) papers on the strength of materials, etc., delivered by Hodgkinson to the Manchester Society, see Rawson, *op. cit.* See also R. W. Bailey, 'The Contributions of Manchester Researches to Mechanical Science', *Minutes of Proceedings of the Meeting of the Institution of Mechanical Engineers in Manchester, 25th June 1929*, for a good account of the work of Fairbairn and Hodgkinson.   [2] *Manchester Guardian*, 15 and 22 May 1830.

[3] *British Association, Third Report* (1833), *Fourth Report* (1835), *Fifth Report* (1835), and *Seventh Report* (1837).

[4] *Royal Society Transactions*, May 1840 and June 1857. He was awarded the Society's gold medal for the first paper, on the strength of cast-iron pillars.

[5] Rawson, *op. cit.*, pp. 147 and 166–7. See also the account of Hodgkinson's exhaustive experiments in the *Report of the Royal Commission on the application of Iron to Railway Structures* (1849).

[6] Rawson, *op. cit.*, p. 155.

IV

It would appear, from our investigations in Manchester, that there was a closer association between science and industry in the early Industrial Revolution than has hitherto been thought. Evidence from other cities and towns indicates that scientific and technical education was fairly widespread in this period. Most of the books and periodicals to which we have referred had a national market. The itinerant lecturers in natural philosophy visited most of the chief towns; some of them, such as Adam Walker, had schools in London. Philosophical societies were also widely established. And many of the leading manufacturers in different parts of the country displayed a keen interest in science and in its possible industrial applications.

London was still the acknowledged centre of scientific thought and education, but surprisingly little is known about the contribution of London's industries to the Industrial Revolution, let alone about the education of its artisans and its manufacturers. It had been an educational centre at least since the sixteenth century, and as the demands of the time changed from legal to commercial education and again to industrial training so the metropolis found teachers to satisfy those demands. Dr. Hans has described some of the new academies of eighteenth-century London: notably the Little Tower Street Academy, growing out of Thomas Watts's private school for young clerks at Abchurch Lane and transferring in the 1740s to Poland Street; Soho Academy, founded in 1717-18 and lasting for a century; and John Rule's Academy at Islington.[1] The first of these institutions had a distinguished record of scientific teaching, in which Thomas Watts himself (fl. 1715-27) and his co-principal Benjamin Worster (fl. 1717-1726), James Stirling, F.R.S. (1692-1770), William Vream (fl. 1710-27) and Archibald Patoun, F.R.S. (fl. 1725-39) all played their parts.[2] Desaguliers was closely connected with this school, through his association both with Stirling and with Vream, the instrument-maker, who assisted him in scientific lectures and experiments. Desaguliers and Stirling both used to lecture in the Bedford Coffee House, Covent Garden, at which Fellows of the Royal Society were accustomed to assemble.[3] Martin Clare, the founder of the Soho Academy, was also a F.R.S. In 1735 he published his *Motion of Fluids*, in which he acknowledged the assistance of Desaguliers, who was probably connected also with this school. He was succeeded apparently by John Barrow, author

[1] Hans, *op. cit.*, pp. 82-93.

[2] See Hans, *op. cit.*, and also Taylor, *Math. Pract. of Hanov. England.* See also above, p. 41.   [3] See above, pp. 37, 41, 58.

of the *New Universal Dictionary of Arts and Science* (1753), another work in the encyclopaedic line established by Harris and Chambers. By 1800 the Soho Academy was acknowledged to be one of the first schools in London, with a special fame for producing painters such as Rowlandson and Turner. Hackney Academy and Kensington Academy are similar examples of schools in London with a strong emphasis upon modern languages, mathematics, and science, opening their doors to a wide variety of pupils.[1]

Most of these academies had a strongly vocational character, training their pupils not only for the professions, the Army and Navy, but also for trade and industry. They catered mainly for the middle classes, but some, like the Islington Academy, undertook to instruct 'artificers of every sort . . . in every part of science, requisite for their several professions'.[2] To them can be added a host of other, less prominent, academies and schools in London noted by Dr. Hans, which provided similar commercial-technical education. They included some of the first evening schools for mechanics and craftsmen, such as the 'Philosophical Institute' in Berwick Street, Soho, conducted by Peter Nicholson (1765–1844), mathematician and architect, who 'came to London in 1789 and established an evening school for carpenters and mechanics, where he taught applied mathematics and published textbooks for carpenters and joiners'; he also provided evening courses in natural philosophy.[3]

Though something is known about schools and academies in London, the information concerning lectures for adults is a good deal thinner. William Whiston, Desaguliers, James Stirling, Peter Shaw, Benjamin Martin, James Ferguson, Dr. Demainbray, Adam Walker, Dr. Henry Moyes, Dr. Bryan Higgins, Frederick Accum, and several others, however, are known to have lectured to the general public on natural philosophy, chemistry, and similar subjects in the eighteenth century.[4] That knowledge is not so full as it is likely to become when the records are more fully investigated, yet it is already sufficient to demonstrate the interest shown in scientific lectures by members of all classes of society. John Harrison (1693–1776), of marine chronometer fame, is said to have 'got hold of a copy of a concise course of lectures on natural philosophy and started inventing mechanical instru-

---

[1] Hans, *op. cit.*, pp. 70–9.

[2] *Ibid.*, p. 93, quoting from an advertisement of 1766.

[3] *Ibid.*, p. 158. See above, p. 114, and below, p. 160, for the evening school of Henry Clarke in Salford and for that attended and later conducted by Charles Hutton in Newcastle.

[4] See above, pp. 37–42 and 51–9, for the scientific lectures given in London in the first half of the century. We shall be concerned here with the second half.

ments'.[1] Apothecaries, physicians, and surgeons regularly attended courses of lectures not only on medical subjects by such men as George Wallis, M.D., Frank Nicholls, M.D., William Hunter, Henry George Clough, M.D., William Smellie, James Douglas, M.D., and the Gresham Professor, Henry Pemberton, but also lectures in chemistry. For example, George Fordyce (1736–1802), M.D. Edinburgh 1758, a pupil of Cullen, 'settled in London in 1759 and began a course of lectures on chemistry which he continued for 30 years, materia medica and medicine being added in 1764'.[2] Fordyce became F.R.S. in 1776 and communicated several papers on chemistry, assaying, etc. to the *Philosophical Transactions.* He was a member of a philosophical society to which John Hunter, the surgeon, and William Small, member of the Lunar Society, also belonged.[3] Fordyce was also a friend of James Keir of the Lunar Society, until he attempted with his brother, Alexander Fordyce, the former banker, to obtain concessions from the government in order to manufacture alkali from salt, for which Keir had already developed a process.[4]

Bryan Higgins's school of practical chemistry in Greek Street, Soho, established in July 1774, was also the resort of both amateurs of science and of those professionally interested in its applications.[5] In 1775 he published *A Syllabus of Chemical and Philosophical Enquiries,* which was reprinted with slight changes in the following year. The edition of 1776 describes the syllabus as being 'composed for the Use of the Noblemen and Gentlemen who have subscribed to the Proposals made for the Advancement of Natural Knowledge'. A further syllabus published in 1778 is entitled *Syllabus of Dr. Higgins's Course of Philosophical, Pharmaceutical, and Technical Chemistry.* This course was detailed and very practical in its content, dealing with all the acids then used in industry and with the principal metals. Higgins, in fact, as Dr. Hans has pointed out, established a school of practical scientific research, attended by knowledgeable people, 'mostly physicians, practical chemists, manufacturers and the like, who wanted to improve their knowledge in research methods of chemistry'.[6] Indeed, London

[1] Hans, *op. cit.,* p. 151.
[2] J. R. Partington, *A History of Chemistry,* Vol. III, pp. 692–4. Partington adds that he has seen one set of MS. copies of Fordyce's lectures consisting of 745 pp. 8vo in 4 vols. They were on chemistry and delivered between Aug. 1784 and Dec. 1785. [3] J. Hunter to W. Small, 5 April 1771, A.O.L.B.
[4] See below, pp. 360–6.
[5] For Higgins's career, see *D.N.B.;* W. K. Sullivan, *Quarterly Journal of Medical Science,* N.S., Vol. VIII (1849), pp. 483–7; Hans, *op. cit.,* pp. 149–50; Partington, *op. cit.,* Vol. III, pp. 727–36; F. W. Gibbs, 'Bryan Higgins and his Circle', *Chemistry in Britain,* Vol. I (1965), pp. 60–5. Higgins graduated M.D. at Leyden in 1765.
[6] Hans, *op. cit.,* p. 150. Higgins attracted students from the provinces, such as young Thomas Henry, of Manchester (see below, p. 245).

I

lacking a university, Higgins played there, to some extent, the role of Dr. Black in Scotland, at Glasgow and Edinburgh Universities. Dr. Priestley attended his lectures in 1775 and also bought some chemicals for his own use from Higgins. The relationship led to Higgins accusing Priestley of stealing some of his discoveries, a charge indignantly repudiated by the latter.[1] Out of Higgins's classes also grew, in the 1790s, a 'Society for Philosophical Experiments and Conversations', presided over by Henry Seymour Conway and assisted in its experiments by Thomas Young.[2] Ambrose Godfrey and Charles Gomond Cooke, manufacturing chemists of Southampton Street, London, were subscribers to this society.[3] Higgins also applied his chemical knowledge to practical commercial purposes: he sold chemical and pharmaceutical preparations; he analysed mineral waters, etc.; he apparently analysed materials for Wedgwood (whose London showrooms were in Greek Street); he took out a very successful patent (1779) for cement, as a result of his researches (following those of Dr. Black) on 'calcareous' substances such as lime; he also, like Fordyce, patented a process (1781) for making synthetic soda from brine or sea-water; and in 1796–1801 he acted as a consultant in the West Indies on sugar and rum manufacture.

Bryan Higgins was assisted at his Greek Street Academy by his nephew, William Higgins,[4] as Henry Moyes was assisted by his nephew, William Nicol. Dynasties of lecturers in mathematics and science are by no means uncommon in the eighteenth century. Before coming to London to assist his uncle, William Higgins may have assisted Thomas Beddoes when he was Reader in Chemistry at Oxford.[5] William Higgins published in 1799 an *Essay on the Theory and Practice of Bleaching*, a subject on which he became an important authority in Ireland after his appointment as Professor of Chemistry and Mineralogy in Dublin in 1796.[6] He made a tour of factories in England in 1785 in the fashion of William Lewis, Samuel More, and others.

William Higgins's patron was his fellow-Irishman, Richard Kirwan (1733–1812), F.R.S., who during the decade 1777–87 resided at 11 Newman Street, London, and was familiar with most of the leading

[1] J. Priestley, *Philosophical Empiricism: containing Remarks on a Charge of Plagiarism respecting Dr. H——s interspersed with various Observations relating to different kinds of air* (1775).
[2] *Minutes of the Society for Philosophical Experiments and Conversations* (1795). See below, p. 166, for Young's earlier attendance at Higgins's lectures.
[3] T. S. Wheeler and J. R. Partington, *The Life and Work of William Higgins, Chemist (1763–1825)* (1960), pp. 36–7, n. 53. For Godfrey and Cooke's earlier links with the famous Robert Boyle, see above, pp. 56–7.
[4] See Wheeler and Partington, *op. cit.*, for the career of William Higgins. See also below, pp. 185–7.   [5] *Ibid.*, p. 4.   [6] See below, pp. 320–1 and 327.

scientists of the day, both in Great Britain and in the world at large.[1] To his house came many industrialists, scientists, and members of the nobility. He numbered Dr. Priestley, James Watt, Matthew Boulton, John Whitehurst, and James Keir among his friends, frequently visited Birmingham and may have attended Lunar Society meetings. He was in addition the chairman of the philosophical society meeting at the Chapter Coffee House in London.[2]

The meetings of scientists taking place in private houses are a feature of London society in the second half of the eighteenth century. Besides Kirwan's house, there was also Sir Joseph Banks's home, and we are told that: 'His collections were freely accessible to all scientific men of every nation, and his house in Soho Square became the gathering-place of science.'[3] William Higgins visited Banks's house about April 1785 to show his method of reduction of nitric acid to ammonia by means of tin.[4] Joshua Gilpin, an American paper-manufacturer visiting England, was offered an introduction to a meeting at Banks's.[5] Both Boulton and Watt visited Soho Square and corresponded frequently with him in his private capacity and as President of the Royal Society. He was consulted on mechanical matters and gave his verdict against James Keir's copper alloy bolts for fixing the sheathing to the hulls of Royal Navy warships.[6] Banks was a great clubman: elected in 1766 a Fellow of the Society of Antiquaries, he also became in 1774 a member of the convivial Society of Dilettanti;[7] he was an officer of nearly a dozen scientific societies, a member of the Turks Head Tavern Club of Dr. Johnson and Sir Joshua Reynolds, which came to be known as the Literary Club, and many other societies. No scientific gathering was complete without him, and no social gathering with him was completely unscientific. Thus Mrs. Montagu, the blue-stocking, wrote to a

---

[1] For Kirwan's career, see *D.N.B.*; J. Reilly and N. O'Flynn, *Isis*, Vol. XIII (1930), pp. 298–319; W. H. Brindley, *Nature*, Vol. CXXII (1933), pp. 957–8; P. J. McLaughlin, *Studies*, Dublin, Vol. XXVIII (1939), pp. 28, 461, 593, and Vol. XXIX (1940), pp. 29, 71, and 281; Partington, *op. cit.*, Vol. III (1962), pp. 660–71; Wheeler and Partington, *op. cit.*, pp. 30–3, where other references are given.
[2] See also below, p. 186, for Kirwan's contributions to textile chemistry in Ireland.   [3] *D.N.B.*   [4] Wheeler and Partington, *op. cit.*, p. 30, n. 17.
[5] Journals and Notebooks, 1790–1801, 1830–3, of Joshua Gilpin (1765–1840), Division of Public Records, Harrisburg, Pennsylvania (microfilm in the possession of Friends' House, London): entry under 8 Jan. 1796 in Vols. VI and VIII of Journal for 5 Dec. 1795–11 Jan. 1796 and March 13–20, 1796.
[6] J. Keir to M. Boulton, 15 Jan. 1780, A.O.L.B. Banks denied that he had condemned the metal. See Banks to Keir, 12, 14, and 17 Feb. 1780, Archives of the Royal Society.
[7] J. Evans, *A History of the Society of Antiquaries* (Oxford, 1956), pp. 93 and 149.

friend: 'To convince you that my health is not impaired I will inform you that the next day I did the honours to a large party composed of Sir Joseph Banks, the Bishop of Llandaff and all our distinguished Philosophers.'[1] But the centre of his scientific life was his house in Soho Square,[2] described by his biographer in the following terms:

Every morning the library, the fine reception rooms and the Museum were open to visitors. Everywhere bright fires burnt; on polished tables lay the latest works on scientific subjects . . . Once a week, on Thursdays, he held a formal reception, at which some particular subject, selected beforehand, was demonstrated or discussed. Distinguished foreigners, Jussieu, perhaps, or Broussonnet, Camper, the younger Linnaeus, Cuvier or Fabricius mingled with their English fellow-workers Cavendish, Priestley, Wollaston, Maskelyne, John Hunter, John Ellis and the rest. There too were to be found the explorers and travellers in the cause of science . . . and with them other men of all sorts and conditions, agriculturists, engineers, archaeologists, men of letters . . .[3]

Banks's correspondence reflects the variety of his acquaintance.[4] Though the majority of his correspondents were pure scientists, a considerable minority were cartographers, mathematicians, instrument-makers, engineers, apothecaries, etc. Many of Sir Charles Blagden's letters to him are concerned with industrial processes;[5] Boulton corresponds with him about coining and steam engines; the borough-reeve of Manchester about bleaching; Henry Seymour Conway about the manufacture of saltpetre; John Rennie about locks, canals, docks, and water supply, and so the story could be continued. Besides the

---

[1] Quoted by H. Cameron, *Sir Joseph Banks K.B., P.R.S., The Autocrat of the Philosophers* (1952), p. 76. The Bishop of Llandaff was Richard Watson, previously Professor of Chemistry at Cambridge University (see below, pp. 167–171).

[2] In the 1790s meetings of scientists took place at the home of Maxwell Garthshore, M.D., F.R.S., 'every Saturday evening at 8 o Clock . . . whilst Sir Joseph Banks is in the country but at his house when he is at home': entry for 31 Aug. 1793 in the Diary of Robert Jameson. See J. M. Sweet, 'Robert Jameson in London, 1793', *Annals of Science*, Vol. XIX (June 1963), pp. 81–116.

[3] Cameron, *op. cit.*, pp. 126–7. See Faujas de St. Fond, *Travels through England and Scotland to the Hebrides in 1784* (revised edition of the English translation, 2 vols., Glasgow, 1907), Vol. I, pp. 2–3.

[4] W. R. Dawson (ed.), *The Banks Letters* (1958).

[5] The Blagden Papers in the possession of Mr. James Osborn, Yale University, also reveal Blagden's industrial interests. In July 1785, for example, Blagden visited ironworks in Glamorganshire. He was accompanied by the Hon. Henry Cavendish and the objects of the expedition were partly to verify some barometric researches but principally to study methods of iron manufacture and the geology of mining. In May 1791 he was asking his brother, John Blagden, to send him information about Homfray's process of iron manufacture. He also took steps to send another brother, R. Blagden, who was engaged in a dyeing business, 'the translation of Berthollet's Elements of Dying' (Charles Blagden to John Blagden, 11 Nov. 1791).

meetings at his own house, Banks presided over the Royal Society Club, the best contemporary description of which exists in Faujas de St. Fond's memoirs.[1] The club originated in October 1743; similar meetings for dining and talk were held from the beginning of the Royal Society itself.[2] Both Pepys and Robert Hooke record in their diaries attendances at 'the Society's Club'. Edmund Ward in *The Secret History of Clubs* (1709) refers to the Virtuoso Club of the Royal Society meeting in 1701, but by 1709 consisting of a few men meeting informally. Meetings took place from 1743 to 1780 at the Mitre Tavern in Fleet Street and then moved to the Crown and Anchor in the Strand, where it continued to meet until 1847. Joshua Gilpin recorded in his diary a meeting which he attended there in 1796, when the club consisted of '40 Members of the Society elected by ballot and [who] meet regularly to dine on the day of the Royal Society's meetings'.[3] Many men of technical and scientific skill who were not Fellows of the Royal Society attended this club as visitors.

The houses of professional lecturers also were naturally enough often used as the venue for their lectures. Thus Dr. Frank Nicholls charged forty guineas for a series of thirty-nine lectures delivered at his home on the corner of Lincoln's Inn Fields.[4] Adam Walker took a house in George Street, Hanover Square, 'where he has read every Winter to numerous and genteel audiences'.[5] He shared his room with Dr. Henry Moyes, and the house was also used for meetings of the Chapter Coffee House Society.[6] It is interesting to note that John Whitehurst, another member of that Society, died at his house in Bolt Court, Fleet Street, where James Ferguson had previously lived and died.

More information has recently come to light about the philosophical society meeting sometimes at the Chapter Coffee House, sometimes at the Baptist's Head Coffee House and sometimes at Adam Walker's house in the 1780s.[7] It can now be clearly seen that this is one and the

[1] *Op. cit.*, Vol. I, pp. 46–9.
[2] Sir Henry Lyons, F.R.S., *The Royal Society, 1660–1940* (Cambridge, 1944), pp. 170–2.
[3] Joshua Gilpin, *op. cit.*, Vols. VI and VIII, under 17 March 1796.
[4] J. Kobler, *The Reluctant Surgeon, The Life of John Hunter* (1960), p. 41.
[5] The *European Magazine and London Review*, June 1792.
[6] Minute Book, Gunther MS. 4, Oxford Museum for the History of Science. See also Forbes MSS., Register House, Edinburgh, letter addressed to Dr. Garthshore and 'Received 4 April 1783': 'Dr. Moyes Lectures at Plaisterers Hall, Aldermanbury on Mondays, Wednesdays and Fridays at seven in the Evening—at Walker's Rooms, Hanover Square, Tuesdays, Thursdays, Saturdays at Twelve o'c Noon'. A Note from E. C. Maitland to Forbes, dated Thursday 17 April 1783, invites him to dinner at 4 o'clock on the following Monday to meet Moyes at King's Arms Yard. Blagden also attended Moyes's lectures.
[7] Gunther MS. 4.

same society, not two or more societies as has sometimes been suggested by earlier writers. References to it occur in both the Boulton and Watt correspondence and the Wedgwood correspondence.[1] John Playfair also attended it as a visitor when he was introduced by Benjamin Vaughan.[2] Boulton, Watt, Wedgwood, Priestley, and James Keir from the midlands were honorary members, but the hard core of the Society were London residents, such as Richard Kirwan, John Whitehurst, J. H. de Magellan, Edward Nairne, William Nicholson, Benjamin Vaughan, and Adam Walker. The minute-books of the Society extend from 1782 to 1787 and describe the discussions and experiments entered into by the members, ranging from the properties of electricity to the analysis of common air. Physicians, instrument-makers, industrial entrepreneurs, and gentlemen with scientific interests all met in this important society, and correspondence from scientists in other parts of Britain and on the continent was frequently read out and discussed. J. H. Magellan (1722–90), F.R.S., is a particularly interesting member of this society since his presence demonstrates the relationships established with continental scientists. An expert on mathematical instruments, he was a disseminator and popularizer of scientific and technical information, and possibly an industrial spy. He was known to Benjamin Franklin, Lavoisier, Priestley, Watt, and indeed to almost every distinguished scientist or engineer of the period. Besides publishing several pamphlets, he appears to have made scientific instruments and sold them to continental buyers.[3] He is credited with being the first to use the term 'specific heat' (*chaleur specifique*).[4]

The coffee-houses from their beginnings at the Restoration were the meeting-places of scientific clubs in London.[5] Dr. Benjamin Franklin's Club of Thirteen, which included Wedgwood and his partner, Thomas Bentley, Daniel Solander, Banks's librarian, R. E. Raspe, the geologist and author of *Baron Munchausen*, and John Whitehurst, met at Old Slaughter's Coffee House.[6] R. L. Edgeworth was introduced by James Keir to a scientific club which met at Jack's Coffee House and later Young Slaughter's Coffee House, and which included Sir Joseph Banks, Solander, Sir Charles Blagden, Dr. George Fordyce,

---

[1] See E. Robinson, 'The Origins and Life-Span of the Lunar Society', *University of Birmingham Historical Journal*, Vol. XI, no. 1 (1967), p. 8, quoting Valentine Gardner, M.P., to M. Boulton, 26 July 1803, A.O.L.B., and Wedgwood's 'Commonplace Book 1', p. 158, Wedgwood MSS., Barlaston.

[2] John Playfair, *Works* (Edinburgh, 4 vols., 1822), Vol. I, p. xxxv.

[3] See Taylor, *Math. Pract. of Hanov. England*, pp. 210–11; Partington, *op. cit.*, Vol. III, pp. 248–9.

[4] Partington, *op. cit.*, p. 156.   [5] See above, pp. 57–8.

[6] See E. Robinson, 'R. E. Raspe, Franklin's Club of Thirteen, and the Lunar Society', *Annals of Science*, Vol. XI (1955).

John Smeaton, Jesse Ramsden, and others.[1] Matthew Boulton attended a London philosophical society in 1766 which included Whitehurst, James Keir, Wedgwood, Bentley, and others.[2] It is not always possible to separate one club from another because it was common for interested persons to be members of several different clubs. But what is common to most of these clubs is a membership that includes industrialists, engineers, physicians, and instrument-makers, besides aristocratic amateurs of science.

Just as the Royal Society Club met at Garaway's Coffee House in Cornhill in the 1670s, so the Society for the Encouragement of Arts, Commerce and Manufactures had its formal origins at Rawthmell's Coffee House, Henrietta Street, Covent Garden, on 22 March 1754. Rawthmell's was favoured by Dr. Richard Mead, F.R.S., and other Fellows of the Royal Society, so that in this respect there is a close link between the Royal Society and the Society of Arts. Four of the eleven founder-members of the Society of Arts were also Fellows of the Royal Society: Dr. Stephen Hales, Henry Baker, Gustavus Brander, James Short.[3] Other founder-members included Viscount Folkestone, Lord Romney, John Goodchild, Nicholas Crisp, watch-maker, Charles Lawrence, Husband Messiter, a surgeon, and William Shipley, a drawing master. The early membership was remarkable for its distinction and its variety, but for the purpose of establishing the link between the Society and the growing technology of the period it is only necessary to name Joshua Ward, the pioneer manufacturer of sulphuric acid; Professor John Robison, Watt's friend; Peter Shaw, the chemist; Robert Mylne, architect and engineer; Captain Thomas Desaguliers, Dr. J. T. Desaguliers' son; Thomas Yeoman, the engineer; Robert Dossie; and John Dollond, the optical instrument-maker. But even men whose fame rested in quite other fields were interested in the Society's practical work. Thus Dr. Kippis testified that he 'once heard Dr. Johnson speak in the Society of Arts and Manufactures, upon a subject relative to mechanics, with a propriety, perspicuity and energy,

---

[1] *Memoirs of R. L. Edgeworth Esq., begun by himself and concluded by his Daughter, Marie Edgeworth* (2 vols., 1820), Vol. I, p. 188.

[2] M. Boulton to Mrs. A. Boulton, 23 Oct. 1766, A.O.L.B.

[3] Sir Henry Trueman Wood, *A History of the Royal Society of Arts* (1913), p. 12; and D. Hudson and K. W. Luckhurst, *The Royal Society of Arts, 1754–1954* (1954), p. 6. Whilst our book was in the press, Mr. D. G. C. Allan produced his life of *William Shipley Founder of the Royal Society of Arts* (1968), which provides interesting new evidence, including references to links between Shipley, Dr. Stephen Hales, and Thomas Yeoman (see below, chap. XI. Mr. Allan also mentions an earlier effort, in 1722, at founding a 'Chamber of Arts, For the Preserving and Improvement of Operative Knowledge, the Mechanical Arts, Inventions; and Manufactures', and in 1738 'A Proposal for the Encouragement of Arts and Sciences by the Royal Society', both foreshadowing the Society of Arts.

which excited general admiration'.[1] As for the general work of the Society in promoting technical advances and technical education, it has been well described in the two standard histories of the Royal Society of Arts, but one can safely say that there was hardly a single industry in England untouched by the Society's endeavours. Much of the early work of the Society came to be known from publication in the *Gentlemen's Magazine* and other periodicals; then the *Museum Rusticum et Commerciale*, appearing monthly from 1764 to 1766, published parts of the Society's proceedings. When this failed, Robert Dossie began his *Memoirs of Agriculture and other Œconomical Arts*, which appeared in three volumes in 1768, 1771, and 1782. (Arthur Young claimed to have originated the idea of publishing the *Transactions* in 1783.) Professor Schofield's comment on these activities is that 'However influential the Society of Arts may have been in encouraging (and spreading information about) technological improvements, it cannot seriously be maintained that there was much of a scientific nature in its work.'[2] But it is difficult to see a different set of principles at work in the Lunar Society from the Society of Arts, and what may be claimed for the one may equally well be claimed for the other. Too narrow a definition of science may lead the historian to disbelieve his own eyes everywhere in England in the eighteenth century, whether he be contemplating a meeting of the Royal Society or a lecture in Adam Walker's parlour. C. C. Gillispie's evaluation of the place of science in these social activities is more helpful. In his view, 'when scientists turned to industry, it was to describe the trades, to study the processes and to classify the principles . . . the eighteenth-century application of science to industry then, was little more and nothing less than the attempt to develop a natural history of industry.'[3] His quotation from Berthollet is particularly apt:

We are frequently able to explain the circumstances of an operation which we owe entirely to a blind practice, improved by the trials of many ages; we separate from it everything superfluous; we simplify what is complicated; and we employ analogy in transferring to one process what has been found useful in another. But there is still a great number of facts which we cannot explain, and which elude all theory: we must then content ourselves with detailing the processes of the art; not attempting idle explanations, but waiting till experience throw greater light upon the subject.[4]

[1] Hudson and Luckhurst, *op. cit.*, p. 29.
[2] R. E. Schofield, 'The Industrial Orientation of Science in the Lunar Society of Birmingham', *Isis*, Vol. 48, no. 154 (Dec. 1957), p. 408.
[3] C. C. Gillispie, 'The Natural History of Industry', *ibid.*, p. 405.
[4] Berthollet, *Elements of the Art of Dyeing* (Edinburgh, 1792), p. 17, quoted by Gillispie, *op. cit.*, p. 405.

As Gillispie explains, 'the eighteenth-century scientists do not write of the application of theory. What they say is that science illuminates the arts, that it enlightens the artisans.'[1] In this sense both the Society of Arts and the Lunar Society demonstrate the operations of science.

The same purpose can be observed in the pamphlet published by Benjamin Thompson, Count Rumford, in 1799, containing:

Proposals for forming by subscription, in the Metropolis of the British Empire, a Public Institution for diffusing the knowledge and facilitating the general introduction of Useful Mechanical Inventions and Improvements, and for teaching, by courses of Philosophical Lectures and Experiments, the Application of Science to the Common Purposes of Life.

Thompson intended to create 'relations between philosphers and work-men' directed to 'the improvement of agriculture, manufactures, commerce, and the augmentation of domestic comforts'. The result was the Royal Institution, whose first formal meeting of managers was held, appropriately enough, at the house of Sir Joseph Banks on 9 March 1799. The founder intended that there should be 'an industrial school for artisans; a collection of models of fireplaces, grates, stoves, steam-engines, spinning wheels, etc.; a professor was to be appointed and provided with a well-equipped lecture room; and a convenient club with a restaurant and school of cookery was to be included'.[2] Thomas Garnett was appointed to the Chair of Chemistry and Natural Philosophy in 1799, but resigned in 1801. In that year Humphry Davy was first appointed assistant lecturer then lecturer in chemistry, and finally, in May 1802, Professor of Chemistry. Thomas Young was appointed Professor of Natural Philosophy in August 1801 but resigned in 1803, since his lectures were not successful with the majority of the audience to which they were directed, although of great value to informed scientists. The success of the Royal Institution was primar-ily due to Davy, whose lectures became popular social occasions in London attracting very large audiences.[3]

After Thompson's departure in 1802 the school for mechanics, the

---

[1] Gillispie, *op. cit.*, p. 404.

[2] A. Wood, *Thomas Young, Natural Philosopher, 1773–1829* (Cambridge, 1954), p. 118.

[3] See J. Kendall, *Humphry Davy* (1954), and G. A. Foote, 'Sir Humphry Davy and his Audience at the Royal Institution', *Isis*, Vol. 43, no. 131 (April 1952), pp. 6–12. Also Thomas Martin, *The Royal Institution* (1942). Lives of Davy include J. A. Paris, *The Life of Sir Humphry Davy* (1831), J. Davy, *Memoirs of the Life of Sir Humphry Davy* (2 vols., 1836), and T. E. Thorpe, *Humphry Davy, Poet and Philosopher* (1896). For a recent study, see Sir H. Hartley, *Humphry Davy* (1966).

workshops and kitchens were dropped and everything depended on the lectures and on the *Journal* which was to be

exclusively devoted to the diffusion of the knowledge of new and interesting scientific discoveries and of useful improvements in mechanics, arts and manufactures; and particularly of making known all such new inventions and contrivances as tend to facilitate labour and render it more productive; to promote domestic economy and increase the conveniences, comforts and enjoyments of life.[1]

The practical slant of the lectures was testified to by Lady Holland at the same time as she criticized the Institution: it was, she said, 'a bad imitation of the *Institut* at Paris; hitherto there is only one Professor, who is a jack-of-all-trades, as he lectures alike upon chemistry and ship-building'.[2] Others, like the young Robert Peel, who attended the natural science lectures in the 1805 season, may have been more sympathetic.[3] Certainly every effort was made by the Managers to provide lectures that would be useful to the tradesmen of London and other parts. Thus in July 1801 they passed the following resolutions:

Resolved—That a Course of Lectures on the Chemical Principles of the Art of Tanning be given by Mr. Davy. To commence the second of November next; and that respectable persons of the trade shall be recommended by Proprietors of the Institution, to be admitted to these lectures gratis.

Resolved further—That Mr. Davy have permission to absent himself during the months of July, August, and September, for the purpose of making himself more particularly acquainted with the practical part of the business of tanning, in order to prepare himself for giving the above-mentioned course of lectures.[4]

A Syllabus of A Course of Lectures on Chemistry[5] delivered at the Royal Institution by Davy in 1802 has one section concerned with 'the Chemistry of the Arts', with lectures on Agriculture, Tanning, Bleaching, Dyeing, Metallurgy, Glass and Porcelain, Preparation of Food and Drink, and the Management of Heat and Light artificially produced. When, on the 21 January 1802, Davy delivered his introductory lecture he declared that:

in consequence of the multiplication of the means of instruction, the man of science and the manufacturer are daily becoming more assimilated to each other. The artist, who formerly affected to despise scientific principles, because he was incapable of perceiving the advantages of them, is now so far enlightened as to favour the adoption of new processes in his art, wherever they are evidently

[1] Minute of the Managers, 31 March 1800.
[2] Earl of Ilchester (ed.), *The Journal of Elizabeth, Lady Holland* (1791–1811) (2 vols., 1908), Vol. II, p. 52, cited by Foote, *op. cit.*, p. 9.
[3] C. S. Parker, *Sir Robert Peel* (2 vols., 1899), Vol. I, p. 16, cited by Foote, *op. cit.*, p. 9.   [4] Kendall, *op. cit.*, p. 65.
[5] There is a copy of this syllabus in the University of Manchester Library.

connected with a diminution of labour; and the increase of projectors, even to too great an extent, demonstrates the enthusiasm of the public mind in its search after improvement.[1]

Though the impression generally created of Davy's lectures at the Royal Institution is that they were given before an upper-class audience of notable figures in society, one wonders how far the tradesmen of London also frequented them. Michael Faraday, destined in 1813 to become Davy's assistant, certainly attended Davy's lectures at the Royal Institution as well as John Tatum's lectures at the City Philosophical Society while still a humble apprentice to G. Rieban, a bookbinder and bookseller at 2 Blandford Street, London.[2]

In speaking of Faraday we are already beginning to stray beyond the broad chronological limits that we have set to our inquiry, but Faraday clearly continues the tradition of 'self-education' in science which was the hall-mark of the early nineteenth as well as of the eighteenth century. In this respect Faraday differs very little from Humphry Davy, John Dalton, and William Herschel, or from Watt, Priestley, and Wedgwood. Like his earlier counterparts he attended lectures, made his own experiments with equipment of his own devising, conversed and corresponded with like-minded friends, and above all read the books that were available and visited the technological marvels of his time. His master, G. Rieban, left on record how the young Faraday spent his time:

After the regular hours of business, he was chiefly employed in Drawing and Copying from the Artists Repository a work published in numbers which he took in weekly—also Electrical Machines from the Dictionary of Arts and Sciences and other works which came in to bind . . . He went on early walks in the morning visiting always some Works of Art or searching for some Mineral or Vegetable curiosity—Holloway Water Works, Highgate Archway, W. Middlesex Water Works—Strand Bridge—Junction Water Works, etc. etc. Sketching the Machinery, Calculating the force of Steam Engines etc. his mind ever engaged . . . he much admired Henry's Chemistry in 4 Vols. he bought and interleaved great part of it, occasionally adding Notes with Drawings and Observations.[3]

We know also that Faraday was greatly influenced by Mrs. Marcet's *Conversations on Chemistry* and by the scientific articles of the *Encyclopaedia Britannica*.[4] Mrs. Marcet had attended Davy's lectures and got

[1] Quoted by Paris, *op. cit.*, p. 88.

[2] L. Pearce Williams, 'Michael Faraday's Education in Science', *Isis*, Vol. 51, no. 166 (Dec. 1960), pp. 515–30.

[3] Quoted by L. P. Williams, *op. cit.*, p. 516. See below, pp. 247–8, for W. Henry's *Chemistry*.

[4] H. Bence-Jones, *Life and Letters of Faraday* (2nd edn., 2 vols., 1870), Vol. I, p. 11.

much of her material from him, while the *Britannica's* articles had often been written by men like John Robison who combined a knowledge of the best science of their time with a real familiarity with industrial practice. The boyhoods of Faraday and Davy have similarities when it comes to intellectual experience. Davy's earliest biographer relates how Davies Gilbert accompanied the young Davy to Hayle Copper House and there introduced him to Dr. Edwards, later the chemical lecturer in the school of St. Bartholomew's Hospital. Dr. Edwards had a fine laboratory and library which entranced Humphry Davy and helped to confirm him in his scientific enthusiasms.[1] Davies Gilbert, Bingham Borlase, Dr. Edwards, Gregory Watt, Thomas Beddoes—such were the friends of Davy's youth and they represent many of the typical enthusiasts for science and technology in their day, though none of them were in the first ranks of scientific or technical fame. All these friends Davy acquired in the west country in the 1790s. Fifteen years later, in the metropolis itself, much wider opportunities were available to those who had faith in the unity of science and industry.

This unity, as we have previously observed, was particularly evident among the scientific instrument-makers and clock-makers of London.[2] The links between scientists and such craftsmen, illustrated by the example of Boyle and Tompion in the earlier Scientific Revolution, were strengthened in the eighteenth century. We have already referred to the associations of the two Hauksbees, William Vream, and other instrument-makers with scientists such as Desaguliers, Whiston, and Shaw in the early eighteenth century. Similar examples could be multiplied in the later part of the century: the names of Graham, Sisson, Dollond, Harrison, Ramsden, Adams, Bird, and Nairne became illustrious in that period.[3] They and many others continued both to provide instruments for the scientists with whom many of them associated in the Royal Society and also to further the diffusion of scientific knowledge by their own publications and lectures. As we shall see, moreover, instrument-making and clock-making made very important contributions to the early development of mechanical engineering.[4] Unfortunately, no detailed study has yet been made of the development of engineering in London during the Industrial Revolution, but there is no doubt that the metal-working tools—the lathes, drills, gear-cutting engines, etc.—of these London craftsmen were fundamental in engineering evolution. Some of the well-known London engineering firms, such as Holtzappfel, originated from such

---

[1] Paris, *op. cit.*, p. 34.   [2] See above, pp. 13–25, 28, 37, 40–2.
[3] See Hans, *op. cit.*, pp. 150–2, and Taylor, *Math. Pract. of Hanov. England,* chap. V.   [4] See below, chap. XIII.

tool-making activity, with its mathematical precision.[1] London, like Lancashire, was able to build a powerful engineering industry on this foundation—indeed London was a reservoir of skilled craftsmanship on which the early provincial engineering firms drew very heavily.[2]

The development of chemical manufactures in London has been similarly neglected, but again there is evidence of links between these and the scientific activities which we have been surveying. The chemical-industrial interests of Shaw, Lewis, and Dossie in the early and middle eighteenth century were continued by Fordyce, Higgins, Kirwan, Davy, and others in the later eighteenth and early nineteenth centuries. As in engineering, however, little research has been carried out into chemical manufacturing firms of the metropolis. We have previously noted the alkali patents of Fordyce and Higgins.[3] John Collison, of Battersea, who took out another in 1782, was a chemical manufacturer with some scientific knowledge; he corresponded with Dr. Black regarding analysis of his alkali.[4] We also know that he provided technical assistance to the firm of Livesey, Hargreaves & Co., of Manchester, in the early application of chlorine bleaching;[5] Thomas Henry said that he 'appeared well informed on Chemical subjects'.[6] Also in the field of textile chemistry, Dr. Edward Bancroft, F.R.S., of London, was renowned both for his chemical knowledge and also for his discoveries in dyeing; his *Experimental Researches concerning the Philosophy of Permanent Colours* (1794, 1813) was a classic work.[7]

The development of sal-ammoniac (ammonium chloride) manufacture in London and elsewhere in the eighteenth century has recently been traced.[8] Unfortunately, little could be discovered about most of the individual manufacturers engaged in it. Professor Multhauf points out that 'virtually all of the chemical data applying to the various processes for producing sal ammoniac can be found in the scientific literature prior to the effective foundation of the European sal ammoniac industry', and he considers it 'very probable that the obscure men who were primarily responsible for the success of that industry were beneficiaries of the literature of popular science which flourished in the mid-eighteenth century'.[9] But he is inclined to think that they

[1] See above, pp. 63 and 78, for Holtzappfel.
[2] See below, pp. 437-8, for the examples of Boulton & Watt and various Manchester engineers.
[3] See above, pp. 121 and 122. See also below, pp. 360-6.
[4] Clow, *op. cit.*, p. 98. See also below, p. 365.   [5] See below, p. 288.
[6] T. Henry to J. Watt, 25 Feb. 1788, Dol.   [7] See below, pp. 340 and 349.
[8] R. P. Multhauf, 'Sal Ammoniac: A Case History in Industrialization', *Technology and Culture*, Vol. VI (1965), pp. 569-86. Sal ammoniac was at this time extensively used by dyers, braziers, and tinplate makers (Clow, *op. cit.*, p. 246).   [9] Multhauf, *op. cit.*, p. 581.

were characterized by their manufacturing 'ingenuity and a spirit of enterprise', rather than by their scientific knowledge. Professor Multhauf, however, appears to have underestimated their scientific attainments. William Menish, for example, one of the 'obscure inventors' whom he briefly mentions as having taken out a patent for sal-ammoniac manufacture by the sulphate process in 1792,[1] was apparently a graduate of Glasgow University and attended Dr. Black's lectures there in the winter of 1760–1.[2] He was a son of William Menish, a surgeon of Wellington, Somerset, and his younger brother, Henry, graduated M.D. at Glasgow in 1763. Both Dr. Black and James Watt refer to William Menish junior as Dr. Menish; Watt stayed on one occasion at his house at Harrow. Either he or his younger brother, Henry, practised as a surgeon and apothecary in Chelmsford (*Medical Register*, 1780). William Menish is listed in the *Universal British Directory* (1791) as 'sal-ammoniac manufacturer, Whitechapel Road', and in *Holden's Directory* (1805) as 'Menish, J.[sic], hartshorn manufactory, 111 Whitechapel Road'. He died on 6 April 1813 at Bromley, Kent, in his seventy-ninth year.[3] Clearly he should be associated, by virtue of his scientific background, with professional men of science such as Dr. Hutton who engaged in sal-ammoniac manufacture.[4]

So, too, should another of the 'obscure men' whom Professor Multhauf mentions, Hector Campbell, who is referred to as 'Dr.' by Samuel Parkes[5] and who was also apparently a Scottish medical graduate. He was able to apply his chemical knowledge to papermaking, taking out a patent for an improved method of bleaching rags with 'oxygenated muriatic acid' (chlorine) in 1792 (no. 1922).[6] In the

[1] Multhauf, *op. cit.*, p. 577.　　[2] J. Black to J. Watt, 15 March 1780, Dol.
[3] *Gentlemen's Magazine*, Vol. 83, pt. i (1813).
[4] Dr. James Hutton (1726–97), of course, the famous Scottish geologist, who had studied medicine at Edinburgh, Paris and Leyden universities, is a well-known example of the links between science and industry, especially in his sal-ammoniac factory, established in 1756. So, too, are the Earl of Dundonald and Charles Macintosh, who were also pioneers in this manufacture. (See, in addition to Multhauf's article, Clow, *op. cit.*, pp. 246 and 419–23, for their development of sal-ammoniac manufacture in Scotland.) It seems very probable, from what we know of the relationships between Dr. Black and these pioneers of Scottish chemical manufactures, that Professor Multhauf is incorrect in stating that Black 'did not know how they produced sal ammoniac'. Professor Multhauf himself refers to the great secrecy surrounding this manufacture, and we know that Black, who was a particularly close friend of Hutton, was not the sort of man to betray confidences (cf. below, pp. 365–6 and 369).
[5] Parkes, *Chemical Essays* (2nd edn., 1823), Vol. I, p. 451.
[6] *Repertory of Arts and Manufactures*, Vol. I (1794), pp. 156–62; *Abridgments of Patent Specifications* (Paper, &c., 1858). He referred to this method as 'invented by the French'. See below, p. 334, n. 3.

patent specification he described himself as a 'Chemist', of Bermondsey; he had started this manufacture in the previous year. His firm appears in the 1796 supplement to the *British Trade Directory* as 'Campbell, Hector & Co., Chymists, 27 Horsleydown Lane', Bermondsey. Later, from 1806 onwards, he appears as Hector Campbell, M.D., of 17 St. Peter's Hill, Doctors Commons.[1] Two papers by Campbell, who signed himself 'M.D.', of Fleet Street, London, clearly demonstrate his application of chemical knowledge to industrial processes.[2] In the first, comparing paper-making in England and France, he refers to the works of Scheele, Berthollet, and Chaptal on bleaching with 'oxymuriatic acid' and mentions having himself bleached rags by this method in the early 1790s. The second paper is concerned with Campbell's method of producing not only sal ammoniac (muriate of ammonia), but also sulphate of soda and carbonate of magnesia, from a mixed solution of sulphate of magnesia, carbonate of ammonia, and muriate of soda, 'principally by the Agency of Frost'. This was evidently the method referred to by Parkes, which proved commercially unsuccessful, but Campbell's paper clearly demonstrates his very considerable knowledge of contemporary chemical science and his experience in chemical analysis and experiment.

Similar links between science and industry are discernible in other branches of chemical manufacture in London. The refining of platinum, for instance, usually credited to the famous analytical chemist, William Hyde Wollaston, was apparently first developed by Thomas Cock, who, as an amateur chemist and metallurgist, first succeeded in working the metal in William Allen's laboratories at Plough Court and later showed his method at Allen's request to Dr. Wollaston.[3] Wollaston, moreover, collaborated in the industrial development of this process with Samuel Parkes, of the Haggerston Chemical Works, London, notable for his *Chemical Essays* (5 vols., 1815).[4] Like earlier chemists, such as Shaw, Lewis, and Dossie, Wollaston took a direct and practical interest in industrial processes. 'Each summer, when activities at the Royal Society were suspended, he was in the habit of taking an extended trip in Britain or abroad; he was particularly interested in seeing

[1] Information kindly supplied by the Guildhall Library, London.
[2] Nicholson's *Journal of Natural Philosophy*, May 1802, pp. 6–9, and June 1802, pp. 117–20.
[3] M. E. Weeks, 'The Chemical Contributions of William Allen', *Journal of Chemical Education*, Vol. 35 (1958), p. 70. For a recent assessment of Wollaston, see P. T. Hinde, 'William Hyde Wollaston: The man and his "equivalents" ', *ibid.*, Vol. 43 (1966), pp. 673–6. We shall deal with William Allen below, pp. 137–8.
[4] Hinde, *op. cit.*, p. 674. This process, perfected about 1805, is said to have yielded Wollaston £30,000.

machinery and manufactures, or in making geological observations.'[1]

Samuel Parkes, on the other hand, himself a manufacturing chemist, took a profound interest in applied chemistry, in which, as his *Chemical Essays* amply demonstrate, he was immensely knowledgeable.[2] Not merely was he widely read in the works of all leading chemists, British and foreign, but he, too, had 'for many years been in the habit of visiting the principal manufactories of these kingdoms',[3] while, at the same time, he had frequently been engaged at his own chemical works in 'experimental investigations'. He confirmed that the publications of chemists such as Watson, Bergman, and Scheele, relating as they did to many industrial processes and materials, had 'contributed in no small degree to the information of the public mind, and to that growing taste for chemical pursuits which is one of the characteristics of the present age'.[4] His own *Essays* were chiefly intended to inform 'the chemist or artisan', with a view 'to the improvement of the manufactures of the kingdom', and the first chapter was devoted to 'the importance and utility of chemistry'.[5] He emphasized that almost all manufactures were 'dependent upon chemistry for their improvement and successful practice', as he proceeded to demonstrate in copious detail.

In the pharmaceutical manufactures of London, we have seen how the firm of Godfrey and Cooke were early associated with Robert Boyle, and in the later eighteenth century with Bryan Higgins.[6] It seems reasonable to suppose that they were able to develop useful applications of the chemical knowledge they acquired. Another more

---

[1] Hinde, *op. cit.*, p. 674. See below, pp. 170 and 173–4, for the similar interests of the Cambridge professors Richard Watson and William Farish.

[2] Parkes, a zealous Unitarian, had been a pupil of Dr. Stephen Addington, a well-known Dissenting minister and educationalist, at whose school 'modern' subjects were taught. Hans, *op. cit.*, p. 61.

[3] Preface to the *Chemical Essays*.

[4] *Ibid.* It is interesting to note how Dr. Edward Cullen, who translated the *Physical and Chemical Essays* (1784) of the Swedish chemist, Törbern Bergman, strongly emphasized Bergman's concern that his researches should be 'immediately applicable to the uses of life. He despises barren and unproductive speculation; he considers science as desirable only so far as it is beneficial . . . Works like those of Mr. Bergman should have an universal circulation; they are written not to the scholar alone, but to the artist [artisan], to the manufacturer, and to the world in general.' The works of most of the leading foreign chemists were similarly translated and read by scientifically-minded industrialists, as we have already noted in Manchester.

[5] He referred approvingly to Bancroft's work and to the lectures of Farish at Cambridge and Ure in the Andersonian Institution, Glasgow, with their 'popular descriptions of processes, and working models of machinery'. See below, pp. 167–170, 172–5, and 349.
[6] See above, pp. 56–7 and 122.

famous London pharmacist of the late eighteenth and early nineteenth centuries was the Quaker William Allen (1770–1843),[1] who acquired an interest in chemistry while at Rochester School.[2] After working for several years in his father's silk-manufacturing business, he entered the Plough Court Pharmacy of J. G. Bevan in Lombard Street, founded in 1715, and on Bevan's retirement he took over the business in 1795, together with a partner. In 1793 he attended Higgins's lectures, where he learnt about 'the new system of chemistry' and began a planned course of studies. In the following year he became a member of the Chemical Society at Guy's Hospital, where for many years he gave lectures on chemistry and natural philosophy to the medical students; in 1795 he also joined the Physical Society there. In 1797 he acquired a new partner, Luke Howard (1772–1864),[3] also a Quaker, who, after a mainly classical education at a private school, had been apprenticed to a retail chemist in Stockport, where he taught himself scientific chemistry, being impressed particularly by Lavoisier and the French school. Whilst there, in the late 1780s and early 1790s, he had experimented with chlorine bleaching at the request of local manufacturers and became expert in this field. In 1793 he commenced business in London, near Temple Bar, before joining Allen, with whom he was in partnership from 1797 to 1806; his main responsibility was for the manufacturing part of the business established at Plaistow, in Essex.[4] He shared Allen's scientific interests, especially in the 'Askesian Society', which they and some friends formed at Plough Court in 1796 for scientific experimentation and discussion; some of the papers read before the Society were later published in Tilloch's *Philosophical Magazine*. They were also active in the Linnaean Society and the British Mineralogical Society, and both were eventually elected F.R.S. Howard became particularly famous for his work in meteorology, especially cloud formations; he maintained a lifelong friendship and correspondence with John Dalton, with whom he shared these

[1] See *D.N.B.* and, for a recent study, Weeks, *op. cit.* See also Anon., *Life of William Allen* (2 vols., Philadelphia, 1847); E. C. Cripps, *Plough Court. The story of a Notable Pharmacy, 1715–1927* (Allen & Hanbury Ltd., 1927); D. Chapman-Houston and E. C. Cripps, *Through a City Archway. The Story of Allen & Hanbury, 1715–1954* (1954).

[2] One of the first grammar schools to introduce scientific subjects. Hans, *op. cit.*, p. 39.

[3] See *D.N.B.* See also below, p. 325, for Howard's connections with chlorine bleaching.

[4] When the partnership was dissolved at the end of 1806, Howard carried on with the chemical works, which were removed to Stratford, London. Howards of Ilford is the modern descendant. See A. W. Slater, 'Autobiographical Memoir of Joseph Jewell 1763–1846', *Camden Miscellany*, Vol. XXII (1964), pp. 113–78. Jewell became Howard's partner.

meteorological as well as chemical interests. Allen became a close friend of Humphry Davy, whose lectures he attended at the Royal Institution, and it was at Davy's request that he himself began in 1804 to lecture there on natural philosophy. In the following years he carried out many scientific experiments; he was also elected to the Institution's board of management. Thus he was associated with all the leading scientific figures in London, while he also had friends, including Professor T. C. Hope of Edinburgh, among Scottish scientists, as well as foreign ones such as Professor Pictet and the famous Swedish chemist Berzelius.[1] Allen and Howard thus provide a remarkable metropolitan example of scientific-industrial interests.

V

Throughout the eighteenth century London remained intellectually pre-eminent, but its predominance was gradually reduced by the establishment of numerous provincial literary and philosophical societies. The earliest of these were founded, however, not in the rapidly developing industrial towns of the midlands and the north, but in the more purely agricultural eastern counties. The first was apparently the 'Gentlemen's Society' at Spalding, originating in 1709 and formally constituted in 1712, and it was soon followed by others in Stamford, Peterborough, Wisbech, Boston, Lincoln, and Doncaster.[2] These other societies, however, do not appear to have proved very successful or long-lived: those at Stamford and Peterborough, for example, had 'sunk into mere taverns and clubs' by mid-century.[3] That at Spalding, however, managed to survive. Its main interests, as

---

[1] He became first president of the Pharmaceutical Society of Great Britain in 1841, but long before then his interests had been gradually drawn from science to philanthropic movements: agitation against the slave trade and slavery; education; Robert Owen and the New Lanark Mills; prison reform, etc. The business of Allen & Hanbury's was incorporated in 1893.

[2] 'An Account of the Gentlemen's Society at Spalding', in J. Nichols, *Bibliotecha Topographica Britannica*, no. XX (1784), repeated by R. Gough and J. Nichols in J. Nichols, *Literary Anecdotes of the Eighteenth Century*, Vol. VI, pt. i (1812), pp. 1-162; W. Moore, *The Gentlemen's Society at Spalding: Its Origin and Progress* (1851), also repeats much of this article. These sources also refer to early societies in Worcester and Dublin (for the latter, see below, pp. 185-6).

[3] Gough and Nichols, *op. cit.*, pp. 4-5. Dr. William Stukeley had tried to revive the Stamford Society in the 1730s and 1740s, with interests in astronomy and meteorology, as well as in literary and antiquarian pursuits. Among the members were John Grundy, the land-surveyor and engineer, John Macklin, mathematician and secretary of the Royal Society, Edward Laurence, surveyor, Tycho Wing, astronomer, as well as members from the clerical, medical, legal, and other professions (*ibid.*, pp. 4-5 and 427, and Stukeley's *Memoirs*, Surtees Society Publications, Vol. LXXIII (1882, for 1880), p. 123.)

of the other societies, appear to have been literary and antiquarian. Maurice Johnson, its founder, barrister and member of the Inner Temple, and steward to noble estates in the Spalding area, was also a founder-member of the Society of Antiquaries in London, with which he maintained a continuous correspondence. The early meetings of the Spalding Society, at Younger's Coffee House in Abbey-yard, were mainly occupied in reading and discussing the *Tatler*, the *Spectator*, and other literature, while most of its members also belonged to the Society of Antiquaries. But the Society's original rules show that it was established for 'Improvement in the Liberal Sciences', as well as in 'Polite Learning', and according to a letter of Maurice Johnson in 1745/6, '[we] put ourselves under the protection of the Royal Society of London, from our first fixing; and had the happiness of their regard. Then Sir Isaac Newton held their chair, and my [former] tutor Dr. [James] Jurin was their secretary, with whom I kept correspondence.'[1] 'We deal', he said, 'in all arts and sciences',[2] and the surviving minute-books show lectures and experiments not only on medicine, natural history, and astronomy, but also on mathematics, mechanics, and natural philosophy in general, with a particular interest in fen-drainage, so important in that area. Indeed it appears that, under the influence of John Rowning, M.A., of Magdalene College, Cambridge—afterwards rector of Anderby, Lincolnshire, author of *A Compendious System of Natural Philosophy* (1738)[3]—who was 'a constant attendant of the Meetings of the Society',[4] it became primarily concerned with 'Experimental Philosophy',[5] for which it equipped itself with suitable apparatus, in addition to its museum and library.

Unfortunately the surviving membership lists[6] do not distinguish between 'regular' and 'honorary' members, except for the early founders, but it seems certain that such distinguished philosophers as Newton, Desaguliers, and many others were in the latter category. That men of such eminence, however, should have given their patronage to the Society shows the status it achieved. Moreover, among those whom we may regard as almost certainly regular members were some interesting figures, in addition to Rowning. Captain John Perry (1670–1732), for example, famous as the 'Comptroller of the Maritime works to Czar Peter in Russia' and for his stopping of the Dagenham Breach on the River Thames, was in his later life engineer to the 'Adventurers for Draining Deeping Fen' and apparently resided in Spalding, where,

---

[1] Gough and Nichols, *op. cit.*, pp. 2–3, footnote. See also below, pp. 159–60, for Jurin. [2] *Ibid.*, pp. 6–7. [3] See above, p. 34.
[4] Gough and Nichols, *op. cit.*, p. 109. [5] *Ibid.*, p. 124.
[6] In Gough and Nichols, *op. cit.*, pp. 69–122, and Moore, *op. cit.*, pp. 19–60.

on 16 April 1730, he became a member of the Gentlemen's Society, which still possesses manuscript material relating to his drainage works.[1] He was obviously a man of some mathematical and surveying ability. He was succeeded as engineer to the Adventurers of Deeping Fen by John Grundy, who joined the Society on 10 June 1731.[2] Grundy was employed by the Duke of Buccleuch to survey the manor of Spalding, and on a map of the town which he presented to the Society in 1732 described himself as a 'Land Surveyor & Teacher of the Mathematicks';[3] he taught mathematics at Spalding Grammar School. A native of Bilston, in Leicestershire, he appears to have been self-taught, for there is an inscription on his tomb in St. Mary's Church, Congerstone, in that county, where he died in 1748, to the memory of this engineer, 'late of Spalding, who, without the advantage of a liberal education, had gained by his industry a competent knowledge in several sciences'. Gough and Nichols state that he was 'much employed in drawing and surveying the navigations in the counties of Chester, Lancaster, and Lincoln'.[4] In this work it is clear that he was applying principles which he had acquired from publications on river hydraulics, for in the appendix to a pamphlet which he wrote in 1745, replying to criticisms by Daniel Coppin of his scheme for improving the River Witham, Grundy referred to works by Guglielmini, Castelli, Westerdyke, and others.[5] This, as Mr. L. E. Harris has pointed out, is 'important as showing that Grundy was acquainted with two of the most significant books of that time on the behaviour of rivers', viz. Benedotto Castelli's *Della Misura dell'Acque Correnti* (1628)[6] and Domenico Guglielmini's *Della Natura de' Fiumi* (1697), both published, of course, in Italy, where the theory of hydrodynamics had been most closely studied. Some years earlier, in fact, writing about the plans for improving the River Dee at Chester, in the mid-1730s, he supported

[1] See *D.N.B.;* Smiles, *Lives of the Engineers* (1861), Vol. I, pp. 73–82; S. B. Hamilton, 'Captain John Perry, 1670–1732', *Newcomen Soc. Trans.*, Vol. XXVII (1949–51), pp. 241–53.

[2] Gough and Nichols, *op. cit.*, p. 86; Mrs. E. C. Wright's contribution to the discussion on S. B. Donkin's paper, 'The Society of Civil Engineers (Smeatonians)', *Newcomen Soc. Trans.*, Vol. XVII (1936–7), p. 64, and her own paper, 'The Early Smeatonians', *ibid.*, Vol. XVIII (1937–8), p. 103; L. E. Harris's contribution to the discussion on A. W. Skempton's paper, 'The Engineers of the English River Navigations, 1620–1760', *ibid.*, p. 51.

[3] He similarly described himself in a pamphlet of 1736, *An Examination and refutation of Mr. Badeslade's new-cut canal* (at Chester, for the River Dee Navigation).

[4] *Op. cit.*, p. 86. See also [R. Gough,] *British Topography* (1780), Vol. I, pp. 259–60 and 266*, for references to his survey and drainage work on the River Welland and Deeping Fen and also on the River Dee. See also Willan, *op. cit.*, pp. 81–3.   [5] Harris, in Skempton, *loc. cit.*

[6] Translated by Thomas Salusbury, *Discourse of the Mensuration of Running Waters* (1661).

his case with *Philosophical and Mathematical Reasons*, in which, as Professor Willan has pointed out, he emphasized that

special qualifications were necessary for dealing with tidal, as opposed to non-tidal rivers. Grundy maintained that the engineer for the former must be a good mathematician, able to take surveys and levels and draw maps; he must understand natural philosophy and the laws of motion; he must be able to account for the rainfall and the force of backwater; finally, he must be capable of inventing engines.[1]

Thus, though self-taught, Grundy may be regarded, within the limits of his day, as a scientific engineer.[2]

In his survey of the Witham, published in 1744, he was assisted by his son, John Grundy junior (1719-83), who followed in his father's footsteps as mathematician, surveyor, and engineer, and also as a member of the Gentlemen's Society at Spalding (from 27 December 1739), residing there until his death in 1783.[3] He succeeded his father as 'surveyor and agent for [the] Adventurers for Deeping Fens', and also published further plans for the Witham in 1753 and 1761 (the latter in conjunction with John Smeaton and Langley Edwards); he was also involved in various other river improvements, in the drainage of Holderness, and construction of Hull docks. Smeaton, who reported on some of Grundy's schemes, clearly had great respect for his engineering ability. It is not therefore surprising to find Grundy participating with Smeaton, Mylne, Yeoman, and other leading engineers in the formation in 1771 of a Society of Civil Engineers, later known as the Smeatonian Society and forerunner of the Institution of Civil Engineers established in 1818. The members of this Society, as previously pointed out, were interested in mathematics, mechanics, hydraulics, and natural philosophy generally,[4] like the earlier Spalding Society, since schemes of fen-drainage, river improvement, canal building, and water power were based to a considerable extent upon scientific principles as well as practical experience. *Omnia in Numero Pondere et Mensura*— the motto of the Smeatonians—was not that of illiterate or innumerate empiricists.

The Spalding Society also included Humphry Smith (d. 1742), another 'engineer for draining the Fens' who collaborated with John

[1] Willan, *op. cit.*, pp. 81–2. See above, pp. 45–6, however, for Professor Willan's doubts regarding the practical importance of scientific theory in river engineering at that time.
[2] Harris, in Skempton, *loc. cit.*, where Professor Skempton agrees with this view. See also above, p. 46.
[3] Gough and Nichols, *op. cit.*, p. 87; Nichols, *British Topography*,Vol. I, pp. 580–1; Donkin, *op. cit.*, especially Mrs. Wright's contribution, p. 64; Wright, *op. cit.*, especially pp. 103–5; and Harris, *loc. cit.*   [4] See above, p. 74.

Grundy senior, as well as a plentiful smattering of surgeons, physicians, and apothecaries, along with clerics and teachers. In its heyday it must have been a lively and interesting body. In the second half of the century, however, it became moribund, sinking 'to the condition of a country club'.[1] The economic and intellectual centre of gravity was shifting from the south and east to the midlands and the north, where similar societies were now rapidly springing up.

VI

It was in the rapidly expanding urban centres of the Industrial Revolution that the alliance between science and technology was to develop most strongly. The Manchester Literary and Philosophical Society was paralleled in Birmingham by the famous Lunar Society. In Birmingham there was also, as Professor Ashton has pointed out, 'much coming and going between the laboratory and the workshop, and men like James Watt, Josiah Wedgwood, William Reynolds, and James Keir were at home in the one as in the other'.[2] They doubtless benefited a good deal from their regular intercourse with such scientists as Erasmus Darwin, Richard Lovell Edgeworth, and Joseph Priestley.[3] 'Never before in history was there such an advantageous syncretism of pure science and advancing industry as in the Lunar Society.'[4]

This orientation of science towards industry in the deliberations of the Lunar Society has been examined more closely in recent years,[5] and though it is clear that the industrial emphasis did not debar the 'Lunatics' from engaging in pure research in electricity and the collection and identification of new gases, at a time when their potential applications to industry were not known, yet a good deal of the correspondence between friends in the society has a very practical flavour. Keir, Watt, and Black shared an interest in a patent for the manufacture of alkali, and the course of their interests is described very fully in correspondence between Watt and Black;[6] Keir and Boulton wished to exploit com-

[1] Moore, *op. cit.*, p. 14; Gough and Nichols, *op. cit.*, pp. 135 and 161–2.

[2] Ashton, *op. cit.*, p. 16.

[3] *Ibid.*, p. 21. See also P. Mantoux, *The Industrial Revolution in the Eighteenth Century* (1928), pp. 387–8. The stimulus and benefits were mutual. See E. Robinson, 'The Lunar Society and the Improvement of Scientific Instruments', *Annals of Science*, Vol. XII (1956), pp. 296–304, Vol. XIII (1957), pp. 1–8.

[4] Clow, *op. cit.*, p. 614.

[5] R. E. Schofield, 'Industrial Orientation of Science in the Lunar Society of Birmingham', *Isis*, Vol. 48 (1957), pp. 408–15. Since our article originally appeared, Professor Schofield has produced his book on *The Lunar Society of Birmingham* (1963).

[6] See below, chap. IX. Mr. Robinson and the late Professor McKie are bringing out an edited version of the Black-Watt papers.

mercially an alloy, discovered by Keir, which had almost the same composition as Muntz's metal;[1] Wedgwood supplied chemical utensils for Priestley's experiments, while Priestley analysed minerals of possible utility in pottery;[2] Small, Boulton, Keir, and Watt, with the advice of John Whitehurst, F.R.S., the Derby watch-maker, were involved in a plan to mass-produce clocks, while the tools of the clock- and instrument-making trade assisted Boulton and Watt in their development of engineering technique.[3] Even before Watt came to Birmingham, the members of the Lunar Society were interested in the design of steam engines. Boulton had invented a steam engine of his own, which he discussed with Erasmus Darwin and Benjamin Franklin. Boulton's 'toy' business also caused him to be interested in metallurgy. In these pursuits he was aided by William Withering, who translated Bergman's *Outlines of Mineralogy* (Birmingham, 1783), and by James Keir. Similar examples of collaboration in the industrial applications of science could be multiplied.

Besides those men of industry and science who are known to have been members of the Lunar Society, there were others who long maintained close relationships with members of the Society. Josiah Wedgwood is the first in importance among some very important men. He was a F.R.S. and a member of the Derby Philosophical Society, as well as of numerous other organizations, including Franklin's 'Club of Thirteen'.[4] The interplay between industry and science in Wedgwood's life is best seen in his scheme to combine with other Staffordshire potters to set up a research establishment.[5] He also encouraged scientific interests in his own district, by patronizing the lectures of John Warltire and engaging him to instruct his own children and those of Erasmus Darwin in chemistry.[6] Wedgwood's own laboratory assistant was Alexander Chisholm, a chemist of no mean ability.[7]

Like Wedgwood, John Roebuck and Samuel Garbett cannot be recognized as full members of the Lunar Society, though they were certainly on very close terms with Boulton, Watt, and others. The refining laboratory and the vitriol works operated in Birmingham by

[1] H. W. Dickinson, *Matthew Boulton* (Cambridge, 1937), pp. 101–3.
[2] E. Robinson, 'The Lunar Society and the Improvement of Scientific Instruments', *Annals of Science*, Vol. XIII (1957), pp. 1–2.
[3] *Ibid.*, Vol. XII (1956), pp. 296–304.
[4] E. Robinson, 'R. E. Raspe, Franklin's Club of Thirteen, and the Lunar Society', *Annals of Science*, Vol. XI (1955).
[5] R. E. Schofield, 'Josiah Wedgwood and a Proposed Eighteenth-century Industrial Research Organization', *Isis*, Vol. 47 (March 1956).
[6] D. McKie, 'Mr. Warltire, a good chymist', *Endeavour*, Vol. X (Jan. 1951).
[7] See above, pp. 54 and 78.

Garbett and Roebuck are *loci classici* of the union between chemistry and industry in the development of eighteenth-century Birmingham. Among the occasional visitors to the Lunar Society were John Smeaton, the engineer, Henry Moyes, the itinerant lecturer, Pieter Camper, the Dutch scientist, and R. E. Raspe, geologist, assayer, and man of science.[1] There must have been many others whose visits cannot be substantiated by written evidence, since the Lunar Society kept no records, but from those who are known to have attended meetings, we can compile a most impressive record of the relationships between scientists and industrialists. Indeed many of the industrialists have some claim to scientific prominence in their own right. We should not overlook the fact, moreover, that the 'Lunatics' spread their influence by their membership of other societies. All of them were Fellows of the Royal Society and many were members of the Society for the Encouragement of Arts, Manufactures and Commerce.[2]

In view of these varied scientific-industrial interests in Birmingham, it is not surprising to find itinerant lecturers operating there apparently with the same frequency as in Manchester, though no thorough survey of their visits has yet been undertaken. References to them occur quite frequently in Matthew Boulton's correspondence. It is probable that Boulton's own interest in electricity, which gained him the friendship of Benjamin Franklin, had first been aroused by the lectures in Birmingham of Thomas Yeoman[3] and Benjamin Martin.[4] Dr. John Taylor, the oculist, also regularly lectured in the town at an early date.[5] We have already referred to Boulton's contacts with Adam Walker;[6] Samuel Galton (1753–1832), Quaker gunsmith and member of the Lunar Society, also attended Walker's lectures in Birmingham in 1776 and 1781.[7] There is also a letter from Joseph Priestley recommending

[1] R. E. Schofield, in his valuable paper, 'Membership of the Lunar Society of Birmingham', *Annals of Science*, Vol. XII (1956), denies that Raspe and Pieter Camper ever attended any meeting of the Lunar Society, but their visits can be substantiated from James Watt's pocket diaries at Doldowlod (e.g. under 13 June 1779: 'Sunday at Soho with the Lunar Society. Mr Raspe was there & Mr Edgeworth'). They were not, however, regular attenders.

[2] R. E. Schofield, 'The Society of Arts and the Lunar Society of Birmingham', *Journal of the Royal Society of Arts*, June and Aug. 1959.

[3] *Aris's Birmingham Gazette*, 20 and 27 Oct. 1746, 3 Nov. 1746, 29 Dec. 1746, 5 and 12 Jan. 1747.

[4] *Ibid.*, 13 and 17 Aug. 1747.

[5] See for example, *Aris's Birmingham Gazette*, 16 Jan. 1743, 7 and 14 Dec. 1748, and 30 May 1748.

[6] See above, pp. 105–6. See also Gibbs, *op. cit.*, p. 115, for Walker's lectures in Birmingham in 1781.

[7] Karl Pearson, *The Life, Letters and Labours of Francis Galton* (2 vols., Cambridge, 1914), Vol. I, p. 47.

John Warltire to Boulton,[1] and it was from Birmingham that Warltire wrote to Priestley about their experiment of passing an electric spark through a closed glass vessel.[2] Warltire lectured there in 1776, 1780, 1781, and 1782.[3] He also became the teacher of Josiah Wedgwood's children at Etruria and the assistant of Dr. Darwin. John Roebuck asked Boulton to collect subscribers for Moyes,[4] who, like Warltire, quickly became known to several members of the Lunar Society. Moyes had intended to settle in Birmingham at one time as a 'teacher of Mathematicks Astronomy and Music'. In September and October 1782, Moyes was lecturing in Birmingham,[5] and it was at this time that he attended a meeting of the Lunar Society and became involved in a lengthy argument with John Smeaton, the engineer, much to Watt's disgust.[6] Moyes, nevertheless, established friendly links with several of the members; in the following year, for example, he acknowledged receipt of a copy of Bergman's *Mineralogy* translated by Dr. William Withering of the Lunar Society.[7] A few years later we find a Mr. Burton giving a course of lectures on natural and experimental philosophy, with a 'very valuable and extensive Apparatus'.[8] At the same time a Mr. Long, recommended to Watt from London, was also lecturing there.[9] *Aris's Birmingham Gazette* shows that James Ferguson, James Booth, and many others also lectured in Birmingham during this period.

That the lectures of such men were considered to be an essential part of the education of every young business-man, can be seen from the concern of John Gilbert, the Duke of Bridgwater's steward, that his son should accompany Boulton to a course of lectures by John Arden in Birmingham, while the young man was Boulton's guest.[10] Boulton

[1] J. Priestley to M. Boulton, 28 Sept. 1776, A.O.L.B. See also *Aris's Birmingham Gazette*, 4 Nov. 1776, announcing a third course of lectures by Warltire 'upon the same plan as the two preceding Courses'.
[2] Dated 21 April 1781; Priestley, *Experiments and Observations* (1781), Vol. II, p. 398. [3] Gibbs, *op. cit.*, pp. 114–16. See also below, p. 146.
[4] John Roebuck to Matthew Boulton, 3 Oct. 1777, Dol. Roebuck had earlier written to Watt on the same errand, saying: 'I wrote to you sometime agoe desiring your recommendation in favour of Mr. Moyes. Since then I have procured him the Countenance and favour of some principal gentlemen in Yorkshire near Leeds', 1 Jan. 1777, Dol. See J. Roebuck to Watt, 14 Aug. 1777, Dol.
[5] See *Aris's Birmingham Gazette*, 30 Sept. 1782 and 14 Oct. 1782. See also Gibbs, *op. cit.*, p. 116. [6] J. Watt to Boulton, 28 Oct. 1782, A.O.L.B.
[7] H. Moyes to W. Withering, Easter Monday 1783, Royal Society of Medicine Library. [8] *Aris's Birmingham Gazette*, 30 Oct. and 13 Nov. 1786.
[9] *Ibid.*, 20 Nov. 1786. Annie Watt to James Watt, 13 Nov. 1786, B.R.L. Both Burton and Long lectured in Manchester in the 1780s (see above, p. 107).
[10] He thought it 'of the Greatest Consequence' that his son should attend these lectures. J. Gilbert (Worsley) to M. Boulton (Birmingham), 7 Sept. 1765, A.O.L.B. *Aris's Birmingham Gazette*, 2 and 9 Sept. 1765, contains advertisements of Arden's lectures.

himself arranged for his own son, M. R. Boulton, to attend Dr. Beddoes's and Dr. Hornsby's lectures in Oxford.[1] The taste remained with young Boulton in adulthood, since his pocket diary refers to his attendance at John Stancliffe's lectures on galvanism.[2] As in Manchester, the advertisements of these lectures often stressed their practical application to local industries. Thus John Warltire's lectures on chemistry would give Birmingham manufacturers 'an Opportunity of informing themselves of the particular Properties of the various substances imployed in the different Processes they are necessarily conducting every Day; and consequently a Means of improving them to *their own Advantage*'.[3] When in 1797 Dr. Thomas Garnett of Glasgow proposed to embark on a lecture tour through some of the large towns of England, he sent a copy of his prospectus to James Watt, telling him that Birmingham seemed to be a natural place to visit 'as many of the manufactures are connected with Chemistry and other branches of Natural Philosophy'.[4] His proposals were very ambitious:

I would commence the Lectures on the first of June, and continue to give them for 4 months. I would propose to give 2 Lectures on Chemistry and one on Experimental Philosophy . . . Besides these, I propose to read a course of Lectures on the Arts and Manufactures connected with Chemistry and Natural Philosophy. This course would be complete and scientific, and would require a Lecture every day to complete it in 4 months. In this course the principles of Natural Philosophy would be both demonstrated Mathematically and illustrated by experiments. The different machines would be explained by working models, and the various processes of the arts would be performed before the audience. This is the plan which I follow in this City, and which has been attended with success beyond the most sanguine expectations of the managers of this institution. If you think that such a plan would be acceptable to the inhabitants of Birmingham, and if it was probable that 100 Subscribers could be procured to the first Course, 50 to the 2nd and 30 to the last, I should be induced to pay you a visit.

Watt's physician, Dr. Carmichael, however, considered that Garnett was being rather too ambitious for Birmingham, but thought that he might well give the first two courses and return for a third.[5] It seems, however, that this would not have been Garnett's first visit to Birming-

---

[1] A. Noble Brown (Balliol College, Oxford) to M. Boulton (Soho), 25 Feb. 1791, A.O.L.B.
[2] M. R. Boulton's Pocket Diary, 4 and 5 March, 5 April 1801, A.O.L.B.
[3] John Warltire's advertisement in *Aris's Birmingham Gazette*, 11 Oct. 1779.
[4] T. Garnett, to J. Watt, 24 Jan. 1797, Dol. See below, p. 181, for Garnett's lectures in Anderson's Institution, Glasgow. See also above, p. 107, for his lectures in Manchester.
[5] J. Carmichael to J. Watt, 9 Feb. 1797, Dol.

ham and that he had lectured in the town in 1796, for there is a letter of that year from James Watt to Thomas Walker of Manchester describing Garnett's success as a lecturer.[1]

VII

In the area of Sheffield, Rotherham, and Doncaster, there is similar evidence of scientific activity, though most of the evidence so far assembled is fuller for the first decade of the nineteenth century than it is for the late eighteenth century. This can largely be accounted for by the imperfect survival of the town's eighteenth-century newspapers, but it should be noted that as early as May 1761, John Arden was giving a course of twenty lectures in the town hall of Rotherham: 'in which will be exhibited the Experiments necessary for the Explanation of Natural Philosophy, Mechanics, Astronomy, Geography, Hydrostatics, Pneumatics and Optics'.[2] There is also some evidence of an interest in mathematics centred in Sheffield round John Eadon, master at the Free Writing School, who besides being a schoolmaster was a land-surveyor and author, publishing in 1766 *The Arithmetician's Guide*. Another Sheffield publication of the 1760s was *The Lady's and Gentleman's Scientific Expositor*, to which a number of local people submitted mathematical problems and solutions. In 1783, the gentlemen of Sheffield instituted a dining-club called 'The Monthly Club', which met at the Angel 'on the Thursday nearest the full moon', but it seems to have been of a more convivial character than its counterparts in Birmingham and Derby.[3] By the winter of 1800–1, however, there was a society in existence in Sheffield which had a distinct scientific cast to it. This was a 'Society of Gentlemen' which commissioned a course of lectures by the Rev. Thomas Olivers Warwick, M.D.[4] The members included the Rev. Benjamin Naylor, successively partner in a silver-plating concern and with James Montgomery in the Sheffield newspaper the *Iris*, who left Sheffield in 1805 to run a Manchester cotton-mill belonging to his family;[5] John Moorhouse, surgeon to the Overseers of the Poor, Sheffield; J. and G. Shore, bankers of Norton; Robert Brightmore, a master cutler; the Rev. Astley Meanley, Presby-

[1] J. Watt to T. Walker, 18 May 1796, Boulton and Watt Letter Book (Office), Sept. 1795–Aug. 1796, B.R.L.   [2] *Sheffield Public Advertiser*, 26 May 1761.
[3] R. E. Leader, *Sheffield in the Eighteenth Century* (Sheffield, 1901), p. 117.
[4] M. Brook, 'Dr. Warwick's Chemistry Lectures and the Scientific Audience in Sheffield (1799–1801)', *Annals of Science*, Vol. XI (1955), pp. 224–37.
[5] *The Commercial Directory for 1816–17* (Manchester, 1816) gives under 'Cotton Spinners', p. 230, Benjamin Naylor and Co., Chorlton Row. In 1787 Naylor was the minister of the Upper Chapel in Norfolk-Street, Pinson-Lane, Sheffield.

terian minister at the nearby village of Stannington; John Aldred, who was later partner with his brother-in-law Dr. Warwick in a firm at Rotherham manufacturing dye-stuffs; Mr. Aldam, probably a member of the well-known Quaker family of south Yorkshire; William Staniforth, surgeon and oculist; and Dr. Hugh Cheney, who was first surgeon of the Sheffield Central Infirmary founded in 1797. Messrs. Naylor, Moorhouse, Staniforth, and Cheney were among those who signed a circular in January 1804 calling for a meeting to form a literary and philosophical society. Such a society was not created until later in the nineteenth century, but there was already in existence a Society for the Promotion of Useful Knowledge. To the members of the Society of Gentlemen listed above should be added the name of Joseph Hunter, the historian of Sheffield, who came to be the secretary of the Society for Useful Knowledge. Among the members of both societies there are not only links with trade, and with the medical profession, but also with Nonconformity. Naylor, a Presbyterian minister, had been educated at Warrington Academy; Staniforth was a Presbyterian; Moorhouse was a 'Rational Dissenter'; Meanley was a Presbyterian minister and had been educated at Daventry and Warrington; and Aldam was a Quaker. Joseph Hunter, who had been at school with Dr. Roebuck at the Free School, Sheffield, before the latter went on to Northampton, went to the Unitarian College, York, in 1809 and there studied for the ministry. The lecturer himself, Dr. Warwick, came to Rotherham as minister of the Presbyterian chapel in 1793, and had been educated at the Dissenting Academy at Daventry and Northampton, which had been Dr. John Roebuck's school.[1]

It should be remembered, in this connection, that Rotherham boasted its own Dissenting Academy, which provided like other academies a considerable amount of scientific education for its pupils. It was instituted in November 1795 under the guidance of the Rev. Edward Williams, D.D. (1751–1813),[2] whose reports make mention of lectures on astronomy, the mechanical powers, hydrostatics and pneumatics (1799), on optics (1801) and philosophy, particularly pneumatics and electricity (1802). In 1803 a course of ten lectures on natural philosophy was delivered at the academy by John Warltire. Walker's *Philosophical Lectures* were possessed by the school, together with various scientific instruments—an orrery, a galvanic trough, globes, telescope, quadrant, etc.

[1] See J. Hunter, 'Autobiographical Miscellany', pp. 54–5, MS. in Sheffield Public Library.
[2] H. McLachlan, *op. cit.*, pp. 199–208. It should be noted that there was a Williams who was a member of the Society of Gentlemen at Sheffield.

Sheffield and Doncaster also had schools in which mathematics and the sciences were cultivated. We have already mentioned the Free Writing School in Sheffield run by John Eadon, the mathematician.[1] The Rev. Peter Inchbald conducted a school in Doncaster at which he offered 'instruction in the principles of general Science'.[2] He shared with his friend Thomas Asline Ward of Sheffield a keen interest in science and belonged also, as an honorary member, to the Society for the Promotion of Useful Knowledge, to which he sometimes took one or two of his pupils, in the same way that Dr. Doddridge invited his pupils to the Northampton Philosophical Society.[3] Ward describes a visit by Inchbald and two of his pupils, Peter and Paul Wynch, to Ward's house on 28 June 1804, and adds that they 'are on their way to Manchester and Liverpool . . . examining those mechanical inventions and improvements which have so greatly conduced to the commercial greatness of this kingdom'.[4] Peter Inchbald was responsible for inviting Dr. Henry Moyes to lecture at Doncaster in October and November 1807, when the blind scientist had a great success, sometimes lecturing to as many as 120 people at a time. When Moyes died at Doncaster on 10 December 1807, Inchbald was one of the pall-bearers. He also entertained William Nicol, Moyes's demonstrator and nephew, when the younger man, continuing his profession, came to Doncaster in the following year.

It is from the diaries of Joseph Hunter (1783–1861)[5] and Thomas Asline Ward (1781–1871)[6] of Sheffield that we learn much more about science in this area. Hunter tells us, for example, in his 'Autobiographical Miscellany': '1800. At the beginning of this year began to attend Dr. Warwick's lectures on Chemistry and read all the Chemical books within reach. Got a little kind of apparatus & made experiments. I was very eager about it.'[7] But Warwick was lecturing in

[1] See above, p. 147.

[2] *Doncaster, Nottingham and Lincoln Gazette*, 29 Jan. 1802. See J. A. Harrison, 'Blind Henry Moyes, "An Excellent Lecturer in Philosophy" ', *Annals of Science*, Vol. XIII (1957), pp. 109–25.

[3] See Thomas A. Ward, 'Diary' (2 June 1805–13 July 1806), p. 1, for 5 June 1805, Sheffield Public Library.

[4] See T. A. Ward, 'Diary', for that date.

[5] See Joseph Hunter, 'Autobiographical Miscellany', Sheffield Public Library; and 'Journal for 1800', B.M. Add. 24,880. Hunter is well known as the antiquarian of Yorkshire. He spent his life from 1783 to 1805 in Sheffield; from 1805 to 1809 in York; from 1809 to 1833 in Bath, and from 1833 to 1861 in London. He was Minister of Trim St. Chapel, Bath, for the whole of his time in that town.

[6] 'Diaries' and 'Notebooks' of T. A. Ward, Local History Archives, Sheffield Public Library. Ward was one of the leading cutlers and citizens of Sheffield.

[7] Hunter, 'Miscellany', p. 162, Also pp. 31 and 55.

Rotherham and Sheffield the year before[1] and his lectures were advertised as 'peculiarly interesting to the manufacturers of this place'. The reports of his lectures[2] indicate that they were, in fact, on acids, metallurgy, alloys and assaying, and many other matters which show the doctor's claims to have been justified. Since Warwick had been a pupil of Robert Cleghorn at Glasgow University he was in a direct line of descent from Dr. Joseph Black, and he had also attended lectures in London which may have brought him into contact with George Fordyce, Bryan Higgins, and others interested in the applications of chemistry to industry.

Thomas Asline Ward tells how, on 8 January 1803, he 'supped at Mr. E. Wilson's with Rev^d. B. Naylor, D^r. Younge, Messrs Thompson, Ernest, Moorhouse, Wade, Daniels and Favell it being their Club night'.[3] On 8 November 1804, he wrote:

I attended the fortnightly meeting of a society of which Inchbald and I were unanimously received as members the last meeting. Its aim is the promotion of useful knowledge. Each in turn reads a paper on any subject he prefers, except religion and British politics, which, as they might cause disputes, are wisely avoided. The members are Dr. [Hugh] Cheney, Mr. [Daniel] Barnard, Dr. [David Daniel] Davies, Mr. [Richard] Sutcliffe, [Mr. Hall] Overend, [Mr. Robert] Earnest, [Mr. Charles] Sylvester, [Mr. Joseph] Hunter, [Mr. Edmund] Wilson, Rev. B[enjamin] Naylor, Mr. [Samuel] Lucas, [Mr. John] Moorhouse, [Mr. John] Favell, [Mr. Thomas] Beilby, and [Mr. Samuel] Hobson, besides an honorary member, Mr. [White] Watson.[4] The paper read this evening was by Mr. Sutcliffe, and attended by experiments. It is well known that the white lead manufactory is very pernicious to those employed in it. Mr. S. hopes he has discovered a manner of preparing the article which will obviate this dreadful effect and be quite as cheap an operation.[5]

---

[1] *Sheffield Iris*, 30 Aug. 1799.

[2] Brook, *op. cit.*, pp. 232–7.

[3] MS. 'Pocket Book', 1803, Local History Archives, Sheffield Public Libraries. Since this is obviously the same society to which Inchbald and Ward were elected in 1804, and since that society was known as the Society for the Promotion of Useful Knowledge, the suggestion, made by Mr. M. Brook in his paper cited above, that the circular of Jan. 1804 calling for the creation of a literary and philosophical society refers to the Society for the Promotion of Useful Knowledge, hardly seems acceptable.

[4] From 'The Minute Book of a Society for the Promotion of Useful Knowledge' (6 Jan. 1804–27 Nov. 1805), Sheffield Public Libraries, may be added the names of the Rev. William Whitelegg, Rev. Samuel Catlow (honorary member), of Mansfield, and Mr. J. Biglow (honorary member), of Rossington, near Doncaster. Inchbald was an honorary member. He was elected with Ward on 25 Oct. 1804. See also Ward's 'Diary', 2 June 1805–13 July 1806, for 5 June 1805.

[5] A. B. Bell (ed.), *Peeps into the Past Being Passages from the Diary of Thomas Asline Ward* (Sheffield, 1909), p. 55; Ward, 'Diary' (1 Jan. 1804 to 26 May 1805), p. 119.

Beilby was an optician and instrument-maker.[1] Richard Sutcliffe was an apothecary and druggist, who on 8 May 1805 lectured to the society, with experiments, on mercury. Among his apprentices were Mr. Hall Overend and Mr. Jonathan Barber senior. John Favell was another surgeon; Charles Sylvester was himself a public lecturer on natural philosophy who had risen from the ranks of the artisans, having been formerly a plated-wire worker.[2] On 24 April 1805 he lectured to the Sheffield Society on Galvanism, his usual subject, and Ward wrote in his diary: 'He is now in the midst of a Course at Rotherham and has been at Leeds, York, Hull, Chesterfield, Mansfield, Nottingham, &c besides lecturing 2 or 3 times in Sheffield.'[3] Ward himself had heard him give two public lectures in Sheffield on Galvanism on 8 and 12 March 1804, paying a half-crown for each lecture.[4]

Various other meetings of the Society are recorded in Ward's diaries: 27 February 1805, when 'Mr. Moorhouse gave a sketch of the various opinions which have been promulgated respecting the cause of the present appearance of the Earth' and Ward decided to follow

[1] *Gales and Martin's Directory of Sheffield, 1787*, ed. S. O. Addy (Sheffield, 1889), p. 72: '*Proctors and Beilby*, telescope, microscope, perspective, reading glass, and spectacle makers, Milk-Street'.

[2] Sylvester's later career demonstrates the close connection between the life of a public lecturer in science and the professions of chemist and mechanical engineer. In the 'Private Memoirs' of Edward Strutt of Derby, son of the cotton-spinner, William Strutt, and himself a distinguished man of science (typescript 'Memoir of William Strutt', Derby Public Library, pp. 39–40), we are told of Charles Sylvester that:

> He had scarcely any education before he was sixteen and was entirely self-educated. Yet he became an eminent practical chemist. . . . He first came to Derby to give a course of scientific lectures; and liking the Society of the place (particularly my father), he removed there and continued till his death to be an intimate friend of my father, who showed him much kindness. He gave me instruction in chemistry and mathematics, of which he made much use in his scientific investigations . . .'

In 1809 Sylvester published in Liverpool *An Elementary Treatise on Chemistry*, in which he quotes with approval Fourcroy's statement that '*Tout homme qui recoit une education liberale, compte aujourd'hui la chimie parmi les objets les plus indispensables de ses etudes*'. His book, *The Philosophy of Domestic Economy* (Nottingham, 1819), is concerned with the heating equipment, washing-machine, W.C., etc., designed by William Strutt, F.R.S., and installed by Sylvester, for the Derby General Infirmary, and informs us that Sylvester first met William Strutt in Derby in 1807. Similar installations were also placed in other public institutions and private houses. Sylvester stresses the industrial applications of the stoves for drying cotton yarn and calicoes in Strutt's mills and their potential application to paper-making, pottery, sugar-making, etc. After leaving Derby, Sylvester installed himself in London as a specialist in heating engineering.

[3] Ward, 'Diary' (1804–5), p. 172, See also 'Notebook', same date.

[4] Ward, 'Pocket Book', 1804, under 'Cash'.

Kirwan's Neptunist theories;[1] 13 March 1805, when Mr. Lucas gave a paper on the explosion of heated copper, when poured into water;[2] 27 March 1805, when climate was discussed;[3] 10 April 1805, when they talked about agriculture;[4] and so on. On 12 April 1804 Overend had read a paper 'on the changes which take place during the conversion of barley into malt'.[5] Ward was also an inveterate attender of public lectures on science. Besides Sylvester's lectures already referred to, he also attended two lectures at the Tontine Inn, Sheffield, on the 14 and 18 December 1802 by John Warltire: the first was on the properties of heat, light, and the change of colours, and the second on minerals.[6] On 17 March 1806 he attended a lecture by Dr. Stancliffe on 'the Rise & progress of Agriculture' and 'the effects of Lime &c as manures'.[7] He supped afterwards with the lecturer and heard a long story from him about the annoyance that his success as a lecturer in Newcastle-upon-Tyne had inflicted upon the actor, Stephen Kemble.[8] On 23 February 1807 he subscribed a guinea for Dr. Moyes's course of 22 lectures, to which he had been recommended by Inchbald, and religiously attended every lecture from that night until the last lecture on the 17th April.[9] On 18 September of the same year, Ward listened to Mr. Richard Dalton lecturing on 'Pneumatics, pneumatic chemistry, hydrostatics, and electricity',[10] and in March and April 1809 he listened to a further twenty-two lectures by William Nicol, Moyes's nephew.[11]

These lecturers and others appeared also in Doncaster. Stancliffe, who described himself as 'Bachelor of Physic of the University of Cambridge; Fellow of the Linnean Society; Lecturer on Chemistry at the Middlesex Hospital and Leverian Museum, London; Honorary Member of the Literary and Philosophical Society, Newcastle-on-Tyne etc.', announced 'Twelve Experimental Lectures on Chemistry, Galvanism, Agriculture, the Arts and Manufactures of the United Kingdom'.[12] Richard Dalton, of York, lectured in Doncaster, August 1807, and lectured also in York in 1807, 1812, 1813, 1816, 1825, and 1826. William Nicol, besides appearing with Moyes, lectured on his own account in Doncaster, in October 1808. He is also found in Sheffield in 1809, 1810, and 1818, and in Leeds in 1810. John Banks, who des-

---

[1] Ward, 'Diary', 27 Feb. 1805, p. 160. Also 'Minute Book'.
[2] 'Diary', p. 161.   [3] *Ibid.*, p. 165. Favell gave this paper.
[4] *Ibid.*, p. 170.   [5] 'Minute Book'.
[6] Ward, 'Pocket Book', 1802.
[7] *Ibid.*, 1806: see also 'Diary', p. 110.
[8] 'Diary', p. 110.   [9] 'Pocket Book', 1807.
[10] *Ibid.*   [11] *Ibid.*, 1809.
[12] *Doncaster, Nottingham and Lincoln Gazette*, 7 Feb. 1806.

cribed himself as 'Lecturer in Philosophy to the College of Arts and Sciences in Manchester', advertised a course of lectures at Doncaster on 'Mechanics, Pneumatics, Hydrostatics, Hydraulics, Optics, Electricity, Magnetism and Astronomy' in March 1787.[1]

VIII

From Leeds, too, there is similar evidence. We have already mentioned the scientific interests of John Marshall, the great flax-spinner,[2] but they merit closer scrutiny. His surviving notes of 'Philosophical Lectures & Extracts',[3] for example, reveal the same pattern as we have found in Manchester, Birmingham, and Sheffield. His first notes were on James Booth's 'Philosophical Lectures' in December 1790, including pneumatics, hydrostatics, hydraulics, astronomy, optics, pneumatic chemistry, electricity, mechanics, and miscellaneous topics, including 'Water Wheels and Steam Engines', all these lectures being illustrated by numerous experiments and models. Later, in 1804, Marshall also heard lectures by Warltire and in 1805 by Moyes (whose works he had referred to in his notes on Booth's lectures). Moyes's lectures were mostly geographical and geological, and the remainder of Marshall's notebook is filled with references from numerous books and journals on these subjects, including geological works by such men as Hutton, Bruce, De Luc, and many others; these reveal a particularly strong interest in coal measures. Marshall also wrote notes from 'Leslie on Heat', 1804.[4] Another subject in which he was obviously very interested was 'the new invention for bleaching' by means of 'dephlogisticated marine acid' (chlorine), which 'discharges all colours from vegetable substances [and] . . . whitens a p[iec]e of cloth in a few hours'. These notes were taken from Booth's lecture on pneumatic chemistry, which included explanations of the preparation and use of the new bleaching agent. Moreover, at the end of these lecture-notes, Marshall added

[1] *The Yorkshire Journal and General Weekly Advertiser*, 10 and 24 March 1787. Since the above was written, we have been informed of an undated handbill in the Sheffield Public Reference Library (Miscellanous Papers, M.P. 415 S.), in which Banks advertised a lecture on water-wheels, illustrated with experiments. We are grateful to Dr. A.J. Pacey, of the University of Manchester Institute of Science and Technology, for giving us this reference. Dr. Pacey is interested in tracing the influence of Galileo's *Discorsi* in England and suggests that Banks derived his theoretical knowledge on the strength of materials, for instance, from the works of William Emerson, who had expressed Galileo's theory in Newtonian terms. See Banks, *On the Power of Machines* (1803), pp. 73–4.
[2] See above, p. 78, and Rimmer, *op. cit.*, pp. 14, 50–3, 103 and 105.
[3] In the Brotherton Library, University of Leeds, Marshall MS. 42.
[4] The work referred to was by Professor John Leslie, of Edinburgh University: *An Experimental Inquiry into the Nature and Propagation of Heat* (1804).

extracts from Scheele's *Chemical Essays* on the preparation and properties of 'dephlogisticated muriatic [or marine] acid'.

Some years later, after becoming well established as a flax-spinner,
Marshall must have recalled these early chemical notes, when he commenced a series of experiments on chlorine bleaching, which formed
the subject of another very interesting notebook, compiled between
1797 and 1802.[1] With the details of these experiments we shall deal
later, in the chapter on bleaching, but we wish to emphasize here
how thoroughly Marshall investigated all published scientific works
on this subject, especially those of French chemists such as Berthollet,
Lavoisier, Pajot des Charmes, and Chaptal, and also of British chemists;
he also obtained practical assistance from French chemists employed
in the Lancashire cotton industry. At the same time, his own thorough
experiments reveal a rational, scientific procedure, and a not inconsiderable scientific knowledge.

The same is also true of his investigations into mill construction,
textile machinery, spinning and weaving processes, etc., which form
the subjects of other notebooks.[2] Marshall, it is clear, acquired all the
best practical knowledge available from the late 1780s onwards, from
leading millwrights and machine-makers, the emergent mechanical
engineers,[3] but his notes also reveal his strong interests in applied
science. Thus, writing on the design and construction of gear-wheels,
he refers to a book on *The Theory of Wheels* (1795) by 'M.M.', in
trying to determine mathematically 'the true form of the teeth', i.e.
their correct pitch and depth.[4] Similar mathematical calculations are
included on the sizes (lengths and diameters) of bobbins and on the
dimensions, speeds, etc. of rollers, spindles, gearing, etc. in spinning
and roving frames.[5] Mathematical formulae are likewise employed
in regard to the shapes and sizes of iron beams, in expansion tables
(taken from the *Philosophical Transactions*), and concerning the 'Power
required for working Machines', together with 'Apparatus for Experiment'.[6] He also noted the chemical formula of the cement patented
by the chemist Bryan Higgins,[7] doubtless for use in factory-building.
Much the same impression is given by the notebooks on spinning and
weaving, with carefully controlled experiments, machine-drawings,

---

[1] In the Brotherton Library, University of Leeds, Marshall MS. 55. The first
page of these notes (crossed out), shows that Marshall had, in fact, begun bleaching experiments with caustic potash in 1788, but apparently discontinued them.
   [2] In the Brotherton Library, University of Leeds, Marshall MSS. 53, 54, 56, and
57.   [3] See below, pp. 399–400 and 443–5.
   [4] MS. 57, pp. 24–5 and 38.   [5] *Ibid.*, pp. 34–7 and 48–53.
   [6] *Ibid.*, pp. 40–1, 46.
   [7] *Ibid.*, p. 70. See above, p. 122, for Higgins's patent.

and mathematical calculations regarding sizes, speeds, etc.[1] What non-sense this all makes of traditional notions about the universality in the textile industry of unlettered, empirical craftsmen, operating by rule-of-thumb!

Benjamin Gott, the equally famous Leeds woollen manufacturer, was likewise scientifically minded. In dyeing, for example, he made similar use of applied chemistry. Dr. and Mrs. Clow even go so far as to say that he was 'the first dyer to attempt scientific control of his industrial processes', though his priority in this field is doubtful.[2] He did, however, spend thousands of pounds on systematic experiments with steam dyeing, in which he corresponded lengthily with John Southern, of Boulton and Watt's Soho factory, who supplied him with scientific and technical advice, including a copy of Fourcroy's *Philosophy of Chemistry*. His sons, too, like those of Boulton and Watt, were given a scientific training, including, for example, practical experiments from William Henry's *Elements of Experimental Chemistry*.[3]

Benjamin Gott and his sons were not the only ones in Leeds to utilize textile chemistry. Benjamin Wilson (1721–88), F.R.S., outstanding both as a painter and natural philosopher, gave scientific help to James Berkenhout, son of a Leeds merchant, in developing a scarlet dye for cotton:[4] '. . . he pointed out to him what authors he should read, made some chemical experiments for him, and explained the philosophy on which they seemed to depend, and the reason why animal substances were more apt to receive the scarlet dye than vegetable substances'. As a result, Berkenhout made considerable advances and in 1779, with the support of Wilson, Sir George Savile, and others, succeeded in obtaining a £5,000 reward from the government for his discoveries.[5]

---

[1] MSS. 53, 54, and 56.   [2] Clow, *op. cit.*, p. 220, Cf. below, chap. IX.

[3] W. B. Crump, *The Leeds Woollen Industry 1780–1820* (Leeds, 1931), pp. 272–312; Clow, *op. cit.*, pp. 220–3. These statements are based on the Bean Ing Mill Note-Book, 1800–25, and the Pattern Book of William Gott, Benjamin's son, dated 1815, both in the Brotherton Library, University of Leeds, Gott MS. XIV (i) and MS. XVI respectively.

[4] R. V. Taylor, *Biographia Leodiensis* (1865), p. 646. Berkenhout's process was for dyeing cotton red with cochineal. More information about Wilson is provided by G. L'E. Turner in *Notes and Records of the Royal Society*, Vol. 22 (1967), pp. 105–12. Son of a once-wealthy Leeds merchant, he became best known as a painter, but also developed an interest in experimental philosophy through Dr. (later Sir) William Watson, F.R.S. He was also on friendly terms with John Smeaton, the engineer. It was apparently for his published works on electricity that he was elected F.R.S. in 1751, but he also obviously knew a good deal of chemistry.

[5] Taylor, *loc. cit.* See also *Journals of the House of Commons*, Vol. XXXVII, pp. 392–3 and 422; E. Bancroft, *Experimental Researches concerning the Philosophy of Permanent Colours* (2nd edn., 1813), Vol. I, pp. 536–8; A. P. Wadsworth and

Scientific interests in Leeds had been stimulated some years earlier by the itinerant lecturers in natural philosophy,[1] and especially by Joseph Priestley's ministry at the Mill Hill Chapel.[2] Priestley's successor, William Wood, sustained these interests,[3] and there appear to have been the same links between Nonconformity, science, and industry in Leeds as in Manchester, Birmingham, and Sheffield. Similar links also existed between medical men and scientifically-minded industrialists, though Dr. William Hey (1736–1819), F.R.S., the Leeds surgeon, who played a role in Leeds similar to that of Dr. Thomas Percival in Manchester, was an Anglican.[4] Hey's early education was at Heath Academy, near Wakefield, run by Joseph Randall, 'whose attainments in the mathematics and the science of calculation was [sic] considerable',[5] and among the tutors there was John Arden (later, as we have seen, well known as an itinerant lecturer),[6] from whom Hey acquired 'a taste for natural philosophy'. After his medical education at St. George's Hospital, London, Hey returned to Leeds in 1759 to commence medical practice; he played a leading part in founding Leeds Infirmary in 1767–8 and in establishing a local medical society and library. About this time Hey became very friendly with Priestley, then at Mill Hill Chapel; he shared the latter's chemical interests, and it was with Priestley's support that he was elected F.R.S. in 1775. It is not surprising, therefore, to find Hey playing a leading role in the establishment in 1783 of a Leeds Philosophical and Literary Society, of which he became president.[7] Unfortunately, the records of

---

J. de L. Mann, *The Cotton Trade and Industrial Lancashire, 1600–1780* (Manchester, 1931), p. 181, n.1. Berkenhout's process, however, does not appear to have proved very successful. See below, p. 343, n. 4.

[1] We have not been able to make a detailed study of the Leeds newspapers, but sampling reveals frequent visits by these lecturers, as in Manchester and other towns.

[2] Gibbs, *Joseph Priestley*, chaps. 4–6, *passim*; W. L. Schroeder, *Mill Hill Chapel Leeds 1674–1924* (1924); F. Beckwith, 'The Beginnings of the Leeds Library', *Thoresby Society Publications*, Vol. XXXVII (1945), pp. 145–65. Priestley was in Leeds from 1767 to 1773, and played a leading role in founding the Library in 1768.   [3] Schroeder, *op. cit.*, pp. 50–1.

[4] J. Pearson, *The Life of William Hey, Esq. F.R.S.* (1822).

[5] *Ibid.*, pp. 3–4. See also Hans, *op. cit.*, pp. 94–5. George Gargrave (1710–85), the mathematician (see *D.N.B.*), also taught there. The academy failed, however, in 1754, whereupon Randall established a school at York, while Gargrave set up a 'Mathematical School' at Wakefield. In 1765 Randall published *An Introduction to so much of the Arts and Sciences more immediately concerned in an excellent education for trades in its lower scenes and more genteel professions and for preparing Young Gentlemen in Grammar Schools to attend lectures in the Universities.*

[6] See above, p. 104. Hey later met him, in 1778, giving a course of philosophical lectures in Bath. Pearson, *op. cit.*, p. 49.

[7] *Ibid.*, pp. 51–4. Hey was also elected a member of the recently founded Literary and Philosophical Society in Manchester.

this Society have disappeared and little is known of its activities,[1] but local newspapers provide some interesting references. Its formation was announced in March 1783: 'Some Gentlemen in Leeds being of Opinion that the Establishment of a Society for promoting Natural Knowledge would be a Source of Improvement and Entertainment, have agreed to request the Concurrence of such Gentlemen and Others in this Town and Neighbourhood as are desirous of encouraging this Scheme.'[2] An inaugural meeting was to be held in the Circulating Library on the following Thursday evening, 6 March. It was then agreed to form a 'Philosophical and Literary Society', at the first meeting of which, on Wednesday, 9 April 1783, 'a Discourse on the Utility of Philosophical Inquiries' would be delivered.[3]

Further information about this Society is provided by a later account of its origins,[4] which, after referring to Priestley's experiments in Leeds on gases and electricity and to his friendship with Hey in the late 1760s and early 1770s—until he left the town in 1773—went on to relate how such informal philosophical contacts led to the formation of a regular society.

Besides his friend Mr. Hey, there were several in the vicinity of Leeds with whom Priestley enjoyed congenial intercourse. Mr. Mitchell, of Thornhill, the discoverer of the method of making artificial magnets, Mr. Dawson, of Royds Hall, the founder of the Low-moor Iron-works, and Mr. Smeaton, of Austhorpe, formed with him no contemptible philosophic coterie, [which] soon afterwards matured into a regular association. Ten years after Dr. Priestley left Leeds—in 1783—a society called the Leeds Philosophical and Literary Society was organized, partly from the materials of this early association. Mr. Hey was its first president, and Mr. Sheepshanks its secretary; and it consisted of from twenty-five to thirty members. Amongst the more distinguished of these were the successor of Dr. Priestley at Mill Hill, the learned and scientific Mr. Wood, Dr. William Hird, the biographer of Samuel Fothergill, a physician of great skill, and a man of exquisite taste, the Rev. Miles Atkinson, and Mr. Gamaliel Lloyd. The experimental operations of this society were greatly facilitated by the ingenuity of Mr. Aspdin, a watch-maker in the town, who made the greater part of its apparatus.[5]

Thus the medical and clerical professions were strongly represented in the Society, which brought together men of different religious denominations. Dr. William Hey, an Anglican with Methodist

---

[1] Some of Hey's papers have been reprinted in the *Proceedings of the Leeds Philosophical Society*, Vol. VI, pt. ii (1952), pp. 65–77.

[2] *Leeds Intelligencer*, 4 March 1783.   [3] *Ibid.*, 11 March 1783.

[4] *Leeds Mercury*, 13 Oct. 1832.

[5] Other members mentioned were 'Mr. Blayds and Col. Smithson'; the former was probably John Blayds, a Leeds banker.

sympathies, and Dr. William Hird, apparently a Quaker,[1] were both associated for many years with the Leeds General Infirmary. The Rev. William Sheepshanks and the Rev. Miles Atkinson were Anglican clergymen,[2] and so, too, was the Rev. John Michell (1724–93), rector of Thornhill, well known for his discoveries not only in magnetism but also in optics, astronomy, and geology, and a friend of Priestley;[3] Michell was also closely associated with the founders of the Lunar Society of Birmingham, Dr. William Small, Dr. Erasmus Darwin, and Matthew Boulton, visiting them at their houses in Birmingham and Lichfield, and discussing scientific matters, such as electrical experiments, with them. William Wood, as we have previously mentioned, was a Nonconformist minister, successor of Priestley at Mill Hill Chapel; a founder-member of the Linnaean Society, he wrote numerous botanical contributions for the *Annual Review* and Rees's *Cyclopaedia*.[4] Among the industrial members, John Smeaton, the famous engineer, was outstanding; his works were at Austhorpe, near Leeds.[5] Since engineering was closely associated with clock- and instrument-making, Smeaton, himself trained as an instrument-maker and responsible for several improvements in scientific instruments, would appreciate the skills of the clock-maker, Aspdin, who made the Society's apparatus. Another outstanding industrial member was Joseph Dawson (1740–1813), founder of the famous Low Moor iron-works, near Bradford.[6] Born in humble circumstances, but of considerable intellectual ability, he was enabled by Nonconformist charity to be educated at Daventry Academy and then at Glasgow University. In 1768 he was ordained minister of Upper Chapel, Idle, near Leeds, where he became the intimate friend of Dr. Priestley, whose religious opinions and scientific interests he shared.[7] All accounts agree on his scientific enthusiasm and ability, and on his application of scientific knowledge to industry. At first he supplemented his small ministerial

[1] He wrote a *Tribute to the Memory of Dr. Fothergill*, the Quaker. See R. V. Taylor, *Biographia Leodiensis* (1865), p. 336.

[2] See Taylor, *op. cit.*, pp. 239–42 and 242–7 respectively.

[3] See *D.N.B.* and cuttings in Leeds Central Library. 'Michell' is the usual spelling of his name, but 'Mitchell' occurs in contemporary and later literature.

[4] Taylor, *op. cit.*, pp. 232–9, and Schroeder, *op. cit.*, pp. 50–1.

[5] Pearson, *op. cit.*, p. 52, states that Smeaton attended some of the Society's meetings, but that it is uncertain whether he was a regular member or an occasional visitor. See above, pp. 73–4, for Smeaton's scientific and technological interests.

[6] J. James, *The History and Topography of Bradford* (1841), p. 328; J. H. Turner, *Nonconformity in Idle* (1876), pp. 46–8; R. V. Taylor, *Yorkshire Anecdotes* (1883), pp. 129–30; [J. Burnley, ed.], *Fortunes Made in Business* (1884), Vol. I, pp. 87–128.

[7] James, *op. cit.*, stated that some of Priestley's scientific apparatus was still preserved at Royds Hall, Dawson's residence.

stipend by running a private school, but then started some coal mines near to the chapel and eventually, in about 1790, played a leading role in establishing the Low Moor ironworks. His knowledge of geology, mineralogy, metallurgy, and chemistry is said to have helped him considerably in exploiting these mineral resources. When he became president of the Yorkshire and Derbyshire Ironmasters' Association, on its establishment in 1799, he proposed that its activities should include the reading of papers on iron-manufacturing processes, and that a gold medal should be awarded annually for the best contribution; he himself set the ball rolling with a paper on 'The Effects of Air and Moisture on Blast Furnaces', and others followed.[1]

The Philosophical and Literary Society also included some commercial men, such as Gamaliel Lloyd, woollen merchant, mayor of Leeds in 1778, a founder of the city library and advocate of general education.[2] For unknown reasons, however, the Society lasted only three years,[3] and the same fate befell two subsequent efforts to found a philosophical society in Leeds.[4] Edward Baines, in his Leeds directory of 1817, lamented the fact that 'Philosophical researches are not much cultivated in Leeds', except—but it is a noteworthy exception—for those 'which have an immediate reference to Commerce and Manufacture'. In the following year, however, a Philosophical and Literary Society was permanently established in the city, including among its members leading industrialists such as John Marshall (flax-spinner), Benjamin and John Gott (woollen manufacturers), John Bischoff (woollen manufacturer), Edward S. George (chemical manufacturer), James Fenton and Matthew Murray (engineers), as well as medical men, including William Hey and his son, and several merchants.[5] Many of the papers read to the Society were of scientific-industrial interest.

IX

In Newcastle upon Tyne, in this same period, we come across similar evidence. James Jurin (1684–1750),[6] a product of Christ's Hospital School[7] and Cambridge graduate, who eventually became a reputable

---

[1] T. S. Ashton, *Iron and Steel in the Industrial Revolution* (1963), pp. 179 and 222. These researches and papers may be compared with those of David Mushet at this same time in Scotland (see below, pp. 184–5).

[2] Taylor, *Biographia Leodiensis*. His father, George Lloyd (d. 1783), was a F.R.S.

[3] *Leeds Intelligencer*, 13 Oct. 1832; Pearson, *op. cit.*, p. 52.

[4] E. Kitson Clark, *The History of 100 Years . . . of the Leeds Philosophical and Literary Society* (Leeds, 1924), p. 2.   [5] *Ibid.*, chap. II.

[6] *D.N.B.* and Hans, *op. cit.*, p. 39. See also above, p. 139.

[7] Which, as Hans, *op. cit.*, pp. 38–9, has pointed out, began to introduce mathematical and scientific subjects in the second half of the seventeenth century.

scientist, a Fellow and Secretary of the Royal Society, was a master at Newcastle Grammar School between 1709 and 1715 and lectured on experimental philosophy. These scientific interests were maintained in the area by John Horsley (1685–1732),[1] son of a tailor, educated at Newcastle Grammar School and Edinburgh University (M.A. 1723), who became a Presbyterian minister at Morpeth, where he established a private school. Elected F.R.S. in 1730, he was in the following year advertising in the *Newcastle Courant*, a course of experimental philosophy to be delivered in Morpeth, and the extensive scientific apparatus which he accumulated was acquired after his death by Dr. Caleb Rotheram, of Kendal Academy, and was eventually transferred to Warrington Academy.[2] Likewise, his *Short and General Account of . . . fundamental Principles of Natural Philosophy* was revised and 'adapted to a Course of Experiments perform'd in Glasgow' by John Booth in 1743.[3]

Horsley's place in Newcastle was soon filled by Isaac Thomson, who established a regular course of natural philosophy there in 1739, assisted by Robert Harrison (1715–1802).[4] Together they published in 1757 *A Short Account of a Course on Natural and Experimental Philosophy*, which they were to deliver as evening lectures in the Cordwainers' Hall—fifteen lectures in all, including mechanics, hydrostatics, pneumatics, optics, and astronomy. In 1757 Harrison became master of the Trinity House School in Newcastle.

Another outstanding figure in Newcastle was Charles Hutton (1737–1820), F.R.S., son of a miner, who became one of the most famous mathematicians of his day.[5] The utilitarian aims of his Mathematical School, which he took over from a Mr. James whose evening classes he had previously attended, are illustrated by the following advertisement which he published in 1760:[6]

To be opened on Monday, April 14th, 1760, at the Head of the Flesh Market, down by the Entry, Newcastle, A Writing and Mathematical School, where persons may be fully and expeditiously qualified for business and where such as intend to go through a regular course of Arts and Sciences may be completely grounded therein at large, viz. in Writing according to its latest and best improvements: Arithmetic in all its parts; Merchant's Accounts (or the true Italian method of Book-keeping); Algebra; Geometry, elemental and practical; Mensuration; Trigonometry, plane and spherical; Projection of the Sphere;

[1] *D.N.B.*; Hans, *op. cit.*, p. 59; Taylor, *Math. Pract. of Hanov. England*, pp. 8 and 130.　　　　　　　　　　[2] See above, pp. 90 and 103.

[3] F. W. Gibbs, *Ambix*, Vol. VI (1960), p. 112.

[4] Hans, *op. cit.*, p. 144. For Harrison, see also *D.N.B.*

[5] *D.N.B.*; Hans, *op. cit.*, pp. 109–10 and 189; Taylor, *op. cit.*, pp. 34 and 235.

[6] Hans, *loc. cit.*

Conic Sections; Mechanics; Statics and Hydrostatics; the Doctrine of Fluxions, etc. Together with their various applications in Navigation, Surveying, Altimetry and Longimetry; Gunnery, Dialling, Gauging, geography, Astronomy, etc. etc. Also the use of Globes, etc. Likewise shorthand according to new and facile characters never yet published by C. Hutton.

His school was also a centre of public lectures on natural philosophy, by such men as Rotheram and Ferguson.[1] He wrote many works on mathematics and natural philosophy, including *A Treatise on Mensuration* (1767)[2] and *A Mathematical and Philosophical Dictionary* (1796).

In 1773 Hutton was appointed Professor of Mathematics at Woolwich Military Academy, but scientific interests continued to flourish in Newcastle. We come across Henry Moyes, Gustavus Katterfelto, and others lecturing on natural philosophy, and in 1793 the Newcastle Literary and Philosophical Society was founded, following the Manchester and Birmingham precedents.[3] A Unitarian minister, the Rev. William Turner, who had been educated at Warrington Academy, played a leading role in the establishment of this Society, similar to that of the Rev. Thomas Barnes in Manchester. Like the Manchester Society, that in Newcastle was broad in its scope, with moral and literary as well as scientific purposes, but from the first it showed a marked interest in the application of natural philosophy to industrial processes, particularly to the local industries of coal-, lead-, and iron-mining, metal, chemical, and glass manufactures.[4] The papers on coal-mining were, as might be expected, most numerous of all. Indeed, the Newcastle Literary and Philosophical Society was the cradle of the North of England Institute of Mining Engineers, eventually established in 1852.

Regular courses of scientific lectures were instituted by the Newcastle Society in 1803, when the Rev. William Turner was appointed to 'a permanent Lectureship in the several branches of Natural and Experimental Philosophy, Chemistry, etc.',[5] a post which he held until 1833.

[1] Hans, *op. cit.*, pp. 110, 148, and 159. Hans confuses Dr. Caleb Rotheram (1694–1752), of Kendal Academy, with his son Caleb (1738–96), who was a friend of Priestley's. [2] See above, pp. 111–12, for its advertisement in Manchester.
[3] R. S. Watson, *The History of the Literary and Philosophical Society of Newcastle-upon-Tyne* (1897). The references to Moyes and Katterfelto are on pp. 203–6.
[4] See especially Turner's paper on the objects of the proposed society (*ibid.*, pp. 35–40) and the brief survey of papers read at the monthly meetings (*ibid.*, chap. vii). When Joshua Gilpin, the American paper-manufacturer, visited the Society in 1795, in company with William Turner, he was shown, among other things, a 'Collection of the Strata to 135 fathoms' ('Journal', Vol. IV, 5 Oct. 1795).
[5] Watson, *op. cit.*, pp. 208–16. Previously, it was pointed out, scientific education could be acquired only 'by expensive residence in a distant capital', or by 'the precarious arrival of occasional lecturers' (*ibid.*, p. 217).

'Without mechanics and chemistry', it was pointed out, 'how wretched would be our manufactures, and how worthless the amount of our mineral possessions! Exactly in proportion to the state of these sciences is the condition of the one and the value of the other.' Turner was provided with a lecture room and a 'philosophical apparatus' formerly belonging to Dr. Thomas Garnett (recently deceased) of the Royal Institution. He greatly stressed in his introductory discourse 'the application of Philosophy to the arts' (manufactures), especially to local industries, and in formulating his courses he was careful 'to apply the [scientific] principles laid down to the explanation of the machines, etc. most useful in this district'.[1]

The Newcastle Literary and Philosophical Society was in touch with the Lunar Society through Richard Chambers's correspondence with Matthew Boulton. The plans for the society were submitted to Boulton in 1793 and he was also asked to honour them with his 'Observations upon any Subject connected with Coal or the Coal Trade—particularly if you could furnish us with an account of the Wedgeburgh mines ...'[2] As early as 1784, Chambers had introduced Henry Cort to Boulton. In April 1795 Chambers wrote on behalf of the Literary and Philosophical Society to Boulton requesting him to distribute some enclosed letters about the coal trade 'among the most intelligent Gentlemen concerned in Coal Works in your Neighbourhood'.[3] It is clear that the philosophical societies could be a means of diffusing applied science throughout the country.

X

The scientific-industrial activities in large cities such as London, Manchester, Birmingham, Sheffield, Leeds, and Newcastle were matched by similar interests in smaller towns. Derby, for example, had a philosophical society begun in 1783, though formally inaugurated in 1784.[4] The moving spirit in it was Erasmus Darwin, who had previously had a hand in founding the Lichfield Philosophical Society and the Lunar Society. One of the first letters relating to the Derby Philosophical Society is an invitation from Darwin to the 'Lunatics' to hold joint meetings.[5] The members of the society were varied. They included local worthies like Richard Leaper, C. S. Hope, and John Crompton,

---

[1] Watson, *op. cit.*, p. 223.

[2] R. Chambers to M. Boulton, 6 March 1793, A.O.L.B.

[3] R. Chambers to M. Boulton, 30 April 1795, A.O.L.B.

[4] E. Robinson, 'The Derby Philosophical Society', *Annals of Science*, Vol. IX (1953), pp. 359–67. See below, chap. IV.

[5] E. Darwin to M. Boulton, 17 Jan. 1784, A.O.L.B.

all of whom became mayors of Derby; the local aristocracy such as Lord Belper and Sir Brooke Boothby; Josiah Wedgwood and his inventor-cousin, Ralph Wedgwood; a number of local medical men; and last, but not least, local industrialists, such as William Strutt, William Duesbury, and Dr. Robert Bage of Elford.

William Strutt, son of Jedediah Strutt, was a co-founder with Erasmus Darwin of the Derby Society, and on Darwin's death in 1802 succeeded him as President of the Society.[1] Through Darwin, Strutt became a close friend of R. L. Edgeworth, another member of the Lunar Society. He also knew Peter Ewart, with whom he discussed his scientific writings, and Ewart's friend, George Lee, while Thomas Hewes built two of the wheels for the Strutts' West Mill;[2] thus he had close links with scientifically-minded industrialists in Manchester as well as in Birmingham. Strutt is typical of the second generation of educated industrialists: 'Although it was Strutt's mechanical genius that secured his election as Fellow of the Royal Society in June, 1817, when he was 60, he maintained throughout life an unflagging interest in many branches of pure science. His reading was wide—it included the works of Newton, Euler, Blair, Priestley, Landen and Vince . . .'[3] He was no doubt assisted in achieving this wide range of reading by the fine collection of books bought by the Derby Philosophical Society, which the members of the Society were entitled to borrow.[4] William Duesbury, another of the industrialist members of the Derby Society, was well known as a porcelain manufacturer of the period, and Robert Bage, the paper-manufacturer and friend of William Hutton, also belonged to the Society.

The Derby Philosophical Society was probably typical of the societies in Leicester, Northampton, Exeter, and other middle-sized towns. John Coltman (1727–1808), one of the leading worsted spinners and hosiers in Leicester, was a man of wide culture. Educated at the Nonconformist academy at Kibworth (forerunner of Warrington Academy), under the Rev. John Aikin, he was later in the habit of meeting with various friends 'for the discussion of literary and philosophical subjects every week' and established the first Book Society in the town. 'Mr. Coltman, too, ever anxious for the promotion of science, invited to his house, Mr. Waltyre [Warltire], a lecturer on Electricity, Hydraulics, etc.'[5] Dr. Priestley also visited him.[6] Later, in

---

[1] R. S. Fitton and A. P. Wadsworth, *The Strutts and the Arkwrights, 1758–1830* (Manchester, 1958), p. 175.

[2] *Ibid.*, pp. 175–7, 207*n*., 215*n*. For Ewart, Lee, and Hewes, see above, pp. 69–71. and 98–100.　　　[3] *Ibid.*, pp. 170–1.　　　[4] See below, pp. 194–6.

[5] Catherine H. Beale, *Catherine Hutton and Her Friends* (Birmingham, 1895), pp. 44 and 64–5.　　　[6] *Ibid.*, p. 87.

1790, Coltman joined with Richard Phillips, bookseller and proprietor of the *Leicester Herald* (subsequently founder of the famous London publishing house), and with others such as Dr. Thomas Arnold, a well-known physician and writer on insanity, and Robert Brewin, another worsted spinner, who was related to Dr. Priestley, in the establishment of a literary and philosophical society in Leicester.[1] At this same time Phillips also established an Adelphi Society, a group of young men interested in the study of science and Radical politics.[2] These activities were stimulated in Leicester, as in other towns, by courses of lectures on 'natural and experimental philosophy, astronomy, chemistry, etc.', given in the town hall by various itinerant philosophers.[3]

These scientific-industrial interests were not confined to the northern and midland provincial towns. In April 1757, for example,

We hear from Bath, Bristol, Frome, Warminster and other places that great numbers of Gentlemen, Ladies, etc. who have lately attended Mr. [William] Griffis's Lectures upon Experimental Philosophy, Chemistry, Fortification, etc. expressed the highest satisfaction at the great Variety of Experiments which he performs with a very large and noble Set of Instruments.[4]

Now he was to visit Salisbury, where he would give lectures, illustrated with '40 models', on Newtonian natural philosophy, mechanics, hydrostatics and hydraulics, pneumatics, and the 'application of these Principles to the Improvement of Trade in several Arts and Professions', followed by lectures on optics, military architecture (fortification) geography, and astronomy—the whole for a subscription of one guinea.

During the later eighteenth century, Griffis was followed in the south-west by all the well-known itinerant lecturers. Advertisements appeared in the Bristol newspapers for courses by John Arden, James Ferguson, John Warltire, Henry Moyes, Adam Walker junior, and Henry Clarke, while a local personality, Benjamin Donne (or Donn), gave several courses each year from 1765 to 1798. On several occasions during this period there were two lecturers operating simultaneously within the town, while Donne was often holding courses in the same

---

[1] Beale, *op. cit.*, p. 110; A. Temple Patterson, *Radical Leicester* (1954), pp. 67–8.
[2] Patterson, *op. cit.*, p. 68. The anti-Radical attacks of 1793–4, however, brought about its collapse, for 'experiments with electricity had become confounded in the public mind with the construction of infernal machines' (*ibid.*, p. 73).
[3] *Ibid.*, p. 13.
[4] *Salisbury Journal*, 12 April 1757. (We are very grateful to Miss Julia de Lacy Mann for this reference.) As we have previously mentioned, this 'celebrated Philosophic Apparatus' of Griffis (or Griffith) was later acquired by Adam Walker (see above, p. 105).

week both at Bristol and Bath. Donne, who published *An Epitome of Natural and Experimental Philosophy* in 1769, came of a mathematical family, his father, George Donn, having kept a mathematical school at Bideford, Devon, in the early eighteenth century.[1] Benjamin established a similar academy in Bristol, in which

> Young Gentlemen are boarded and taught Writing, Arithmetic, Book-keeping, Navigation and Geography. Also the elements of Algebra, Altimetry, Architecture, Astronomy, Chances, Conics, Decimals, Dialling, Fluxions, Fortification, Gauging, Geometry, Gunnery, Hydraulics, Hydrostatics, Level-ling, Mechanics, Mensuration, Optics, Perspective, Pneumatics, Ship-building, Surveying, Trigonometry Plane and Spherical, with the use of Mathematical and Philosophical Instruments.[2]

In addition to being a teacher and lecturer, Donne was also an instrument-maker, surveyor, and map-maker.

Bristol was also fortunate in numbering among its citizens Dr. Thomas Beddoes, formerly Reader in Chemistry at Oxford,[3] and founder of the Bristol Pneumatic Institute, where Humphrey Davy began his scientific career. Beddoes was a close friend of James Watt, Erasmus Darwin, and other members of the Lunar Society, and he was related by marriage to R. L. Edgeworth. The Wedgwoods knew him and were in the habit of staying in Clifton. He assisted the Oxford engineer and aeronaut, James Sadler, in his experiments upon steam engines in 1792–3.[4]

While Warltire was lecturing in Bristol, his host was Joseph Fry, who corresponded with James Watt about Warltire's attempts to obtain a soft metal from manganese.[5] Fry shared the chemical interests of Warltire and Watt, being engaged at this time, together with John Collison, in the establishment of a works at Battersea, London, for making alkali from common salt for use in his soapworks at Bristol.[6]

Wherever one cares to look—in large industrial cities or smaller

---

[1] For Benjamin Donn(e), see *D.N.B.*; Hans, *op. cit.*, pp. 99–101 and 146; and E. Robinson, 'Benjamin Donn (1729–98), Teacher of Mathematics and Navigation', *Annals of Science*, Vol. XIX, no. 1 (March 1963), and 'Dr. Beddoes and Benjamin Donn on the teaching of Mathematics', *Mathematical Gazette*, May 1965, note 2603.

[2] Advertisement in his *Epitome*, quoted by Hans, *op. cit.*, p. 100.

[3] See below, pp. 176–7.

[4] A. Raistrick, *Dynasty of Ironfounders. The Darbys of Coalbrookdale* (1953), pp. 158–9.

[5] Joseph Fry to James Watt, 1st Month 16, 1783, Dol.

[6] L. Gittins, 'The Manufacture of Alkali in Britain, 1779–1789', *Annals of Science*, Vol. XXII, no. 3 (Sept. 1966), p. 177. See also above p. 133, and below, pp. 361–2, 364–6.

market towns[1]—one comes across similar evidence of the diffusion of scientific and technical knowledge in lectures, schools, books, etc. The ways in which such knowledge was diffused, at various levels of society, are also demonstrated by the experiences of Thomas Young, the eminent natural philosopher.[2] Whilst a boy, he was given the run of a neighbouring land-surveyor's library, in which he found 'many books relating to science, and particularly a Dictionary of Arts and Sciences, in three volumes, folio'. He also had the use there of 'several mathematical and philosophical instruments'. Whilst at boarding school in the early 1780s, at Compton, Dorsetshire, though receiving a predominantly classical education, he learnt a great deal informally from Josiah Jeffrey, the school usher, who not only loaned him Benjamin Martin's *Lectures on Natural Philosophy*, but also taught him how to use the lathe, to make telescopes and other optical instruments, to grind and prepare colours, and to bind books. Privately educated between 1787 and 1792, in company with Hudson Gurney, grandson of David Barclay, the brewer and banker, at the latter's country house at Youngsbury, near Ware, he was able, in addition to the formal classical and linguistic studies, to read widely on his own in mathematics, astronomy, natural philosophy, and botany, including Newton's *Principia* and *Opticks*, the botanical works of Linnaeus, works on chemistry by Black and Lavoisier and on medicine by Boerhaave. In London during winter vacations at David Barclay's town house, he attended Dr. Bryan Higgins's lectures on chemistry. All this was before studying for the medical profession at London and Edinburgh, and then at Göttingen and Cambridge, and before being appointed in 1801 to the Chair of Natural Philosophy at the Royal Institution.

XI

The Industrial Revolution also made some impact upon the universities of Oxford and Cambridge,[3] though these have often been portrayed as backwaters of classical and theological studies in this

[1] In York, for example, we have found numerous newspaper advertisements of visits by itinerant lecturers and of schools providing instruction in mathematics and in commercial and technical subjects.

[2] A. Wood and F. Oldham, *Thomas Young Natural Philosopher 1773–1829* (1954). See also the older work by G. Peacock, *Life of Thomas Young* (1855).

[3] Dr and Mrs. Clow, while rightly emphasizing the great part played by the Scottish Universities and the Dissenting Academies in the 'Chemical Revolution', make only a cursory survey of developments in Oxford and Cambridge (*op. cit.*, pp. 600–3).

period. Joseph Priestley, for example, in his *Letters to Pitt* (1787), referred to them as 'pools of stagnant water', by comparison with the Dissenting Academies. And the *Edinburgh Review* later condemned Oxford particularly as a place 'where the dictates of Aristotle are still listened to as infallible decrees, and where the infancy of science is mistaken for its maturity, the mathematical sciences have never flourished; and the scholar has no means of advancing beyond the mere elements of geometry'; while at Cambridge, though 'mathematical learning is there the great object of study', the method of study was antiquated and other scientific subjects were neglected.[1] There is, of course, some truth in these contemporary statements, which have often been echoed by later critics, but they are greatly exaggerated and ignore new developments which were taking place in those ancient seats of learning. In the late seventeenth and early eighteenth centuries, as we have seen, mathematics, natural philosophy, and chemistry had become subjects of study there, with some energetic professors, who had introduced experimental methods.[2] It is true that in the second half of the eighteenth and early nineteenth centuries, apart from Henry Cavendish, the names that occur readily to our minds as representing Britain's contribution to science—such as Joseph Black, Joseph Priestley, and Humphry Davy—were not products of Oxford and Cambridge. Nevertheless, science was being studied there and increasing attention was being devoted to its industrial applications. It is significant, for instance, that Richard Watson (1737–1816), elected Professor of Chemistry at Cambridge in 1764, 'should have decided that the application of chemistry to the arts and manufactures was the most fitting theme at that time for a university course'.[3] Yet Watson is often taken as an example of the low depths to which chemistry teaching at Cambridge had sunk.[4] It is true, as he himself admitted,[5] that at the time of his appointment he 'knew nothing at all of Chemistry, had never read a syllable on the subject; nor seen a single experiment in it'; but he made 'extraordinary exertions' to remedy this ignorance, sent to Paris for a skilled operator (Hoffmann), buried himself in his laboratory, and in fourteen months was able to read a course of chemical lectures 'to a very full audience'. He also succeeded in obtaining for the Professor of Chemistry a stipend of £100 annually from the Crown, and in the following years continued to read lectures

---

[1] *Edinburgh Review*, Vol. XI (1808), p. 283.
[2] See above, pp. 31–6.
[3] F. W. Gibbs, *Annals of Science*, Vol. VII (1951), p. 236.
[4] As by Turner, *History of Science Teaching in England*, p. 46.
[5] *Anecdotes of the Life of Richard Watson, Bishop of Landaff; Written by Himself* (2nd edn., 2 vols., 1818), Vol. I, pp. 45–6.

'to very crowded audiences'. His published works,[1] closely examined by Dr. Coleby,[2] show that, although he kept in touch with the current researches of Black, Priestley, Cavendish, Macquer, Lavoisier, and other great contemporary chemists, and used them in his lectures, 'questions of fundamental theory were not his main interest . . . He was much more concerned with problems of physical chemistry . . . and with the application of chemical research to metallurgy and industry.' Dr. Sherwood Taylor has come to the same conclusion:[3] 'The experiments which he narrates as his own are chiefly concerned with manufacturing processes rather than with the advancement of pure science.' A 'pure' scientist may perhaps turn up his nose at such utilitarian concerns, but from our point of view they are of great interest.

In his *Plan of a Course of Chemical Lectures* (1771) he devoted considerable attention to the metallurgy and assaying of the principal metals, and industrial processes are given for green vitriol, alum, nitre, common salt, sal ammoniac, plate glass, porcelain, enamels, etc.; he also referred to mordant dyeing, giving Macquer's explanation, to the researches of Margraaf and Macquer on alum, and to those of Margraaf on the extraction of sugar from roots and on the fusible salt of urine (nitre or saltpetre). Watson's best-known work, however, was his *Chemical Essays*, which became one of the most widely read and appreciated books on chemistry of the age; they were intended, like his lectures, for those who had little or no previous knowledge of the subject—though they were constantly referred to in the works of the most knowledgeable scientists—and they proved very popular, over 2,000 copies of the first three volumes being sold in five years and the eleventh and last edition being reached by 1800.[4]

Watson's accounts of manufacturing and mining processes are based on first-hand knowledge obtained by actual visits and investigations. Thus he says, 'I have had the curiosity to go to the bottom of some of the most famous mines in England; such as the copper mine at

---

[1] *Institutiones Chemicarum . . . Pars Metallurgica* (Cambridge, 1768); *An Essay on the Subjects of Chemistry and their General Division* (Cambridge, 1771); *A Plan of a Course of Chemical Lectures* (Cambridge, 1771); *Chemical Essays* (5 vols., Cambridge, 1781–7). He also read five papers to the Royal Society (of which he was elected a Fellow) and one to the Manchester Literary and Philosophical Society.

[2] L. J. M. Coleby, 'Richard Watson, Professor of Chemistry in the University of Cambridge, 1764–71', *Annals of Science*, Vol. IX (1953), pp. 101–23. See also Wordsworth, *Scholae Academicae*, pp. 189–90, and *D.N.B.*

[3] *Philosophical Magazine*, Commemoration Number, July 1948, p. 163.

[4] Even after the revolution in chemistry in the late eighteenth and early nineteenth centuries, they were still highly regarded by chemists such as Davy and Thomson, who were themselves keenly interested in applied science. Coleby, *op. cit.*, p. 114.

Eaton; the coal-mines at Whitehaven and Newcastle; the cannel coal mines in Lancashire; and a variety of lead mines in Derbyshire and other places; . . . the rock salt mines at Northwich.'[1] He made innumerable similar visits to manufacturing establishments, which enabled him to write with practical authority as well as scientific insight. He had no doubts about the utilitarian possibilities of applied chemistry.[2]

The uses of chemistry, not only in the medicinal but in every economical art are too extensive to be enumerated, and too notorious to want illustration . . . It cannot be questioned that the arts of dying, painting, brewing, distilling, tanning, of making glass, enamels, porcelane, artificial stone, common-salt, sal-ammoniac, salt-petre, potash, sugar, and a great variety of others, have received improvement from chemical inquiry and are capable of receiving much more.

He particularly emphasized that the extraction of metals from their ores, and their combination into new alloys, could be greatly developed by chemical experiments; he pointed, by way of example, to the recent discovery of platinum, cobalt, and nickel, and chemical experiments upon them, and to Margraaf's work on extraction of zinc from calamine, which had made possible the establishment of the copper works near Bristol, to supply the Birmingham metal trades. He therefore proposed that a 'Chemical Academy' should be set up, with a view to 'the improvement of metallurgy, and the other mechanic arts dependent on chemistry'.

There were, of course, as he realized, impediments to the development of industrial chemistry. On the one hand, philosophers were too often cloistered academics. Watson emphasized that 'It is not enough to employ operators in this business; a man must blacken his own hands with charcoal, he must sweat over the furnace, and inhale many a noxious vapour before he can become a chemist.' He must also acquire industrial experience. 'On the other hand, the artists [artisans] and manufacturers themselves are generally illiterate, timid, and bigotted to particular modes of carrying on their respective operations. Being unacquainted with the learned, or modern, languages, they seldom know anything of modern discoveries, or of the methods of working practised in foreign countries.' Manufacturers were also wary of risking their capital in experimental processes.

Watson, however, was not content simply to describe existing industrial operations, or merely to exhort others to apply chemistry, but himself suggested improved processes. When, for example, he was consulted by the government about improving the power of

[1] *Chemical Essays*, Vol. II, pp. 39–40.
[2] See especially *Chemical Essays*, Vol. I, pp. 39–48.

M

gunpowder, he 'suggested to them the making charcoal by distilling the wood in close vessels', which was put into execution at Hythe, in 1787, using iron cylinders, and was said to have resulted in a saving to the Government of £100,000 annually.[1] As Dr. Coleby comments: 'The grant of £100 a year to Cambridge in aid of chemistry had soon paid the Government a handsome dividend.'[2] Similarly he made suggestions[3] for saving lead smelters thousands of pounds annually by condensing the sulphur hitherto wasted in roasting the ore; and after he had 'conversed with some of the principal lead smelters in Derbyshire', and practical proof had been provided, his suggestions were adopted. He likewise suggested[4] to copper smelters that they should consider 'whether a *manufactory* of *green vitriol* might not be established at this [copper works in Anglesey] and all other places where *copper* is obtained by *precipitation*' with iron, instead of throwing away the waste liquors. This suggestion also was adopted in the following years, by the Mona Vitriol Company, in which young Thomas Henry, chemist, of Manchester, was involved.[5] Watson also suggested that coke instead of charcoal could be used not only for making pig-iron but also for converting pig-iron into bar-iron, if only research were done on the different kinds of coke obtained by distillation in retorts at different degrees of heat.[6] Here he was continuing the earlier interests of William Lewis,[7] and it is not impossible that Henry Cort and others, who successfully developed the puddling process in coke-fired furnaces, may have been acquainted with his writings.[8] Watson also threw out another pregnant suggestion,[9] regarding distillation of the volatile products from coke-ovens, but it was many years before this came to birth. There is no doubt that professors such as Watson, when visiting ironworks, etc., gave scientific advice to manufacturers, in return for being permitted to observe industrial processes, in the same way as Scottish professors—Home, Cullen, Hutton, Black, Ferguson, and Anderson.[10]

[1] *Anecdotes*, Vol. I, pp. 240–2. He included several essays on saltpetre and gunpowder (including its use for blasting in mines) in the first two volumes of his *Chemical Essays*.   [2] *Op. cit.*, p. 119.

[3] In a paper on 'Chemical Experiments and Observations on Lead Ores', dated 13 June 1777, read to the Royal Society on 9 July 1778 and published in the *Philosophical Transactions* for that year; reprinted *in Chemical Essays*, Vol. III, Essay 6, where it is followed by another 'On the Smelting of Lead Ores as practised in Derbyshire'.   [4] *Chemical Essays*, Vol. I, Essay 6, on 'Vitriols'.

[5] See below, pp. 245–6, and references there quoted.

[6] *Chemical Essays*, Vol. II, Essay 9, on 'Pit-coal'.   [7] See above, p. 54.

[8] As were engineers such as Smeaton and Watt. (See above, p. 75, for Smeaton.)

[9] In his Essay on 'Pit-coal'.

[10] As the Clows and others have pointed out. See also above, pp. 75–7, 79–80, and below, pp. 178–85, 256–7, and chap. X.

In 1771 Watson moved to the highly prized Regius Chair of Divinity, and was succeeded as Professor of Chemistry by Isaac Pennington, who held this Chair till 1793 but never lectured. Pennington, however, found an able deputy in Isaac Milner (1750–1820), F.R.S., one of Watson's pupils, who, after an interval of some years, resumed the chemical lectures.[1] Meanwhile, interest in natural philosophy and its practical applications was being revived at Cambridge, after a backward period from the mid-century, in which successive science professors—John Colson and Edward Waring, Lucasian Professors of Mathematics, and Antony Shepherd, in the Plumian Chair of Astronomy and Natural Philosophy—were intellectually not very distinguished and also irregular lecturers. Their deficiencies were to some extent made up for by non-professorial Fellows such as George Atwood, of Trinity, who lectured in natural philosophy, including numerous experiments in mechanics, hydrostatics, optics, astronomy, magnetism and electricity.[2] Similarly the Rev. T. Parkinson, at Christ's, published *A System of Mechanics* and *A System of Mechanics and Hydrostatics* in 1785–9.[3]

In the later eighteenth century the development of industry and the growing interest in science throughout the country caused a renewed emphasis on the teaching of applied science, including engineering as well as chemistry.[4] The Jacksonian Chair of Natural Philosophy was established under the will of the Rev. Richard Jackson (d. 1782), who specifically requested that the professor should 'adhere to the plain facts, both in the history and narrative of experiments without adding any hypothesis (unless after the manner that is done by Isaac Newton at the end of his Optics)'; his design was 'the promotion of real and useful knowledge', by experimental methods.[5] Isaac Milner, who was appointed first Jacksonian Professor in 1783, had already, as we have just noticed, proved an able deputy to Pennington, the non-lecturing Professor of Chemistry. His interests in applied science may have derived from his industrial background, as a one-time apprentice in a woollen mill. In addition to pursuing his chemical studies, he also communicated mathematical papers to the Royal Society, of which he was elected a Fellow in 1780. From 1784 to 1792, when he retired from

---

[1] L. J. M. Coleby, 'Isaac Milner and the Jacksonian Chair of Natural Philosophy', *Annals of Science*, Vol. X (1954), pp. 234–57; T. J. N. Hilken, *Engineering at Cambridge University 1783–1965* (Cambridge, 1967), pp. 33–8.

[2] *A Description of the Experiments intended to illustrate a Course of Lectures on the Principles of Natural Philosophy* (1776), republished in enlarged form in 1784 as *An Analysis of a Course of Lectures on the Principles of Natural Philosophy*. See also Hilken, *op. cit.*, p. 35.

[3] Wordsworth, *op. cit.*, p. 74, refers to his tutoring students in mechanics.

[4] Hilken, *op. cit.*, pp. 12–13.   [5] *Ibid.*, pp. 15–16.

the Jacksonian Chair, he proved a very popular lecturer on chemistry and natural philosophy,[1] speaking to as many as 200 students in one course, aided by the entertaining German demonstrator, Hoffmann.[2] He talked about steam engines, as well as air-pumps and other scientific instruments, and spent a considerable amount of money on apparatus. He also demonstrated his practical interests by fitting up at Queen's a large room which he used as workshop and laboratory: 'He installed lathes, furnaces, work-benches, grindstones, bellows, blow-pipes, electrostatic appliances, etc., and there, either alone or in collaboration with friends with similar interests, carried out chemical, electrical and mechanical experiments.'[3]

On appointment as Dean of Carlisle in 1791 and as Vice-Chancellor of the University in 1792, Milner resigned the Jacksonian Chair, but in 1798 was elected to the Lucasian Professorship of Mathematics and continued to take an active interest in experimental philosophy and engineering. In 1801 he was asked by the government to advise on the construction of a new bridge over the Thames, and in his report strongly recommended his colleague, William Farish, as a man of great mathematical and mechanical knowledge.[4]

William Farish (1759–1837), of Magdalene College, who had applied for the Jacksonian Chair on Milner's retirement in 1792 but was defeated by F. J. H. Wollaston, succeeded Pennington as Professor of Chemistry in 1794. The titles of chairs, however, provide an unreliable guide to the actual interests and lectures of the professors. Samuel Vince, for example, Plumian Professor from 1796 onwards, continued the earlier practice of combining astronomy with natural or experimental philosophy;[5] Wollaston, therefore, the Jacksonian Professor in the latter subject, eventually devoted himself to chemistry, while

[1] See *A Plan of a Course of Chemical Lectures* (1784), and *A Plan of a Course of Experimental Lectures introductory to the Study of Chemistry and other branches of Natural Philosophy* (undated).

[2] Coleby, *op. cit.*, p. 241; Hilken, *op. cit.*, pp. 35–6.

[3] Hilken, *op. cit.*, p. 36.   [4] *Ibid.*, pp. 37–8.

[5] *Ibid.*, p. 45; Wordsworth, *op. cit.*, pp. 193, 244, and 254. That Vince's lectures were not without interest to practising engineers is shown by the example of John Southern, Boulton & Watt's manager at Soho, Birmingham, whose interest in Vince's lecture 'On the Motion of Bodies affected by Friction', *Phil. Trans.*, Vol. LXXV (1785), pp. 165–89, reprinted in the *Philosophical Magazine*, Vol. XVII (1803–4), pp. 47–58, 113–19, led him to carry out experiments of his own 'on actual and heavy machinery' (*Phil. Mag., ibid.*, pp. 120–5). Earlier, Southern had himself read a paper on this subject to the Royal Society on 24 Dec. 1801 (not printed in *Phil. Trans.*, but see *Phil. Mag.*, Vol. XI (1801–2), pp. 377–8), in which he had also referred to Vince's work. This, and another paper by Southern, 'On the Equilibrium of Arches', especially in bridge-building (*Phil. Mag.*, Vol. XI, pp. 97–107), strikingly demonstrates his very high ability in applied mathematics. William Strutt also read Vince's works (see above, p. 163)

Farish, Professor of Chemistry, struck out a new line for himself in 'The Application of Natural Philosophy, Natural History, Chemistry, etc. to the Arts, Manufactures and Agriculture of Great Britain'.[1] Farish's course has very recently been described, by the historian of engineering at Cambridge, as probably 'the first serious attempt in any British university to study the industrial implications of a rapidly advancing technology'.[2] Captain Hilken also refers to Farish as 'the first man to teach the construction of machines as a subject in its own right instead of merely using mechanisms as examples to illustrate the principles of theoretical physics or applied mathematics'.[3] This may well be true of Cambridge[4] and Oxford, but Farish had been preceded in this field by many itinerant lecturers in natural philosophy, and also, as we shall see, by Professor John Anderson at Glasgow University. Nevertheless, his course is of very great interest:

He made a study of the machines in common use, the production of raw materials and all branches of civil engineering. He read widely on these subjects, but was not content with second-hand information. During vacations he travelled over the country, visiting factories, harbour works and mines and discussing technical matters . . . and thus amassing a prodigious store of miscellaneous knowledge.[5]

Captain Hilken's account of Farish's work is based mainly on the *Plan of a Course of Lectures on Arts and Manufactures, more particularly such as relate to Chemistry*, issued by Farish in 1796. This outline we have been able to supplement by means of notes on these lectures by John Hey, son of William Hey, the Leeds doctor, who entered Magdalene College in 1794 and graduated in 1799; he was then elected a Fellow, but died in January 1801.[6] These notes are much too copious for us to

---

[1] Hilken, *op. cit.*, p. 40; Wordsworth, *op. cit.*, pp. 190, 193, 244. F. J. Hyde Wollaston was eclipsed in reputation by Dr. William Hyde Wollaston, F.R.S., who, after studying medicine at Caius, soon became one of the outstanding British analytical chemists of the late eighteenth and early nineteenth centuries.
[2] Hilken, *op. cit.*, p. 40.   [3] *Ibid.*, p. 38.
[4] Though at Cambridge we do hear that 'about 1750 Dr. Charles Mason, one of the senior fellows of Trinity (B.A. 1722), a good mathematician, used to be much interested in practical mechanics, working at his lathe, &c.', Wordsworth, *op. cit.*, p. 190, n. 4.   [5] *Ibid.*, p. 39.
[6] For William Hey, see above, pp. 156–9. For John Hey, see Pearson's *Life of William Hey* (1822), pp. 44–5; J. A. Venn, *Alumni Cantabrigienses* (1947), p. 351: *Gentleman's Magazine*, Jan. 1801, p. 89. John Hey's notes consist of (i) jottings on the interleaved pages of Farish's printed *Plan* referred to above, and (ii) a separate book of detailed lecture-notes, described as 'Part of Mr. Farish's Lectures on the Various Arts and Manufactures—at Cambridge' (Hey MSS. 183 and 184, Brotherton Library, University of Leeds). Farish is incorrectly referred to as 'Parish' in 'The Hey Papers', *Leeds Phil. Soc. Proceedings*, Vol. VI, pt. ii (1952), p. 65.

deal with here. They confirm, however, the excellent summary of Farish's lectures in the Cambridge *University Calendar* of 1802,[1] especially Farish's wide knowledge of actual industrial practices, his use of mechanical models and drawings, and his linking of theory with practice.

After having taken an *actual* survey of almost everything curious in the manufactures of the Kingdom, the Professor contrived a mode of exhibiting the operations and processes that are in use in nearly all of them. Having provided himself with a number of *Brass Wheels* of all forms, and sizes, such, that any two of them can work with each other, the *Cogs* being all equal; and also with a variety of Axles, Bars, Screws, Clamps, &c., he constructs at pleasure, with the addition of the peculiar parts, *working Models* of almost every kind of Machine. These he puts in motion by a *Water Wheel,* or a *Steam Engine,* in such a way, as to make them in general do the actual work of the real Machine on a small scale; and he explains at the same time the chemical and philosophical principles, on which the various processes of the Arts exhibited, depend.

In the course of his lectures he explains the theory and practice of *Mining* and of *Smelting* metallic Ores—of bringing them to nature—of converting, purifying, compounding, and separating the Metals, and the numerous and various Manufactures which depend upon them, as well as the Arts which are more remotely connected with them, such as *Etching* and *Engraving*. He exhibits the method of obtaining *Coal* and other *Minerals,* the processes by which *Sulphur, Alum, common Salt, Acids, Alcalies, Nitre,* and other *saline* substances are obtained, and in which they are used; the mechanical process in the formation of *Gunpowder,* as well as its theory and effects. He shews the arts of procuring and working *Animal* and *Vegetable* substances; the great staple manufactures of the country, in *Wool, Cotton, Linen, Silk*; together with the various chemical arts of *Bleaching,* or *Preparing* Cloth, of *Printing* it, of using *adjective* and *substantive* colours, and *Mordants* or *Intermediates* in Dying. He explains in general the nature of Machinery: the moving powers, such as *Water-Wheels, Windmills,* and particularly the *agency* of Steam, which is the *great* cause of the modern improvement and extension of manufactures. He treats likewise on subjects which relate to the carrying on, or facilitating, the commerce of the country, such as *Inland Navigation,* the construction of *Bridges, Aqueducts, Locks, Inclined Planes,* and other contrivances, by which *Vessels* are raised or lowered from one Level to another; of *Ships, Docks, Harbours,* and *Naval Architecture.* On the whole, it is the great design of these Lectures to excite the attention of persons already acquainted with the principles of Mathematics, Philosophy, and Chemistry, to *Real Practice*; and by drawing their minds to the consideration of the most useful inventions of ingenious men, in all parts of the kingdom, to enlarge their sphere of amusement and instruction and to promote the improvement and progress of the Arts.

[1] Quoted in Wordsworth, *op. cit.,* pp. 190–2. See also Hilken's digest of the 1796 *Plan, op. cit.,* pp. 40–2.

These lectures were very popular and attracted large audiences, especially on account of the fascinating variety of model machines employed. Farish did not, however, introduce these 'new methods' of scientific-technological illustration,[1] which, as we have seen, had long been practised by itinerant lecturers. He succeeded Wollaston to the Jacksonian Professorship in 1813 and continued his lectures successfully almost until his death in 1837.[2]

Similar, though less far-reaching, developments took place at Oxford during this period, especially in chemistry. The connections between chemical and medical studies remained very close, but this does not mean that the chemistry taught had no wider applications. Dr. John Parsons (1742–85), who became the first Matthew Lee Reader in Anatomy in 1767, lectured on chemistry, both 'Philosophical' and 'Philosophical and Practical'.[3] His successor, William Thomson (1761–1803), a pupil of Joseph Black, also shared wide interests in chemistry.[4]

In 1781 Dr. Martin Wall (1747–1824), son of Dr. John Wall (1708–1776), of Merton, founder of the Worcester Porcelain Company (1751), became Public Reader in Chemistry at Oxford. There is in the Bodleian *A Syllabus of a Course of Lectures in Chemistry read at the Museum, Oxford, by Martin Wall* (Oxford, 1782),[5] a course of which something more is known through the *Letters of Richard Radcliff and John James of Queen's College, Oxford, 1755–83* (Oxford, 1888). In a letter to his father the younger John James gave an account of Wall's introductory lecture: 'This lecture treats first of the uses, and then of the history of the art. Its uses are various. It mingles itself with all the arts in which fire or mixture are employed; the dyer, refiner, painter, smith, and the cook himself are all in some degree chemists. But its most eminent powers are reserved for medicine.'[6] It is clear that, in addition to hearing from Wall about 'the chemical theories and experiments of Boyle, Margraff, Priestley, Scheele, Boerhaave, Lavoisier, and Cavendish',[7] students also learned something of their practical applications. It is true that, as young James stated, a large part of the class consisted 'either of clergymen, or men intended for the Church',[8] but he referred to the precedent of

---

[1] As stated by Hilken, *op. cit.*, p. 42.

[2] He was succeeded by Robert Willis (1800–75), his most brilliant pupil, who continued and developed his methods, thus solidly laying the foundations of engineering at Cambridge. Hilken, *op. cit.*, pp. 50–7.

[3] Sinclair and Robb-Smith, *op. cit.*, p. 39.

[4] Black visited him in Oxford in 1788: Joseph Black to James Watt, 7 Aug. 1788, Dol.

[5] See also his *Dissertation on Select Subjects in Chemistry and Medicine* (Oxford, 1783).

[6] *Op. cit.*, p. 164. John James jun. to John James sen., 31 Oct. 1781.

[7] Turner, *op. cit.*, p. 50. [8] *Op. cit.*, p. 177.

Dr. Stephen Hales and the example of Dr. Richard Watson, of Cambridge, 'who is certainly a respectable divine, [and] is at the same time an excellent chemist'. He also mentioned 'Dr. Adams, the Head of Pembroke College, a distinguished divine in Oxford, [who] is said by Dr. Wall to be considerably deep in chemistry'. Divines, Anglican as well as Nonconformist, as we have seen, were very active in provincial philosophical societies and occasionally in industrial enterprises, and commonly associated with scientifically-minded industrialists.

Martin Wall, like Richard Watson, communicated a paper to the Manchester Literary and Philosophical Society, on the 'vegetable fixed alkali' (potash) and nitre,[1] which demonstrates his practical interests. After explaining the processes of production, and referring to 'the extensive employment of the Vegetable Alkali and Nitre in many of the principal operations of Chemistry', he emphasized the importance of 'any new experiments by which the qualities and preparation of articles so important may be perfectly investigated and explained, and the great expence attending the present mode of preparing or importing them diminished'.[2]

Dr. Thomas Hornsby (1733–1810), appointed Savilian Professor of Astronomy in 1763 and Sedleian Professor of Experimental Philosophy in 1782, gave lectures on experimental philosophy and chemistry and was abreast of the recent discoveries relating to gases, as surviving advertisements of his lectures[3] show. His lectures on natural philosophy were so good that Matthew Boulton arranged for his son, who was not an undergraduate, to attend them.[4] He owned an apparatus for illustrating his lectures which was valued by Edward Nairne at about £375.[5] He was somewhat eclipsed, however, by Dr. Thomas Beddoes (1760–1808),[6] whose lectures were so popular that the attendance 'exceeded anything known in the university since the thirteenth century'.[7] Beddoes taught chemistry every day except Saturday and Monday when he worked in the library, and he brought pressure to bear upon the Bodleian to improve its collection of scientific publica-

[1] *Manchester Lit. & Phil. Soc. Memoirs*, Vol. II (1785), pp. 67–79. The paper was read on 19 Nov. 1783.

[2] *Ibid.*, pp. 67–8. He stated that £150,000 was paid out annually on imports of potash from Russia and other foreign countries.

[3] In the Oxford Museum of the History of Science. See the advertisement for a course of lectures on 'The different Kinds of Air, Natural and Factitious', 2 June 1785.

[4] Taylor, *Math. Pract. of Hanov. England*, pp. 80 and 234.

[5] Gunther, *Early Science in Oxford*, Vol. I, p. 195.

[6] See E. Robinson, 'Thomas Beddoes, M.D., and the Reform of Science Teaching in Oxford', *Annals of Science*, Vol. XI (1955), pp. 137–41; F. W. Gibbs and W. A. Smeaton, 'Thomas Beddoes at Oxford', *Ambix*, Vol. 9 (1961). See also above, p. 165.   [7] Clow, *op. cit.*, p. 603.

tions, pointing out the need to instruct 'the youth of a great commercial State in the principles of commerce and manufactures'. Beddoes held his Readership in Chemistry from 1788 to 1792 and his biographer reported that

The time of Dr. Beddoes's residence in Oxford was a brilliant one in the annals of the University. Science was cultivated more than it has been since, and I believe that I may say the same of the period which preceded. Dr. Thomson's lectures on anatomy and mineralogy, and Dr. Sibthorp's on botany, were delivering at the same period; and produced a taste for scientific researches which bordered on enthusiasm.[1]

As a close friend of James Watt, Josiah Wedgwood, Erasmus Darwin, and James Keir, as well as son-in-law of Richard Lovell Edgeworth, he may be described as a member by adoption of the Lunar Society of Birmingham. This social context ensured that Beddoes would have very strong interests in the application of science to everyday life. Though his pro-French political views eventually led to his expulsion from Oxford, he was not without friends while he was there, and Christ Church, at the instigation of their Dean, the Rev. Cyril Jackson, elected him a member of its Common Room. It was probably Beddoes and William Thomson who attended the Lunar Society of Birmingham with Jackson on 3 August 1789.[2]

As his assistant Beddoes had the services of James Sadler (1753–1828), the son of a local confectioner. In 1789 Sadler himself lectured in Oxford on 'philosophical fire-works', and this was 'no doubt a perfectly serious subject, since six years later he was conducting gunnery trials at Woolwich'.[3] He was described as 'a clever, practical and experimental manipulator in chemistry'. In August 1795 trials were carried out at Woolwich Arsenal on Sadler's 'new Pattern Twelve-pounder Piece of Ordnance' with favourable results. He also patented a thirty-two pounder which included his invention of a non-recoil mounting. His interest in steam engines may have started as early as 1786. Coalbrookdale erected engines after his designs in 1792 and 1793; in 1796 there was a Sadler engine working on Garlick Hill, London, for Sutton, Keen & Co., mustard-makers; while in 1805 there was a Sadler engine in use at Portsmouth Dockyard. In 1796 he moved with his family to London and started a mineral-water factory near Golden Square. For a man who had made his first balloon ascent on 4 October

[1] J. E. Stock, *Memoirs of the Life of Thomas Beddoes, M.D.* (Bristol, 1811), p. 24.
[2] See M. Boulton to M. R. Boulton, 3 Aug. 1789, A.O.L.B.
[3] G. L'E. Turner, 'James Sadler, Oxford Engineer, Chemist and Aeronaut', in *Oxfordshire Roundabout*, Vol. I, no. 1 (June 1965), pp. 24–5 and 52.

1784, the connection with mineral-water manufacture, and with pneumatic chemistry, is not far to seek. In 1796 also he was appointed chemist to the Board of Naval Works at a salary of £400 a year. In later years he described himself as 'late Member of the Board of Naval Works, and Inspector of Chemistry to the Admiralty'. He carried out researches into the copper-sheathing of ships, the distillation of sea-water, improvements in brewing on ships, the seasoning of timber, the combustion of gunpowder for fire-ships, improvements in air-pumps, the construction of signal-lights, impurities in the nitrate of ammonia, and later made several more balloon ascents.

Beddoes was succeeded by Robert Bourne (1761–1829), whose *Syllabus of a Course of Chemical Lectures read at the Museum, Oxford, in 1794* declares that 'the Course will be practical rather than theoretical'.[1] As a physician he obtained the unenviable sobriquet of 'the bourn whence no traveller ever returns',[2] but he is not without interest for us. His 'Introductory Lecture to a Course of Chemistry read at the Laboratory in Oxford, on February 17, 1797', deals chiefly with the utility of chemistry and explains why Oxford undergraduates might be expected to have an interest in chemistry. First of all, 'many of [the students] will become M.P.s and have to decide questions highly interesting to our Manufactures and Commerce; others may have a fondness for agriculture . . .', while 'those Gentlemen of Fortune whose property consists of metallic mines' will naturally be interested. Then, too, it ought not to be forgotten that 'a physician ignorant of chemistry cannot be well skilled in his profession'.

XII

The association between science and industry was particularly strong in Scotland, where there occurred in this period a brilliant flowering of scientific thought and technology.[3] In Edinburgh and Glasgow Universities, great scientists such as William Cullen, Joseph Black, James Hutton, and Francis Home made important contributions to industrial technology. Their students included later famous industrialists such as John Roebuck, James Keir, and Charles Macintosh, and they were in close contact with scientists and manufacturers in Birmingham, Manchester, and other English industrial centres. The association between Black, Roebuck, and Watt in the development of the steam engine is famous, but there was important scientific-industrial collaboration in other fields, such as bleaching and the production of synthetic

---

[1] Gunther, *op. cit.*, Vol. I, p. 69.   [2] *Ibid.*, p. 40.
[3] See Clow, *op. cit.*, especially chap. XXV, for a detailed account.

soda.[1] Edinburgh and Glasgow also had philosophical and chemical societies and academies in which science was taught outside the universities.

Dr. and Mrs. Clow have dealt at such length with these Scottish developments that there is little need for us to elaborate upon them here. It is, however, worthwhile to emphasize the role of Dr. John Anderson, Professor of Natural Philosophy in Glasgow University from 1757 to 1796 and founder of Anderson's Institution, in the dissemination of scientific knowledge among the artisans of that city.[2] Anderson was closely associated with James Watt and many other industrialists, and 'delighted in visiting the workshops of the artisans and mechanics, and giving them such information as was likely to benefit them in their respective arts, receiving in return a knowledge of those which he could not otherwise have obtained'.[3] This evidence is supported by another near-contemporary statement,[4] which confirms that Anderson visited mechanics

in their workshops, [and] instructed them even there in the principles of Science . . . The shop of James Watt, a common working Mechanic, was a favourite lounging place of his, as of most ingenious and scientific characters about Glasgow. He gave Watt an old model of Newcomen's steam engine to repair, and from this simple circumstance arose the most wonderful invention of modern times.

Though it hardly seems appropriate to call Watt 'a common working Mechanic'—indeed he was an uncommonly intelligent and highly skilled craftsman, with a scientific outlook[5]—there is no doubt that Anderson, like Robison, Black, and others, did frequently visit him in his workshop for philosophical as well as mechanical discussion, and that it was Anderson who asked Watt to repair a model of a Newcomen engine which he was using for demonstration purposes in his natural philosophy classes.[6] (It is significant, incidentally, that Watt's steam-engine experiments should have been upon a small model, of the type commonly used by itinerant lecturers as well as by Anderson at Glasgow and by Farish at Cambridge.) Watt's association with Black

---

[1] See below, chap. IX, for the Black–Watt collaboration on synthetic soda.

[2] See J. Muir, *John Anderson* (Glasgow, 1950). This we have supplemented with contemporary evidence.

[3] 'Memoir of Professor Anderson', *The Glasgow Mechanics' Magazine; and Annals of Philosophy*, Vol. III (1825), p. vii.

[4] W. G., 'On the Origins of Mechanics' Institutions', *The Scots Mechanics' Magazine and Journal of Arts, Sciences and Literature*, Vol. I, no. iii (March 1825), pp. 97–104. We are grateful to Mr. Gerard Turnbull, of Glasgow University, for obtaining copies of these articles for use.   [5] See above, pp. 79–80.

[6] See Arago, *op. cit.*, pp. 524–5, 543; Muirhead, *op. cit.*, pp. 58–9, 62–3, 75; Williamson, *op. cit.*, pp. 159–65.

is famous, but that with Anderson has generally been very inadequately appreciated, though Williamson pointed out, well over a century ago, that Anderson's brother and Watt were schoolfellows and that when Watt was appointed instrument-maker to Glasgow University he became very friendly with Professor Anderson, frequently spending evenings at his house, where scientific works were made freely available to him.

Anderson opened his classes at the University to workmen of all kinds, for whom he provided a special evening course 'of a more popular nature, illustrated by experiments', for which he collected an extensive apparatus. 'The late Professor Anderson . . . gave a Course of Experimental Lectures, to Mechanics of every description who chose to attend. This Course continued under the Professor for about forty years, and a great many Journeymen Mechanics attended, many of whom are now eminent Masters still living.'[1] This statement was confirmed by James Crichton, another associate and former pupil of Anderson's, who stated that these lectures had begun at least as early as 1770 and that his father had attended them and had even been boarded in the professor's house. He also added: 'In 1790 when I attended the Professor's Lectures there were about 200 auditors, of whom about *the half were people of all descriptions belonging to the Town*, that is those who belonged to no other class in the College. Many of these were *workpeople*, some of whom the Professor had *met with casually and invited* to attend his lectures.'[2]

William Clydesdale, when he was an apprentice with Mr. Ninian Glen of Glasgow, attended Anderson's lectures on mechanical subjects in about the year 1781, paying one guinea for the same course for which university students paid two and a half guineas: 'These Lectures were frequented by Apprentices and Journeymen of all descriptions, and generally triple the number attended that Class, than did the one which he conducted at 11 o'clock, forenoon.—We were also often invited to his Morning Class, when the Experiments were Optical, or otherwise interesting to the Mechanics, and required day-light.'[3]

Anderson himself in February 1792 described the work done in his 'Experimental Philosophy' course. Beginning with only half a dozen students, the numbers 'gradually increased, so that for many years they have been from 150 to 200 (including gown students)', among whom

---

[1] W. G., *op. cit.*, p. 99, quoting from a letter from John Parsell (Anderson's 'Amanuensis and Operator' for nine years) to John and Robert Hart of Glasgow, 9 Feb. 1825.

[2] *Ibid.*, pp. 99–100, quoting a letter from Crichton to R. Hart, 11 Feb. 1825.

[3] *Ibid.*, p. 100, quoting a letter from Clydesdale to J. and R. Hart, 9 Feb. 1825.

were '*Town's People of almost every Rank, Age, and Employment*'.[1] In his will he was able to claim that, as a result of his efforts, 'the Manufacturers and Artificers in Glasgow have become distinguished in a high degree, for their General Knowledge, as well as for their Abilities and Progress in their several Arts'.[2]

The main outcome of Anderson's will was the establishment in 1796 of 'Anderson's Institution, or New School of Arts, Manufactures, &c.', forerunner of the Royal Technical College, now the University of Strathclyde, in Glasgow. The main aim of this Institution, in the words of Dr. Thomas Garnett, its first professor (whom we have previously mentioned among itinerant lecturers in natural philosophy[3]), was 'to promote useful knowledge and improvement in Science and Philosophy applied to the various branches of trade and manufactures carried on in this populous City and neighbourhood'.[4] Garnett had a thoroughly scientific background.[5] Born in Westmorland in 1766, articled at the age of fifteen to the celebrated John Dawson of Sedbergh, surgeon and mathematician, he had eventually entered Edinburgh University, where he was particularly interested by Black's lectures, and graduated M.D. in 1788. After completion of his medical training in London, he practised at Bradford, Knaresborough, and Harrogate, but was not apparently very successful and was in Liverpool preparing for emigration to America in 1795 when he was invited to deliver a course of lectures on chemistry and natural philosophy. The success of this course, which was afterwards repeated at Manchester, Birmingham, and elsewhere, was so great that he was offered the new professorship at Anderson's Institution. The attendances at his lectures there, as the late Professor Muir has remarked, were 'astonishingly large': in the first session 972 students regularly attended, including 528 for his philosophical lectures and 344 for those on chemistry.[6] In 1799, however, he moved to the Royal Institution in London, established on very similar lines to Anderson's.[7]

Dr. George Birkbeck was appointed his successor, and proved equally popular. In 1801–2 more than five hundred operatives attended

---

[1] *Ibid.*, p. 98. He gave free admission to professional people, such as teachers, doctors, etc., and also 'gardeners, painters, shopmen, porters, founders, bookbinders, barbers, tailors, potters, glass-blowers, gunsmiths, engravers, brewers, and turners' (*ibid.*, pp. 98–9).

[2] Muir, *op. cit.*, p. 11.   [3] See above, pp. 107 and 146.

[4] Muir, *op. cit.*, p. 25. See also *ibid.*, pp. 98–9, for the outline of Garnett's first course of lectures, in which the practical industrial applications of natural philosophy and chemistry were strongly emphasized.

[5] See *D.N.B.* and his obituary in Nicholson's *Journal of Natural Philosophy*, Vol. III (1802), pp. 62–4.

[6] Muir, *op. cit.*, pp. 99–101.   [7]See above, p. 129. He died in 1802.

his course on 'Mechanic powers', which was 'calculated chiefly for working tradesmen'.[1] Birkbeck after a few years also followed Garnett to London, where he took the lead in establishing Mechanics' Institutes and was associated with the founding of University College. The first Mechanics' Institute, established in Glasgow in 1823, on Birkbeck's advice, was an offshoot from Anderson's Institution, and was followed a year later by that in London, the first in England. Birkbeck, meanwhile, had been succeeded at Anderson's in 1804 by Dr. Andrew Ure, whose lectures continued to attract hundreds of students annually throughout the first quarter of the nineteenth century.[2] Many of these students were to become prominent in industry, including, for example, J. B. Neilson, of the newly established Glasgow gas-works and later of 'hot-blast' fame in the iron industry.[3] Another was James Young (1811–83), F.R.S., who, as a workman, attended evening classes at Anderson's; he became Professor Thomas Graham's assistant, and then, in the late 1830s and 1840s, a chemical works manager, first with Muspratt's at Newton-le-Willows and later with Tennant's in Manchester, before founding the shale-oil or paraffin industry.[4] A contemporary of his at Anderson's, and a lifelong friend, was Lyon Playfair (later Lord Playfair), whose scientific-industrial achievements hardly need emphasizing here.

The ready availability of instruction in industrial chemistry during the late eighteenth century is also evidenced by a letter from Dr. Joseph Black to his brother Alexander, secretary of the British Cast Plate Glass Manufactory, dated 20 February 1788, recommending a young lad as an apprentice: 'You will judge on trial how fast he may be brought forward, and when he is fixed and in a train, it will be easy to give him an opportunity of attending a Course of Lectures on Chemistry.'[5] Professors such as Black and Robison in Scottish universities were in close contact with industrialists,[6] giving scientific advice on chemical or engineering matters, and, in return, themselves acquiring much knowledge of industrial techniques. This relationship is well

---

[1] Muir, *op. cit.*, p. 107. See also T. Kelly, *George Birkbeck* (Liverpool, 1957).
[2] *Muir, op. cit.*, pp. 108–12. Before beginning this course, Ure 'visited all the Principal Manufactories and Mines in the Kingdom, he also visited the Workshops of the Artizans and Laboratories of the Chemists', to see the latest manufacturing improvements (W. G., *op. cit.*, p. 103).
[3] Clow, *op. cit.*, p. 606.   [4] *D.N.B.*
[5] D. McKie and D. Kennedy, 'On some letters of Joseph Black and Others', *Annals of Science*, Vol. XVI (1960), p. 149. See above, pp. 5–7, for Black's chemical lectures.
[6] See, for example, above, pp. 75–7, 79–80, and below, pp. 256–7 and chap. X, for their relations with Telford, Rennie, Watt, Roebuck, and Keir. Dr. and Mrs. Clow have provided many similar examples.

illustrated by a letter from Robison to Watt at the end of 1796,[1] in which he outlined his proposed plan for

printing a work on the employment of the Civil Engineer Millwright &c. such as might be useful to the *mere Workman*, might incite some to a further scientific, but elementary knowledge of their profession, and furnish the means for their proficiency . . . This I thought might be accomplished by a Book in the following form. The running text should be intirely practical, containing no science, but only the results of scientific investigation, connected (when I am able) by some palpable reasoning. [The text, however, would be accompanied by explanatory notes.] . . . The Work should contain the principles of Mechanics, Hydrostatics, Hydraulics, & Pneumatics—Strength of Materials—Strains—proper disposition of both—Friction—Resistance of Mediums—of sand and of soft or ductile bodies—Theory of Machines—including both the theory of their operation and of their best construction.

Robison proposed that this work should be on the same level as his articles in the *Encyclopaedia Britannica* on such topics as philosophy, physics, pumps, resistance of fluids, rivers, roofs, rope-making, seamanship, sound, steam engine, strength of materials, etc., but a book would provide a better opportunity of integration. He also thought that

It wd. be an agreeable thing to myself, & would be some Credit to the work, to found the whole on a series of original and appropriated experiments made by myself . . . I have a good Situation for Hydraulic Experim[en]ts at a Mill within 500 yards of my house at Boghall . . . Other Exper[imen]ts on the strength of Matter, on Friction &c, I shoud have great difficulty in, for want of great Apparatus and Workmen. For these I should apply to you—or perhaps come & place myself beside your Workmen—I should hope for such extracts from the myriads [of experiments] which you have made as you could with propriety communicate.[2]

He would be especially grateful for information and drawings on the development of the steam engine, from Newcomen's to Watt's latest improvements. Robison similarly obtained help from Rennie,[3] and in 1797 he published *Outlines of a Course of Lectures on Mechanical Philosophy*, followed in 1804, just before his death, by *Elements of Mechanical Philosophy*, of which only the first volume was published. Ill health, it would appear, prevented him from bringing his ideas fully to fruition, though David Brewster later collected his articles from the *Encyclopaedia Britannica* and elsewhere into *A System of Mechanical Philosophy* (Edinburgh, 4 vols., 1822).

[1] J. Robison to J. Watt, Dec. 1796, Dol.
[2] See above, pp. 75–6 and 77, 117–18, for similar scientific-industrial collaboration between Telford and Barlow and between Fairbairn and Hodgkinson.
[3] See above, pp. 76–7.

The Scottish iron industry, as well as the Scottish engineering and chemical industries, also appears to have derived some benefit from applied science. John Roebuck's experiments at the Carron ironworks in the 1760s provide an early example. The historian of this company has recently referred to Roebuck's efforts in 'trying to apply the latest scientific knowledge at Carron', in his experiments with blast-furnace blowing cylinders, etc., and on producing bar-iron and steel with coke fuel.[1] Professor Campbell has stressed the importance of Roebuck's scientific training, his early education at the Sheffield Grammar School and at Doddridge's Dissenting Academy at Northampton, followed by medical and chemical studies at Edinburgh and Leyden Universities. After developing the lead-chamber process of vitriol manufacture first at Birmingham and then at Prestonpans in the late 1740s, he turned to iron manufacture, coal-mining, etc., and the Carron ironworks, which he helped to found, reaped the benefit of his scientific approach to industrial matters:

A pioneering concern such as Carron Company could not fail to benefit from such centres of intellectual discussion as Edinburgh and Glasgow then were. But the Company gained in a more special way through the intimate personal contacts of leading partners with the best minds in Edinburgh and Glasgow. Roebuck in his own right could claim to be one of the brilliant group of Edinburgh intellectuals . . .[2]

Unfortunately, however, although Professor Campbell refers to Roebuck's numerous experiments in iron manufacture and to some of his attempted improvements in smelting and forging processes, he provides no details of the ways in which Roebuck was actually applying scientific knowledge or techniques. Moreover, Roebuck's efforts appear, for the most part, to have ended in failure.

At the turn of the century, David Mushet provides a much more striking and more fully documented example of applied chemistry and metallurgy in the Scottish iron industry. First at the Clyde Valley and then at the Calder Ironworks, Mushet produced a remarkable series of papers in the *Philosophical Magazine*, from 1798 onwards, over a period of more than ten years,[3] in which, whilst deploring on the one hand the general lack of scientific knowledge among ironfounders and on the other hand the equal lack of practical ironworking experience

---

[1] R. H. Campbell, *Carron Company* (1961), chap. II. See also P. Mantoux, *The Industrial Revolution in the Eighteenth Century* (1928), pp. 299 and 309–10; T. S. Ashton, *Iron and Steel in the Industrial Revolution* (1963), pp. 51, 87*n*., and 88.

[2] Campbell, *op. cit.*, p. 3.

[3] These were eventually collected and published together as *Papers in Iron and Steel* (1840).

among chemical and metallurgical philosophers, he himself demonstrated in brilliant fashion how chemistry and mineralogy could be usefully applied in iron smelting and forging. He had a profound knowledge of the works of French chemists of the Lavoisierian 'oxidation' school and of the works of mineralogists such as Bergman and Kirwan, and he utilized this knowledge in carrying out and interpreting innumerable experiments on all aspects of iron manufacture. These researches included chemical analyses of raw materials such as charcoal, coal, and ironstone, and of pig-, cast-, and wrought-iron products, and he also investigated the effects of varying temperature, pressure, humidity, and velocity in furnace blasts. His knowledge of mineralogy and his analyses of ores enabled him quickly to appreciate the possibilities of the famous black-band ironstone which he discovered in about 1800.[1] He met with much benighted opposition, but it is not unreasonable to suppose that some other ironmasters may have benefited from his published researches, just as the earlier works of Lewis, Dossie, and Watson in England were not without practical metallurgical applications.[2] Neilson's similar application of chemical knowledge, as we have mentioned, was soon to revolutionize the Scottish iron industry with the hot-blast process.[3]

XIII

Ireland also participated in these scientific-industrial developments. A Dublin Philosophical Society had been founded as early as 1683 and the (Royal) Dublin Society had existed continuously since 1731. A Medico-Philosophical Society was formed in 1756 and the Irish Academy in 1783 (chartered 1786), the first volume of its *Transactions* appearing in 1787. Towards the end of the eighteenth century the Dublin Society felt 'that the time had come when the formation of schools of science, in which qualified professors might lecture . . . in accordance with the example and precedent set in such matters in England and Scotland, [would be] more likely to further the purposes for which the Society had originally been founded'.[4] Hence William Higgins (1763–1825) was appointed professor of chemistry and mineralogy to the Society

---

[1] See Clow, *op. cit.*, pp. 342, 344, 351–2, and 357.
[2] See above, pp. 53–5 and 167–70. See also above, pp. 158–9, for the similar researches of Joseph Dawson, of the Low Moor Ironworks, near Bradford.
[3] See above, p. 182.
[4] H. F. Berry, *A History of the Royal Society of Dublin* (1915), p. 159. That some of the itinerant lecturers in natural philosophy visited Ireland is shown by the example of Adam Walker (see above, pp. 105–6).

in 1795,[1] equipped with a laboratory, and instructed 'to make such Experiments on Dying [*sic*] Materials and other Articles, wherein Chymistry may assist the Arts'.[2] He also began to provide a 'course of Experimental Chemistry', which was advertised in *Saunders' News Letter* and the *Hibernian Journal* of 1797 and later years; his 'chymical lectures', according to Kirwan, were attended 'by professional men or Mechanics'.[3] Professors of other scientific subjects were also appointed by the Dublin Society: in 1795 Dr. Walter Wade became professor and lecturer in botany, and in 1800 James Lynch, a Dublin optician, was appointed to a similar post in hydraulics, mechanics, experimental philosophy, etc.[4] The most eminent scientist in the Society, however, was Dr. Richard Kirwan (1733–1812), F.R.S., one of the most notable chemists in Europe at this time, whose activities in London we have previously mentioned;[5] he returned to Ireland in 1787 and was president of the Royal Irish Academy between 1799 and 1812.

Many of the books and papers published by Kirwan and Higgins were of considerable importance to industrial chemistry, especially to the Irish linen industry. It is very probable that Kirwan was the first to introduce chlorine bleaching into Ireland,[6] and he also carried out researches into alkaline substances used in bleaching.[7] Higgins in addition to holding a chair in the Dublin Society, was also chemist to the Irish Linen Board between 1795 and 1822, during which years he carried out numerous researches into bleaching and other branches of textile chemistry, with important practical applications.[8] His work as

---

[1] He was a nephew of Dr. Bryan Higgins (see above, pp. 121–2), with whom he had studied chemistry in London in the 1780s. He had also spent about two years at Oxford. In 1792 he returned to Ireland, to become chemist to the Apothecaries Hall, but resigned this position in 1795, when he transferred to the Dublin Society. See *D.N.B.*; W. K. Sullivan, *Dublin Quarterly Journal of Medical Science*, N.S., Vol. VIII (1849), pp. 487–95; T. S. Wheeler, *Endeavour*, Vol. XI (1952), pp. 47–52, and *Studies*, Dublin, Vol. 54 (1954), pp. 78–91, 207–15, 327–31; J. R. Partington, *A Short History of Chemistry* (1951), pp. 272–6, and *A History of Chemistry*, Vol. III (1962), pp. 736–54; T. S. Wheeler and J. R. Partington, *The Life and Work of William Higgins, Chemist, 1763–1825* (1960).

[2] Berry, *op. cit.*, p. 355; Wheeler and Partington, *op. cit.*, pp. 17–21. The content of Higgins's lectures is shown in *A Syllabus of a Course of Chemistry for the year 1802* (Dublin, 1801).

[3] R. Kirwan to Sir J. Banks, April 1797, quoted by Wheeler and Partington, *op. cit.*, p. 33.   [4] Berry, *op. cit.*, pp. 159–60.

[5] See above, pp. 122–3.   [6] See below, pp. 259–60 and 319–20.

[7] *Transactions of the Royal Irish Academy*, Vol. III (1789), pp. 3–47; *Repertory of Arts and Manufactures*, Vol. V (1796), pp. 255–70, 330–49, 396–417. Kirwan carried out these researches 'solely with a view to the utility of the public', and his paper was packed with invaluable information for practical bleachers on chemical analysis and testing of their alkaline materials.

[8] See below, pp. 320–1 and 327. As early as 1716 the Linen Board had procured the assistance of William Maple, 'Chymist', at an 'experiment of making Pot-

chemist to the Dublin Society also included research into other industries such as brewing and tanning.

James Muspratt (1793–1886), pioneer of synthetic alkali or soda manufacture in Great Britain,[1] was born in Dublin and studied at a commercial school in that city, before being apprenticed to a wholesale druggist there. We are told that he 'rejoiced to devote what time he could spare to the study of such books as the Dictionary of Chemistry by Nicholson, and a translation of the works of Guyton de Morveau, and also to the making of experiments'.[2] This training and knowledge enabled him, after a period of war service, to establish a chemical manufactory in Dublin, with a partner named Abbott, producing hydrochloric acid and prussiate of potash, and eventually, in 1822–3, to establish a works in Liverpool for the manufacture of soda by the Leblanc process.

Whilst in Dublin he had become acquainted with a fellow-Irishman, Josias Christopher Gamble (1776–1848), who shared his chemical interests and who was to join with him in pioneering soda manufacture.[3] Gamble, of Scottish Presbyterian extraction, was a graduate of Glasgow University (M.A., 1797), who had become a Presbyterian minister, but who, inspired by Dr. Cleghorn's lectures in chemistry at the University, had become keenly interested in chemical experiments and particularly in chlorine bleaching.[4] He acquired knowledge of the bleaching-powder process patented by Charles Tennant, of the St. Rollox chemical works, Glasgow, and decided to leave the ministry and start up as a chlorine-bleach manufacturer, first in County Monaghan and then near Dublin, where he produced not only chlorine-bleach powder, but also his own sulphuric acid, Glauber's salt (sodium sulphate), potash and alum. Eventually, in 1828, he followed Muspratt to England and partnered him in a Leblanc soda works

ashes'; Maple later carried out useful work on tanning and was for many years curator and registrar to the Dublin Society. Higgins's predecessor as chemist to the Linen Board, first permanently appointed in 1782, was James Clarke, who had also carried out experimental researches on potash. See Wheeler and Partington, *op. cit.*, p. 12.

[1] J. Fenwick Allen, *Some Founders of the Chemical Industry* (1906), pp. 69–100; D. Reilly, 'The Muspratts and the Gambles—Pioneers in England's Alkali Industry', *Journal of Chemical Education*, Vol. 28 (1951), pp. 650–3.

[2] Allen, *op. cit.*, p. 71. For Nicholson, see above, p. 83. Guyton de Morveau was one of the earliest French chemists to attempt the manufacture of soda from common salt (*ibid.*, p. 79; see also below, p. 370).

[3] *Ibid.*, pp. 39–66, 82–4, and Reilly, *op. cit.*

[4] Allen, *op. cit.*, pp. 44–5. According to Reilly, *op. cit.*, p. 651, Gamble became interested in chemistry as a result of attending lectures at the Royal Academical Institution in Belfast; but he was probably continuing interests first aroused at Glasgow.

in St. Helens. Both Muspratt and Gamble, of course, found their main market in the textiles industry, firstly in Irish linen and then in Lancashire cotton, and they were clearly aided and inspired by their chemical knowledge not only of the English and French chemists previously cited, but also, very probably, of the Irish chemists Kirwan and Higgins, with their strong textile-chemical interests.

From this survey of the United Kingdom, it is clear that there was a widespread interest in applied science in the late eighteenth century. Many scientists or 'natural philosophers' had industrial interests, and many industrialists were interested in science. It may be objected, however, that these interests were too empirical and too fragmentary to be called 'scientific'.[1] This criticism can be met by considering what science was at this period. It is true that many of the scientific theories of the day—the whole complex of phlogiston theory, the theories of chemical combinations, of dyeing, bleaching, etc.—were later to be proved erroneous, but they were useful concepts at the time: modern science has evolved through a series of discarded hypotheses. The scientific theories of the late eighteenth century ought to be criticized only in the light of the knowledge then available. At that time, as a contemporary, William Nicholson, pointed out, scientists were 'not sufficiently advanced in any part of the knowledge of nature, to reason with safety, without constant recurrence to the test of experiment'.[2] This widespread adoption of what has been called 'the scientific method'—the numerous attempts 'to improve natural knowledge by experiment', the original aim of the Royal Society—was of considerable importance to industry, for much of the experimenting was of a practical, even utilitarian kind, in which scientists and industrialists had mutual interests.

A modern chemist may easily take an unhistorical view of the comparatively primitive attempts made by such men as Watt, Keir, and Henry to analyse chemical processes too complex for the chemical knowledge of the period. But their efforts were undoubtedly scientific in the sense of being based on rationally ordered experiments. Watt and Keir, for example, certainly thought in terms of controlled experiments in a laboratory when they were considering the manufacture of alkali or the chlorine process of bleaching; so, too, did Thomas Henry, Charles Taylor, and Thomas Cooper in their developments of bleaching and dyeing, and the same is true, as we have seen, of many other

[1] Such a criticism was made by Dr. D. W. F. Hardie at the Economic History Conference in 1960.
[2] W. Nicholson, *The First Principles of Chemistry* (1790), pp. 413–14.

manufacturers, in various industries, ranging from Josiah Wedgwood in pottery to John Marshall in textiles.

No doubt the mass of manufacturers in the late eighteenth century had little scientific knowledge, but a substantial minority, including some of the most significant figures, were passionately interested in it, for utilitarian as well as intellectual reasons, and there seems little doubt that such knowledge contributed to the industrial changes of that period. We do not wish to push this thesis too far—to underestimate the contributions of unscientific though intelligent practical craftsmen—but the evidence appears to necessitate some modification of the traditional view of the Industrial Revolution.

IV The Derby Philosophical Society[1]

When Peacock described the wordy evenings at Headlong Hall or
Crotchet Castle, he had a sound pattern of fact on which to work—
the little coteries established all over England of men ready and anxious
to discuss the latest scientific discoveries, the most recent 'improvements'. Samuel Smiles said that these provincial coteries 'were usually
centres of the best and most intelligent society of their neighbourhoods,
and were for the most part distinguished by an active and liberal spirit
of enquiry'.[2] Probably the greatest of them was the Lunar Society at
Birmingham which met at Boulton's residence on Handsworth Heath,
and certainly one of the liveliest members of that very distinguished
society was Dr. Erasmus Darwin of Lichfield. The spirit of the man
breathes strongly in his letter of 5 April 1778 to Matthew Boulton
excusing himself for being unable to attend one of the Lunar meetings:

> Lord! what inventions, what wit, what rhetoric, metaphysical, mechanical,
> & pyrotechnical, will be on the wing, bandy'd like a shuttlecock from one to
> another of your troop of philosophers! while poor I, I by myself I, imprizon'd
> in a post chaise, am joggled, and jostled, & bump'd, & bruised along the King's
> high road, to make war upon a pox or a fever![3]

Darwin was one of the founders of the Lunar Society, and even earlier
had founded a botanical society at Lichfield which published a translation, under his direction but with the advice of William Withering,[4]
of Linnaeus's *Genera Plantarum*. When he was obliged to leave Lichfield
for Derby in 1782, he was cut off from regular meetings with the

---

[1] Based on an article by Eric Robinson in *Annals of Science*, Vol. IX (1953),
pp. 359–67.   [2] S. Smiles, *Boulton and Watt* (1865), p. 367.
[3] A.O.L.B., *Darwin* Box. Quoted by R. V. Wells, 'The Lunar Society', *School
Science Review*, Nov. 1951.
[4] See T. Whitmore Peck and K. Douglas Wilkinson, *William Withering of
Birmingham* (Bristol and London, 1950).

Birmingham philosophers and felt the loss deeply, as Priestley did later when he was forced to find sanctuary in America. 'I am here cut of [*sic*] from the milk of science, which flows in such redundant streams from your learned lunations,'[1] Darwin wrote from Derby to Boulton. He was not the man, however, to languish under such conditions and he very soon founded his third philosophical society, in Derby, and thus extended even further that Midlands community of scientific gentlemen in which the members corresponded with one another and with the leading scientists of Europe and America. One of Darwin's first letters after the inauguration of the Derby Society was of course to Boulton, inviting the members of the Lunar Society to join with the Derby philosophers in a meeting at Darwin's residence in Derby, and proposing a return visit to Birmingham: 'We have establish'd an infant philosophical Society at Derby, but do not presume to compare it to your well-grown gigantic philosophers at Birmingham. Perhaps like the free-mason societies, we may sometime make your society a visit, our number at present amounts to seven, and we meet hebdomidally.'[2] A year later the Derby Society sent a message by airballoon to their fellow-philosophers in Birmingham: 'You heerd we sent your society an air-balloon, which was calculated to have fallen in your garden at Soho; but the wicked wind carried to Sr Edward Littletons.'[3] These societies did in fact represent a new freemasonry, and as freemasonry is reputed to be a movement of some importance in professional circles today, so this scientific movement in the eighteenth century linked together manufacturers, scientists, and men of letters and speeded the technological advance of the Industrial Revolution.

We have not discovered when the first meeting of the Derby philosophers took place, but it was certainly before the formal inauguration at Darwin's house on 18 July 1784 when Darwin gave an address to them.[4] The letters quoted above show that such a society existed in 1783, and the first meetings probably took place in the early part of that year: Darwin's letter to Boulton is dated 4 March 1783, and on 13 March 1783, Susannah Wedgwood wrote to her father, Josiah Wedgwood: 'The Philosophical Club goes on with great spirit, all the ingenious gentlemen in the town belong to it, they meet every saturday night at each others houses. . . .'[5] Susannah Wedgwood was

[1] A.O.L.B., *Darwin* Box, 26 Dec. 1782.   [2] *Ibid.*, 4 March 1783.

[3] *Ibid.*, Erasmus Darwin to Matthew Boulton, 17 Jan. 1784.

[4] Draft of an *Address to the first meeting of the Literary Society at Derby*, 18 July 1784, in the 'Darwin Commonplace Book', the Charles Darwin Museum, Down House, Downe, Kent. There is an identical printed version, of the same date, in the Derby Collection, Derby Borough Library.

[5] Wedgwood letters, Darwin Museum, Downe, 13 March 1783.

staying in Derby at the house of Robert French, a man who had the courage to tell Sir Joshua Reynolds that he thought Fuseli's 'Nightmare' was the best picture in the Academy exhibition of 1782. His name appears second in the list of the members of the Philosophical Society. The meetings of the Society were social occasions, often of a fairly light-hearted character, but the desire for knowledge was genuine. What was wanted were '*gentlemanlike* facts', a phrase which Darwin used when asking James Watt for information about his steam engine.[1] On occasion, in the early days of the Society, the ladies had to be interested:

> The last meeting was at Mr. Struts The Miss S——s keeps [*sic*] their brothers house, & consequently were obliged to make tea & preside at the supper table, they did not like this at all, but Doctor D—— with his usual politeness made it very agreeable to them by shewing several entertaining experiments adapted to the capacities of young women; one was roasting a tube, which turned round itself. . . .[2]

But Dr. Darwin's inaugural address to the Society makes it quite clear that the drawing-room experiments had a serious scientific purpose, and in a typical metaphor drawn from contemporary books of travel he stresses that theme of scientific freemasonry which we have already mentioned:

> These bright examples of the numerous literary societies already established, we cannot do better than to imitate; and by collecting their annual publications, together with those of other ingenious philosophers, and forming a permanent, increasing, and valuable library, we shall create a kind of band of Wampum, a chain of concord, which may hold our own Society together; and I hope, at some distant time, perhaps not very distant, by our own publications we may add something to the common heap of knowledge, which I prophesy will never cease to accumulate so long as the human footstep is seen upon the earth.

The idea of a scientific library was discussed probably quite frequently by the members of the Lunar Society because a similar collection was formed in Birmingham.[3] Dr. Withering, the botanist, wrote to Boulton on 25 June 1784: 'We meet on Tuesday to make Laws for the New Library of Scientific Books.'[4] The date suggests that the letter might refer to the Derby library, but Withering is not mentioned in the records of the Derby Society and was living at that time in Birmingham. We assume therefore that the philosophers at Derby and at Birmingham both established scientific libraries within a month of

---

[1] Printed in J. P. Muirhead, *Mechanical Inventions of James Watt* (1854), Vol. II, p. 231.   [2] Susannah Wedgwood to Josiah Wedgwood, 13 March 1783.
[3] A.O.L.B., Box *Birmingham 1*.   [4] A.O.L.B., under *Withering W.*

each other. The rules for the Derby Society were drawn up on 7 August 1784,[1] and it was decided that '*Books of Natural History and Philosophy* in English, French, or Latin, shall be ordered by the President at every Meeting, if approved by a majority of the members present; and shall be purchased and duly circulated by the Secretary'. The Society met at the King's Head in Derby at 6 p.m. on the first Saturday in every month, thus changing the earlier habit of meeting weekly.

From the catalogue preserved in Derby Borough Library we find the names of the resident and non-resident members. Dr. Darwin was President, and presumably in order to strengthen the connections between the two societies he was made an honorary fellow of the Literary and Philosophical Society of Manchester on 28 April 1784.[2] The certificate is signed by Thomas Percival, F.R.S., among others. Three years later, 22 November 1787, he was elected a member of the Medical Society of London. He also became a Fellow of the Royal Society. Apart from Josiah Wedgwood, F.R.S., who was a non-resident member, the other members are much less well known. Robert Darwin, F.R.S., Erasmus's son, Ralph Wedgwood, Josiah's inventor-cousin, and William Strutt, F.R.S., were members. Strutt was a member of the important family of Derby mill-owners, and designer of the Derby Infirmary.[3] Although John Whitehurst, F.R.S., the Derby watch-maker and Darwin's fellow 'lunatic', and Abraham Bennet, F.R.S., the inventor of the electric doubler, are not mentioned in the Society's catalogue as being members,[4] it seems very probable that they attended at least some of the early meetings. Other members were local worthies; Richard Leaper, C. S. Hope, and John Crompton were all mayors of Derby. William Duesbury was a porcelain manufacturer who did business with Boulton[5] and Dr. Bage of Elford was the owner of a paper mill. The medical profession supplied the greatest number to the Society, and the aristocracy were represented by Lord Belper, and by Sir Brooke Boothby who had been a member of the Lichfield Society. One other member, Mr. Archdale, we know something of because Darwin introduced him to Boulton as 'a gentleman of classical abilities . . . who is going to make speeches in the Irish parlament—& wishes to see all things visible on the earth'.[6] Later in

[1] 'Laws or Regulations Agreed upon by the Philosophical Society at Derby, At their first regular meeting, 7 Aug. 1784', Derby Borough Library.
[2] Darwin Commonplace Book, Darwin Museum, Downe.
[3] See the *Strutt Collection*, Derby Borough Library.
[4] F. W. Shurlock in *Science Progress* (1925), pp. 452–65, maintains that Bennet was a member, but his quotations from the catalogue only show that Bennet's book on electricity was purchased by the society (Item 213).
[5] Boulton Letter Book, 1776–9, p. 454, A.O.L.B.
[6] 22 Oct. 1789, *Darwin* Box, A.O.L.B.

its history, in the nineteenth century, George Spencer, Herbert Spencer's father, was secretary of the Society, and Herbert Spencer himself did some work in connection with it. But for the most part its members seem to have been gentlemen of moderate means and moderate scholarship. Darwin's wish that the world of learning should be enriched by contributions from the Derby Philosophical Society seems to have been disappointed. The Society did not publish a collection of papers, but it gathered together quite a large library and scientific lectures were given under its auspices until 1857, when the Society was amalgamated with the Derby Town and County Museum.

The list of books bought is particularly interesting because the books were ordered by Darwin in his official capacity as president, and because they give a picture of the range of reading to which the late eighteenth-century philosopher aspired. The catalogue is in the form of a book about 8 inches by 12 inches beginning with a list of the books purchased, entered merely in order of acquisition, followed by a printed section which is a loan register. There is no classification and in order that we may describe the collection the following headings seem useful:

(1) Proceedings of other scientific and literary societies
(2) Chemistry
(3) Geology and Mineralogy
(4) Electricity
(5) Botany
(6) Medicine
(7) Travel
(8) Politics, Economics, and Sociology

There are 275 items recorded in the catalogue which has been preserved, but many of these items comprise five or six volumes. The volumes were lent to the members for a certain period and the secretary, Richard Roe, kept a record of the loans, paid carriers, and collected fines. The secretary was paid twelve guineas a year for these services.

The publications of all the leading scientific societies of Great Britain and France were bought. For example the publications of the Royal Societies of London and Edinburgh, the Society of Arts, the Royal Irish Academy, the London Medical Society, and the societies of Manchester and Bath were taken. France is represented by the proceedings of the Académié Royale des Sciences in the cities of Paris and Toulouse, the Société Royale d'Agriculture de Paris, and the Academy of Dijon. The Physical Society of Lausanne and the American Academy of Arts and Sciences at Boston extend the international range of the publications. It is clear that every effort was made to keep abreast of the researches of other philosophical societies, not only through these publications, but also through the books published by leading scientists at home and abroad. Books by Darwin himself, John White-hurst, Priestley, James Keir, De Luc, and John Ash, all members of the Lunar Society, appear in the catalogue, and the one gift to the library is James Keir's life of another Lunar Society member, Thomas Day, the book being presented by Dr. Darwin. Swediaur of the Manchester Society, Beddoes of Bristol, Monro and Cullen of Edinburgh are all to be found in the list.

In *The Chemical Revolution* (London, 1952), A. and N. Clow demonstrate the importance of chemistry in the Industrial Revolution, and it is true that the Derby philosophers seem to have been much more interested in chemistry than in the mechanical sciences. All the leading works on both sides of the phlogiston controversy, including of course Priestley's and Lavoisier's, were bought and some of the titles also indicate an interest in the industrial applications of chemistry.[1] It is a little refreshing to see that the gentlemen of Derby were alive to the dangers of the pollution of the atmosphere by industrial smoke.[2] Through their proximity to Matlock and Buxton they were also interested in the analysis of the spa waters. They read James Keir's *Dictionary of Chemistry*, that book which Boulton described as 'the most compleat Book ever yet published'[3] when he sent it to his son in Germany. Since the famous Blue John quarries were near Derby and also since most of the members had geological interests, keeping fossil cabinets and foraging for specimens, numerous works on geology, including those of Hutton, another intimate of Darwin, were bought for the library. In botany they were, like all their contemporaries, admirers of Linnaeus, and the catalogue is scattered with translations of Linnaeus and descriptions of his system, crowned by Darwin's own *Loves of the Plants*. There are two copies of the Lichfield Society edition of Linnaeus's system of classification, Pulteney's accounts of the development of botanical studies in England, and volumes by famous botanists such as William Curtis, Thomas Martyn, and Smith, first president of the Linnean Society. This was the age of the amateur botanist, when

[1] Richard Watson's *Chemical Essays*; Bergman, *Usefulness of Chemistry*; Berkenhout, *First Lines of the theory and practice of philosophical chemistry*; Keir, *Dictionary of Chemistry*; Rigby, *Chemical observations on Sugar*; W. Richardson, *Chemical principles of the metallic arts: designed chiefly for the use of Manufacturers*, etc., etc.

[2] They bought Dr. Franklin, *On smoky chimnies*, and Benjamin Taylor, *Lecture on the atmosphere of London*.

[3] Matthew Boulton to M. R. Boulton, 26 Oct. 1789, *M. R. Boulton* Box, A.O.L.B.

Horace Walpole complained of the hours that Gray spent reading Linnaeus instead of writing poetry.

A typical page (p. 12) of the catalogue covers these items:

Item No.	Title		No. of volumes
203	Clark on the diseases of horses		1
204	Pilkinton's hist. of the present state and antiquities of Derbyshire	8vo	1
205	Paley's principles of moral & political philosophy	8vo	2
206	Loves of the plants, a poem	4to	1
207	a. Annales de Chymie par Levoisier, &c	8vo	2
208	b. do.	tome 3	1
209	Essai analytique sur l'air pur et les differentes especes d'air, par M. de la Metherie	8vo	2
210	Traite elementair de chymie par Levoisier	8vo	2
211	Lecture on the atmosphere of London, by Benj. Taylor	4to considered as 12mo	1
212	Medical communications, vol. 1	8vo	1
213	Bennet on electricity	8vo	1
214	a. Memoires de la societe des sciences physiques de Lausanne vols. 1 and 2	4to	2
215	Letters from Barbary, France, Spain, Portugal, &c, by an English officer	8vo	2
216	Travels through the interior parts of America, in a series of letters, by an officer	8vo	2
217	Higgins's comparative view of the phlogistic and anti-phlogistic theories	8vo	1
218	Keir's dictionary of chemistry, &c		
	a. part 1 ———————	4to	1
219	Brook on electricity, the air pump, and the barometer	4to	1

This page gives a good impression of the variety of books collected. It contains for example two of the travel books which were bought by the Society.[1] The members' general reading was extended by Adam Smith's *Wealth of Nations*, Howard's *On Prisons*, and Paley's *Moral and Political Principles*.

At first political differences within the various literary and philosophical societies do not seem to have mattered, but as the excitement in France increased and the revolution grew nearer, many scientific men looked to France for a new spirit in government. Priestley in Birmingham became a conspicuous political figure, the subject of violent cartoons, of frequent misrepresentation, and an obvious target for anti-Jacobin feeling. Matthew Robinson Boulton brought the latest news of France to a meeting of the Lunar Society[2] and liberal opinion undoubt-

[1] Other travel accounts were Ulloa, *Voyages to South America*; Sparman, *Voyage to the Cape of Good Hope*; Governor Phillip's *Voyage to Botany Bay*; and Bruce's *Travels into Abyssinia*.

[2] *Life of Mary Anne Schimmelpenninck (née Galton)*, ed. C. C. Hankin (London, 1859).

edly grew strong in that Society.[1] Then came the Priestley Riots on 14 July 1791 and the destruction of Priestley's laboratory. The Derby Philosophical Society, unlike its counterpart at Manchester, sent a message of sympathy to Priestley, but like the society at Manchester suffered a minor revolution within its own ranks. The address to Dr. Priestley was not phrased in the most diplomatic terms

An ADDRESS to Dr. PRIESTLEY

Agreed upon at a Meeting of the Philosophical Society at Derby, Sept. 3 1791.

Sir,

We condole with yourself and with the scientific world on the loss of your valuable library, your experimental apparatus, and your more valuable manuscripts: at the same time we beg leave to congratulate you on your personal safety in having escaped the sacrilegious hands of the savages at Birmingham.

Almost all great minds in all ages of the world, who have endeavoured to benefit mankind, have been persecuted by them; Galileo for his philosophical discoveries was imprisoned by the inquisition; and Socrates found a cup of hemlock his reward for teaching 'there is one God'. Your enemies, unable to conquer your arguments by reason, have had recourse to violence; they have halloo'd upon you the dogs of unfeeling ignorance, and of frantic fanaticism; they have kindled fires, like those of the inquisition, not to illuminate the truth, but, like the dark lantern of the assassin, to light the murderer to his prey. . . .

R. Roe, Secretary.

The Derby Philosophical Society asked Priestley to 'leave the unfruitful fields of polemical theology' and pursue his scientific work. Priestley replied on 21 September 1791, thanking them for their encouragement, but saying 'Excuse me, however, if I still join theological to philosophical studies, and if I consider the former as greatly superior in importance to mankind to the latter. . . .' In the *Derby Mercury* for 6 October 1791, Mr. Hope, a member of the Society, published an advertisement[2] pointing out that the Address had been 'agreed to and fabricated by only five members of the Society out of 37' and that the general annual meeting of Saturday, 1 October, had decided that a fortnight's notice had to be given of any intended 'Act of Publicity'. This was not the end of the matter. On 10 October 1791, Richard Roe, the secretary, issued a further statement saying that Mr. Hope had been asked to resign, 'An advertisement, misrepresenting a transaction of

---

[1] Erasmus Darwin to James Watt, 19 Jan. 1790, B.R.L., Case 25, Parcel A: 'Do you not congratulate your grand-children on the dawn of universal liberty? I feel myself becoming all French both in chemistry & politics . . .'

[2] Dated 3 Oct. 1791.

the Philosophical Society at Derby, having been inserted in last week's newspapers'. Roe goes on to say:

That of thirty-seven members, thirteen only are resident in the Town, and that the address to Dr. Priestley was voted unanimously at a regular monthly meeting, at which was present the usual number of attending members, and that as it contain'd no reference to the Doctor's *political* opinions, and even recommended him to decline those theological controversies which seem to have provoked the vengeance of his adversaries, it was conceived that no man of a liberal mind would object to congratulating him on his escape from the violence of an enraged mob; and that there could be no member of a *Philosophical* Society who did not regret the demolition of his valuable laboratory and manuscripts; . . .

Nevertheless, we can judge that there was a current of opinion within this as within other scientific societies not unfavourable to the French Revolution in its early stages.

The Society weathered the storm and lived until 1857. Erasmus Darwin continued for some years to stimulate scientific interests among the membership. Thus we find him writing to James Watt in 1794 that he had 'invited all our Derby Philosophers to tea this afternoon, Sunday, to see your apparatus'.[1] And shortly before his death he was still writing enthusiastically to William Strutt, 'I wish on Sunday morn to see the grand effects of your electrical apparatus', and suggesting a series of investigations into electrical phenomena, including the construction of 'galvanic piles' and the electrical decomposition of water.[2]

Darwin died in 1802. With his vast intellectual energy, wit, and good humour, he had been the Society's mainspring and his death must have been a great loss. The growing attraction of the metropolis, the centralizing tendency in learned publications, and improved methods of transport also weakened the Society. The Manchester Literary and Philosophical Society was eventually overwhelmed by the Royal Society,[3] and the Derby Philosophical Society had never been so strong as its Manchester counterpart. Nevertheless, through its half-century of remaining life the Derby Philosophical Society continued to foster scientific activity and to organize lectures; and it bequeathed the library for which its amateurs of science had subscribed. Less distinguished

[1] Erasmus Darwin to James Watt, 17 Aug. 1794. Dol. The apparatus referred to was apparently a pneumatic apparatus for experiments on gases.

[2] Erasmus Darwin to William Strutt, 6 Aug. 1801, quoted in 'William Strutt, A Memoir' (c. 1830), typescript f. MSS., no. 3524, p. 16, in Derby Borough Library.

[3] See H. J. Fleure, 'The Manchester Literary and Philosophical Society', *Endeavour*, Vol. 6 (1947), pp. 147-51.

than the Lunar Society or the Manchester Literary and Philosophical Society, it had links with both and helped to keep alive that spirit of inquiry on which England's industrial and scientific greatness was built. It was also a monument to the great social qualities and the insatiable scientific interest of its founder, Erasmus Darwin.

V Training Captains of Industry[1]

The Education of
Matthew Robinson Boulton (1770–1842)
and James Watt, Junior (1769–1848)

One of the most powerful firms in England during the Industrial Revolution was the Boulton and Watt partnership in Birmingham. Eventually the business passed into the hands of the sons, and, according to Wedgwood's biographer, went into 'a sleepy decline, . . . upheld rather by the traditions of the past than by any genius or business capacity of those who had succeeded'.[2] Even a superficial examination of the correspondence at Birmingham reveals, however, that Matthew Robinson Boulton and James Watt junior were both of them able and intelligent men, and that the business transacted by the firm was much greater than it had been in their fathers' day. This is the view held by Erich Roll: 'The fathers had been builders, the sons were organisers; and, although the older generation had laid the foundations, the new built a superstructure of such unique elaboration that it becomes difficult to balance the merit of the two generations . . .'[3] But the fathers built not only machines but also men. They took the greatest care to make their sons fit to succeed them and their views on education reflect a great deal of light on the ideas held by the midland philosophers as a group. R. L. Edgeworth and his daughter Maria, Erasmus Darwin, Thomas Day, Samuel Galton junior, and Joseph Priestley, all wrote books on education and were all close friends of Matthew Boulton and James Watt.

James Watt confessed himself to be a self-made man, but was proud of his status as a 'gentleman' which his genius had obtained for him. He had, of course, been educated at Greenock Academy and his grand-

---

[1] This chapter is based on an article by Eric Robinson in *Annals of Science*, Vol. X, no. 4 (Dec. 1954).
[2] Eliza Meteyard, *A Group of Englishmen (1795–1815)* (1871), p. 2.
[3] Erich Roll, *An Early Experiment in Industrial Organization* (1930), p. 270.

father, Thomas Watt, had been a teacher of mathematics, so that he was not without formal education. It is probable indeed that his education had been sounder than that of most grammar-school boys in England at the same period. During his tenure of office as maker of scientific instruments to the University of Glasgow he extended his self-education and acquired a knowledge of German and Italian that enabled him to read scientific works in those languages. He also knew some French as he shows by writing sentences in that language in some of his letters to Joseph Black, but his occasional quotations from Latin in the same correspondence show him to be not very sure in his knowledge. Nothing much is known about the elder Boulton's education except that he went to the Rev. John Hausted's academy in Deritend, Birmingham, and that he had a slight acquaintance with Latin and an insufficient knowledge of French, and that the English of his earliest business letters is not beyond reproach. It also seems that, though he was always an interested and able scientist, he had enjoyed little formal instruction in science, because James Keir remarked: 'Mr. B. is a proof how much scientific knowledge may be acquired without much regular study.'[1] The letters which passed between these men and their sons, however, reveal that they had very distinct views about the sort of education required by boys who were to become gentlemen, manufacturers, and philosophers.

The first record of M. R. Boulton's schooling seems to be some letters from the Rev. Henry Pickering at Winson Green, then 'near Birmingham'. In 1777 M. R. Boulton was a day boy at this small academy, but two years later Mr. Pickering advised Boulton to send his son to him as a parlour boarder, because as a day pupil he would 'necessarily acquire a vicious pronunciation & vulgar dialect'.[2] How many Birmingham parents are still worried by those same speech habits! The schoolmaster also says: 'I do not mean to say that we are calculated to finish his Education on that liberal & genteel Plan that you I know wish & intend to adopt—but I think at his age we can do as well for him as Schools in general at a greater distance.'[3] The adjectives 'liberal' and 'genteel' are an interesting comment on Boulton's plans, and so is the fact that, when the boy was aged only nine, these plans had already been conceived for him. Mr. Pickering was obviously speech-conscious and it is quite probable that he made young

[1] James Keir, *Memoir of Matthew Boulton* (City of Birmingham School of Printing, 1947), p. 8.
[2] Rev. Henry Pickering to M.B., 5 Dec. 1779, A.O.L.B. (In footnote references James Watt and Matthew Boulton are abbreviated to J.W. and M.B. respectively.)
[3] *Ibid.*

Boulton take part in a little performance described in *Aris's Birmingham Gazette* for 20 December 1779:

> On Wednesday Evening last, previous to their breaking up for the Holidays, the Young Gentlemen educated under Mr. Pickering, at Winson Green, near this Town, delivered several select Orations, from various Authors, with a Propriety and Ease, which reflected great Credit on the Master's Care and Attention, and afforded the utmost Satisfaction to a genteel Audience, assembled on the Occasion.

From Mr. Pickering's the boy proceeded in 1780 to Twickenham where he came under the care of Mr. L. M. Stretch. Only the bills for the period 1780–4 remain, except for a letter from M. R. Boulton to his father dated 6 October 1783, which begins: 'As usual, at the beginning of the month, I present you with a Specimen of my writing.' Both fathers badgered their sons throughout their schooldays about writing. James Watt sounds amusingly Scottish: 'I observe you always send me a whole sheet of paper though you write only on one page of it which is bad economy & a needless piece of ceremony towards me—You also have lately written in too large a character and your lines too far distant for Letter writing . . .'[1] One can imagine young Jimmy Watt trying to cheat a little only to meet with his father's solemn rebuke. The thousands of letters hand-written by Boulton and Watt do show nevertheless how essential it then was to write a good hand. The copying machine invented by Watt, which, by the way, was later taken over by the sons, reduced the labour but made it even more important to write the master-copy well.

James Watt junior had been sent to the same school at Winson Green, Birmingham, as Matt. Boulton. There is a letter of 5 August 1780, apparently addressed to a local schoolmaster, from James Watt, complaining that his son has fallen into habits of 'insolence sauciness and disobedience', and that consequently he had sent for him home and given him 'a severe correction'.[2] He asks the schoolmaster to continue the treatment. There is no doubt that young James and his stepmother did not get on at all well together and the insolence may have been towards her, though Watt only spoke of insolence to other boys and to the servants.[3] He was also slovenly and even more careless of his clothes than his father was: 'and you will own that *was* bad enough,

[1] J.W. to J.W., jun., 13 July 1784, B.R.L.
[2] J.W. to an unnamed schoolmaster, 5 Aug. 1780, Dol.
[3] J.W. to Mrs. Marr, 13 Aug. 1780, Dol. See also Watt to his cousin, Nancy, 15 June 1780, Dol., where Watt says that he believes James to be better fitted for a lawyer or a merchant than an engineer. He complains of James's insolence and unreliability in his 'word of Honour'.

not to say *is* yet, be that as it may I have seen the evil of it and am resolved to cure him of it if possible'.[1] It was for these reasons and not primarily because the boy did not progress in his learning that Watt decided to change his son's school. He wrote to the Rev. Mr. Deane at Shiffnall:

Having a son about 12 years of age at Winson green School whom I want to remove to some other school where his learning and manners may be more particularly attended to than they can be in so numerous a school my friend Doctor Withering has done me the favour to mention you as a person to whose care I may safely entrust his education . . .

The boy is now attending writing and arithmetic in which latter he is advanced as far as to the extraction of the Cube root. He also reads latin, has read Caesar a first time and is now reading Virgils Aeneid. In a fathers opinion he is no despicable scholar for his years.

I want him to be further instructed in Writing, Arithmetic, and Latin, and to be radically taught Euclids Geometry, with such of the dependent Sciences as his time admits of.

But what I most desire is that a strict attention be paid to his manners and morals these being the most essential and also the prevailing reason with me to remove him from a public school.

When you do me the pleasure to answer this letter shall be obliged to you to lett me know if you have any persons who teach French & dancing to your Scholars . . .[2]

The stress upon moral education is heard in all Watt's letters about his son and in one he says that he does not mind what profession his son chooses so long as he proves to be 'an honest industrious man'.[3] Nevertheless he has firm ideas about the subjects that he expects his son to study.

Young James does not seem to have gone to Shiffnall and in January 1781 Watt was still making inquiries. He writes to the Rev. Mr. Palmer saying that Mr. Walker's school is too expensive for his son and one wonders whether this may have been a school run by Adam Walker, the itinerant lecturer in science.[4] At the age of fifteen, Jimmy was sent to John Wilkinson's ironworks at Bersham where he was to study practical book-keeping, geometry, and algebra in his leisure hours, and work three hours a day in the carpenter's shop. During his year at Bersham, Jimmy was bombarded with questions and advice from his father. The first thing to be noticed is the insistence upon some practical and manual education. Watt asks after his son's

[1] *Ibid.*   [2] J.W. to the Rev. Mr. Deane, 2 Nov. 1780, Dol.
[3] J.W. to Mrs. Marr, 15 June 1780, Dol.
[4] J.W. to the Rev. Mr. Palmer, 12 Jan. 1781, Dol. For Adam Walker, see above, pp. 104–6.

'progress in carpenter work' and asks him for his drawing of a furnace,[1] but some of James's drawing was of an ornamental character as well as a practical.[2] About some drawings the father observed that they failed 'more in the shading than in the perspective'.[3]

The boy was trained to 'a punctual and regular correspondence', but he was to avoid a stilted manner in his letters since: 'in general in Letter writing he succeeds best who writes like a correct Speaker in common conversation, & it is for this reason that women of any tolerable education write more easily than men; because they have not such a confusion of words in their head & study less'.[4] In addition the boy was to study merchants' accounts in the afternoon 'from Dinner time to 6 at night, along with your Arithmetic & algebra'.[5] The father kept a particularly watchful eye on his son's study of mathematics, especially in the solution of mechanical problems. He sets him a problem on weights and pulleys and is most disappointed with Jimmy's failure to provide the solution. 'I thought you would easily have given me the solution of the problem on the weights & pullies as it is particu[la]rly Exemplified in the 1st Book of the Octavo Edition of s'Gravesande in the chapter of oblique powers . . .'[6] The book was the classic translation of the Dutch mathematician s'Gravesande's work by Dr. Desaguliers, which became one of the standard school text-books in mathematics in Britain. James Watt himself had used the same text-book as a boy. The emphasis upon practical subjects while the boy was in Mr. Turner's care at Bersham meant that he would have to cut down his classical studies, but his father thought that he should still contrive to do some classical reading every day. While he was at Bersham the boy was warned against taking the Lord's name in vain, exhorted not to be too talkative when travelling home in the coach, to 'spend no more than what is necessary and avoid drinking strong liquors, but pay your share of the reckoning'.[7] He was also given advice about tipping and other points of social etiquette. In October 1784, when the boy was fifteen, Watt thought of sending him to university but instead decided to send him abroad.

Matt's education, at first, was rather less practical. He was now at the school of the Rev. S. Parlby, at Stoke by Nayland, near Colchester. His main study seems to have been Latin, and with some attention to

[1] J.W. to J.W., jun., 13 July 1784, B.R.L.
[2] A drawing of cherubs by James Watt junior, n.d., Dol. Exhibited in the Lunar Society Bicentenary Exhibition, Birmingham Museum and Art Gallery, Nov. 1966. [3] J.W. to J.W., jun., 1 Aug. 1784, B.R.L.
[4] J.W. to J.W., jun., 13 July 1784, B.R.L.
[5] J.W. to J.W., jun., 1 Aug. 1784, B.R.L.
[6] *Ibid.* [7] J.W. to J.W., jun., 17 Oct. 1784, B.R.L.

Latin pronunciation. Mr. Parlby hoped 'to qualify him so far in the Latin Tongue by Xstmass, that he may be enabled to read a Latin poet or Historian with tolerable fluency of construction & certainty of Judgmt. as to the Author's sense'.[1] He continued to declaim 'memoriter' passages from English and Latin poetry 'to *the room*', though 'With respect to his defect in Pronunciation, I fear it will not be soon or easily remedied'. He adds: 'I seldom fail to correct him whenever he omits the aspirate, his most inveterate provinciality in conversation.'[2] However, Mr. Parlby recognized the superiority of his home background: 'Indeed for me to pretend to lead him into any philosophical pursuits wd be absurd, as the advantages he may reap in that way from his Friends at home both in Theory & practice are so infinitely superior to any I cd pretend to hold out to him.'[3] The philosophical attitude was even maintained, to some degree, betwen father and son, because when Boulton wanted his son to learn Greek he bowed to Matt's objection that it was then too late to begin.

There is a greater similarity in the education of the two boys once they begin their further education. Both fathers decided upon a private education for their sons, on the continent, where they would be able to learn the French and German languages. A great merchant house like that of Boulton and Watt sold its goods to the whole of Europe and had agents and customers in every country. Watt kept up a critical commentary on his son's linguistic progress, and advised him to make the most of his opportunities in Geneva: 'As there are two German Gentlemen in the house I hope you will improve that circumstance so as to acquire some knowledge of the pronunciation of that language, which is a very necessary one both to merchants and men of science.'[4] Young Watt was sent to Geneva to be under the eye of the scientist, J. A. de Luc, and to attend the lectures at the Academy. At home he continued his geometrical and algebraical studies with M. Duvillard. He was to attend the lectures on natural philosophy if he found himself able to understand them, but chemistry had to wait for the moment so as not to interfere with 'your more important studies, which are mathematicks, natural philosophy, and drawing'. 'The parts of mathematicks that are most essential are Geometry comprehending Conic sections and the Doctrine of curves, and Algebra with the whole science of calculation and the application of it to Geometry.'[5]

[1] Rev. S. Parlby to M.B., 28 Feb. 1785, A.O.L.B.
[2] Rev. S. Parlby to M.B., 14 Feb. 1786, A.O.L.B.
[3] Rev. S. Parlby to M.B., 28 Feb. 1785, A.O.L.B.
[4] J.W. to J.W., jun., 16 Jan. 1784 [in error for 1785], B.R.L.
[5] J.W. to J.W., jun., 13 March 1785, B.R.L.

Pictet, the educationist and, be it remembered, translator of the Edge-worths' *Practical Education*, invited young James to attend his course of lectures for nothing, but the father insisted upon paying. There is no doubt that the boy was in a position to learn a lot from men like De Luc, Pictet, and De Saussure, with whom he was in contact. Apparently he found De Saussure pompous, but his comment upon this was severely rebuked by his father: 'Your remark in relation to Mr. De Saussure may be just, but you show too much self importance in saying that you will not trouble him with your visits, as he is one of the richest and most powerful men in the City and is also a Gentleman of much learning, science, & genius . . .'[1] Poor James was kept in very severe check. His father attributed his absence of mind to reading too many novels; he was told that he went to the theatre too often and that was bad for his morals; he was told in every letter to be more economical; and he was to be sure to rise in plenty of time to begin his studies at six in the morning. His mathematical studies became more advanced: 'You should divide your lesson into 2 parts, one to be dedicated to Geometry and the other to the higher parts of Algebra such as cubic & Biquadratic Equations, the summation of series, the squaring of curves, and other applications of Algebra to geometry.. . .'[2] And at the same time, being now reasonably proficient in French, he was to start German. Only one relief he seemed to enjoy. Joseph Priestley junior was in Geneva studying science and French, and it so happened that he was taking lessons in fencing. Presumably Watt learnt this from Dr. Priestley in Birmingham, and then thought it would be a good idea for James 'to stand firm on your legs, and put your body into graceful attitudes'. However, two months' lessons were not to cost more than one louis per month.

By July 1785 James Watt decided that his son had benefited suffi-ciently from Geneva and that it was time for him to make a move. He wrote to his son saying that he intended that Jimmy

should leave Geneva about the begginning [*sic*] of Octr and go to some part of Germany to acquire that language and to finish your studies, but I am indeter-mined yet whether I shall send you to a merchants Counting house in Ham-burgh, Koningsburgh [*sic*] or some other great trading town to initiate you into the nature of commerce, or if I shall send you to Gottingen university to pursue the study of usefull science. I fear that in the latter case you will by following pursuits principally literary contract a dislike for the serious and uniform business of a Counting house . . . You know my intention has hitherto been to breed you to the business of an Engineer if you should possess that

[1] J.W. to J.W., jun., 16 Jan. 1784, B.R.L.
[2] J.W. to J.W., jun., 1 May 1785, B.R.L.

Ingenuity and attention to mechanical pursuits that I did at your time of life . . .[1]

In addition Watt, who was in the midst of litigation over the patent for his steam engine, feared that an engineer's life without patents was not worth while: 'I therefore wish your Education to be such as may qualify you for a manufacturer as well as an Engineer so that your dependence may be on yourself and not on the caprices of Judges & Juries in case of any reverse of fortune.'[2] The word 'literary' in connection with the boy's studies at Bersham and Geneva now sounds a little odd, but it does emphasize how important it was thought to be to have a practical education. The only entry italicized in the index to the Edgeworths' *Practical Education* reads: '*Education practical, never ends!*'[3] In the same volume we read that 'Much of the time that is spent in teaching boys to walk upon stilts might be more advantageously employed in teaching them to walk well without them.'[4]

About August 1785, Watt decided to send his son to Herr Reinhard at Eisenach in Saxony. After he had learnt German there, he was to go 'to Freyberg in Saxony where there is a School for teaching the miners Mechanicks mathematicks & metallurgy, and one Charpentier a first rate Mathematician teaches that science'.[5] If this plan failed he might go into a counting-house at Leipzig. The priority of Jimmy's education at Eisenach had been succinctly described by his father: 'As to your studies they are to be in the first place the German Language and writing, secondly the pursuing your Mathematical studies, thirdly merchants accounts as soon as you understand Mr. Reinhard well enough to begin them fourthly drawing and dancing . . .'[6] Watt, while making it clear that mathematics was the key science, laid stress on orderliness in correspondence and note-taking and the surviving papers of father and son show that the lesson was taught by a master in the art to an apt pupil: 'let your memorandums be made in clear and orderly method so that by an Index you may easily refer to them, and keep your letters, and other loose papers, regularly folded up and docketed'.[7]

Matthew Robinson Boulton also found his way to Herr Reinhard's and, since the account in the Boultons' correspondence is fuller, we shall now revert to the story of Matt's education. Young Boulton's

---

[1] J.W. to J.W., jun., 17 July 1785, B.R.L.   [2] *Ibid.*
[3] Maria and R. L. Edgeworth, *Essays on Practical Education* (2 vols., 1789), Index.
[4] *Ibid.*, Vol. II, p. 408. This remark is made about speaking in public.
[5] J.W. to J.W., jun., 12 Aug. 1785, B.R.L.
[6] J.W. to J.W., jun., 2 Sept. 1785, B.R.L.
[7] J.W. to J.W., jun., 13 March 1785, B.R.L.

education, like young Watt's, went on at home as well as at school. As a small child he played with balloons filled with hydrogen and helped his father with his experiments in the laboratory at Soho. In 1782 his father wrote to him from Cornwall telling him how, in his rides to the mines, he was always stopping to pick up mineral curiosities which he assayed in his private laboratory on his return to Truro.

There is nothing could add to this pleasure so much as the having your assistance in making solutions precipitations filtrations evaporations & chrystalizations; but previous to the exercising of any art or science it is necessary to learn it, and previous to that, it is necessary you should learn certain languages such as your own, the latin, the french; and then go to the university to learn science & then you come to the more agreeable part: viz: the application of that knowledge . . .[1]

Besides helping his father in his experiments, Matt was encouraged to take an interest in the family business. In February 1785 he was taken round the Albion Mill by the engineer, John Rennie, who was then managing the mill for Boulton. In a similar way Jimmy Watt, two months earlier, had been told to visit Perier's steam engine in Paris and report on it to his father. The two boys were expected to see the practical application of their scientific studies.

In 1785 Matt was taken to France in the company of his father and James Watt, who were negotiating with Perier about the Paris steam engine. There he was left with Mons. H. Bourdon at Versailles. The boy's day is mapped out in Bourdon's prospectus, written in copperplate, but a fuller description is given by the boy himself:

I have endeavoured to dispose the days of the different masters in such a manner as to have employment for every day. Monday, I rise at 7 o Clock, when the hairdresser pays his visit, after that till Breakfast at 9 o Clock I generally spend in reading at ten I learn my dialouges [*sic*] & translate some English piece into french till twelve the Hour of the fencing Master who rests an Hour, I draw till two when I prepare for dinner a quarter past, we rest at Table an Hour; after dinner we walk or take other amusements as the weather will permit at five Mr. Bourdon corrects my translation &c I read to him to get the pronunciation as you wished.

At seven I return into my Chamber, for the evening I prepare the sums & questions which the mathematical master gives me who I attend every tuesday, thursday, & Satur[da]y from half past eight to ten; the same day the drawing master comes at six & rest[s] an Hour; Thursday, Friday Saturday the drawing master, who will suit your taste, large heads in Crayons, I have the fencing master the rest. Any alterations that you may point out I will with pleasure adopt . . .[2]

[1] M.B. to M.B., jun., 28 Oct. 1782, M.B. Letter Book 1781-3, A.O.L.B.
[2] M.B. jun., to M.B., 20 March 1787, A.O.L.B.

It was no narrow apprenticeship that Boulton desired for his son. Boulton was as insistent as Lord Chesterfield himself that his son should have all the appearance and attributes of a gentleman. He was to have lessons in riding, to dance a minuet gracefully and elegantly, and above all to be informed on the political movements of the day in France. The 'Assembly of the Notables' was being held at Versailles in 1787, and Matthew Boulton hoped that his son had seen the procession in which the king rode. The boy wrote back in a truly critical spirit:

> The last assembly of Notables was held yesterday when the king made a speech, wherin [sic] he mentions the result of the meetings & the reforms he intends to make, but it appears that the business which cost a large sum & for which they have taken such a long time might have been transacted in a week, but nothing can be done here without great pomp & magnificence.[1]

Meanwhile, in the middle of these great events, the more sober matters of Matt's education were proceeding. He was reading the works of Voltaire, the *Contes et Moraux* of Marmontel ('to form the style and to give the tone of conversation in the first ranks of society'), Bossuet's *Orations Funèbres*, Bruyère's *Characters*, the Abbé Raynal's *Philosophical and Political History*, and M. De Genlin's *Theatre*, with an English translation at the side. Boulton, more liberal than Watt, recommended *Telemachus* and Fanny Burney's *Evelina*. Yet the real core of his studies was mathematics and science. On the Versailles prospectus Matthew Boulton has pencilled the following remarks:[2]

Not to dispute but may argue

Euclid perfectly	
Algebry	
Use of Logarithms	} Geometry
plain and Spericle [sic] Trigonometry	
Conic Sections	

Fancy	
Architecture	} Drawing
Perspective	

(The injunction at the head of these notes may be a reference to a remark made by Benjamin Franklin to a young lady: 'So, you see, I think you had the best of the argument; and, as you gave it up in complaisance to the company, I think you had also the best of the dispute.'[3]) The mathematics speaks for itself, while it is natural that Boulton, who had been obliged to found a drawing-school for his

---

[1] M.B. jun., to M.B., 27 May 1787, A.O.L.B.
[2] H. Bourdon to M.B., Emploi de la Journée, 1786.
[3] Edgeworth, *op. cit.*, Vol. I, p. 167.

Soho employees, should insist on a knowledge of drawing. The Edge-worths also would have approved: 'To understand prints of machines, a previous knowledge of what is meant by elevation, a profile, a section, a perspective view, and a (vue d'oiseau) bird's eye view is necessary.'[1]

M. Bourdon does not seem to have fulfilled all Matthew Boulton's expectations, and in 1787 the boy was removed to M. Manuel in Paris. His literary education continued with more Voltaire and M. Thomas's 'sublime and lofty letter to the people'. He visited the theatre with M. Manuel because 'le théatre nous donnera de bonnes leçons de morale et de gout', and his tutor observed that he listened with approval to the lines from 'hamlet',

> on retrouve un ami, son épouse, une amante:
> mais un vertueux père est bien précieux
> qu'on ne tient qu'une fois de la bonté des dieux[2]

M. Manuel also says that mathematics are to Matt's taste. His father indulged him by sending him arithmetical problems about steam engines and their consumption of coal. In another letter he writes: 'I wish you to read to study & to understand the Six Books of Euclid. I wish you to understand Conick Sections also Trigonometry, Algebra, the use of Lograthims [sic] & the doctrine of Ratios . . .'[3] But Matt was also to study science:

> I wish you to attend philosophical Lectures when you have laid the founda-tion aforesaid & then you will easily learn those sciences which depend thereon such as Opticks—Hydrolicks—Hydrostaticks—Pnumaticks etc. At the same time you are improving your self in Mathematicks you may attend some publick Lecturer in Metalurgy & Chymistry in General but above all things persue constantly something that is laudable & never pass an idle moment . . .[4]

After such a course, the final warning seems hardly necessary. It was a pity that Matt had no great interest in science, because he had to spend three whole mornings a week 'dans le trop magnifique laboratoire de M. Sage', and was expected to send news of any scientific discovery to his father and to buy for him any important scientific publications. Boulton père was a chemist rather than a mathematician, and some of his comments upon chemistry are the most interesting in the corres-pondence. 'A man', he tells his son, 'will never make a good Chymist unless he acquires a dexterity, & neatness in making expts, even down to the pulverising in a Morter, or blowing the Bellows, distinctness, order, regularity, neatness, exactness, & Cleanliness are necessary in the

[1] Edgeworth, *op. cit.*, Vol. II, p. 98.
[2] Manuel to M.B., 25 Nov. 1787, A.O.L.B.
[3] M.B. to M.B., jun., 1 Dec. 1787, A.O.L.B.   [4] *Ibid.*

Laboratory, in the Manufactory, & in the Counting house.'[1] The laboratory, the factory, and the counting-house—the three are indissolubly linked in Boulton's mind. It is characteristic that, in the same letter in which he tells his son to use a pestle and mortar, he asks him to keep his personal accounts carefully and to learn for this purpose the Italian or double-entry method of book-keeping. Boulton also tells his son:

I wish you would endeavour to acquaint your Self with the knowledge of Exchanges between one Commercial Country and another, with the Theory of Money and the rules of calculating Exchanges or the relative Value a given quantity of Gold or Silver in one Country bears to the like quantity in another and the Causes of the Fluctuations.[2]

Matt is to keep personal accounts at all times, to date his letters in an orderly way and to keep a diary.[3] Above all he must learn the first lesson of a man of business—to keep his mouth shut about business matters.[4] Matthew Boulton was a firm believer in the middle class and the middle-class virtues. He even breaks into song, echoing the Miller of Mansfield, in one letter:

> How happy a State does the Miller possess
> Who would be no greater nor fears to be less
> On his Mill & himself he depends for support
> Without Bowing or serving or Cringing at Court.[5]

For the aristocratic Englishman in Paris he had supreme contempt and strongly warned his son against having anything to do with them. The boy was, however, to observe the fashions in buttons in Paris and to report his observations to Birmingham. One begins to notice that Matt is being introduced to business slowly but purposefully. He undertakes little errands to Droz,[6] the engraver, and is told about the state of the business at home. He also acts as a channel for scientific information between Franch scientists and the Lunar Society.

But what of the moral and spiritual education of young Boulton? It is noteworthy that Boulton, like the Edgeworths in *Practical Education*, never once mentions religion in his letters of advice. This was not because he lacked affection for the boy: 'I long to hear from you, pray

[1] M.B. to M.B., jun., 19 Dec. 1787, A.O.L.B.
[2] M.B. to M.B., jun., 1 Dec. 1789, A.O.L.B.
[3] M.B. to M.B., jun., 4 March 1789, A.O.L.B.
[4] M.B. to M.B., jun., 11 March 1788, A.O.L.B.
[5] M.B. to M.B., jun., 2 Oct. 1787, A.O.L.B.
[6] M.B. to M.B., jun., 11 March 1788, A.O.L.B. 'I beg youl begin to be a man of business. Be silent about everything that passes between me & Mr. Droz.' There are several letters on this topic from Nov. 1787 to May 1788.

write often, tell me your feelings & what may pass in your mind.'[1] Nor
was it because he lacked interest in the boy's moral welfare. He warns
him about the dangers of fashion, of debt from card-playing, and of the
necessity to be punctual, neat, and honest. One must believe him when
he says to his son: 'There is nothing on Earth I so much wish for, as to
make you a Man, a good Man, a usefull Man, and consequently a happy
Man.'[2] But none of these things apparently required religious instruction, nor is there any trace of strong religious feeling elsewhere in
Boulton's correspondence. The failing of character which seems to
have disturbed Boulton most is his son's lack of industry.

In 1788 Matthew Boulton returned to Birmingham for a short
holiday. It was probably then that Mary Anne Schimmelpenninck saw
him at a meeting of the Lunar Society:

> Amongst the habitual family routine to which we now returned, was that
> of receiving the Lunar meetings. The first of these was marked by Mr. Boulton's
> presenting to the company his son, just returned from a long *séjour* at Paris. I
> well remember my astonishment at his full dress in the highest adornment of
> Parisian fashion; but I noticed, as a remarkable thing, that the company (which
> consisted of some of the first men in Europe) all with one accord gathered
> round him, and asked innumerable questions . . .[3]

The degree of Matt's fashion may be exaggerated by this Quaker lady,
but not the degree of interest which the Lunar Society showed in all
things French. James Keir, Erasmus Darwin, and Joseph Priestley would
have been all agog to hear the latest French news. But the holiday was
not very long, and Matt, who had already been taking German
lessons in Paris, was sent off to Staedfeld in Haute Saxe, to the Rev. M.
Reinhard. James Watt junior had returned from there and Boulton was
greatly impressed by his progress. Young Startin, the son of another
business acquaintance, was also under Reinhard's care. Reinhard states
the terms on which he will take young Boulton:

> Si Mons: Boulton le père voudrait me confier son fils unique, j'en serais bien
> content. La pension serait la même que Vous, Monsieur, m'avez donné; savoir
> 40£ sterl. pour l'instruction dans la langue allemande, Geographie, Arith-
> metique et dans la Musique, pour la table etc. Vous m'avez donné 45£ et j'ai
> fait faire blanchir le linges [*sic*] de Mr. Votre fils, j'ai payé le frisseur, et j'avais
> promis de faire venir deux fois la semaine les maitres à designer et à danser . . .[4]

At first all went well at Staedfeld. According to Reinhard, Matt
showed genius in mechanical things, he learnt in nine months more

[1] M.B. to M.B., jun., 23 Jan. 1786, A.O.L.B.
[2] M.B. to M.B., jun., 19 Dec. 1787, A.O.L.B.
[3] *Life of Mary Anne Schimmelpenninck (née Galton)*, ed. C. C. Hankin (1860),
p. 125.   [4] Reinhard to J.W., 9 Aug. 1787, A.O.L.B.

German than Jim Watt had learnt in twelve, and he was progressing in mineralogy and learning the technical terms used by the local miners. All that worried his father at this time was whether he could send him some soft fine Welsh flannel to wear under his shirt because of the cold. Unfortunately, however, Matt was now nineteen and chose this moment to fall in love with a skittish young aristocrat, the Baroness de Wangenheim. Reinhard packed him off out of the district and immediately told Matt's father about the affair, giving one account in the letter which he allowed Matt to see and a different one in a letter which he sent secretly. Matt, having decided perhaps that a patrimony was better than a dowry, wished to retire to Freiburg and to study under Werner, who was then teaching mineralogy there, but his father preferred him to work under J. C. Wiegleb at Langensalza on the grounds that the Cornish mines were beginning to fail and Matt was unlikely to have much to do with mining:

> My present fears are that you will have taken your departure to Fryberg & if that is the case you may stay so long as you can gain any knowledge & improvement and then I submit to Mr Reinhards advice & your own Judgement whether you shall join Mr Collins or go to study Chymistry under Mr Wiegleb at Langensalza or go into a Counting house for 6 or 12 Months either at Leipzig or Hamburgh ... A Knowledge of Mechanicks, Mathematicks, Chymistry, Mechanick Arts, & Commerce; joined to the Character of an Honest Man is what I am very anxious you should possess & thereby continue to embelish, extend, & enrich the House I have layd the Foundation of.[1]

If Matt passed through Nuremburg, he was to be careful to see their manufactures, especially their methods of making brass tinsel and gold leaf. Matt was also given various commissions connected with his father's new Mint, distributing sample coins and inserting advertisements in the German papers.

Johan Christian Wiegleb had studied pharmacy under Sartorius at Dresden, and he was a friend of Jacquin, the botanist, and indeed of most German men of science. As a defender of the phlogistic school of chemistry, he was the more acceptable to the followers of Priestley, but his chief attraction to Boulton was that he was 'a good practical Chymist & particularly in all those branches which relate to the *Arts* in general'.[2] His text-book, *Manuel de chimie générale appliquée aux arts* (Berlin and Stettin, 1779) was highly thought of at this period. Boulton advised his son to make drawings of all Wiegleb's furnaces and utensils, just as young James Watt had been told to do at Bersham. In order to keep Matt in touch with developments in chemistry at

[1] M.B. to M.B., jun., 3 Aug. 1789, A.O.L.B.
[2] M.B. to M.B., jun., 8 March 1790, A.O.L.B.

home, Boulton sent him the first volume of Keir's *Dictionary of Chemistry*, and proposed sending him Dr. Black's lectures in five MS. volumes. He sent him Bennet's book on electricity, a thermometer, some coins, a box of fossils, and referred him to Withering's paper on *terra ponderosa*, and to Dr. Crawford on heat.

The year is 1790. Matthew Boulton is twenty years of age, and his father is beginning to feel himself infirm and overloaded with business. Boulton and Watt have lost several thousands of pounds through the Cornish mines and Matthew Boulton feels that he must get to know his son better. All the more so, because James Watt and his son have quarrelled, so that Boulton has been unable to find a suitable partner for his son. The young Watt has gone to Taylor and Maxwell, fustian makers, in Manchester, where he is fast coming under the influence of the republican, Thomas Cooper. That connection with Cooper nearly proved to be his ruin, for he was a fugitive for three or four years from the English government. In 1794 he returned from Europe to take his place in the family business. His education for the family business has ended in 1790, and so has the formal education of Matt Boulton, though it was not his father's intention at the time:

> You may this Summer get a little introduction & light into business & next winter (if God permitt) you may go to Edinburg, attend Dr. Blacks lectures through one course of Chymistry in your own language, also Dr. Robison, through one course of Natural, Experimental, & Mechanical Philosophy; at the same time you may study Rhetorick & the Belle Lettres which are essential embelishments to the Character of a Gentleman. Then return to business for a time after which you may make a carefull tour through Germany . . .[1]

It so happened that this ambitious plan was never completed, but even without these refinements Matthew Robinson Boulton's education had been thorough in the extreme. The main core of his studies and those of the younger James Watt had been modern languages, mathematics, and science. Their studies in the various branches of science had throughout a practical application. A small point substantiating this is the fact that we constantly read of the bags of tools which they took around with them on their travels. Though young James Watt's education was narrower than young Boulton's, yet they both had some knowledge of Latin and some acquaintance with the liberal arts. In history Matt was brought up on Priestley and the French historians; in science they were both reared on Priestley and the phlogiston school. Matt was also encouraged to be a gentleman—well but not gaudily dressed, able to dance a minuet elegantly, acquainted with music, the theatre, and

[1] M.B. to M.B., jun., 8 March 1790, A.O.L.B.

painting. James was educated as an engineer, and his more liberal thoughts and tastes were severely repressed by his father. The education of both after the age of fifteen was private. In the next generation Matt sent his son to Eton. Their finest education probably came from the great men with whom they were in constant touch at home and abroad, and as a consequence of that education their outlook was European and not insular. It is hardly surprising that they expanded their fathers' business beyond recognition, and became large-scale manufacturers of a sort that became the pattern of the Industrial Revolution. Within a few years of his entry into the business Matthew Robinson Boulton valued his time at £2,000 per annum. Nor were either of them confined entirely to their business; James Watt junior corresponded with Keir on geology and chemistry and maintained a long friendship with Wordsworth whom he met in Paris,[1] while Matthew Boulton junior corresponded with Beddoes, and with such artists as Sir Thomas Lawrence and John Flaxman. If these were two of the captains of industry, how many others had a similarly interesting education?[2]

[1] 'A day with Wordsworth', *Blackwood's Magazine* (Jan. 1927), quoted by F. W. Bateson, *William Wordsworth*, p. 96.

[2] That Watt influenced others to send their sons to be educated abroad is evident from surviving letters such as that from James Watt junior to his father (2 March 1790, Dol.), referring to an inquiry from a Mr. Weaver, apparently of Macclesfield, about the Freiburg Academy, and another from D'Ewes Coke, of Brookhill Hall, near Mansfield, to James Watt (3 Feb. 1793, Dol.), about his son's education under Wiegleb at Langensalza and then at Freiburg.

VI The International Exchange of Men and Machines 1750-1800[1]

As seen in the business records of Matthew Boulton

Eighteenth-century industry was conducted in an atmosphere of secrecy. The newspapers of Manchester, Birmingham, and other industrial centres, during the 1770s and 80s, contain frequent references to foreign spies who were snooping in factories and warehouses to learn the trade secrets of the area and to entice away the workmen who knew them. Committees were formed to protect these trade secrets by warning the locality about foreigners and by enforcing the various acts against the exportation of tools and the enticing of artisans abroad, so that every manufacturer became spy-conscious and perhaps more deliberately secretive than he already was. But this secretive attitude was already strong in British industry, not merely with regard to foreigners, but also with neighbours or potential competitors at home. Despite the efforts of such societies as the Society for the Encouragement of the Arts, which sought to make inventions freely available to the community, there was no general sympathy for the liberal exchange of knowledge about machines and skills. This is abundantly clear in the way in which Matthew Boulton dealt with Birmingham and Sheffield manufacturers who tried to seduce his skilled workmen.[2] During Arthur Young's visit to Birmingham to see the industries, he encountered the same kind of obstructions that foreign spies could expect to meet.[3] This caginess, strong enough towards neighbours, formed an

---

[1] This chapter is based on an article by Eric Robinson in *Business History*, Vol. I, no. 1 (Dec. 1958), pp. 3–15.

[2] Matthew Boulton Letter Book, 1768–73, A.O.L.B., Matthew Boulton to John Taylor, 23 Jan. 1769; Hodges Box, A.O.L.B., John Hodges to Matthew Boulton, 30 Nov. 1780; Matthew Boulton Letter Book, 1796–8, A.O.L.B., John Hodges to Mr. Betson, 10 April 1800.

[3] A. Young, *A Six Months Tour through the North of England* (2nd edn., 1771), Vol. 3, p. 279: 'I was no where more disappointed than at Birmingham; where I could not gain any intelligence even of the most common nature, through the excessive jealousy of the manufacturers. It seems the *French* have carried off

almost impenetrable barrier to dealings with foreigners. The object of this paper is to inquire how far this system of secrecy and protection was consistently adopted by Matthew Boulton in his dealings with foreigners, and how far, by the end of the eighteenth century, it had come to be replaced by a more liberal attitude. It will be seen that patriotic duty in this matter often conflicted with self-interest, but that there was a good deal of confusion in Boulton's mind about what his duty was and where his interests lay.

There was a fundamental moral weakness in a system which made it the manufacturer's patriotic duty to entice the foreigner's workmen and to acquire his secrets while sitting Cerberus-like over his own. This attitude is expressed very clearly in a letter from Messrs. Autran and Ador fils at Pforzheim to Matthew Boulton:

> We know you are not unacquainted with the establishment of our hardware factory; no less are we acquainted with the bonds you wish to place on it. You fulfilled, Sir, the duty of a good citizen in persecuting us, and when you wished to do us the most harm, we considered you moved by patriotic zeal rather than ill-will. . . . But at the same time we hope that after thus having done you justice, your own sense of what is fair will do us the grace of rendering what is just in its turn. Our reasons for building our factory were exactly those which made you oppose it, that is patriotic zeal. . . .[1]

Self-interest was in fact reinforced by patriotism, and since it was considered a loss to the private manufacturer if someone else knew his trade secrets, it seemed equally obvious that the nation should also keep her trade secrets from other nations. Then, as today, however, there was some confusion in people's minds. The international community of science was growing stronger; the typical figure of the age was Benjamin Franklin, member of almost every respectable scientific society in Europe—and even the more conservative Boulton and Watt had many scientific correspondents in France and Germany. Internationalism in science could conflict with patriotic duty or self-interest. J. H. van Liender might offer scientific honours to Boulton and Watt but they were obliged to reply that they were business-men trying to exploit a mechanical invention.[2] Moreover it had to be recognized that the foreigner also might have something to teach, and how were you to learn from him without teaching him something in return, especially if you were tied to him by business interests? It is also the object of this

---

several of their fabricks, and thereby injured the town not a little: This makes them so cautious, that they will shew strangers scarce anything; . . .'

[1] A.O.L.B., Autran et Ador fils to Matthew Boulton, 8 June 1770.

[2] Erich Roll, *An Early Experiment in Industrial Organization: being a History of the Firm of Boulton & Watt, 1775–1805* (1930), pp. 47–8.

P

paper to emphasize that the traffic in men, machines, and methods was not by any means all one way. The result was that there was a good deal of conflict between publicly expressed principles and private behaviour, and this can be substantiated in the records of Matthew Boulton. T. S. Ashton has noted in the State Papers Domestic some occasions when Birmingham artificers were enticed to Sweden and Austria.[1] Those incidents are, therefore, not described here but only referred to in sufficient detail to introduce new material from the Assay Office Papers, which Ashton was unable to consult when his book was published.

Perhaps the first point to establish is that it is no mere local patriotism for the citizen of Birmingham to visualize his city as already internationally famous well before the end of the eighteenth century. It seems to have been, with the neighbouring towns of Wolverhampton, Walsall, and Bilston, one of the greatest hardware centres in the world by the middle of the eighteenth century. Foreign buyers made twice-yearly visits to the area and there were a number of factors in the city engaged in the export trade. Many Birmingham merchants had foreign experience, such as Boulton's partner, John Fothergill, who had served his apprenticeship to a relation in Königsberg and had also travelled in Italy. The links with the continent were, therefore, very close and the firm of Boulton and Fothergill even had two secret foreign partners: John Friedrich Bargum, founder of the Royal Danish Guinea Company, and John Herman Ebbinghaus, a German hardware merchant of Iserlohn. Thus there were frequent opportunities for exchanging trade secrets, such as the occasions when Boulton asked Ebbinghaus to discover for him the method of manufacturing the white metal of Saxe-Gotha or told Bargum how to prepare German spelter for the English market. The connection with Solomon Hymen, a merchant of Paris, was particularly valuable. In a letter of 1772 Boulton asked Hymen to find out the French secret of making *or moulu* and authorized him to pay a few guineas for it.[2] It was the second attempt made through Hymen's help to find this out and Boulton describes how he had paid a guinea to some friend of Hymen's in a garret to watch the process but that it had proved to be unsatisfactory.

[1] T. S. Ashton, *Iron and Steel in the Industrial Revolution* (1924), pp. 200–5.

[2] Matthew Boulton Letter Book, 1771–3, A.O.L.B., Matthew Boulton to Solomon Hymen, 17 June 1772, 'In France *or moulu* was the name given to gold which had been ground up finely with mercury to form an amalgam used in gilding ornamental brass and other objects. In England ormolu is the name given to brass of high purity with a high content of zinc cast in ornamental forms and gilded, for the enrichment of other objects such as furniture.' H. W. Dickinson, *Matthew Boulton* (1936), p. 54.

During a short visit to Paris in November 1765, Boulton had witnessed the French method of silvering and their way of boiling brass work in some kind of sauce to give it a bright golden colour.[1] He also inquired in London of S. M. Diemar's[2] German workman:

> si le dit ouvrier sait quels sont les Matériaux dont on fait usage en France pour la dorure commune, & de quoi les francois se servent pour donner ensuitte à cette dorure ce Lustre & cette belle Apparence qui eu relêvé l'Eclat. Si Votre Ouvrier sait donner les informations nécessaires a ce sujet et sait en faire les Opérations lorsqu'on en aura besoin, dans ce Cas je pourrois peut-être procurer moyen de luy donner de l'Occupation. . . .[3]

Peter Hulphens of Amsterdam warned Boulton that a certain Frenchman was in Birmingham, pretending that he knew how to gild with a kind of varnish equally well as with gold but was contracting large debts in the meanwhile.[4] Fothergill, Boulton's partner, suggested craftsmen in London who might know the secret, and so, by inquiring of workmen at home and abroad, Boulton was eventually able to break the monopoly held by the French of *or moulu* manufacture. In the letter to Hymen, quoted above, Boulton tentatively proposed that Hymen should obtain workmen for him in Paris but added that he would not press the matter since he knew it would be against Hymen's principles. Though the suggestion is tactfully made, there is little doubt that it was not a random shot. In March 1775 Boulton asked Hymen, this time without apology, to find him in Paris 'an Excellent Workman in the Silversmith trade'.[5] The same agent also supplied him with tools for gold lace, which had to be obtained from a Monsieur Carpentier, a master gilder, at *rue du Four, faubourg St. Germain à côté de la rue de l'Egout*.[6] He also contacted a Monsieur Didelot, another master gilder in Paris, to obtain the tools for Didelot's son who was then working at Boulton and Fothergill's factory.[7] Naturally enough, other customers performed similar services. Thomas Winckelmann of Brussels, an old customer who had patronized Boulton's father and

[1] A.O.L.B., Matthew Boulton to Ann Boulton, 27 Nov. 1765.
[2] S. M. Diemar was presumably a toy-maker. His address is given by Boulton as: 'faceing the new Lying-in Hospital, the Surry Side of Westminster Bridge'.
[3] Boulton and Fothergill Letter Book, 1771–3, A.O.L.B., Matthew Boulton to S. M. Diemar, no date.
[4] Matthew Boulton to Peter Hulphens, 28 Sept. 1771, Boulton and Fothergill Letter Book, 1771–3, A.O.L.B.
[5] Letter Book 1774–7, A.O.L.B., Matthew Boulton to Solomon Hymen, 13 March 1775.
[6] Boulton and Fothergill Letter Book, 1771–3, A.O.L.B., Boulton and Fothergill to Solomon Hymen, 7 March 1772.
[7] Boulton and Fothergill Letter Book, 1771–3, A.O.L.B., Matthew Boulton to Solomon Hymen, ?1772.

often visited Birmingham, purchased a lathe for Boulton, and Peter Hulphens of Amsterdam forwarded brazier's tools. The latter also recommended a Flemish workman in the gilding line.[1] It cannot be said then that Boulton had any scruples about enticing foreign workmen or bribing them to part with their secrets or buying foreign tools and machinery, all of which he appeared to find very reprehensible when the traffic was reversed.

We know in fact that throughout Boulton's lifetime he recruited skilled workmen wherever he could find them. R. E. Raspe wrote to Boulton, 2 April 1791: 'For engravers I have wrote to Vienna, Hanau and Sweden—as also to Berlin—without compromising you—simply requesting to let you know whether and how they can and will engage with you.'[2] Adam Afzelius[3] also assisted Boulton in his search for foreign engravers. The two most famous in the annals of Soho were Jean Pierre Droz (1746–1832)[4] and Conrad Heinrich Küchler (1740?–1810),[5] a Frenchman and a Fleming. In 1771 the following Frenchmen were employed at Soho: Didelot, already named above; Marriset, dismissed that year but continued to work in Birmingham for John Taylor and others; Jean de la Fontaine, a buckle-maker from Wolverhampton; and most expert of them all, Alexandre Tournant, an expert mathematical instrument-maker and gilder, who had worked for François Thomas Germain (1726–91), the famous French goldsmith, and for the Berlin Academy of Sciences, and who later received a royal appointment from Louis XVI. Nor was Boulton the only Birmingham manufacturer to employ foreign hands: John Taylor was renowned for engaging French workers at high wages.[6] There was also the case of the German, Kern, who was taken on by Boulton and Fothergill, but who had previously been employed by one, Orsel.[7] William Matthews was asked to board Kern and was told:

He is a very level sensible well behav'd Man he was enticed from Germany by one Orsel of Birmingham who has behaved to him in a very scandalous shabby dirty manner he is an exceeding good workman in inlaying gold and silver in steel, for the Art or secret Mr. Orsel agreed to give him a certain sum of money to come to England and teach and work for him and him only but

[1] Boulton and Fothergill Letter Book, 1771–3, A.O.L.B., Matthew Boulton to Peter Hulphens, 28 Sept. 1771.
[2] A.O.L.B.
[3] Adam Afzelius [1750–1837], pupil of Linnaeus and famous Swedish naturalist.
[4] L. Forrer, *Biographical Dictionary of Medallists* (6 vols., 1904).
[5] *Ibid.*
[6] *Reminiscences of Thomas Henry Ryland*, ed. W. H. Ryland (1904), p. 46.
[7] He may have been connected with Orsel, Père et fils, of Aix-en-Provence with whom Boulton and Fothergill traded.

when Orsel had learnd all he cou'd he wou'd not pay him according to Agreement but playd him many dirty tricks . . .[1]

This workman from Saxony was so good that Boulton and Fothergill wanted him to be kept from contact with all other jewellers in London lest they should lose his services. Frenchmen are mentioned in the firm's letter-books in connection with original designs and with the copying of designs. John Taylor and Samuel Garbett testified before a Select Committee of the Commons in 1759: 'That there are Two or Three Drawing Schools established at Birmingham for the Instruction of Youth in the Arts of Designing and Drawing, and 30 or 40 *Frenchmen*, and *Germans* are constantly employed in Drawing and Designing. . . .'[2] Thus, in one way and another, continental craftsmen made quite a considerable contribution to the Birmingham hardware industry and Boulton never ceased to turn to foreign aid if he wanted something he could not obtain as well at home. As late as 1790 he bought a lathe from J. B. Dupeyrat (1759–1834), an engraver and lathe-maker, and a friend of Benjamin Franklin. This lathe, which was a kind of reducing machine, Dupeyrat was to come to Soho to instal.[3] It is true to say that Boulton was always alert to French inventions in the luxury trades. The Genevese, Aimé Argand, ordered instruments on Boulton's behalf if he saw any that were not already used at Soho.

This is one side of the exchange. The better known is the way in which foreign countries sent their spies to England and tried to entice British manufacturers and workmen abroad.[4] Garbett and Taylor declared in 1759: 'That many Attempts have been made by Foreigners to decoy our Manufacturers from hence; and that some have been prevented going, but some are gone, particularly one Master Workman from *Birmingham*, who used to employ 3 or 400 Men is gone, and took some of his Men with him. . . .'[5] This person was probably a Birmingham button-maker, Michael Alcock, who was a principal figure in the Industrial Revolution in France. The French biographer of Holker has said: 'La biographie d'Alcock, précurseur de Wilkinson, Wendel, Tubeuf, est à écrire,' but though there seems to be material for such a biography in France, there is little known about him in

[1] Boulton and Fothergill Letter Book, 1771–3, A.O.L.B., Boulton and Fothergill to William Matthews, 30 Oct. 1771.
[2] *Commons Journals*, Vol. XXVIII, p. 496.
[3] Letter Book, 1789–92, A.O.L.B., Matthew Boulton to G. Foucaut, 5 March 1790.
[4] The story of J. R. Holker and others who helped to develop the French textile industry is told in W. O. Henderson, *Britain and Industrial Europe, 1750–1870* (1954). I shall content myself with dealing with some Birmingham craftsmen.
[5] *Commons Journals*, Vol. XXVIII, p. 496.

Birmingham. He left England for France in the 1750s and settled in Saint-Omer. Shortly afterwards he moved to La Charité-sur-Loire, where he began a new factory in partnership with a Miss Willoughby, and in 1758 he promised D. C. Trudaine to make Saint-Etienne and La Charité, 'Birminghams en France'.[1] Both Miss Willoughby and Alcock returned to England to recruit workers, and Alcock was paid 2,400 livres for settling eight English artificers at Saint-Etienne.[2] In 1762 he founded works at Roanne and Villefay-en-Charolais. His sons, Michael and Joseph, later managed the establishment at Roanne which was described by Thomas Ingram:

> That at Roanne established by Les frères Alcock is conducted with greater skill & to more advantage, [than la Charité] these two young men are such that you would admire, & as far as my decernment will carry me, I never saw men better calculated for the undertaking they are engaged in .... Mr. Alcock asked much after your works & confessed their interest was naturally opposite to the trade of Birm & said they were taking ev'ry method to outdo us. . . .[3]

A few years later, in 1780, Fothergill told Boulton that Michael Alcock wished to open a shop in Paris for the sale of Soho mechanical paintings, *or moulu* frames, mirrors, etc.,[4] though nothing seems to have come of the project. That year Boulton was trading with a William Alcock of George Town, Minorca, as well as Joseph Alcock of Roanne. Michael Alcock senior died in Paris at the age of seventy-one.[5] Some of the workmen taken abroad by Alcock migrated still further and one was reported by one of Boulton's correspondents to be working in a Portuguese button factory at Oporto in 1773.[6]

We also learn from the Boulton papers about emigrants from Birmingham to Vienna. On 1 August 1769 a certain Robert Hickman applied to Boulton for work, having been dismissed by Capper, one of Boulton's neighbours on Snow Hill, Birmingham. Hickman was probably related to Josiah Hickman who supplied Boulton with candle snuffers[7] and we know that the Hickmans were quite a large family, all engaged in the hardware trade. The next we hear of the Hickmans is that Robert and William Hickman have established a

[1] A. Rémond, *John Holker* . . . (1946), p. 37.
[2] W. O. Henderson, *op. cit.*, p. 38.
[3] A.O.L.B., Thos. Ingram (Lyons) to Matthew Boulton, 11 March 1774; Thomas Ingram became High Bailiff in 1784, a Guardian of the Assay Office, Birmingham, and died in Bewdley, 1817.
[4] Fothergill Box, A.O.L.B., John Fothergill to Matthew Boulton, 7 Sept. 1780.
[5] *Aris's Birmingham Gazette*, 4 June 1785, B.R.L.
[6] A.O.L.B., Samuel Aislabie (Lisbon) to Matthew Boulton, 11 Aug. 1773.
[7] Boulton and Fothergill Letter Book, 1764–6, A.O.L.B., Matthew Boulton to Josiah Hickman, 2 July 1765.

metal and iron goods factory in Vienna, in 1779.[1] Seven years later, in 1786, they moved their factory to Ebersdorf, but in 1785 Robert Hickman suddenly appeared back in Birmingham, having attained prosperity. This is how Samuel Garbett described to Isaac Hawkins Browne, M.P., the return of the prodigal:

> That in January last, Robert Hickman, an ingenious artist who had left Birmingham some Years since to work at his Trade in Vienna, and had great Encouragement from the Emperor, returned to this Neighbourhood where he hath a Father & three Brothers, all ingenious Workmen—Upon Information given to me that there was reason to suppose he was tempting Workmen to go to Vienna, I sent a civil card desiring to see him. He came in a coach & was elegantly dressed. Upon my acquainting him that I had reason to suppose his Business here was to procure Workmen, he assured me he was returned to live in his native Place upon the Property he had acquired. I expressed my doubts & the reason we had to fear from him and his family—that Eyes were upon them—and that we shou'd advertize a Reward of fifty Guineas for Information of Offers made to our People to go abroad, and that if he returned to Vienna, we shou'd solicit his Majestys Ministers to outlaw him & many of his Servants at Vienna. We accordingly advertized & Hickman instantly left Town: but since then his family, who are not People of unquestionable Property, have delivered Commissions for great quantities of Hardware, & I don't doubt for Tools, & several of the family are likely to go with them by way of Ireland to the Continent. . . .[2]

But this was not by any means the first occasion on which attempts had been made to decoy workers from Birmingham to Vienna. In 1766 Garbett tried to have British workers in Vienna outlawed, and he enlisted the services of an English jockey to track down three British workmen who had gone there.[3] On 6 September 1768 the Graf von Zinzendorf, from Austria, visited Birmingham and was shown round the Soho factory by Matthew Boulton himself. According to von Zinzendorf, the very first thing Boulton did was to express his anxieties because '*on avait débauché des ouvriers pour les envoyer à Vienne*'.[4] It is worth observing, however, that von Zinzendorf was allowed to see several factories in Birmingham, including Baskerville's japan-works, Clay's papier-mâché factory, and other well-known places. He was even given a letter of introduction to Garbett at Carron though Garbett was not there when he made his visit. On such occasions as von

---

[1] Johann Slokar, *Geschichte der österreichisen Industrie unter Kaiser Franz I* (1914).
[2] A.O.L.B., Samuel Garbett to Isaac Hawkins Browne, M.P., 16 July 1785.
[3] T. S. Ashton, *op. cit.*, p. 204.
[4] 'Observations du Comte Charles de Zinzendorf pendant des voyages par la Grande Bretagne et Irlande: l'an 1768', microfilm, University of Manchester from Hof und Stats-Arkiv, Vienna.

Zinzendorf's visits some of the anomalies of the situation may be observed. He kept a diary in which, for example, he noted in detail various processes in the textile industries, and was evidently a careful observer, yet he seems to have made his way, unhampered, through the industrial centres of Great Britain, presumably because he was a man of society with influential friends. As for the workmen who allowed themselves to be enticed abroad, it was Samuel Garbett's belief that they had best be outlawed so that the example would deter others. There is little doubt that these threats did alarm some who had already emigrated and were frightened to return. This was the case with some Birmingham workmen whom N. Vyse found in the small Swiss town of Thun.[1] They were employed by Messrs. Ponnier and Stellie and had been brought over to carry on a manufacture of gilt and steel trinkets under the protection of the Canton of Berne. As their skills had been learnt, their pay was reduced and they were encouraged to move on, but they were timid of competing with the Swiss workmen and feared to return to England because they might be prosecuted. The workmen named by Vyse were: '*Pearson*, a gilder who work'd formerly with Mr. Taylor, *Benjamin Price*, watch chain maker, work'd with Prat on Mount Pleasant. *George Cashmore*, Chain Maker, *Harrison*, his father lives at the Cottage of Content.' Boulton tried to discover from Lord Dartmouth what the government wished to be done about these workmen and others like them. He did not receive a great deal of encouragement. If the manufacturers of Birmingham and Sheffield cared to band together and subscribe the return fares of the workmen, and guarantee them work upon their return, then the government would not prosecute.[2]

Of all the Birmingham manufacturers, Matthew Boulton himself would have been the greatest prize for a foreign nation to allure. It is well known that Russian offers were made to his partners, James Watt and William Small, and Boulton himself was invited to erect a factory in Sweden. Professor Ashton has described how, in 1753, the King of Sweden offered special privileges to foreigners who would engage in the iron and steel industries in Sweden, and how, ten years later, a passenger from Birmingham to London left behind in the fly in which he had travelled, a pocket-book with papers concerning some English workmen who were going to erect an iron foundry in Sweden.[3] A fortnight later, Samuel Garbett gave information that Matthew Boulton had received lucrative offers from a Swedish Intendant of Com-

[1] A.O.L.B., N. Vyse (Berne) to Matthew Boulton, 9 Sept. 1767.
[2] A.O.L.B., Lord Dartmouth to Matthew Boulton, ?1767.
[3] T. S. Ashton, *op. cit.*, pp. 202 *et seq.*

merce.[1] In 1765 Lovell Stanhope made inquiries about Boulton's dealings with Dr. Solander.[2] James Farquharson, the chief clerk for Roebuck, and Garbett, replied that Solander had recommended two or three Swedish gentlemen to visit the manufacturers of Birmingham in 1763 but that he had not been in Birmingham in 1765. Farquharson also pointed out that Boulton had invested large sums in his factory buildings at Soho and was not likely to be considering emigration.[3] This he confirmed in a further letter of 19 September 1765.[4] Unfortunately the Home Office records do not tell us more about the events of 1763, but certain papers at the Assay Office, Birmingham, clarify the situation. The Intendant of Commerce[5] was John Westermann and he visited Soho in company with John Alstroemer.[6] Both these men are known to have been recruiting English ironworkers for Sweden.[7] Westermann wrote in French, for reasons of secrecy, from Cologne on 16 September 1763.[8] A most interesting feature of this letter is that it suggests that the idea of going to Sweden was first proposed by Boulton and not Westermann. Since the letter was obviously not intended for anyone to see except Boulton, this must have been how the matter appeared to Westermann:

Comme Vous m'avait entendre Monsieur, que Vous etiez, presque determiné de Vous etablir dans quelque Pays etranger,[9] si vous pouviez trouver la facilité et des-

[1] Home Office Papers, 1760–5, no. 1359, 26 June 1765.
[2] For accounts of Solander's activities as a recruiting agent for Swedish industry, see Riksarkivet, Huvudafdeling, Stockholm, Diplomatica, Anglica 1, Svenska Sändebuds Skrivelser till K.Mt, Brev till Kanslipresidenten, and *ibid.*, Brev till Kanslikollegium, and *ibid.*, Brev fran Kanslikollegium och andre ambetsvarkl Sverige, and *ibid*, Brev fran Myndigheter och Enskilda samt strödda Beskickningshandlingar, and Utrikes Expeditions Registratur. I am indebted to Mr. H. S. Kent of the University of Adelaide for these references. Mr. Kent has also published two valuable articles on Anglo-Scandinavian commercial relations in the eighteenth century in *The Norseman*, Vol. XI, no. 3, and Vol. XII, no. 6.
[3] Home Office Papers, 1760–5, no. 1821, 6 July 1765.
[4] *Ibid.*, no. 1919, 19 Sept. 1765.
[5] To the best of my knowledge not previously identified. He was also known by the name of Liljecrantz.
[6] Johan Alströmer, 1742–86. See *Svenskt Biografiskt Lexikon*, which, however, does not mention this visit to England.
[7] See Rydberg, *Svenska, Studiereser till England under Frihetstiden* (1951), pp. 137–8. Rydberg also mentions the suspicion displayed by Birmingham manufacturers to Angerstein who visited Birmingham in 1754 because they were jealous of their machines and artificers, pp. 178–9. There are valuable shorter accounts in English by M. W. Flinn, 'The Travel-Diaries of Swedish Engineers of the Eighteenth Century as sources of Technological History', a paper read to the Newcomen Society, 8 Jan. 1958, and by A. Birch, 'Foreign Observers of the British Iron Industry during the Eighteenth Century', *Journal of Economic History*, Vol. XV (1955).
[8] A.O.L.B., John Westermann to Matthew Boulton, 16 Sept. 1763.
[9] My italics.

avantages convenables pour entreprendre et conduire avec succès une telle Manufacture que la votre, j'ai réflechi ulterieurement sur cette affaire, qui me paroit d'une nature à pouvoir devenir également utile pour Vous et pour la Suède.

If Boulton had been pretending an interest merely in order to implicate Westermann, there is no reason why he should have delayed his replies to Lovell Stanhope, or why he should have been still under suspicion two years later. Though Boulton and Garbett were acquainted with each other in 1763, they were not then the close friends they were to become later, and it is difficult to believe that Boulton confided in Garbett about the Swedish affair. There is, moreover, a possible reason why Boulton may have spoken in this way to Westermann in the summer of 1763. During that summer he quarrelled with his partner, Fothergill, who came to see him one Sunday morning and declared that he proposed to terminate their partnership and had given instructions accordingly to his solicitor.[1] Westermann may have caught Boulton in a dispirited mood during Fothergill's two months' absence. By 1765 Boulton's financial position had improved as a result of his brother-in-law's, Luke Robinson's, death. It is interesting to note that among the inducements offered to Boulton by Westermann was a moneyed partner.

The terms which Westermann was going to propose to the Swedish government on Boulton's behalf were:

(*a*) £500 for the travelling expenses of Boulton and his workmen,
(*b*) liberty to establish his manufactory anywhere in Sweden, though Westermann hoped he would choose Norrköping,
(*c*) an advance of £1,500 for his water-wheel and tools, which was to become a gift as soon as the factory was in working order,
(*d*) a bounty of twenty to twenty-five per cent on everything he exported,
(*e*) exemption from duty on all English coals he imported,
(*f*) a moneyed partner,
(*g*) special terms of discount for his bills of exchange received from overseas customers.[2]

Westermann added that all depended upon Boulton being able to persuade sufficient workmen in each branch of the manufacture to accompany him.

In November 1763, Westermann confirmed that the Swedish government would accept these terms, and asked Boulton to reply

---

[1] Fothergill Box, A.O.L.B., Statement of affairs, 1763–80.
[2] A.O.L.B., John Westermann to Matthew Boulton, 16 Sept. 1763.

under cover to Spalding and Brander,[1] merchants, in Cornhill, White Lion Court, London.[2] Apparently the negotiations went no further. But in later years Boulton came to be on friendly terms with Solander, continually expressed a close interest in Swedish affairs and corresponded with Sir John Goodricke, British ambassador to Sweden. One cannot help wondering too if there is any significance in the fact that from 1765 onwards a certain Jas. Farquharson was living next door to Matthew Boulton on Snow Hill, Birmingham, in a house owned by Boulton.[3] Could this have been the clerk to the firm of Roebuck and Garbett who reported on Boulton and Solander to the government? If so, he had only moved into Boulton's house in the significant year, 1765.

If Boulton was still under suspicion in 1766, he took steps that year to transform himself into a loyal citizen, anxious to assist the government in detecting foreigners who enticed away British craftsmen. Two Genevese gentlemen, called Preponnier and Ador, visited Birmingham that year in search of artificers for a factory in Pforzheim, owned by Messrs. Autran and Ador fils, under the patronage of the Margrave of Baden-Durlach, who also had a financial interest in their firm.[4] Unfortunately for Preponnier and Ador one of the men they had persuaded to join them changed his mind and let out the whole story. The leading manufacturers of Birmingham—Samuel Garbett, John Taylor, John Gimblett, Sampson Lloyd, Matthew Boulton, and others —wrote a letter to the Secretary of State to have the workmen stopped. Boulton and Garbett also employed their own investigator, Charles Sanders, to search out information and in particular to watch some goods deposited in Smithfield steel, which were never called for. Garbett himself went down to London and the whole affair has a distinctly 'cloak and dagger' atmosphere about it. Sanders reported that a workman called Tonks, late of Birmingham, had been offered £150 a year and board, or £200 a year without board, to go abroad.[5] As we have seen,[6] Autran and Ador later resumed a correspondence with Boulton, justified their actions, and wished to sell Boulton and Fothergill's products abroad. In 1773, John Sebastian Clais, 'an ingenious

[1] This was the leading Swedish merchant house in London. Abraham Spalding (1712–82) was a friend of Solander and a member of several English scientific societies. Gustaf Brander (1720–87) was also a well-known natural philosopher. See Rydberg, *op. cit.*, pp. 87–9.

[2] A.O.L.B., John Westermann to Matthew Boulton, 14 Nov. 1763.

[3] Rate Books, B.R.L.

[4] This incident was first described by A. Westwood, 'Le Premier Pas de Pforzheim', *Bulletin of the Birmingham Jewellers' and Silversmiths' Association*, Nov. 1932.

[5] A.O.L.B., Charles Sanders to Matthew Boulton, 29 Oct. 1772.

[6] See above, p. 217.

Mettalurgist from Saxony' who had been introduced to Boulton by J. R. Valtravers,[1] visited Pforzheim and sent a report back to Birmingham. Autran and Ador refused to let Clais see over their factory where they employed two hundred people just in making ornamental chains,[2] but since he had seen Boulton's factory he was sure he would not see a better one. According to Clais, Autran and Ador had enough business on their order books for two years ahead.

John Sebastian Clais, who later rose to high position in Prussia under Frederick II, was another of those useful foreign correspondents with whom Boulton was regularly furnished. He was the inventor of a new type of scales, had interests in a copper work and lead mine in the Isle of Man, was acquainted with Benjamin Franklin, and had a wide fund of knowledge about metals. He informed Boulton about the Nuremberg manufacture of brass, saying: '. . . the Bronze coulours I became acquainted of, in a most easy way, though they are so secret and I must tell you they make of every Louse an Elephant. . . . I have seen on the Rhine at Reuenweed a Steel Manufactory, where they make pretty good work. . . .'[3] It was also through him that Boulton learned of a Mr. Gaubrecht from Königsberg who had received a privilege from Frederick II for button-making, preparatory to a prohibition on imports of Birmingham buttons, and who had visited Birmingham to buy a great many machines.[4] In the same letter Clais sent his regards to Boulton's physician William Small, and said that in a previous letter he had told Boulton how to purify tortoise-shell, a manufacture engaged in at Soho. Contacts of this sort, on what might be called the natural philosopher level, were valuable to both sides, and seem on the whole to have been welcomed by Boulton and by Watt.

W. H. Chaloner has described Marchant de la Houlière's visit to England in 1775:

to investigate whether the superior quality of English coal and the peculiar nature of English iron ores constituted the reason why they were used together, with such a great measure of success in that country, to make cast and wrought iron; to see the methods used for the purpose and to become acquainted with the preparatory treatment given to these minerals, what substances are added to them in order to facilitate smelting, and lastly, the construction of furnaces and bellows.[5]

[1] A.O.L.B., J. R. Valtravers to Matthew Boulton, 29 Oct. 1772.
[2] A.O.L.B., John Sebastian Clais (Karlsruhe) to Matthew Boulton, 26 July 1773.   [3] *Ibid.*
[4] A.O.L.B., John Sebastian Clais to Matthew Boulton, 3 Aug. 1772.
[5] From Marchant de la Houlière's *Report to the French Government* (1775), translated, with an introduction and notes, by W. H. Chaloner, and reprinted from *Edgar Allen News*, Dec. 1948 and Jan. 1949.

Marchant de la Houlière's visit to the leading ironworks of the Midlands, including Coalbrookdale, Wright's and Jesson's at West Bromwich, John Wood's at Wednesbury, William Wilkinson's at Bersham, William Reynold's at Ketley, and John Wilkinson's at Broseley, were largely due to the recommendations of Matthew Boulton,[1] and he seems to have received every encouragement from manufacturers who had good reason to be spy-conscious. The Duc de Chaulnes, a distinguished name among scientific instrument-makers, was also received in a friendly manner at Soho in 1783.[2] Baron Stein aroused Boulton's anger when he attempted to view the engine at Barclay and Perkins's brewery, but was later welcomed at Soho. It seems that where Boulton and Watt did not fear attempts by foreigners to pirate their invention, they were quite prepared to be helpful. In fact it became Boulton's ambition to develop a foreign trade in steam engines. He told his son Matthew Robinson Boulton, who was studying in Germany: 'In the course of all your travils, & acquaintance Pray enquire if any Steam Engines are wanted in Saxony or any other parts of Germany as I wish to cultivate a foreign trade in that Branch . . .'[3] and thus seems to have recognized the fact that it was useless to deny new inventions to intelligent foreigners and better to supply their demands.

The position held by Boulton thus came to conflict with that of other manufacturers in Birmingham. On 12 July 1799 Boulton obtained an Act of Parliament enabling him to export machinery for a mint at St. Petersburg.[4] The manufacturers of Birmingham petitioned the government against that Act on 21 June 1800 but to no effect. Boulton's argument was that rolling mills, steam engines, and other machines were widely known on the continent so that there was no purpose served in denying the Russians these machines made at Soho. In a memorandum to Alexander Baxter, Boulton pointed out that the advantages of the machines far exceeded those to be expected from coining machinery, and even suggested that the steam mill for laminating copper might be useful to the Imperial Navy.[5] Not only that, but several Russians were trained at Soho, including A. Deriabin, who rose to be Chief Director of the Department of the Mints, Mines and Salt Works of Russia.

So the situation arose that by the end of the century a leading Birmingham manufacturer could export valuable machinery with the

[1] A.O.L.B., Marchant de la Houlière to Matthew Boulton, 31 Aug. 1775.
[2] A.O.L.B., James Watt to Matthew Boulton, 5 Sept. 1783.
[3] A.O.L.B., Matthew Boulton to Matthew Robinson Boulton, 26 Oct. 1789.
[4] See Eric Robinson, 'Birmingham Capitalists and Russian Workers', *History Today*, Oct. 1956.
[5] A.O.L.B., Matthew Boulton to Alexander Baxter, 15 Aug. 1796.

assistance of the government and in defiance of other industrial opinion in the midlands. The younger generation of Birmingham industrialists and merchants was well equipped to profit from international friendships but was prevented by the revolutionary wars from so doing. Matthew Robinson Boulton and James Watt junior, in whose hands the future of Soho lay, had been educated in France, Switzerland, and Germany; their cousin, Zachary Walker, had travelled very widely on the continent, as did Gregory Watt after the Peace of Amiens. John Roebuck junior and Peter Capper junior were established as merchants in St. Petersburg. The climate of opinion was about to change. It seems valid to conclude that Matthew Boulton's papers show that the behaviour of Birmingham manufacturers in the later eighteenth century towards the international traffic in men and machines was never very consistent. Boulton himself realized too clearly how much he could learn from the continent as well as how much he could teach. Consequently he shifted his mercantilist stand to suit his interests. Moreover the international links between men of science and men of fashion were such that it was difficult to refuse visits from persons bearing letters of introduction, without causing offence, quite apart from what you might learn from such people and the pleasure you could derive from their company. In general terms, secrecy proved to be bad business, and when the demand for British machinery strengthened, some manufacturers like Boulton saw that it was better business to satisfy the demand rather than resist it. In the conflict between the doubtful patriotic values of mercantilism and the obvious demands of self-interest, self-interest gradually won.

VII Early Industrial Chemists
Thomas Henry (1734-1816)
of Manchester, and his Sons[1]

The well-known names of mechanical inventors figure prominently in all the text-book accounts of the Industrial Revolution. Spinning and weaving machines and steam engines make a direct, concrete, visual appeal, and their effects are fairly easy to grasp. But the concurrent changes in chemical processes are more subtle and more difficult for the non-scientific layman to appreciate. Yet they were equally important: indeed, without them the mechanical inventions would, in many cases, soon have created bottlenecks in production. This is particularly true in regard to bleaching, dyeing, and calico-printing, where applied chemistry made vital contributions. Though the names of the chemists concerned have generally sunk into oblivion, in their day they were justly celebrated and, as a mid-nineteenth-century Manchester observer remarked, chemists such as Thomas Henry 'deserve to be ranked with such distinguished inventors as Arkwright, Kay, and Crompton'.[2]

Henry, as we have previously seen, was not alone in this field of applied chemistry, for there was in Manchester, especially in the Literary and Philosophical Society, a fruitful collaboration between science and industry similar to that in Birmingham and the Lunar Society. But he was undoubtedly one of the most important figures and deserves a niche in the industrial hall of fame.[3] He was not, of

[1] This is a previously unpublished contribution by A. E. Musson. Many of the references from the Boulton and Watt papers have been kindly supplied by Eric Robinson.

[2] H. G. Duffield, *The Stranger's Guide to Manchester* (Manchester, 1851), pp. 169–71. See also J. H. Park and E. Glouberman, 'The Importance of Chemical Developments in the Textile Industries during the Industrial Revolution', *Journal of Chemical Education*, Vol. 9, no. 7 (July 1932), pp. 1143–70.

[3] See W. Henry, 'A Tribute to the Memory of the late President [Thomas Henry] of the Literary and Philosophical Society of Manchester', *Manchester Lit.*

course, a great scientist, distinguished by his original discoveries: he is not to be compared, in this respect, with British contemporaries such as Black or Priestley. His youngest son, Dr. William Henry (1774–1836), was scientifically more eminent.[1] The father's achievements, as Bosdin Leech has pointed out, were 'essentially different. Thomas was utilitarian and for the application to everyday life of the awakening science of chemistry.'[2] He provides, in fact, a good example of the emergent industrial chemist, combining laboratory research with practical application.

He was descended, according to his son, from 'a respectable family, which, for several generations, had resided in the county of Antrim'. His father moved to Wales, married a clergyman's daughter, and set up a girl's boarding school, first in Wrexham and later in Manchester. Thomas was born at Wrexham on 26 October 1734, and after early tutoring by his mother was educated at the local grammar school. His parents being unable to afford to send him to Oxford, he was apprenticed to a Mr. Jones, 'an eminent apothecary of Wrexham', on whose death he was articled to 'a respectable apothecary at Knutsford in Cheshire'.

In neither of these situations did Mr. Henry enjoy any extraordinary opportunities of [intellectual] improvement. The only book, which he remembered to have been put into his hands, by either of his masters, was the Latin edition of Boerhaave's Chemistry . . . His reading was, therefore, entirely self-directed; and, by means of such books as chance threw into his way, he acquired a share of knowledge, creditable both to his abilities and his industry.[3]

Either now or later, there is no doubt, as we shall see, that Henry acquired an extensive familiarity with the works of both British and foreign chemists.

At the expiry of his apprenticeship, he became principal assistant to Mr. Malbon, then the leading apothecary in Oxford, where he was able to form associations with various colleges, including attendance at anatomy lectures at which John Hunter was demonstrator. In

& *Phil. Soc. Memoirs*, 2nd ser., Vol. III (1819), pp. 204–40. This forms the basis of later accounts of Henry's life, e.g. J. Wheeler, *Manchester: Its Political, Social and Commercial History* (1836), pp. 488–93; R. Angus Smith, *A Centenary of Science in Manchester* (1883), chap. vii; E.M.Brockbank, *Sketches of the Lives and Work of the Honorary Medical Staff of the Manchester Infirmary* (1904), pp. 72–82; E. Bosdin Leech, 'Some Manchester Medical Authors and their Works, V. The Henrys', *Manchester University Medical School Gazette* (1927), pp. 74–7 and 108–112; W. Kirkby, 'Thomas Henry, F.R.S. The Inventor of Calcined Magnesia', *The Chemist and Druggist*, 1 Dec. 1934, pp. 674–5; J. R. Partington, *A History of Chemistry*, Vol. III (1962), pp. 690–2.   [1] See below, pp. 246–50.
[2] *Op. cit.*, p. 109.   [3] W. Henry, 'Tribute', pp. 207–8.

1759, however, he moved back to Knutsford, where he married, and five years later was able to succeed to the business of 'a respectable apothecary in Manchester; where he continued, for nearly half a century, to be employed in medical attendance, for the most part on the more opulent inhabitants of the town and neighbourhood'.[1] He was appointed one of the visiting apothecaries to the Infirmary in 1779 and until nearly the end of his life played an important part in its concerns.[2]

The greater part of his time, especially during these early years, must obviously have been taken up by his apothecary's work, and his first researches sprang therefrom.

During his apprenticeship, Mr. Henry had manifested a decided taste for chemical pursuits, and had availed himself of all the means in his power, limited as indeed they were, to become experimentally acquainted with that science. This taste he continued to indulge after his settlement in life; and, after having made himself sufficiently master of what was ascertained in that department of knowledge, he felt an ambition to extend its boundaries.[3]

In 1771, therefore, he communicated to the Royal College of Physicians of London 'An Improved Method of Preparing Magnesia Alba', which was published in the second volume of their *Medical Transactions* for the following year. A year later it was reprinted, along with essays on other subjects, in a separate volume of his *Experiments and Observations* (1773). The existence of this 'earth' (alkaline salt), 'magnesia alba'[4] and its medicinal uses as a laxative, etc. had been discovered many years earlier, and Dr. Joseph Black had carried out important experiments on it, both in its common or carbonate form and when calcined (magnesium oxide).[5] Physicians complained, however, of the unsatisfactory nature of the 'magnesia alba' then commercially obtainable. Henry now revealed improved methods of preparing both 'magnesia alba' and calcined magnesia, and explained their various medicinal properties. At the same time, in an appendix, he passed some 'Strictures on Mr. Glass's Magnesia', revealing its fraudulent preparation, impurity, and medical unsuitability. This involved him in a

[1] He appears in Raffald's 1772 Manchester directory as an apothecary, in St. Ann's Square; in that of 1781 in Marsden Street; in Holme's 1788 directory as both apothecary and 'chymist' in King Street; in Scholes's 1794 and later directories at 41 King Street (his residence), with a shop at 1 Essex Street.

[2] Brockbank, *op. cit.*, gives the date of his Infirmary appointment as 1778, but a pencilled note, taken from the Medical Register by Bosdin Leech, in the Manchester Medical School's copy of Brockbank's book, shows the correct date as 1779. [3] W. Henry, 'Tribute', p. 210.

[4] Magnesium carbonate in modern chemical terminology.

[5] Sir W. Ramsay, *Life and Letters of Joseph Black, M.D.* (1918), pp. 22–30; Partington, *op. cit.*, pp. 135 *et seq.*

Q

controversy,[1] in which he was supported by Dr. Percival and Dr. Aikin and emerged victorious; his researches on magnesia were regarded as sufficiently sound to be incorporated by the chemists Bergman and Macquer in their respective accounts of that 'earth'.

Henry's *Experiments and Observations* also included the results of his researches into various methods of preserving foods and water from putrefaction, a subject then a matter of public concern, especially in the Navy. He experimented with calcined magnesia and lime,[2] and was thus led to investigate 'the Sweetening Properties of Fixed Air', or 'still air' (carbonic acid gas, or carbon dioxide), obtained from lime, magnesia, etc. by reaction with acids. In these investigations into the properties of 'fixed air', he had been preceded by such distinguished philosophers as Black, Cavendish, Bergman, and Priestley, and by lesser ones such as Hales, Brownrigg, and Macbride.[3] Priestley preceded him by a year in publishing his directions for impregnating water with 'fixed air', to preserve it and, at the same time, to provide an additional safeguard against scurvy, whilst also giving it 'the peculiar Spirit and Virtues of Pyrmont Water, and other Mineral Waters'.[4] It seems probable that Henry was conducting his experiments at the same time and independently of Priestley. This conjecture is supported by the fact that the latter included in an appendix to his famous *Experiments and Observations on Different Kinds of Air* an account by Henry of his experiments 'On the Effects of Fixed Air on the Preservation of Plants, &c.'[5] Priestley, together with Sir John Pringle and Benjamin Franklin

---

[1] See *Remarks on . . . T. Henry's improved method of preparing Magnesia Alba . . . By a Physician* (1774) and Henry's reply, *A Letter to Dr. Glass* (1774).

[2] Here again he was following in the footsteps of Black and of others such as Hales, Alston, and Macbride. For the three latter, see Partington, *op. cit.*, pp. 112–123, 136, and 143–4.

[3] For references to their researches, see the appropriate sections in Partington, *op. cit.*

[4] J. Priestley, *Directions for Impregnating Water with Fixed Air, in order to communicate to it the peculiar Spirit and Virtues of Pyrmont Water, and other Mineral Waters of a similar Nature* (1772). See also his *Experiments and Observations on Different Kinds of Air*, Vol. II (1775), pp. 263–303. Priestley's experiments are referred to in Partington, *op. cit.*, pp. 247–8, and F. W. Gibbs, *Joseph Priestley* (1965), pp. 57–60, 66–7, and 69. Partington points out, however, that Priestley had been preceded in the invention of artificial 'aerated waters' or 'mineral waters' by Dr. William Brownrigg, of Whitehaven, and also by foreign chemists such as the Frenchman, Gabriel Venel, and the famous Swedish philosopher, Torbern Bergman (*op. cit.*, pp. 78–9, 124–6, and 189). Priestley, however, appears to have been the first to publicize an apparatus for producing such waters. Bergman's papers on this subject have recently been translated into English by Sven M. Jonsson, *On Acid of Air and Treatise on Bitter, Seltzer, Spa and Pyrmont Water and their Synthetical Preparation*, with a Memoir by Uno Boklund, 'Torbern Bergman as Pioneer in the Domain of Mineral Waters' (Stockholm, 1956).

[5] Priestley, *op. cit.*, Vol. III (1777).

(then in London), backed his admission to the Royal Society, of which he became a Fellow in May 1775.[1]

Priestley, of course, remained essentially a philosopher, but Henry soon saw commercial possibilities in his chemical experiments. His researches on 'magnesia alba' were freely disclosed, for medicinal use, but whilst they were being printed it was suggested to him by a friend that he might produce this article for sale. After obtaining the approval of leading members of the College of Physicians, Henry took out a patent and launched into manufacture of magnesia, which was advertised as 'HENRY'S GENUINE MAGNESIA ... sold in Boxes of One Guinea, Ten shillings and Six-pence, and Six Shillings Price; and CALCINED MAGNESIA in Bottles at Three Shillings, or, with ground Stoppers, at Three Shillings and Six-pence each, at his House in Manchester', and also by Joseph Johnson (his publisher) and James Ridley, of London.[2]

This manufacturing and commercial venture proved a success and its founder eventually became known as 'Magnesia Henry'. At first it was on a small scale, apparently in the Essex Street premises, but eventually, soon after 1810, works were established in East Street, St. Peter's (near the later Central Station), in which Henry was partnered by his son William, and the firm of 'Thomas and William Henry, Manufacturing Chemists', remained in existence there until the early 1930s.[3]

Henry naturally continued to take an interest in magnesia, lime, and other 'earths' and published further papers, particularly in regard to their agricultural or industrial uses. Thus he wrote *On the Action of Lime and Marle as Manures, and the Making of Artificial Manures for the Purposes of Agriculture* (1775),[4] and in 1783 he delivered a paper to the Manchester Literary and Philosophical Society 'On the Natural History and Origin of Magnesian Earth', with observations on some of its hitherto unknown or undetermined properties.[5] After

---

[1] W. Henry, 'Tribute', pp. 212–13. Franklin was later influential in securing Henry's election to membership of the American Philosophical Society.

[2] This advertisement appeared at the back of Henry's *Letter to Dr. Glass* (1774). Advertisements also appeared regularly in the *Manchester Mercury*. Henry also publicized it in *An Account of the Medical Virtues of Magnesia Alba, more particularly of Calcined Magnesia, with Plain Directions for the Use of Them* (1775).

[3] There is a letter in the Manchester University Medical Library, from J. B. Lloyd, Superintendent Pharmacist at the Royal Infirmary, Manchester, dated 3 Oct. 1953, which states that Henry's Magnesia came to have 'a fairly wide sale throughout the world'. Lloyd had come across an old bottle in which it was sold, for 'Heartburn, Headache, Acidity of the Stomach, Biliousness, and Gout'; photographs of the wrapper showed that the business was established in 1772.

[4] Reprinted in A. Hunter's *Georgical Essays* (1777), pp. 489–503.

[5] *Manchester Lit. & Phil. Society Memoirs*, Vol. I (1785), pp. 448–73.

describing the various compounds of magnesia earth, he stated: 'I have collected them under one view, as it may be useful to some artists [artisans], particularly those concerned in the potteries, to know what earths and stones contain it, and in what proportions', since the magnesia prepared for the shops was too expensive for manufacturing purposes. We are reminded here of the links between Priestley's chemical researches and Wedgwood's pottery manufacture.[1] Henry may perhaps have had similar links with Wedgwood; the two certainly communicated with each other.[2]

Henry also continued to take an active interest in the exciting researches then going on into different 'airs' (gases), both in Britain and on the continent. The writings of the celebrated French chemist, Lavoisier, who started a revolution against the phlogiston theory and in favour of a new chemical terminology, were first introduced to English readers in translation—*Essays, Physical and Chemical*—by Henry in 1776, followed by another translation of further *Chemical Essays* in 1783. Henry also developed his own early researches, in a paper delivered in May 1784 to the Manchester Literary and Philosophical Society, dealing with 'Observations on the Influences of Fixed Air on Vegetation',[3] thinking that they might lead to an understanding of the action of manures and thus to 'considerable improvements in agriculture'. Though he remained a supporter of phlogiston, as championed by such chemists as Priestley and Kirwan, he nevertheless appreciated the experimental and theoretical skill of Lavoisier, which clearly called for some revision of existing ideas. Eventually, in fact, unlike Priestley, he was converted to the new chemical theory.[4]

It was again typical of Henry, however, that he did not remain content with philosophizing, but put his researches to industrial application, in developing the manufacture of 'aerated' waters by impregnation with 'fixed air'. In his experiments of the early 1770s, as we have seen, he had been contemporaneous with Priestley in putting forward this idea, principally for preserving meat, fruit, etc., but also for medicinal purposes. Henry, like Priestley, discovered that the amount of

[1] See above, pp. 78 and 143.
[2] It was about this time that Wedgwood supported Henry's scheme for a College of Arts and Sciences in Manchester (see below, pp. 241–2): W. Henry, 'Tribute', p. 226. He had previously made 'zealous personal exertions' to persuade the Admiralty to adopt Henry's method of preserving water at sea (see below, p. 237): *ibid.*, pp. 216–17.
[3] *Manchester Lit. & Phil. Soc. Memoirs*, Vol. II (1785), pp. 341–9.
[4] Thus he wrote to James Watt early in 1790: 'It does not appear that Dr. Priestley has adduced anything very convincing against the anti-phlogistons in the last part of the Philosophical Transactions.' T. Henry to J. Watt, Dol., 25 Jan. 1790.

'fixed air' taken up by water could be increased by use of a compressor. 'On this principle', as Dr. Gibbs has observed, 'a whole new industry—the manufacture of artificial mineral waters—was based'.[1]

The Admiralty were particularly interested in Priestley's proposed method for improving water used at sea 'by impregnating the same with fixed air',[2] but it does not seem to have proved successful. Henry, however, continued his experiments and in 1781 produced a booklet, dedicated to the Lords Commissioners, in which he gave *An Account of a Method of Preserving Water, at Sea, from Putrefaction.* Some years previously Dr. Charles Alston, of Edinburgh, had suggested adding lime to casks of water for preserving it at sea; to get rid of its disagreeable taste before drinking, he proposed adding 'magnesia alba' to the water. But this proved impracticable and Henry now proposed passing 'fixed air'—obtained from the action of vitriol, or sulphuric acid, on limestone or chalk—through the lime-water to precipitate the lime (as carbonate), and so restore the water to its original taste.

At the same time, he went on to describe a 'Method of impregnating Water in large Quantities with Fixed Air, so as to give it the properties of Mineral Water, for the Use of the Sick, on Board Ships, and in Hospitals'. He referred to Dr. Priestley's earlier proposals, admitting that he was only applying 'chemical facts, which were already well known': he was merely describing 'an improvement and extension of modes which have been already practised on a less enlarged plan', i.e. Henry was proposing large-scale commercial production of mineral waters. Since Priestley's original experiments, impregnation of water with 'fixed air' had been 'considerably facilitated by the invention of Dr. Nooth's glass machine, with Mr. Parker's and Mr. Magellan's improvements'.[3] But this machine, 'though admirably contrived for the preparation of such quantities of artificial mineral water as may be necessary in private families', was 'too small' for bigger production. Henry therefore suggested a method 'by which the process may be performed on a much larger scale',[4] in which the 'fixed air' would be

---

[1] *Op. cit.*, p. 69. [2] *Ibid.*, p. 59.

[3] These three had produced small domestic machines for making artificial mineral waters. See Gibbs, *op. cit.*, pp. 101 and 138. For Nooth's machine, see the Royal Society's *Philosophical Transactions*, Vol. LXV (1775), pp. 59–66; for Parker's, see J. Priestley, *Experiments and Observations on Different Kinds of Air*, Vol. II (1775), pp. 293–303; for Magellan's, see his *Description of a Glass Apparatus for making . . . the Best Mineral Waters* (1777; 2nd edn., 1779; 3rd edn., 1783). Magellan stated that he had distributed Priestley's pamphlet on the continent, and that 'many thousands' of these glass machines had been sent abroad.

[4] Kirkby, *op. cit.*, describes this as 'the first piece of apparatus made in this country for the production of carbonated water in bulk sufficient for . . commercial purposes'.

forced into the water by squeezing a pig's bladder fastened inter-
mediately between the flask containing the sulphuric acid and chalk
and the cask containing water. Henry provided recipes for making
artificial Pyrmont and Seltzer waters and 'Mr. Bewley's mephitic
julep'.[1]

But this clearly was still a fairly crude method of production. It was
improved almost immediately, however, by Dr. Haygarth, of Chester,
who freely communicated his idea to Henry. It was described by the
latter in another paper to the Manchester Literary and Philosophical
Society, in November 1781.[2] The principal improvement was the
replacement of the pig's bladder by bellows for compression.[3]

Using this type of apparatus, Henry launched out during the following years into large-scale commercial production of mineral or soda
waters, being one of the first such manufacturers in the country. These
waters were intended primarily for medicinal purposes, and according
to Brockbank, 'occasionally a prescription ordering Henry's soda
water is met with in the publications of the time'.[4] It is not known
exactly when this manufacture was started. Reference to a separate
'mineral water-works' does not appear in the Manchester directories
until 1808-9, when it was conducted in partnership with his son
William at 19 Cupid's-alley, Deansgate.[5] But it was certainly established a good many years earlier, probably in the 1780s, for it had
developed considerably by 1804, when William Henry corresponded
with James Watt junior on the subject.[6] He introduced to the latter a
young man named Thompstone,

who has for some time past been connected with me in the manufacture of
soda water, and other chemical substances. Finding that Schweppe is dispersing
his partners over the country, we begin to fear that he may establish himself at

---

[1] William Bewley, an apothecary of Great Massingham, Norfolk, was another
who had been carrying out experiments on magnesia, lime, and 'mephitic air'
('fixed air'). A friend of Priestley, he was apparently the first to propose the
addition of a small quantity of carbonate of soda to the water, to produce 'soda
water', or 'mephitic julep'. Partington, *op. cit.*, pp. 140, 248, and 251-2.

[2] *Memoirs*, Vol. I (1785), pp. 41-54.

[3] A drawing of a similar model, used by Adam Walker in his popular science
lectures, is reproduced in Gibbs, *op. cit.*, p. 70.

[4] *Op. cit.*, pp. 73-4.

[5] Dean's Manchester & Salford Directory, 1808-9.

[6] Thomas Henry had corresponded with James Watt in the late 1780s on
bleaching, dyeing, and other chemical matters (see below, pp. 243, 299-303). He
had also become friendly with James Watt jun. at that time, whilst the latter was
in Manchester, and had corresponded with him in the 1790s; he had obtained
from him a pneumatic apparatus for experiments on gases. T. Henry to J. Watt
jun., 2 Dec. 1794, B.R.L., Muirhead Box 4 H. This friendly relationship was
continued by William Henry.

Birmingham, and thus deprive us of a very good market for our preparation. I have, therefore, sent Thompstone over to reconnoitre; and to look for a situation in a central part of the town.[1]

He asked for Watt's assistance in providing workmen to fit up their apparatus in Birmingham—as Ewart and Lee had helped them in Manchester—and requested that the utmost secrecy be observed, 'lest the enemy should steal a march on us'.

A few months later he forwarded a copy of a circular letter, issued in Birmingham, announcing that 'we have been encouraged, by the extensive and increasing Demand in this and neighbouring counties for our *Artificial Mineral Waters*, to establish a Manufactory of them in High-street, in this Town'.[2] They would endeavour unremittingly to preserve the superior character for which their artificial waters had 'long been honoured'.

Later that month William Henry wrote to say that they were proposing to establish another mineral water manufactory at Bath, but were afraid of competition from Schweppe's at Bristol. They were also considering the possibility of an establishment at Glasgow. The Birmingham works, 'though very far behind that in Manchester', was going along prosperously.[3]

This expanding business, however, was eventually abandoned by William Henry, who arranged a division of interests with his partner, Samuel Thompstone, taking the magnesia for himself and leaving the mineral waters to the latter.[4]

This was not the only industrial application of Thomas Henry's researches into 'fixed air'. These led him, like Priestley and others, to observe the processes of vinous fermentation and brewing, in the course of which 'fixed air' is given off.[5] And again we find Henry proceeding farther with possible utilitarian applications. In his booklet of 1781 on the preservation of water at sea and preparation of mineral waters, he included a few pages explaining how he had 'repeatedly prepared an artificial yeast, by impregnating flour and water with fixed air, with which I have made very good bread, without the assistance of any other ferment'. He pointed out that by this means 'fresh fermented bread' might be obtained at sea. A few years later he developed these ideas in a paper to the Manchester Literary and Philosophical Society, describing his 'Experiments and Observations on Ferments and

---

[1] W. Henry to J. Watt jun., 4 Feb. 1804, B.R.L., Muirhead Box 4 H.
[2] Dated 17 May 1804, A.O.L.B.
[3] W. Henry to J. Watt jun., 25 May 1804, B.R.L., Muirhead Box 4 H.
[4] Angus Smith, *op. cit.*, p. 127, where the latter's name is mis-spelt 'Thompson'. The businesses are shown separately in Pigot's Directory of 1816–17.
[5] See Gibbs, *op. cit.*, p. 57, and above, p. 85.

Fermentation; by which a Mode of exciting Fermentation in Malt Liquors, without the Aid of Yeast, is pointed out; with an Attempt to form a new Theory of that Process'.[1]

Henry's theorizing, as his son William later pointed out, was not particularly valuable and was soon superseded, but his experiments led him to produce some important practical information.[2]

it was at that time believed that the infusion of malt, called *wort*, could not be made to ferment, without the addition of yeast or barm; but Mr. Henry discovered that wort may be brought into a state of fermentation, by being impregnated with carbonic acid gas. By a fermentation thus excited, he obtained not only good beer, but yeast fit for the making of bread; and, from separate portions of the fermented liquor, he procured also ardent spirit and vinegar, thus proving that the fermentative process had been fully complete.

This paper was, in fact, full of practical hints for brewing and wine-making. The reference to vinegar raises another of Henry's developments at this period—his invention of 'aromatic vinegar'. According to a later letter from his son William,[3]

More than fifteen years ago, during the delivery of a course of lectures, by my father, in this town, he had occasion to notice a property of the radical vinegar, or acetic acid, which had not, to his knowledge, been before observed; viz. its property of dissolving camphor and various essential oils. The compound was found to possess a most pungent and agreeable odour . . .

Thomas Henry suggested the use of this 'aromatic vinegar' as a safeguard against infection and it was adopted by T. B. Bayley, Esq., F.R.S., and other local magistrates, etc., who were frequently 'exposed to the danger of foul and infected air'. It was also through T. B. Bayley that Henry established a connection with Bayley's, a perfumery establishment in Cockspur Street, London, for sale of the new product in the metropolis. This part of Henry's pharmaceutical manufactures became well known, for we find Robert Owen later referring to Henry's works 'for the manufacture of concentrated essence of vinegar'.[4]

All these researches and manufacturing interests were peripheral to the main industrial activities of Manchester in this period, but the rapid development of the cotton industry during the 1780s caused Henry to turn his attention to the possibilities of applying chemistry to bleach-

---

[1] Read 20 April 1785, *Manchester Lit. & Phil. Soc. Memoirs*, Vol. II (1785), pp. 257–77.   [2] W. Henry, 'Tribute', pp. 220–1.
[3] William Henry's letter appeared in Tilloch's *Philosophical Magazine*, Vol. XV (1803), pp. 156–8, and in Nicholson's *Journal of Natural Philosophy*, 2nd ser., Vol. IV (1803), pp. 215–16.
[4] *The Life of Robert Owen, Written by Himself*, Vol. I (1857). p. 56.

ing, dyeing, and calico-printing. On the general relationship between science and industry he spoke strongly in 1781, immediately after the establishment of the Manchester Literary and Philosophical Society,[1] of which he was one of the chief promoters and joint secretary.[2] He referred to the strong prejudices then existing against natural philosophy among merchants and manufacturers, and he admitted that some branches, such as pneumatics, electricity, and optics 'may perhaps be considered rather as amusing and ornamental, than necessary'. But, he went on, 'there are other branches of natural philosophy which may be deemed highly useful and important to commercial men'. Mechanics, hydrostatics, and hydraulics, for instance, were essential to a proper understanding of machinery, steam engines, water-wheels, pumps, etc., while chemistry he considered 'the corner stone of the arts' (manufactures). After referring to its use in metal-refining, glass-making, etc., he particularly stressed the importance of chemistry to the cotton manufactures, though he had to deplore the existing lack of chemical knowledge:

Bleaching is a chemical operation . . . The materials for this process are also the creatures of chemistry, and some degree of chemical knowledge is requisite to enable the operator to judge of their goodness. Quick-lime is prepared by a chemical process. Pot-ash is a product of the same art; to which also vitriolic, and all the acids owe their existence. The manufacture of soap is also a branch of this science. All the operations of the whitster [bleacher]; the steeping, washing and boiling in alkaline lixiviums; exposing to the sun's light, scouring, rubbing and blueing are chemical operations, or founded on chemical principles. The same may be said of the arts of dying [sic] and printing . . . [But] few of the workmen, employed in them, possess the least knowledge of the science . . . The misfortune is, that few dyers are chemists, and few chemists dyers. Practical knowledge should be united to theory, in order to produce the most beneficial discoveries. The chemist is often prevented from availing himself of the result of his experiments by the want of opportunities of repeating them at large: and the workman generally looks down with contempt on any proposals, the subject of which is new to him. Yet under all these disadvantages, I believe it will be confessed, that the arts of dying [sic] and printing owe much of their recent progress to the improvements of men who have made chemistry their study.[3]

Henry's views were strongly supported by other members of the Literary and Philosophical Society and the eventual outcome, as we

[1] 'On the Advantages of Literature and Philosophy in General, and especially on the Consistency of Literary and Philosophical with Commercial Pursuits', read 3 Oct. 1781, *Memoirs*, Vol. I (1785), pp. 7–29.
[2] Later he became a Vice-President and eventually, in 1807, President, which office he held until his death.   [3] On this latter point, see below, chap. IX.

have seen, was the establishment in 1783 of the Manchester College of Arts and Sciences.[1] In the proposed courses, special importance was attached to chemistry because of its application to 'so many of the *arts*, on which our *manufactures* depend', and 'the whole [course] will have a reference to the arts of *Dyeing, Bleaching*, &c. which, depending upon chemical principles, might probably, by the knowledge of those principles, be very greatly extended and improved'.[2] Thomas Henry was appointed to give these lectures, in which he was assisted by his eldest son, Thomas.[3] He delivered several courses, 'to numerous and attentive audiences', during the few years of the College's existence, and continued them afterwards, until deprived of his son's services. In addition to these general lectures on chemistry, Henry also 'delivered a course on the arts of Bleaching, Dyeing, and Calico-Printing; and to render this course more extensively useful, the terms of access to it were made easy to the superior class of operative artisans'.[4] At the same time, Henry obtained practical information from manufacturers such as Charles Taylor, who provided him with dyers' 'receipts' or recipes for use in his lectures.[5] His concern to make chemistry generally accessible and useful is also illustrated by his objections to the new French chemical nomenclature: 'Chemistry being a science so intimately connected with many of the Arts, & consequently necessary to be studied by many illiterate men, should have its language plain and intelligible, & not made up of words compounded from a dead language, which none but men of learning can understand.'[6]

Henry's researches in the mid-1780s were chiefly concerned with dyeing, on which he read a lengthy and very learned paper, in three parts, to the Literary and Philosophical Society in 1786–7: 'Considerations relative to the Nature of Wool, Silk, and Cotton, as Objects of the Art of Dying; on the various Preparations, and Mordants, requisite for these different Substances; and on the Nature and Properties of Colouring Matter. Together with some Observations on the Theory of Dying in General, and particularly the Turkey Red'.[7] The main theoretical interest of this paper may be summarized, very briefly, by saying that, in place of the older 'mechanical' theories of dyeing (with reference to filament structure, porosity, etc.), Henry substituted more purely chemical explanations, based on the theory of 'affinity' or

[1] See above, pp. 91–3.

[2] *Manchester Lit. & Phil. Soc. Memoirs*, Vol. II (1785), p. 39.

[3] *Ibid.*, p. 45, and W. Henry, 'Tribute', pp. 224–5. See below, pp. 244–6, for young Thomas Henry.   [4] W. Henry, 'Tribute', p. 227.

[5] T. Henry to J. Watt, 12 Sept. 1788, Dol.

[6] T. Henry to J. Watt, 25 Jan. 1790, Dol.

[7] *Manchester Lit. & Phil. Soc. Memoirs*, Vol. III (1790), pp. 343–408.

'attraction'.[1] He emphasized his debt to French chemists, such as Hellot, Macquer, d'Apligny, and especially Berthollet, and again stressed the practical importance of chemical knowledge, whereby 'great improvements have been made, within these few years', particularly by John Wilson, of Manchester.[2] His chemical theory, however, was backed up by an enormous amount of practical industrial observation and laboratory experimentation. His work on fixing agents or mordants (or 'bases', as he preferred to call them), such as alum, etc., was particularly brilliant; his opinions on this subject, his son stated thirty years later, were 'still held by the latest and best writers on the principles and practice of Dyeing'.[3]

Henry's paper attracted widespread attention both in this country and abroad, in the original English and also in translation. He was highly regarded not only by eminent philosophical chemists such as Priestley and Berthollet, but also by leading industrial chemists such as James Watt and Theophilus Rupp.[4] Henry's connection with James Watt began with the introduction of chlorine bleaching,[5] and the two men also corresponded freely in regard to dyeing, informing each other in detail of their various chemical experiments on dyes, mordants, etc., as well as writing more generally about the work of other British chemists such as Priestley and Kirwan, or foreign ones like Berthollet and Lavoisier—in regard to the phlogiston controversy, for example, and chemical nomenclature.[6] Henry was also very friendly with James Watt junior whilst the latter was in Manchester, associating with him in the Literary and Philosophical Society, and no doubt helping him to improve his knowledge of textile chemistry.[7] They continued, as we have noticed, to correspond afterwards. We also find Henry introducing foreign scientists to him, such as 'Dr. Prochier of Geneva, who has studied & graduated at Edinburgh, & is well known to Mr. Saussure & the other Philosophers of his country'.[8]

[1] Here again, however, he had been preceded in this theory by such chemists as Dufay, Bergman, and Keir, as was pointed out by Edward Bancroft, *Experimental Researches concerning the Philosophy of Permanent Colours* (2nd edn., 1813), Vol. I, pp. l–lv. But at the time when he produced his paper the 'mechanical' theory, propounded by Hellot, Macquer, and d'Apligny, was still widely held. Berthollet joined with Henry in supporting the theory of chemical combinations in dyeing.

[2] See below, pp. 340–3.   [3] W. Henry, 'Tribute', p. 224.

[4] See above, p. 82, for Rupp's tribute to Henry's researches on dyeing.

[5] See below, pp. 299–303.

[6] See, for example, T. Henry to J. Watt, 12 and 23 Sept. 1788, 7 Jan. 1789, and 25 Jan. 1790, Dol.

[7] In a letter to his father, dated 27 Nov. 1788, Dol., J. Watt jun. says that he proposes to attend Henry's chemical lectures the next year.

[8] T. Henry to J. Watt jun., 27 July 1797, B.R.L., Muirhead Box 4 H.

Thus Henry was contributing to the fruitful interplay between science and industry, especially in dyeing. An equally outstanding contribution was his role in the introduction and dissemination of chlorine bleaching in the late 1780s. As this will be dealt with at length in the next chapter, it need only be stated here that in the development of Berthollet's process Henry was not far behind Watt, with whom he had a prolonged correspondence; he succeeded, after numerous experiments, in introducing several improvements, notably by absorbing chlorine in a lime solution, and he played a leading role in spreading knowledge of the new process among Lancashire bleaching firms. His own business venture in this field, however, proved a miserable disappointment, partly through the duplicity of his partner, John Wilson, but also, it would seem, because he lacked the capital and commercial drive to make a success of large-scale bleaching. The upset caused by Wilson's treatment of him contributed to a serious and prolonged illness in 1789.[1] Moreover, he found that such an extension of his industrial-chemical activities beyond the manufacture of magnesia, mineral waters, etc. was 'inconsistent with his professional employments.'[2] 'His medical occupations had greatly increased', and he was nearly sixty years of age, so that from now on he 'did not embark in new experimental enquiries, yet he continued, for many years, to feel a warm interest in the advancement of science; and to maintain an occasional correspondence with persons highly eminent for their rank as philosophers, both in this and other countries'.[3]

According to his son, Henry had a very considerable 'inventive talent' and 'ingenuity', which enabled him to carry out his researches with rudimentary equipment, for 'at no period of his life, was he in possession of a well furnished laboratory, or of nice and delicate instruments of analysis or research'.[4] Moreover, in addition to carrying on chemical research and managing his chemical business, he was also a busy apothecary. In this profession, too, he kept up with current research and contributed several papers to medical journals.

His hopes of seeing chemistry applied more extensively in industry appeared likely to come to fruition in the achievements of his sons. His eldest son, Thomas, showed early signs of scientific ability, being awarded the Literary and Philosophical Society's silver (or junior) medal in 1785, and earning high praise from Dr. Percival, the President,

---

[1] T. Henry to J. Watt, 25 Jan. 1790, Dol.   [2] W. Henry, 'Tribute', p. 229.
[3] *Ibid.*, pp. 229–30. William Henry referred to 'a considerable collection of letters from persons of this description . . . Many of them are from learned foreigners, with whom he had enjoyed opportunities of personal intercourse during their visits to Manchester.'   [4] *Ibid.*, p. 233.

for a paper reviewing the controversy between Henry Cavendish and Richard Kirwan 'relative to the cause of the diminution of common air, in phlogistic processes'.[1] In his youth he at first assisted his father in the apothecary business and chemical experiments,[2] and apparently had hopes of entering the medical profession, but towards the end of 1788 he decided that he would 'stand a better chance in the Commercial, than the Medical, line'.[3] He intended, as his father informed James Watt, 'to learn the practical part of dying. Being already a tolerable Chemist, if he can join the Practice to Theory, it may be hoped something advantageous may be the Result. . . . This I am convinced of, that no essential improvements can take place in the Art of Dying without such an union.'[4] But first he had to widen his theoretical and practical experience. Thus a few months later Henry wrote to James Watt:

I have been deprived of the assistance of my Son Thomas, who has been in London these six weeks, where he is gone to attend Dr. Higgins's lectures[5] & to mix with the Chemical & Scientific People there, to many of whom I have given him recommendations. You know his Views are directed to the dying business, & if you can give him introductions to any artists in that or the printing line, who may be likely to afford him information; or to any of the Philosophical Gentlemen it will oblige me. I could wish him to see some of the great Chemical Works.[6]

Soon afterwards young Thomas went for training into Potter's, one of the leading Manchester cotton manufacturers and merchants, but his father then became so seriously ill[7] that he was obliged to leave: 'fearing the consequences of leaving my family without a successor to my business, I was induced to give up the plan I had formed for my oldest Son, & to determine on keeping him steady in the medical line'. Young Thomas was therefore placed with Dr. Lyon, a physician and surgeon of Liverpool, for a year's training in surgery and midwifery.[8]

It appears, however, that he soon began to venture into the industrial world again, for there is a reference to his participation in an unsuccessful enterprise for manufacturing vitriol or sulphuric acid: 'The

---

[1] *Manchester Lit. & Phil. Soc. Memoirs*, Vol. II (1785), pp. 510–14.

[2] See above, p. 242

[3] T. Henry to J. Watt, 12 Sept. 1788, Dol. 'Professional merit', his father remarked, 'is so often jostled out of Company by impudence and ignorance' in the medical world.   [4] *Ibid.*   [5] See above, p. 121.

[6] T. Henry to J. Watt, 7 Jan. 1789, Dol. It was mainly through Henry's advice and assistance that James Watt jun. was indentured to Maxwell, Taylor & Co., of Manchester, to learn dyeing and calico-printing. See below, p. 345.

[7] See above, p. 244.

[8] T. Henry to J. Watt, 25 Jan. 1790, Dol. See also J. Watt jun. to J. Watt, 25 May 1789, Dol.

Partnership lately existing between Robert Horne, James Lawson, Andrew Tomlin, Thomas Henry, Jun. and Joshua Parr, carrying on business at Amloch, in the county of Anglesea, under the Firm of the Mona Vitriol Company, was this day dissolved by mutual consent. As witness their hands the 25th of November 1793.'[1]

Young Thomas meanwhile had become very friendly with Thomas Cooper, Priestley, and other Radicals, and this business failure perhaps combined with the 'anti-Jacobin' riots in 1793 to cause him to emigrate, in company with these friends, to America.[2] But he soon became disillusioned there and returned to England in 1796.[3] Unfortunately, however, he died soon afterwards.

Henry's second son, Peter, also had a flair for industrial chemistry, as indicated by a paper to the Literary and Philosophical Society on 16 November 1792, 'On the Action of Metallic Oxydes and Earths upon Oils, in low Degrees of Heat'.[4] He had carried out a series of experiments for removing from whale, linseed, olive, and other oils the colouring matter which rendered them 'unfit for several uses in the Arts'. He, too, it seems, may perhaps have entered dyeing,[5] but as his father communicated his paper to the Literary and Philosophical Society, it appears that he had left Manchester.

The father's hopes were now concentrated on his youngest son, William, whom he took into partnership in 1797, and who, as he justly observed, 'promises to be a very good Chemist and has acquired a considerable degree of professional knowledge, for which he has a most ardent thirst'.[6] He was in fact, the most brilliant of Henry's sons, having just left Edinburgh University, where he studied under Dr. Black.[7] Establishing himself in Manchester as a surgeon and apothecary, and also taking over active management of his father's chemical

---

[1] *Manchester Mercury*, 31 Dec. 1793. For the Mona Vitriol Company, see J. R. Harris *The Copper King* (1964), p. 165, and W. Davies, *General View of the Agricultural and Domestic Economy of North Wales* (1797), pp. 52–3. This venture was perhaps a consequence of young Henry's interest in chlorine bleaching, which greatly stimulated the development of vitriol manufacture.

[2] Gibbs, *op. cit.*, pp. 218, 223, 225, 228.

[3] T. Henry sen. to J. Watt jun., 2 Dec. 1794 and 8 Aug. 1796, B.R.L., Muirhead Box 4 H. [4] *Memoirs*, Vol. IV, Part I (1793), pp. 209–16.

[5] J. Watt jun. to J. Watt, 25 May 1789, Dol.

[6] T. Henry to J. Watt, 20 Jan. 1797, B.R.L., Muirhead Box 4 H.

[7] He was there in 1795–6, but then left to help his father. In 1805–7, however, he returned to Edinburgh and graduated as Doctor of Medicine. For accounts of his life and scientific achievements, see W. C. Henry, 'A Memoir of the Life and Writings of the late Dr. [William] Henry', *Manchester Lit. & Phil. Soc. Memoirs*, 2nd ser., Vol. VI (1842), pp. 99–141; J. Wheeler, *op. cit.*, pp. 495–8, reprinted separately as a *Sketch of the Character of the Late William Henry, M.D., F.R.S. &c.* (1836), by J. Davies; Angus Smith, *op. cit.*; Brockbank, *op. cit.*, pp. 237–40; Bosdin Leech, *op. cit.*; Partington, *op. cit.*, Vol. III, pp. 822–6.

manufactures, he quickly followed in his father's footsteps by giving public lectures on chemistry, in which he emphasized its numerous industrial applications.[1] Indeed, although recognizing the purely philosophical interest of chemistry, he considered that 'it is a conviction of the utility of the science, that can alone recommend it to attentive and persevering study'. He gave many examples of its industrial importance:

The extraction of metals from their ores; the conversion of the rudest materials into the beautiful fabrics of glass and porcelain; the production of wine, ardent spirits, and vinegar; and the dying [*sic*] of linen and woollen manufactures, are only a few of the arts that are dependent on Chemistry for their improvement, and, even for their successful practice.[2]

He recognized, of course, that industrial processes had been improved over the ages by practical empiricism, 'and that they are daily the employment of unlettered and ignorant men', but the achievements of Watt with the steam engine, of Wedgwood in pottery, and of Berthollet in bleaching—to mention only a few illustrious examples—demonstrated the possibilities of applied science. Henry wished to promote 'the union of theory with practice' along explicitly Baconian lines—'to inform the artist [artisan], and to induce him to substitute for vague and random conjecture, the torch of induction, and of rational analogy'.

His lectures were 'illustrated by a great variety of interesting experiments performed by means of an apparatus unusually extensive and valuable'.[3] They formed the basis of his *Epitome of Chemistry* (1801),[4] later developed into *Elements of Experimental Chemistry*, which together

---

[1] *Syllabus of a Course of Lectures on Chemistry* (Manchester, 1799) and *A General View of the Nature and Objects of Chemistry, and of its Application to Arts and Manufactures* (Manchester, 1799). See above, pp. 84 and 107.

[2] *General View*, p. 13. He gave many more examples on pp. 18–35.

[3] Davies, 'Sketch', p. 6.

[4] In the Preface to this work Henry referred particularly to the earlier publications of James Parkinson, *Chemical Pocket Book* (1799), and Professor J. F. A. Gottling, *Description of a Portable Chest of Chemistry* (English trans., 1791), to which his own was very similar. Henry's *Epitome*, in fact, was very largely an up-to-date version of Gottling's work, using the new in place of the old chemical terminology, and he imitated Gottling in supplying 'chemical chests' for carrying out chemical tests and experiments (see below, p. 250). Gottling, who was Professor of Chemistry at Jena in Saxony, wrote his work 'For the Use of Chemists Physicians, Mineralogists, Metallurgists, Scientific Artists, Manufacturers, Farmers, and the Cultivators of Natural Philosophy'. Directions were given for analysing and testing the purity of a wide range of chemical substances. References were also given to works on 'Technical Chemistry' by other Germans such as Gmelin and Succow. There is no doubt that there was considerable German as well as French influence in the 'Chemical Revolution' (cf above, pp. 96–7, for James Watt junior's links with German chemists).

went through eleven editions between 1801 and 1829. Henry also wrote numerous papers in the *Philosophical Transactions* of the Royal Society, in the *Memoirs* of the Manchester Literary and Philosophical Society, and in various scientific periodicals. He became, in fact, one of the leading British scientists of the early nineteenth century, being especially noted for his discoveries in pneumatic chemistry, including 'Henry's Law'.[1] He was a more scientific chemist than his father, but shared his industrial-chemical interests. He is said to have 'greatly extended' the 'lucrative business' established by his father, and to have 'conspicuously shewn that a due and regular attention to business is not incompatible with very high success in science'.[2] Some of his scientific researches were clearly related to his business interests. Thus we find him continuing his father's interests in the analysis and properties of magnesia and other 'earths',[3] while his famous experiments on the 'absorption' or solubility of gases in water[4] were connected with his mineral water manufacture, in which 'fixed air' was dissolved, under pressure, in water. His interest in gases also led him into prolonged researches into coal-gas, which were of immediate practical importance in early gas-works, and which, in turn, encouraged his scientific experiments on carburetted hydrogen, ammonia, and other gases.[5] Closely

[1] That the 'absorption' or solubility of gases in water is proportional to pressure.
[2] Davies, *op. cit.*, p. 6.
[3] 'Experiments on Barytes and Strontites', Nicholson's *Journal of Natural Philosophy*, Vol. III (1800), pp. 168–71; 'Analysis of a native Carbonate of Magnesia from the East Indies', Thomson's *Annals of Philosophy*, new ser., Vol. I (1821), pp. 252–4; 'On the Magnesite discovered in Anglesey', *Edinburgh Journ. Sci.*, Vol. II (1830), pp. 155 ff.
[4] 'Experiments on the Quantity of Gases absorbed by Water, at different Temperatures, and under different Pressures', *Phil. Trans.*, 1803, pp. 29–42 and 274–6.
[5] 'Experiments on carbonated hydrogenous Gas; with a View to determine whether Carbon be a simple or compound Substance', *Phil. Trans.* (1797), pp. 401–15; 'Description of an Apparatus for the Analysis of the Compound Inflammable Gases by slow Combustion; with Experiments on the Gas from Coal', *ibid.* (1808), pp. 282–303; 'Experiments on Ammonia, and an Account of a new Method of Analyzing it, by Combination with Oxygen and other Gases', *ibid.* (1809), pp. 430–49; 'On the aëriform compounds of Charcoal and Hydrogen; with an account of some additional Experiments in the Gases from Oil and Coal', *ibid.* (1821), pp. 136–61; 'On the action of finely divided Platinum on Gaseous Mixtures, and its Application to their Analysis', *ibid.* (1824), pp. 266–89. 'Description of an Eudiometer, and of other Apparatus employed in Experiments on the Gases', *Manchester Lit. & Phil. Soc. Memoirs*, 2nd ser., Vol. II (1813), pp. 384–390; 'Experiments on the Gas from Coal, chiefly with a View to its Practical Application', *ibid.*, Vol. III (1819), pp. 391 ff.; 'Experiments on the Analysis of some of the Aëriform Compounds of Nitrogen', *ibid.*, Vol. IV (1824), pp. 499–517. 'Experiments on the Gases obtained by the destructive Distillation of Wood, Peat, Pit-Coal, Oil, Wax, etc. with a View to the Theory of their Combustion, when employed as sources of artificial Light; and including Observations on

related to these researches were his investigations into fire-damp in
coal mines, as a result of which he emphasized the importance of improved ventilation to prevent gaseous explosions.[1] His connections
with coal mines also led to other practical chemical investigations, e.g.
into the corrosion of cast-iron piping used in the sinking of a pit at
Newcastle-on-Tyne,[2] and into 'Siliciferous Sub-sulphate of Alumine'
found in an Oldham coal mine.[3] Similarly, we find him investigating
the blockage of a pipe by a crystalline deposit in Mutrie's chemical
works at Lloyd-field, Manchester.[4] His researches on muriatic (hydrochloric) acid and 'oxymuriates' and also into potash and other alkalis
were certainly related to local textile-bleaching operations, along the
same lines as his father's, though they also led him into more purely
scientific investigations.[5] From letters which he wrote to Sir Joseph
Banks in 1801[6] we find that he had been engaged for two years in
experiments on the production of 'mineral alkali' (soda) by distillation
of bleachers' waste,[7] that he claimed to have successfully developed a
process for this purpose and had erected a building in which to operate
it. But he ran up against the Excise regulations regarding salt duties and
despite Banks's intervention in the Privy Council's Committee on
Trade and Plantations, he appears to have been unable to get a licence
to carry on this manufacture. Closely related to these chemical-
industrial interests were his researches into common salt, which were

Hydro-Carburets in general, and the Carbonic Oxide', Nicholson's *Journal*, 2nd.
ser., Vol. XI (1805), pp. 65–74.
  See also W. Buckley and A. McCulloch, 'Leaves from an Old Notebook.
A Record of Some Early Experiments on the Carbonisation of Oil and Coal,
by Dr. W. Henry, F.R.S. (1808–24)', *Manchester Lit. & Phil. Soc. Memoirs*, Vol.
LXXIV (1929–30), pp. 64–74. This article demonstrates the close collaboration
between Henry and the firm of Philips and Lee, whose cotton mill was the first
in Manchester to be lit by gas, in 1805. Lee was a close friend of Henry (see above,
pp. 99–100, for his scientific interests). Henry and Dalton also analysed gas from
Manchester public gas-works and from various private establishments.
  [1] 'Experiments on the Fire-damp in Coal Mines', Nicholson's *Journal*, 2nd.
ser., Vol. XIX (1808), pp. 149–53.
  [2] Thomson's *Annals of Philosophy*, Vol. V (1815), pp. 66–68.
  [3] *Ibid.*, Vol. XI (1818), pp. 432–4.
  [4] *Ibid.*, new ser., Vol. XI (1826), pp. 368–71.
  [5] 'An Account of a Series of Experiments, undertaken with a View of decomposing the Muriatic Acid', *Phil. Trans.* (1800), pp. 188–203; 'Additional
Experiments on the Muriatic and Oxymuriatic Acids', *ibid* (1812), pp. 238–46.
'Discovery of a new and improved method of preparing Prussiate of Pot-ash',
Nicholson's *Journal*, Vol. IV (1801), pp. 30–3; 'Further Remarks on the Preparation of Prussiate of Pot-ash. Method of purifying Caustic and carbonated Alkalis
from Sulphate of Pot-ash', *ibid.*, pp. 171–2.
  [6] These letters, listed by Dawson, *The Banks Letters*, p. 407, are all in the
British Museum.   [7] See below, p. 328, n. 4.

R

also motivated by practical commercial considerations.[1] Those on heat and electricity, on the other hand, were entirely in the field of 'pure' research.[2] His gaseous researches, moreover, though having some utilitarian applications, were also more 'pure' than 'applied'.[3] Nevertheless, it is clear that in various fields, in his own chemical works and in other industries, he had many practical chemical interests. He also helped to stimulate both scientific and industrial interest in experimental chemistry by producing 'chemical chests', or 'collections of chemical substances, which I have been induced, by repeated applications of students of this science, to fit up for public sale', for use with his chemical text-books.[4] He supplied one of these, at a price of fifteen guineas, to James Watt junior,[5] with whom he carried on a correspondence for many years. Like his father, he became a Fellow of the Royal Society and was on friendly terms not only with leading chemists, such as Dalton and Davy, but also with the more scientifically-minded industrialists, such as Lee and Ewart.[6] Like his father, too, he introduced other scientists to them.[7] Together, the Henrys certainly made a notable contribution to industrial-chemical developments in the Manchester region during the Industrial Revolution.

[1] 'An Analysis of several Varieties of British and Foreign Salt (Muriate of Soda), with a View to explain their Fitness for different economical Purposes', *Phil. Trans.* (1810), pp. 89–122.

[2] 'A Review of some Experiments which have been supposed to disprove the Materiality of Heat' [Henry was a believer in 'caloric'], *Manchester Lit. & Phil. Soc. Memoirs*, Vol. V (1802), pp. 603–21; 'On the Theories of the excitement of Galvanic Electricity', *ibid.*, 2nd ser., Vol. II (1813), pp. 293–312; 'Experiments on the Chemical Effects of Galvanic Electricity', Nicholson's *Journal*, Vol. IV (1801), pp. 223–6.

[3] In addition to the papers previously mentioned, see those supporting Dalton's atomic 'Theory of the Constitution of Mixed Gases', Nicholson's *Journal*, 2nd ser., Vol. VIII (1804), pp. 297–301, and Vol. IX (1804), pp. 126–8.

[4] *Epitome of Chemistry* (1801), p. v.

[5] W. Henry to J. Watt jun., 3 July and 18 Aug. 1802, B.R.L. Muirhead Box 4 H. Eventually he was unable to cope with the demand for these chests and passed the business to a London manufacturer. Bosdin Leech, *op. cit.*, p. 110.

[6] For example, W. Henry to J. Watt jun., 18 Aug. 1802: 'At Matlock, we met with Davy, and expect a short visit from him in a few days.' W. Henry to M. Boulton, 5 May 1802, A.O.L.B.: 'Our friends, Lee and Ewart, supped with me the other evening.'

[7] W. Henry to J. Watt jun., 26 Aug. 1804, B.R.L., Muirhead Box 4 H, introducing 'Dr. Thomson of Edinburgh, Author of an excellent System of Chemistry'. This was the later well-known Professor of Chemistry in the University of Glasgow, Dr. T. Thomson, whose *System of Chemistry* was first published in 1802 and ran into several editions. He also edited *Annals of Philosophy* (1813–26), a leading scientific periodical, and wrote a *History of Chemistry* (1830–1). See A. Kent, 'Thomas Thomson (1773–1852) Historian of Chemistry', *British Journal for the History of Science*, Vol. II (1964–5), pp. 221–34; J. R. Partington, *A History of Chemistry*, Vol. III (1962), pp. 716–21.

VIII The Introduction of Chlorine Bleaching[1]

Rarely during the Industrial Revolution is it possible to discover precisely how a new manufacturing process was invented and developed. In most industries there are no surviving documents to reveal how inventors and industrialists proceeded, what difficulties they encountered, how they modified their ideas, and how their processes came to be applied generally in industry. We have been very fortunate, therefore, in discovering material which enables us to follow in detail, from original sources, how chlorine bleaching was introduced and developed in this country. The general outline of the story, of course, is already familiar,[2] but from contemporary letters and scientific-technical publications it has been possible to formulate a much fuller, more accurate, and more revealing account of how this revolution in bleaching was brought about. And undoubtedly the most interesting feature which emerges from this research is the close relationship between science and technology, between scientists and industrialists, in achieving this 'break-through'.

Already, some years before this revolution occurred, it is evident that natural philosophers or scientists were making important contributions to bleaching. The traditional processes—the repeated and complicated operations of steeping in water, 'bucking' or 'bowking' (i.e. boiling) in alkaline lyes, 'crofting' or 'grassing' in bleachfields, and 'souring' in buttermilk, with intervening 'washings' and 'soapings'— were originally developed by empirical methods, but from about the middle of the eighteenth century, as Dr. and Mrs. Clow have shown, scientists such as Francis Home, William Cullen, and Joseph Black in

[1] This chapter is a previously unpublished contribution by A. E. Musson. The research on which it is based was jointly undertaken by A. E. Musson and Eric Robinson. Dr. J. L. Moilliet gave helpful advice on chemical matters.
[2] See, for example, S. H. Higgins, *A History of Bleaching* (1924), chap. VI, and Clow, *op. cit.*, chap. IX.

Scotland, James Ferguson, Richard Kirwan, and William Higgins in Ireland, were assisting in such developments as the use of dilute vitriol or sulphuric acid (and later hydrochloric acid) in place of buttermilk in the souring process, and of lime together with alkali in bucking, as well as in analysis of materials such as water, potash, etc. Less is known of such early links between chemistry and bleaching in England, but, as we have seen, chemists such as Lewis and Dossie had carried out important researches into alkalis (e.g. potash and barilla) used in bleaching,[1] while Dr. John Roebuck had developed markets for his vitriol among English bleachers.[2] In Manchester there is evidence that men of science were active in this field. Thomas Henry, for instance, in 1781, emphasized the importance of applied chemistry in bleaching, which he described as 'a chemical operation',[3] and in a paper of the following year Dr. A. Eason, another medical man interested in industrial chemistry, produced his 'Observations on the Use of Acids in Bleaching of Linen', in which he urged bleachers to use muriatic (hydrochloric) acid in place of vitriolic (sulphuric), on account of its superiority for this purpose and since it could now be produced more cheaply.[4]

On the continent, and especially in France, there was a similar interest in applied chemistry, and it is not surprising, therefore, that the original discovery of chlorine and its application to bleaching resulted from the scientific researches of the Swedish chemist, Scheele, and the Frenchman, Berthollet.[5] The latter, as we shall see, was essentially a philosopher, who did not wish to be personally involved in the industrial application of his discoveries, and who, therefore, publicized them freely. He first revealed them in April 1785 in a paper to the Académie Royale des Sciences, published in the *Observations et*

[1] See above, pp. 54–5.
[2] By the 1760s, if not earlier, the textile-manufacturing firm of J. and N. Philips & Co., of Tean, Staffordshire, were purchasing vitriol for use in bleaching. Wadsworth and Mann, *op. cit.*, pp. 178 and 296.
[3] See above, p. 241.
[4] *Manchester Lit. & Phil. Soc. Memoirs*, Vol. I (1785), pp. 240–2.
[5] For a brief but interesting general account of French developments, see C. Ballot, *L'Introduction du Machinisme dans l'Industrie Française* (1923), chap. XI, 'Industries Chimiques'. Scheele's contribution to the introduction of chlorine bleaching, however, has been unduly overshadowed by that of Berthollet. Scheele not merely discovered 'dephlogisticated marine acid' in 1774, but also observed its bleaching properties, as Berthollet himself acknowledged. See *The Chemical Essays of Charles-William Scheele* (translated by Dr. Thomas Beddoes, 1786, from the Transactions of the Academy of Sciences at Stockholm), pp. 90–6. Scheele, however, appears only to have decolorized paper, flowers, and plants with chlorine, and it was Berthollet who revealed its possibilities in bleaching textiles.

Mémoires sur la Physique, sur l'Histoire Naturelle et sur les Arts et Métiers, more commonly known as the *Journal de Physique*;[1] later in the same year he read another longer and more detailed paper to the Academy, which eventually appeared in its *Mémoires.*[2] Berthollet continued his experiments in the following years, producing improvements in both the preparation and application of chlorine, which he again published.[3] His discoveries aroused immediate scientific and industrial interest in France and the new method of bleaching was further developed and applied by other French chemists, including his assistants Bonjour and Welter, and such notable scientists as Chaptal, Décroisille, Fourcroy, and others, whose findings were similarly published in the *Journal de Physique*, the *Annales de Chimie*, the *Mémoires* of the Académie Royale, the *Annales des Arts et Manufactures*, and the *Encyclopédie Méthodique*. By the end of the century, in addition to numerous articles in these scientific and technical journals, long and detailed accounts of the new methods were to be found in various books such as C. Pajot des Charmes, *L'Art du Blanchiment des Toiles* (1798) and R. O'Reilly, *Essai sur le Blanchiment* (1801). By 1790, indeed, Berthollet was able to claim that his invention had been introduced in most of the leading textile-manufacturing towns of France,[4] and by 1801 O'Reilly spoke of 'une révolution complète dans l'art de blanchir',[5] as a result of researches by chemists not only in France, but also in other European countries, such as Great Britain and Germany: 'Nous sommes enfin parvenus à l'époque où les Sciences et les Arts industriels, aidés de leurs ressources réciproques, se portent rapidement vers une amélioration indéfinie.'[6]

Chlorine, the active agent in this revolution, was originally produced by Scheele and Berthollet by the action of marine or muriatic (hydrochloric) acid on black calx of manganese (manganese dioxide), but later by that of vitriol or sulphuric acid on common salt and manganese

---

[1] *Op. cit.*, Vol. XXVI (1785), pp. 321–5. This journal was also sometimes referred to as *Rozier's Journal*, because Rozier was its editor, together with Mongez and de la Métherie. Berthollet's paper was read on 6 April 1785 and published in the May number of the *Journal*.

[2] *Historie de l'Académie Royale des Sciences. Année MDCCLXXV* [1785]. *Avec les Mémoires de Mathématique & de Physique pour la même Année* (1788), pp. 276–95.

[3] See *Journal de Physique*, Aug. 1786 and Sept. 1788, and *Annales de Chimie*, Vol. II (1788), pp. 151–90, and Vol. VI (1790), pp. 204–40. It was his article in the second volume of the *Annales de Chimie*, 'Description du Blanchîment des Toiles et des Fils par l'acide muriatique oxigéné', which provided for the first time a detailed account of the new bleaching methods.

[4] Ballot, *op. cit.*, p. 529.

[5] O'Reilly, *op. cit.*, p. 4.

[6] *Ibid.*, p. 69.

dioxide.[1] Not until Davy's researches in the early nineteenth century, however, was it discovered that chlorine was a separate element.[2] During the previous twenty years chemical theory was in a confused state: supporters of the old 'phlogiston' theory were fighting a rear-guard action against what may be termed the new French 'oxidation' school. The latter, of course, were eventually to carry the day, but in regard to the nature and bleaching action of chlorine both schools were theoretically erroneous. Scheele and other phlogistonians regarded this substance as 'dephlogisticated marine acid', i.e. marine acid deprived by the manganese of its phlogiston, which, they thought, was recovered from coloured materials in the bleaching process. The anti-phlogistonians, on the other hand, including Berthollet, termed the new bleaching agent 'aerated' or 'oxygenated' muriatic acid (or 'oxymuriatic acid'), i.e. muriatic acid enriched from the manganese with 'vital air' or 'oxygen', which, they considered, was released during the bleaching process, to combine with colouring matters, thus decolorizing them.

This latter theory was approaching near to the truth, for the bleaching action of chlorine *is* due to its being a powerful oxidizing agent. Chlorine gas in solution decomposes into oxygen and hydrochloric acid:

$$2Cl_2 + 2HO_2 \rightarrow O_2 + 4HCl$$

Berthollet had observed that, in sunlight, a solution of 'oxymuriatic acid' gave off oxygen, leaving dilute muriatic acid, and he was correct in surmising that when unbleached cloth was immersed in such a solution the colour was removed by the formation of colourless oxidation products. Davy's later discoveries were based on these earlier experimental and theoretical developments.

The fact that many years elapsed before the formulation of a really accurate theoretical explanation of this process does not mean that the researches of chemists such as Scheele and Berthollet, and the new bleaching methods based upon them, were 'unscientific'. These men were among the most outstanding natural philosophers or scientists of the day, and, as we shall see, their methods were applied in

---

[1] The reaction of hydrochloric acid (HCl) on manganese dioxide (MnO_2) produces chlorine, thus:

$$2HCl + MnO_2 \rightarrow MnO + H_2O + Cl_2$$
$$\text{or } 4HCl + MnO_2 \rightarrow MnCl_2 + 2H_2O + Cl_2$$

By using sulphuric acid (H_2SO_4) and salt (sodium chloride, NaCl) with the manganese dioxide, hydrochloric acid is first produced:

$$H_2SO_4 + 2NaCl \rightarrow Na_2SO_4 + 2HCl$$
$$\text{or } H_2SO_4 + NaCl \rightarrow NaHSO_4 + HCl$$

[2] See below, p. 335.

industry by men who were generally well versed in theoretical and practical chemistry. Undoubtedly experiment played a tremendously important part, but it was not blind empiricism: practice and theory progressed together, stimulating each other. There is no doubt that, but for the researches of philosophers in universities, academies, and private laboratories, industrial bleachers would have long continued to use their traditional methods. At the same time, however, our researches have revealed how the practical experience of industrial chemists and bleachers had to be allied with the theoretical knowledge and experimental skills of scientists before the new 'chymical' method could be successfully developed on a large commercial scale.

Berthollet himself succinctly outlined the problems involved at the beginning of 1790:[1]

Il s'agissait de rendre d'une application sûre et facile l'emploi d'un gaz qui est suffocant et qui dévore presque tout ce qu'il touche; il fallait imaginer les appareils, établir les proportions, combiner l'action du gaz avec celle des autres agents, prévenir les pertes et affaiblir les frais; enfin il fallait lutter contre un procédé que l'expérience de tous les siècles paraissait avoir amené à la plus grande simplicité.

The new method of bleaching obviously did not spring from Berthollet's head fully developed and ready for industrial application. Like most important industrial inventions, it required years of patient trial and error before it could be developed from laboratory experiment to large-scale industrial plant. But these trials and tribulations have never been revealed.

The accounts given by Higgins and by the Clows of the early introduction of chlorine bleaching from France into Great Britain are derived mainly from the articles in Rees's *Cyclopaedia*, Parkes's *Chemical Essays*, and the *Annals of Philosophy*.[2] These were based upon information supplied by some of the people directly concerned in these developments, or upon accounts in earlier publications such as those by Berthollet and others, but they are nevertheless second-hand, often confused and conflicting, and lacking in exact details of how chlorine bleaching was developed. Contemporary letters and other evidence, however, enable us to get closer to the actual events and to see more

---

[1] In a memoir written for the Bureau du Commerce, 2 Feb. 1790, quoted by Ballot, *op. cit.*, pp. 531–2.

[2] A. Rees, *The Cyclopaedia; or Universal Dictionary of Arts, Sciences, and Literature* (1819), Vol. IV, article on 'Bleaching', and Vol. XXV, article on 'Oxymuriatic Acid'; S. Parkes, *Chemical Essays* (1815), Vol. IV, 'Essay on Bleaching'; *Annals of Philosophy*, Vol. VI (1815), pp. 42–4, and Vol. VII (1816), pp. 83–4.

precisely how the new method of bleaching was first introduced, how it was gradually developed, and how its use spread in different parts of the country.

First of all, however, let us look more carefully at the evidence in Rees, Parkes, and the *Annals of Philosophy*. Dr. and Mrs. Clow, utilizing these sources, have described how the new bleaching agent was introduced into Scotland by two routes during 1786-7.[1] James Watt, whilst visiting France in the former year and meeting leading scientists, was shown the bleaching action of chlorine by Berthollet. Watt, as the Clows rightly emphasize, was himself 'no mean chemist, and . . . the friend and collaborator of Dr. Black and Dr. Roebuck'. Moreover, his father-in-law, James McGrigor, was a Glasgow bleacher, so Watt had a twofold interest in the new process. But Dr. and Mrs. Clow then go on to state that 'some considerable time elapsed before he could personally supervise a demonstration of the use of chlorine', and they show the delay by referring indirectly to a letter from Watt to Thomas Henry, of Manchester, dated 25 February 1788, saying that 1,500 yards of cloth were then bleaching.[2]

Dr. and Mrs. Clow then proceed, closely following Parkes's account (though not stating so), to give priority to Patrick Copland, Professor of Natural Philosophy at Marischal College, Aberdeen, and Messrs. Gordon, Barron & Co., cotton manufacturers of that town, as having been the first in Great Britain to succeed in large-scale bleaching by the new method, in July 1787. Parkes actually obtained his information from Patrick Milne, at that time M.P. for Cullen, who together with his late brother, being partners in the above firm, had collaborated with Copland in successfully applying chlorine bleaching.[3] Milne had told Parkes how Copland acquired information on the new process from Professor de Saussure, at Geneva, while travelling on the continent with the Duke of Gordon. Dr. and Mrs. Clow have reprinted Copland's confirmatory letter to Parkes, dated 27 April 1814, which he was permitted to publish in the *Annals of Philosophy*, so we shall not reproduce the whole of it here.[4] Several parts of it, however, deserve particular notice. It was 'in the early part of 1787' that Copland met Saussure at Geneva, when he was shown 'the experiment of discharging

[1] *Op. cit.*, pp. 186-8.
[2] Their reference is to Rees' *Cyclopaedia*, article on 'Oxymuriatic Acid'. The letter was also referred to in *Annals of Philosophy*, Vol. VI (July-Dec. 1815), p. 423. The date of the letter was actually 23 Feb. 1788, and it was addressed, not to Thomas Henry, but to Dr. Thomas Percival, who communicated it to Henry. See below, pp. 258 and 285-7.
[3] Parkes, *op. cit.*, Vol. IV, pp. 45-6; *Annals of Philosophy*, Vol. VII (Jan.-June 1816), pp. 98-102.
[4] Their reprint, however, contains numerous small errors.

vegetable colours by the oxymuriatic acid, which though I had met with accounts of (I think in M. de la Metheric's[1] Journal) I had never before seen tried'. Obviously Berthollet's account was already known to leading chemists in this country, who commonly read French and other foreign scientific journals. Copland, being 'well acquainted with the chemical knowledge of the Mr. Milnes', communicated this information to them immediately on his return. This is another example —and by no means the last we shall come across—of the close and fruitful relationships between scientifically-minded industrialists and natural philosophers. They 'instantly' tried it on a hank of yarn, which was bleached in less than an hour. 'To the best of my recollection this was about *the end of July 1787*, and from that time I was frequently informed by Mr. Milne and his late brother that they always continued to use this new mode of bleaching in their manufactory', particularly for urgent orders. 'I also think they were soon enabled to extend its application to larger quantities, by using vessels of white wood in place of glass, as at first. Mr. Milne is, therefore, in my opinion, perfectly correct in stating that *theirs* was the first manufactory in Britain where the new method of bleaching was introduced and continued to be practised.'

Dr. and Mrs. Clow unhesitatingly accept this view, considering that there is 'little doubt that Messrs. Gordon Barron and Company have six months' priority over the large-scale experiments of Watt and MacGregor'. But from careful examination of the evidence provided by Rees, Parkes, and the *Annals of Philosophy*, it is obvious that their priority was not clearly established. Even Copland's letter is by no means so definite and incontrovertible as it appears at first sight. The test conducted with Messrs. Milne in July 1787 (according to Copland's recollection twenty-seven years later) was only on a small scale, and it is hardly to be believed that, when (as we shall see) Berthollet, Watt, and many others experienced such difficulties and delays in developing large-scale production, this Aberdeen firm should have so quickly and so easily achieved success. Indeed Copland himself says that at first they used glass vessels (which they, like others, evidently found impracticable), but that 'soon' they began to bleach larger quantities using wooden vessels.[2] How 'soon', one wonders, and what other development problems did they encounter and how did they overcome them?

Parkes was, not surprisingly, somewhat equivocal in giving credit for the early introduction of the new process. In his *Chemical Essays*

[1] This is a misprint for M. de la Métherie, who was joint editor of the *Journal de Physique*, together with Rozier and Mongez (see above, p. 253).
[2] The word 'soon' is omitted from the Clows' reprint.

he included Berthollet's mention in the *Annales de Chimie* of how he repeated his bleaching experiments in the presence of 'the celebrated Mr. Watt', who, soon after his return to England, wrote to tell him that, in a first trial, he had bleached five hundred pieces of cloth at the bleachfield of his father-in-law, James McGrigor, near Glasgow; Berthollet also mentioned that McGrigor continued (i.e. in 1789) to use the new process.[1] He did not date his demonstration to Watt, but it was obviously not long after his early experiments, and Parkes therefore declared that he had 'no doubt that Mr. Watt was the first person in Great Britain who introduced science into the bleaching process; for before his connection with Mr. MacGregor . . . the whole operation of bleaching was merely the effect of observation and practice'.

Despite this compliment, however, Parkes obviously gave preference to Copland and the Milnes of Aberdeen as being first in Britain to introduce and successfully develop chlorine bleaching. His account therefore produced a protest from William Henry, son of Thomas Henry, who, in a letter to the *Annals of Philosophy*,[2] emphasized the parts played firstly by James Watt and secondly by his father. Henry quoted from the previously mentioned letter by James Watt, dated 23 February 1788,[3] in which Watt had stated not only that 1500 yards of linen were then bleaching under his directions, but also made the following statement: 'I have, *for more than a twelvemonth*, been in possession *and practice* of a method of preparing a liquor from common salt, which possesses bleaching qualities in an eminent degree; but, not being the inventor, I have not attempted to get a patent or exclusive privilege for it.'[4]

It is most surprising that Dr. and Mrs. Clow did not refer to this statement by Watt, nor to William Henry's conclusion from it, 'that all claims for priority that have been hitherto advanced in this country must yield to that of Mr. Watt, whose actual employment of the oxymuriatic acid in bleaching dates from the *beginning* of the year 1787'. Henry also claimed that,

next to Mr. Watt, Mr. [Thomas] Henry was at least equally early with any other person in applying the discovery to practice. In proof of this I might appeal to the general notoriety of the fact in this town and neighbourhood: but I depend chiefly for its establishment, on a number of letters from Mr. Watt

---

[1] Parkes, *op. cit.*, Vol. IV, p. 54; *Ann. Chim.*, Vol. II (1789), p. 160.

[2] *Op. cit.*, Vol. VI (July–Dec. 1815), pp. 421–4.

[3] See above, p. 256, and below, p. 285–7.

[4] Watt merely stated that the inventor was 'an eminent chemist and philosopher at Paris'.

to Mr. Henry, written in the year 1788, which are now before me. They form part of a series, in which each of those gentlemen disclosed, unreservedly to the other, the progress of his experiments in this new art.[1]

In reply to William Henry,[2] Parkes confessed that he had not done his father justice; but by producing Copland's letter he was able to claim 'that oxymuriatic bleaching was employed at Aberdeen in preparing goods for sale many months prior to any such application of it at Manchester, or at any other place in Great Britain, Mr. MacGregor's works in Scotland, where the operations of Mr. Watt were conducted, being alone excepted'. Thus he seems to have admitted Watt's priority.

There were also, it is clear from these publications, other pioneers in the development of chlorine bleaching. Other scientists than Copland would certainly have read the published works of Scheele and Berthollet;[3] in 1786 Beddoes's translation of Scheele's *Chemical Essays* had appeared,[4] and in 1787 there were references to the discoveries of Scheele and Berthollet in the second edition of William Nicholson's *Introduction to Natural Philosophy*. News of such an important innovation must soon have spread in the textile districts, and would be likely to stimulate experiment. Thomas Henry, for instance, 'having received an indistinct account of the new method, but not knowing precisely in what it consisted, immediately set about investigating the steps of the operation'.[5] Personal contacts between scientists—as between Watt and Berthollet, and between Copland and Saussure—were also important. According to Rees's *Cyclopaedia*,[6]

previous to any publication by M. Berthollet, Mr. Scheele communicated to Mr. Kirwan the properties of the dephlogisticated marine acid in whitening vegetable substances, and Mr. Kirwan, then residing in Newman-Street, London, suggested to Mr. C. Taylor [of Manchester], the present secretary to the

[1] We shall later examine some of these letters, which we have recently discovered.

[2] *Annals of Philosophy*, Vol. VII (Jan.–June 1816), pp. 98–102.

[3] See also below, pp. 261–2, 277, 296, 305, for Berthollet's contacts with other British scientists.

[4] Earlier, in 1784, Edmund Cullen had translated the *Physical and Chemical Essays* of T. Bergman, Scheele's great Swedish contemporary, who had also experimented with 'dephlogisticated marine acid'. Educated scientists, of course, would have been able to read the original Latin of both Scheele and Bergman.

[5] W. Henry's 'Tribute' to his father, *Manchester Lit. & Phil. Soc. Memoirs*, 2nd. ser., Vol. III (1819), p. 228. See below, pp. 287–8. According to E. Baines, *History of the Cotton Manufacture in Great Britain* (1835), p. 248, Henry began his investigations 'without knowing any thing of Watt's experiments, but acting merely on the suggestions in Berthollet's papers in the *Journal de Physique*'. His signed copies of the *Annals de Chimie*, from the first volume (1789) onwards, are now in Manchester University Science Library.

[6] Article on 'Bleaching'. Rees's *Cyclopaedia* was originally published in parts, and this article, in Vol. IV, seems first to have appeared in about 1803–4.

Society of Arts, &c., the probability of its use in bleaching; and a whole piece of calico, in the state received from the loom, was, in the spring of 1788, actually bleached white, printed in permanent colours, and produced in the Manchester market ready for sale, having undergone all these operations in less than 48 hours, by the joint efforts of Mr. [Thomas] Cooper, Mr. [Joseph] Baker, and Mr. [Charles] Taylor, which is perhaps the first entire piece, either in France or England, that fully ascertained the real merits of the new mode of bleaching, and a certainty that it might be generally useful in commerce. This experiment was immediately followed by the establishment of a large bleaching concern by Mr. Cooper, Mr. Baker, and Mr. Horridge, at Raikes, near Bolton, in Lancashire, and before any considerable bleaching work was actually at work in France.[1]

The only reference to Watt's contribution in this article was that he was 'the first person who simplified the process of preparing the oxygenated muriatic acid, by means of a mixture of common salt and manganese, previous to the addition of vitriolic acid'.[2] Possibly Rees also intended to give Watt the credit for 'the substitution of large and commodious stills of lead, instead of glass vessels';[3] but his wording is not clear. Henry's work in Manchester he did not even mention.

This article had produced a protest from William Henry identical to that which he later made after publication of Parkes's account.[4] The part played by his father must, he said, have been known to the writer of the article, 'who, at that period, was himself engaged in this town [Manchester] in pursuit of the same object, and was in habits of occasionally communicating with Mr. Henry on the subject'. (The author, it seems, was very probably Charles Taylor, who may perhaps have inflated his own achievements.) As a result of this protest, Rees gave due credit to James Watt and Thomas Henry when he came to write the later article on 'Oxymuriatic Acid'.[5] Nevertheless, it seems very likely that Taylor did acquire knowledge of the process in the way described.[6] The eminent Irish chemist, Richard Kirwan, certainly appears to have obtained early information about chlorine bleaching. Parkes recalled having 'heard that Saussure or Scheele made a similar communication to Kirwan; and this appears probable, from the circumstance of the Irish having taken the lead as proficients in the new mode of bleaching'.[7] Parkes also gives testimony to the achievement of Taylor and Cooper in having 'bleached a whole piece of cotton by

---

[1] These developments will be dealt with in more detail later in this chapter.
[2] See below, pp. 267, n. 2, and 272–3.   [3] See below, p. 278.
[4] In 1815, in fact, he simply sent to *Annals of Philosophy* a copy of his letter to Rees of Dec. 1809, written when the early volumes of the *Cyclopaedia* were about to be reprinted.   [5] *Op. cit.*, Vol. XXV, pt. ii.   [6] See below, p. 289.
[7] Parkes, *op. cit.*, Vol. IV, p. 43, footnote. See below, pp. 319–21, for the introduction of chlorine bleaching in Ireland.

the new process, and printed and calendered it fit for the market in less than three days', and that this success led to the establishment of a large bleachworks at Raikes, near Bolton.

William Henry also admitted their success. At that time his father had been able to produce only half a yard of calico, similarly bleached, but of superior whiteness.[1] These demonstrations were made at a public meeting convened in Manchester early in 1788 to oppose an attempt at securing a Parliamentary monopoly of the new process by Messrs. Bourboulon de Boneuil & Co., of Liverpool, formerly manufacturers of liquid chlorine bleach at Javelle, in France, against whom Berthollet also complained.[2]

These individuals and firms are only referred to very briefly and inadequately by Higgins and by Dr. and Mrs. Clow; much of the evidence is either entirely overlooked or confused. This evidence, moreover, can be supplemented by a great deal of other material, both printed and manuscript, most of which has never previously been utilized.[3]

The earliest information about Berthollet's experiments with the new bleaching agent was conveyed to England in a letter of 31 March 1785, sent by Berthollet himself to Charles Blagden, secretary of the Royal Society.[4] In this he sent prior notice of

quelques expériences que je dois présenter à notre première assemblée [de l'Académie Royale]: elles ont pour objet l'acide marin déphlogistiqué. . .[5]

---

[1] Rees, *op. cit.*, article on 'Oxymuriatic Acid'; *Annals of Philosophy*, Vol. VI (July–Dec. 1815), p. 423.

[2] *Ann. Chim.*, Vol. II (1789), pp. 178–80; Parkes, *op. cit.*, Vol. IV, pp. 57–9 and 61–3, and *Annals of Philosophy*, Vol. VI, pp. 423–4. Boneuil & Co.'s *eau de Javelle* was made by absorbing the chlorine gas in a potash solution. Their activities will figure prominently in the later part of this chapter.

[3] Many letters, for example, in the Boulton and Watt papers, records of various bleaching firms, material relating to Thomas Henry, the letters of Sir Joseph Banks, Sir Charles Blagden, and other scientists, the memoirs of the Manchester Literary and Philosophical Society, the Irish Academy, etc., proceedings in the Tennant bleaching case, the article on bleaching in Brewster's *Edinburgh Encyclopaedia* (1808–30), neglected by modern writers, and also a manuscript by James Rennie, M.A., 'An Essay on the Improvements in the Art of Bleaching by the Application of the Principles of Chemistry' (n.d., but internal evidence shows it to have been written in 1816–17). This manuscript has been very kindly loaned by Major David Gibson-Watt from his collection of papers at Doldowlod. Rennie wrote a series of 'Essays on Bleaching' in the *Glasgow Mechanics' Magazine*, Vols. III and IV (1829). In the early thirties he became Professor of Natural History at King's College, London.

[4] In Mr. James Osborn's Collection, Yale University Library. Blagden had visited France in 1783 and met a number of French scientists, with whom he subsequently corresponded. In another letter to Blagden, of 28 April 1785, Berthollet referred to other scientific researches in France, and in each letter he introduced visitors to England—Baron de Montboissier firstly, and M. de Virly secondly.   [5] He referred Blagden to 'le traité des affinités de Bergman'.

J'ai saturé l'eau distilé du gas de l'acide marin déphlogistiqué à l'appareil de Mr Woulfe et quatre flaccons dont le premier est vide étant environnés de glace. L'eau se sature promptement et alors le gas surabondant prend une forme concrête qui tend à la plus faible chaleur à reprendre l'état gaseux.

He went on to describe how he had mixed alkali and lime with the liquor, and carried out other experiments, concluding that its peculiar properties 'ne sont dues qu'à l'air déphlogistiqué qui se combine avec l'acide marin, mais qui y tient si faiblement qu'il passe très facilement en combinaison avec les substances avec lesquelles il a quelqu' affinité'.[1] This, he went on, explained its bleaching action:

L'acide marin déphlogistiqué produit sur les couleurs tant végétales que minérales des changements semblables à ceux qu'éprouvent les couleurs à une longue exposition à l'air: de sorte que les altérations de ces couleurs dependent d'une combinaison de l'air déphlogistiqué avec les parties colorantes. Lorsque la liqueur a épuisé son action sur les parties colorantes elle reprend les propriétés de l'acide marin ordinaire. Les substances végétales colorées passent donc promptement au blanc, au jaune, ou à une teinte rouge avec ma liqueur, selon qu'elles doivent prendre les teintes étant exposées à l'air.

There is no suggestion in this letter of any use of 'dephlogisticated marine acid' for bleaching textiles, but it seems very probable that Blagden, and other British scientists, soon acquired knowledge of Berthollet's subsequent proposals for its industrial application, by reading his published papers or by personal contact.[2] The first definite evidence, however, does not appear until towards the end of 1786, when Boulton and Watt visited Paris and Berthollet showed them his bleaching experiments. A letter of 30 October 1786[3] to Aimé Argand,[4] in Paris, stated that they would set out for France that week. They did

---

[1] This was a 'phlogistic' explanation; in his papers to the Académie Royale, Berthollet put forward the oxygenation theory.

[2] In 1785, for example, the young British chemist, Smithson Tennant, F.R.S., toured on the continent, visiting Scheele in Sweden, Guyton de Morveau at Dijon, de Saussure in Geneva, and Berthollet in Paris, where he worked in the laboratory of the Academy. W. A. Smeaton, 'Louis Bernard Guyton de Morveau, F.R.S. (1737–1816) and his Relations with British Scientists', *Notes and Records of the Royal Society*, Vol. 22 (1967), p. 117.

[3] In Birmingham Reference Library (B.R.L.)

[4] A Geneva physician and scientist, noted particularly for his invention of the Argand oil-lamp, who had become acquainted with Boulton and Watt in 1783, when he came to England to patent and manufacture his lamp; the patent was obtained in 1784, but was subsequently challenged and the project failed. (See H. W. Dickinson, *Matthew Boulton*, p. 128.) Boulton and Watt, however, remained in friendly contact with him, especially on scientific matters.

not return to England until after mid-January 1787. Meanwhile Watt had written to his wife, Annie, informing her of Berthollet's bleaching discoveries and apparently asking her to pass this information on to her father, James McGrigor,[1] for on 25 December 1786 she replied to him, in Paris, saying, 'I have wrote my Father about the Bleaching it is very wonderful but is your Bleacher not like the sercher after the philosopher['s] stone [who] put in gold to produce gold.' This was an extremely shrewd remark, for, as we shall see, one of the most serious problems facing the early experimenters with chlorine bleaching was the cost of the raw materials used; Annie Watt was to prove herself a very capable helpmate in the endeavours of her husband and father to make the process a practical commercial proposition.

At the time when Berthollet showed his experiments to Boulton and Watt,[2] he had already published accounts of them in the *Journal de Physique*,[3] from which we can see what progress he was making. He had repeated Scheele's experiments in making the 'dephlogisticated marine acid' or 'oxygenated muriatic acid' (as he now termed it) in gaseous form, by the reaction of marine or muriatic (hydrochloric) acid on black calx or oxide of manganese.[4] Scheele had found the gas to be very little soluble in water and had therefore experimented mostly with the gas itself; this was because he had not brought the latter sufficiently into contact with the water, perhaps on account of the difficulties of handling this dangerous vapour. The worst problem, in fact, in dealing with chlorine was its volatility and its blinding and suffocating effects.[5] Berthollet overcame these difficulties by using a Woulfe distilling apparatus, made of glass, in which the gas, generated in a retort, was passed along tubes through a series of bottles or jars containing water, in which it was successively dissolved. Between the retort and the first receiver he placed an intermediate bottle containing a little water, to condense muriatic acid vapours which were liable to pass over.

In his first bleaching experiments, Berthollet used the 'oxygenated' liquor in a highly concentrated form, but found that it badly weakened textiles so treated. When diluted, it bleached safely, but the fabrics subsequently tended to turn yellow. He concluded that treatment with

---

[1] Unfortunately, this letter does not appear to have survived.

[2] That Boulton saw these together with Watt is shown by a letter from him to Berthollet in March 1788, Matthew Boulton Letter Book, 1780–9, A.O.L.B.

[3] See above, pp. 252–3. See also *Ann. Chim.*, Vol. II (1789), in which he makes clear the extent of his researches at the time of Watt's visit.

[4] Simply manganese in common terminology.

[5] It could choke a person, rendering him senseless, or even killing him. In the First World War, of course, it was used as poison-gas.

the liquor alone was inadequate and therefore alternated it with the traditional 'bucking' in alkaline lye, followed by 'souring' in dilute sulphuric acid, whereby he produced a perfect and permanent whiteness. To prevent damage to the cloth and to reduce the liquor's suffocating odour, he also experimented by adding alkali (potash or soda)—thus producing the first hypochlorites[1]—which proved successful, but had two snags: firstly, it weakened the liquor's bleaching effects, and secondly, it added greatly to costs. Berthollet therefore abandoned it,[2] though, as we shall see, hypochlorites were to become the bleaching agents of the future.

His experiments had nearly reached this stage by the time of Boulton and Watt's visit. The latter, as Berthollet observed, instantly grasped the commercial possibilities of his process, and on his return to England at once began, firstly, to consider steps for securing to Berthollet a patent or Parliamentary monopoly, and secondly, to experiment himself in producing the new bleaching agent and, in collaboration with McGrigor, his father-in-law, to make practical trials of it. Meanwhile, secrecy was of the utmost importance, but unfortunately Argand had given information about the new process to friends in London, named Parker and Middleton.[3] In a letter to Argand of 21 February 1787,[4] Watt said that he had approached these two to try to obtain their co-

---

[1] A solution of potash (potassium carbonate, K_2CO_3), for example, is rendered caustic (changed to the hydroxide, KOH) by addition of slaked lime (calcium hydroxide, $Ca(OH)_2$):

$$K_2CO_3 + Ca(OH)_2 \longrightarrow 2KOH + CaCO_3$$

When chlorine gas is passed into the caustic solution, potassium hypochlorite (KOCl) is produced:

$$Cl_2 + 2KOH \longrightarrow KOCl + KCl + H_2O$$

In the bleaching process, the potassium chloride (KCl) is inactive, but the hypochlorite (KOCl) decomposes to produce oxygen.

[2] See below, p. 271.

[3] It is fairly certain that the first-named was William Parker, who appears in the London directories of that period as a 'glass manufacturer' at 69 Fleet Street. We know that he manufactured glass apparatus for making mineral waters, together with eudiometers, etc., at his 'Cut-glass Manufactory' at that address. See above, p. 237, n. 3 (Magellan's pamphlet). He also manufactured lamps for Argand, whom he had first introduced to Boulton in 1783 (Dickinson, *Matthew Boulton*, p. 128). The glass apparatus for production of chlorine would provide him with another market.

The Middleton referred to appears to have been John Middleton, to whom we find Watt writing in 1787 (J. Watt to J. Middleton, 27 Feb. 1787, B.R.L.); they had common French acquaintances such as 'our friends Reveillon & Argand'. There are, however, several people of that name in the London directories of the 1780s, one a black-lead pencil-maker, another a merchant, and a third a colourman.

[4] Boulton & Watt Letter Book (Office), March 1786–Feb. 1788, B.R.L.

operation, and that he had mentioned Berthollet's invention to Mr. Pitt, the Prime Minister.

I proposed to make some experiments myself that I might be able to speak positively to the fact & then to apply boldly to the higher powers for an effectual exclusive privilege being persuaded nothing can be done otherwise. I have now got the app[aratu]s and shall consider it as a debt of honour unpaid until I get it tried. When with Mr. Pitt I mentioned it he said it was a matter of importance & should receive all the *countenance* he could give it.

Four days later he wrote to Berthollet,[1] informing him also that he had mentioned his process to the Prime Minister. At first he had thought of trying to obtain some reward either from the manufacturers, by subscription, or from Parliament, but now considered that the best course was to apply for a patent and at the same time for an Act of Parliament, 'to secure to you the profits of the invention for 15 or 20 years'. To strengthen his case, he hoped actually to provide the Prime Minister with experimental proof of the efficacy of Berthollet's process.

With that view I procured an apparatus & this day set about distilling some of the acid, but from one cause or other have failed in making it strong enough, whether for want of Ice[2] the Gas made its escape through the water in the two Flacons which acted as receivers, or whether I have mistaken the proportions of Manganese & Spirit of Salt [hydrochloric acid], or whether my Manganese was bad, or from what other cause, the first receiver was *very* slightly coloured & the second not at all sensibly, the liquor in both smelt very pungent & tasted acrid like pepper but does [not] whiten linen as I could wish. I shall endeavour to correct my error . . .

He suggested that Berthollet should hold himself in readiness to visit England, to secure the patent, etc., which would cost about £100.

The necessary points to determine are the quantity of [acid] necessary to whiten a given weight of linen or cotton—The probable cost of it—The methods of making it in great [on a large scale]—The substances which can contain it & of [what] vessels may be made to make it in—Please turn these thoughts over in your mind and also all sorts of substitutions which [may] be made for the different ingredients & variations of process.

At the same time Watt wrote to Samuel Garbett,[3] who was going to meet Pitt, asking him to mention the new bleaching invention by

[1] Watt to Berthollet, 25 Feb. 1787, *ibid.*
[2] Berthollet had packed the receiving jars in ice to aid distillation.
[3] J. Watt to S. Garbett, 25 March, 1787, Boulton & Watt Letter Book (Office), Mar. 1786–Feb. 1788, B.R.L. Garbett, of course, was notable for his partnerships with Roebuck in vitriol works at Birmingham and Prestonpans, and in the

s

'a man of Great Merit in France', as being of 'the utmost consequence to our manufactures', but which would 'require still much study and expence to make it practicable in Great'. He was to say that this French chemist would be willing 'to come over here & put it into practice', but only with 'proper encouragement in . . . securing to him effectually the profits of his invention, which cannot be done without an Act of Parliament on purpose or an act regulating patents; for after the dreadfull treatment poor Argand met with in his Lamp patent, what foreigner in his senses will put any trust in them'.

Watt, however, was mistaken in his judgement of Berthollet, who had no desire to engage in the commercial exploitation of his process, but preferred a life devoted to science, as he informed Watt a month later:[1]

Je suis très sensible à l'interêt que vous prenez à un projet qui pourrait m'être avantageux; mais revenir à mon caractère, j'ai entièrement renoncé aux entreprises d'interêt. Quand on aime les sciences on a peu besoin de fortune, et il est si facile d'exposer son bonheur en compromettant sa tranquillité et en se donnant des embarras.

He was continuing with experiments on his process, to make it more practical, and had recently bleached several pieces of cloth so successfully that they could bear comparison with the finest fabrics. He went on to suggest reasons for Watt's failure:

Si vous n'avez pas réussi dans votre premier essai, c'est probablement parceque les luts[2] n'auront pas résisté, et qu'ils auront laissé passer le gaz. D'ailleurs deux flaccons[3] ne suffisent pas et ne produisent pas assez de résistance. Avec quelques tentatives, vous ne manquerez pas de réussir parfaitement. Vous vous rappelez que les proportions sont d'une partie de manganaise contre six d'acide marin et qu'il faut une pinte d'eau dans les flaccons, pour chaque once d'acide marin employé.

Watt did, in fact, soon overcome these early difficulties in producing the bleaching liquid and was able to send a bottle of it to his father-in-

---

Carron ironworks. He, like Boulton, was seasoned in the arts of parliamentary lobbying. See J. M. Norris, 'Samuel Garbett and the Early Development of Industrial Lobbying in Great Britain', *Econ. Hist. Rev.*, 2nd ser., Vol X, no. 3 (April 1958), and E. Robinson, 'Matthew Boulton and the Art of Parliamentary Lobbying', *Hist. Journ.*, Vol. VIII (1964).

[1] C. L. Berthollet to J. Watt, 25 March 1787, Dol. As Matthew Boulton later put it, 'the Character of the Philosopher & love of Science predominated over the Merchant & love of Money'. M. Boulton to C. L. Berthollet, March 1788, Matthew Boulton Letter Book, 1780–9, A.O.L.B.

[2] Lutes to make the apparatus gas-tight. This was one of the constant problems in the distillation process. Pajot des Charmes (*op. cit.*, English trans., pp. 4–5 and 12–18) later laid special emphasis upon the necessity of good luting, which he regarded as 'the soul of distillation'. A kind of putty was used, e.g. made with clay and linseed oil.   [3] Receiving flasks containing water for dissolving the gas.

law, for him to use in bleaching experiments. These were to reveal the numerous and complicated problems involved in applying the new process industrially and commercially. In a long letter of 19 April 1787,[1] James McGrigor acknowledged receipt of the bottle and described his subsequent experiments. He had tried the liquid on various textiles, including coarse and medium linen yarn and cloth, finer linens (lawns and cambrics), and cotton shirting, subjecting them to twice-repeated operations of boiling in alkaline lyes of different strengths, followed by washing, and then steeping in the bleach liquid (for twelve hours the first time and six hours the second), with the result that the cotton shirting was fairly well bleached, but the linen goods were only partially done, the yarn least of all; all the fabrics were 'much impoverished', especially the linen.

McGrigor had then carried out the same operations using a wooden receiving vessel in place of glass ones, since this would obviously be a better commercial proposition. But, although he left the goods in the liquid for two periods of sixteen hours, they came out far inferior. He noticed that the liquid 'lost the stink much sooner in the wood than it did in the Glass Vessels', but did not know whether this was due to the wood or because the liquid had lost some of its strength through being kept and the bottle being often opened.

However as I have by these Experiments seen this whitening power of your liquid and finding by your last Letter that you can make it much Cheaper than you mentioned in your first[2] I have every reason to believe that the liquid under proper management may turn out a very Valuable discovery in Bleaching Providing the goods after undergoing two three or four operations of the Alkali & liquid remain sufficiently strong to stand the work which I apprehend it may be necessary to give them in the common method of Bleaching viz Boiling in weak lyes mixed with sope lying some time on the Grass Rubbing & weak Vitriol sours all these I believe will be necessary in a greater or lesser degree not only to clear the goods perfectly of the liquid but raise their colour & make them fit for the market. From this you will be pleased to observe that I do not think the liquid alone will answer the sole end of Bleaching but the use of it will Expedite the Process greatly & that at not great Expence which in my opinion is a very great matter.

At Watt's request he also offered his opinion as to 'the best plan for your friend the inventor to follow in order to obtain a proper reward for his discovery'. He was not in favour of a patent, 'the more so as you

[1] Dol.

[2] Unfortunately these letters by Watt to McGrigor seem to have disappeared, but this is most likely a reference to Watt having hit on the use of vitriol (sulphuric acid) and salt with manganese (see above, p. 260), so that he no longer needed to purchase the dearer hydrochloric acid.

seem by your Letter to think that a Chimest may easily discover the composition' of the liquid. Nor did he think any great sum could be got for the secret 'till such time as it is very well ascertained how it answers in practice'. If he were the inventor, therefore, he would first make secret trials to perfect the process and then take in goods to bleach, returning them so quickly and so well done that he would soon make a 'noise among the Bleachers, not only here but in every other part in the Country', whereupon he 'would propose to make a dis- covery of the secret either to a few or to a great number upon their paying a certain sum of money which I would leave them to settle among themselves'. He suggested that, if Watt approved of this idea and if he (Watt) was to share with the inventor in the profits of this discovery, he should come to the bleachworks at Clover, near Glasgow, and carry out the necessary large-scale trials, in absolute secrecy.

Having written this, however, McGrigor then added a postscript, replying to an inquiry from Watt as to whether he had ever heard of any such discovery. He had, he said, often heard foolish boasts, while their acquaintance, George Glasgow, had actually altered the colour of a small bit of sheeting very considerably in twenty-four hours, but nothing had come of his efforts. He went on to report, however, that Mr. Walter Beggur, of Edinburgh, and his son had lately been there and 'in the course of conversation on Bleaching the son said that he had heard when at London that some person there could bleach cloth in a few hours'. McGrigor tried to pass this off as another foolish rumour, but, as we shall soon see, it boded ill for their attempted secrecy.[1]

Watt had also wanted to know 'the quantity of liquid necessary to cover all the parts of a given qu[antit]y of Cloth', to which McGrigor replied that, in souring, about five English gallons of vitriol sour were required to each piece (25 yards) of middling-priced linen—six or seven gallons if they were soured when green—and he thought that 'the goods should not be so throng [close together] in the liquid [bleach] as they are often in the Sour'. (It soon became obvious and important that bleachers should know exactly the quantities and also the strengths of the alkaline lyes and bleaching liquors used with different quantities and kinds of cloth, so as to avoid damaging the fabrics and yet ensure that they were properly bleached all over.)

Finally, after asking Watt if he had made any trials of the liquid in wooden vessels, McGrigor concluded by saying that, 'were the liquid to cost the price mentioned in your first letter the discovery would have been worth very little for some of the Bleachers here who use

[1] We have been unable to discover anything further about the activities of either George Glasgow or Walter Beggur.

Lime Bleaches 28 yards of thick Cotton Cloth at 18d. & the people about Manchester does it from 12d. to 14d. the 28 yard piece'. If Watt would send more liquid, he would make more trials.

Again, unfortunately, we do not have Watt's reply, but McGrigor's next letter, of 4 May 1787,[1] indicates further progress. Watt had apparently suggested that McGrigor had used too little of the liquid for the quantity of cloth, whereas the alkaline lye had perhaps not been weak enough; but Watt was sending another and stronger bottle of the bleaching liquid. Moreover, he suggested adding alkali to it, although, as we have seen, Berthollet had decided against this.[2]

Annie[3] in her Letter said I might add a few drops of alkali into the liquid but as she did not mention how much I was affraid I might spoil the Effect . . . so did not do it. But as I observe that you in your last say that when a proper quantity of Alkali was added the Cloth was not weakened but when no Alkali was added 2 or 3 days steeping made the Cloth quite tender pray what quantity of alkali should be added to a given quantity of liquid?

He also took note of what Watt said 'in regard to the Effect the liquid has on dyed Colours. I meant to have tried it upon some printed Goods had not my liquid run short.' This, as we shall see, was an important consideration: bleaching was necessary not merely for whitening grey or brown cloths as they came from the loom, but also for 'clearing' the 'grounds' of printed (especially maddered) goods, without discharging or even impairing the colours, i.e. it would be important to calico-printers as well as to bleachers.[4] But this would obviously be a more difficult operation, requiring more precise knowledge of the strengths of the bleaching liquids employed.

Although McGrigor still had 'some doubts whether the liquid will answer the end of Bleaching in the great', he considered it 'an object well worth investigation' and would readily have joined Watt in a venture to purchase the inventor's right, were it not for the fact—previously rumoured but now revealed in Watt's letter—that the cat had indeed been let out of the bag. McGrigor referred to 'the trans-action your friend [Berthollet] has made with the Gentlemen in London [Parker and Middleton] who I suppose are both made Acquainted with the Secret which puts it in too many hands to remain long a Secret'. Nevertheless, he was obliged to Watt for his intention of asking the inventor's 'liberty for me to practise in the great', which Watt had no doubt of obtaining. He was encouraged by 'the discovery you have made in making the liquid at a much cheaper rate than you

---

[1] Dol. [2] See above, p. 264, and also below, p. 271.
[3] Watt's wife. [4] See below, p. 301.

believe the Gentlemen in London knows anything of which is the great point as coarse goods which is the great Bulk will require a large quantity of liquid'. This fact, combined with Watt's scientific help and his own practical knowledge, would give them a considerable advantage. He therefore asked Watt to arrange for the shipping from Bristol to the Clyde of 'one Ton of the Mineral you mention'—manganese[1]—so that he might begin to produce the bleaching liquid on a large scale.

Berthollet, in response to Watt's request, readily gave his assent to manufacturing experiments on his process, saying that Watt might communicate it to whomsoever he pleased.[2] He also informed Watt of the further progress of his experiments. Like Watt and McGrigor he had found that the practical application of his process was not a simple straightforward matter, but required considerable variations according to type and quality of the textiles treated.

... il m'a paru que les toiles étaient beaucoup plus difficiles les unes que les autres à blanchir et qu'elles exigeaient des quantités très différentes de liqueur; d'ailleurs les lessives plus ou moins fortes, plus ou moins multipliées, contribuent plus ou moins au blanchiment, et font varier les quantités de liqueur nécessaires ...

Je me suis assuré qu'on réussissait beaucoup mieux en employant plusieurs lessives, quatre ou cinq, ou même six pour certaines toiles; il suffit alors de laisser une heure ou deux la toile dans la liqueur entre chaque lessive et il n'y a point besoin que la liqueur soit forte si ce n'est pour la dernière fois; j'étens une liqueur forte, c'est à dire assez colorée en jaune-verd, de deux ou trois parties d'eau, excepté la dernière fois et encore peut-on l'étendre de partie égale d'eau. Une lessive n'a pas besoin d'être longue, un quart d'heure suffit.

In this way, by repeated brief immersions in diluted liquid (instead of longer periods in a concentrated solution), alternated with boilings in alkaline lyes, and with a final souring in dilute sulphuric acid, he succeeded in bleaching any fabric in five to six hours, and the process was not thereby made any more expensive, because much less oxygenated marine (muriatic) acid was necessary. This procedure, as he later pointed out in the *Annales de Chimie*, overcame two major difficulties, which would otherwise have rendered it impracticable for industrial use: firstly, it reduced the danger to workmen from suffocating vapours, and secondly, it prevented the cloth from being weakened.[3] It was still necessary, however, to give the fabrics a thorough final washing, for which he intended to use soap, as Watt had done.

---

[1] See below, pp. 272–3.
[2] C. L. Berthollet to J. Watt, 22 May 1787, Dol. Watt's letter to Berthollet does not appear to have survived.
[3] *Ann. Chim.*, Vol. II (1789), pp. 158–60.

Because of his success with diluted liquid, he no longer found it necessary to add alkali:[1]

Je me suis assuré que lorsqu'on faisait usage d'alkali caustique, l'effet était plus prompt, mais qu'une même quantité de liqueur blanchissait beaucoup moins de toile, parcequ'il se forme un sel qui détonne comme le nitre et dont j'ai fait connaître les propriétés singulières à l'académie des sciences.[2] À présent je ne mets même plus d'alkali dans la liqueur, parceque comme elle est affaiblie je ne crains pas qu'elle attaque la toile.

He hoped that, by mutual exchange of information upon their experiments, they would quickly perfect the process. He urged particularly that, in preparing the liquid, Watt should guard against any gas escaping.

Meanwhile, McGrigor had received another letter from Watt and the bottle of strong liquor, with which he had repeated his previous experiments on linen and cotton fabrics, with alternate boilings in alkaline lye.[3] He put 2½ oz. of this liquor in 2 lb. of water and found that

two operations in the liquid produced a much better Colour than any of my former trials had done with the first Bottle and the goods done [were] not so much impoverished as formerly.

I have therefore better hopes than I formerly had of the liquids answering upon a large Scale & now only wishes to have the inventor's permission & your directions to try it in the great by which I will be much better able to judge

[1] See above, p. 264.
[2] See *Journal de Physique*, Vol. XXXIII (1788), pp. 217–24. See also below, pp. 299–300, for Thomas Henry's similar difficulties. As we have seen (above, p. 264, n. 1), when chlorine gas is passed into a caustic potash solution, hypochlorite (KOCl) is formed (the bleaching agent), together with chloride (KCl, which is inactive). If, however, there is not an excess of alkali present, i.e. if the liquor is saturated with the gas, then chlorate ($KClO_3$) is produced, which has no bleaching action. This was the remarkable new salt which Berthollet had discovered— 'oxygenated muriate of potash' he called it—the 'sel qui détonne': like nitre, it forms an explosive when mixed with combustible materials such as carbon and sulphur. Berthollet found that this salt was not formed if there was an excess of alkali, and that the resultant liquor was an effective bleaching agent, but he supposed that this was due to the oxymuriatic acid remaining uncombined in the solution, or only loosely attached to the alkali, without having surrendered its oxygen to form 'oxygenated muriate (or oxymuriate) of potash'. Later, in 1802, Berthollet's chemical reasoning was carried somewhat farther by Richard Chenevix (*Phil. Trans.* (1802), pp. 126–67), who suggested that 'oxygenized' (or 'oxygenated') muriate existed prior to the formation of what he therefore termed 'hyperoxygenized' muriate of potash (or soda). See also Thomas Hoyle's paper (referred to below, p. 315, n. 7), where the same suggestion had been made. These ideas were eventually to be overthrown by Davy (see below, p. 335), but the terminology lingered on long afterwards. Meanwhile, as we shall see, practical bleachers did succeed in preparing bleach liquors with alkaline solutions.
[3] J. McGrigor to J. Watt, 25 May 1787, Dol.

whether it will be an object worth your attention & mine or not. And if it should appear to be well I shall be extremely happy that you & I should be equaly concerned in the manufacture & sale of the liquid and in case a handsome sum should be obtained from making discovery of the secret to others I should think [it] but right to bring in the inventor for a third share of whatever may be got on that head.

This he thought would be a fair arrangement, since he would be 'at the trouble Expence & risque of making under your direction the Experiments in the Great so as to Ascertain the Value of the discovery'. He asked that, as soon as the inventor's permission was secured, Watt should 'have the whole Apparatus necessary for making the liquid made under your direction at Birmingham & leave as little to be done here as possible and let the Apparatus be such as will make a good large quantity at one time & the cost I shall remit you so soon as you advise me of the amount'.

A few days later Watt wrote to his wife,[1] from London, telling her of Berthollet's permission to make free use of his process, 'so that when I return I shall set seriously about putting your Father in the way of using it, but as he [Berthollet] seems to be pursuing it earnestly I fear there will not be much advantage accrue to us from the use of it, as he will probably make all his discoveries on it publick'.

On his return to Birmingham, Watt wrote a letter to McGrigor, evidently containing detailed directions for producing the 'dephlogisticated marine acid', as he continued to call it. This letter, dated 17 June, does not appear to have survived, but we have McGrigor's lengthy reply of 5 July,[2] in which he said 'your trouble is only to begin for you will very soon perceive by the questions I may put that I am perfectly ignorant in the Process of Distillation ... you will have to drive into my thick scull before you make me Chymist & I will be but a sorry one after all'. For secrecy's sake, he could not seek advice from others and therefore asked Watt to be very explicit in his instructions.[3]

From references in McGrigor's letter to materials used, it is quite clear that Watt had instructed him how to prepare the 'dephlogisticated marine acid' by adding vitriol (sulphuric acid) to salt and manganese. He asked whether he should powder or dissolve the manganese before putting it in the 'still', and he was concerned about the dangerous fumes which would be given off when the vitriol was

---

[1] J. Watt to Annie Watt, 1 June 1787, M. II, B.R.L.   [2] Dol.

[3] For the same reason perhaps he used chemical characters, sent to him by Watt, to avoid naming the materials used. In quoting from these letters, however, we shall translate them into contemporary terminology.

added to the manganese and salt, and about consequent loss of 'dephlogisticated marine acid' gas. There are no details, however, in this letter of the quantities or proportions of these materials to be used. He had not yet heard about the ton of manganese from Bristol; the other ingredients he could probably obtain at Glasgow. He was pleased with Watt's improvement in making the bleaching liquid 'more portable & I suppose cheaper for it is that which will make it answer'. Watt had apparently informed him that he had prepared the last bottle by receiving the gas in a caustic potash solution, thus securing it in a more concentrated form. From his experiments with that liquid, McGrigor thought that cottons could therewith be brought 'very near a finished State', but not linens; he was looking forward to trials 'in the great' with the apparatus which Watt was getting made for him. Watt himself, it appears, was also making experiments, for which McGrigor was to send him brown linen.

McGrigor's views in regard to the difficulties of bleaching linen were confirmed by another letter from Berthollet,[1] in which he gave his opinion that, whilst cotton and hemp could be bleached perfectly by alternate treatment with alkaline lyes and 'dephlogisticated marine acid', linen required something more: 'il lui reste un coup d'œil jaunâtre qu'on ne peut lui enlever que par le moyen d'un savon conformément à ce que vous m'avez écrit: on croit même qu'il faudrait le passer après cela une fois dans le lait aigri'.[2] In this letter Berthollet referred to public trials of the new process, which was now sufficiently proved to be taken up by two large establishments in Flanders. Just as he had freely given Watt and McGrigor permission to exploit it, so he freely gave his aid to its industrial application in France.[3] It was hardly to be expected, therefore, that these revolutionary developments would not soon be carried across the Channel. Watt, however, was seriously alarmed when news now arrived of such a French invasion. This was conveyed to him by Matthew Boulton, who had received information from Samuel Garbett about 'some French men that are come to Liverpool & Manchester to teach the new Art of

[1] C. F. Berthollet to J. Watt, 30 July 1787, Dol.

[2] As traditionally used in the 'souring' process. Dilute sulphuric acid could also be used. *Ann. Chim.*, Vol. II (1789), pp. 159 and 174.

[3] In his famous essay on the new method of bleaching in the *Annales de Chimie* in 1789, Berthollet referred to its establishment, with his assistance, in Valenciennes, Lille, Rouen, Javelle, and Jouy, and to the improvements introduced by other chemists such as Bonjour, Welter, Décroisille, and others. As Ballot rightly observes, however, in France, as in England, practical bleaching experience had to be allied with chemical knowledge to achieve success (*op. cit.*, pp. 529–30).

Bleaching & mean to take out a patent'.[1] They were also intending to exploit an allegedly new process for the manufacture of 'mineral alkali' (soda) from salt, in which Watt was likewise interested.[2] Boulton immediately wrote to Watt, then in London, that he might 'enter a Caveat & take such steps as you may think proper'.

Garbett had been told of these Frenchmen by a friend in Liverpool,[3] who was associated with him in the development of a process whereby they had succeeded in manufacturing a 'Fossil Alkali' or 'Mineral Alkali', i.e. 'an Alkali [made] from Rock Salt for the Soapboilers use', apparently similar to that of the Frenchmen, whose pretensions, therefore, they would oppose. This friend was probably Dr. James Gerard, a Liverpool surgeon, who had approached Dr. Joseph Black in January 1785 regarding the establishment in Liverpool of a works for the manufacture of 'fossil alkali' or soda.[4] Garbett had now received from his friend a sample of

some Mineral Alkali made her[e] from Rock Salt, by two Frenchmen, Messrs. Boneuil & Vallet, who have been at this Place about 2 Months. I am told they actually carry on a Manufactory of this Sort near Paris at this very time under the Firm of Messrs. Alban & Vallet a Javel pres Paris. Their object in coming here, I am further told, was to try whether the Materials of this Country answered equally well in their Process, & at a suitable Expence; & if so to apply for a Patent, or an Act of Parliament. They say their Plan will succeed & they propose going to London upon that Business about the 12th of Sep^r.

They were said to claim that they could sell their alkali cheaper than barilla and that 16 lb. of it would make 1 cwt. of soap, but Garbett's friend very much doubted this. They also possessed 'some Secret in the Art of Bleaching (which they mean to follow) by which the whole Process is finished in a few Hours', and all 'within Doors'. They had given practical demonstrations on small pieces of cloth, but Garbett's friend thought the whole account 'very much in the wonderful'.

Watt, however, knew better and at once contacted Garbett and began to 'take the necessary steps' in London, where he then was.[5] He also wrote to his wife—in Glasgow, helping her father with the bleaching experiments—informing her of the Frenchmen's spurious claims, and of his intentions to frustrate them, particularly in regard to

[1] M. Boulton to J. Watt, 2 Sept. 1787, Box 20, B.R.L. Watt was in London.

[2] For Watt's alkali interests, see below, chap. X.

[3] A copy of this letter, dated 24 Aug. 1787, was enclosed by Boulton in his communication to Watt of 2 Sept.

[4] Sir W. Ramsay, *Life and Letters of Joseph Black, M.D.* (1918), p. 69; Clow, *op. cit.*, p. 99. Gerard had taken out a patent for producing alkali from common salt in 1783. See below, pp. 364–6.

[5] J. Watt to M. Boulton, 10 Sept. 1787, James Watt Box IV, A.O.L.B.

the bleaching process, 'as they have no connection with the inventor', Berthollet.[1] He asked that McGrigor should take measures to secure himself against them in Scotland 'by making the acid before proper witnesses whom he can trust[2] & by using the liquor before some of his confidential servants'. It would be 'necessary to make some of it by using 6 ounces of common spirit of salt [hydrochloric acid] to each oz. of Manganese', in place of the vitriol and common salt, that being the method used by Berthollet. He later suggested that McGrigor should bleach a yard or two of cloth with the liquor and *sell* it to a friend, 'informing him that it is done by a new process in a few hours', but trying to 'avoid making the matter publick'.[3]

Exactly what other steps Watt took to frustrate the French 'pretenders' at this juncture we do not know. In fact the matter appears to have subsided temporarily, until the early months of 1788.[4] The information supplied by Garbett's friend in Liverpool, however, was reasonably accurate. Ballot, in his account of the early development of the French chemical industry, emphasizes the importance of the Javel (or Javelle) firm:[5]

La manufacture de Javel avait été fondée en 1777 sous la protection du comte d'Artois; elle était dirigée par Alban et Vallet; consacrée d'abord à l'huile de vitriol, elle produisit bientôt des eaux fortes, de l'esprit de sel, du vitriol bleu, de l'alum, de la soudre épurée,du blanc de céruse. Alban et Vallet s'interessaient aussi aux aérostats et mirent en pratique le procédé de blanchiment de Berthollet. Cette manufacture était encore prospère en 1790.

Ballot gives no details of their achievements in chlorine bleaching, but from Berthollet's own account we know how he originally provided them with information on his process, and how they then tried to pirate it.[6] He himself had visited Javelle and shown the proprietors how to distil the oxygenated muriatic acid and how to bleach with it. At that time, 'j'employois encore une liqueur concentrée, & j'y mêlois un peu d'alkali'. Soon afterwards the Javelle manufacturers published in different newspapers that they had discovered a new bleaching liquor which they called *lessive de Javelle*. 'Le changement qu'ils avoient fait au procédé que j'avois exécuté en leur présence, consistoit en ce qu'ils mettoient de l'alkali dans l'eau qui reçoit le gaz, ce qui fait que la liquer se concentre beaucoup plus, de manière qu'on peut ensuite l'étendre de plusieurs parties d'eau pour s'en servir.' With this,

[1] J. Watt to Annie Watt, 5 Sept. 1787, M. II, B.R.L.
[2] He suggested 'Mr. [Gilbert] Hamilton & some of yourselves'.
[3] J. Watt to Annie Watt, 10 Sept. 1787, M. II, B.R.L.
[4] See below, pp. 284 *et seq.*   [5] *Op. cit.*, p. 541.
[6] *Ann. Chim.*, Vol. II (1789), pp. 178–80. See also above, p. 261.

as we have seen, Berthollet did not agree[1] and he continued to advocate absorption of the gas in water.[2]

These, then, were the 'pretenders' who established themselves in Liverpool. Garbett's friend referred to them as 'Messrs. Boneuil & Vallet'. Later evidence reveals them more fully as Anthony Bourboulon de Boneuil and Matthew Vallet,[3] of whom the former appears to have been the chief promoter, while Vallet supplied the chemical expertise. Thomas Henry procured information about Bourboulon de Boneuil from 'a Mon'. Browne son to the Inspector of the F[rench] King's gardens who is now here'.[4] 'He says Bourboulon was originally a Man of Fortune, but dissipated; was Treasurer to the C[omte] d'Artois & to the City of Paris; but he afterwards engaged in a Vitriol Work, & became a Bankrupt. Browne seems to consider him an adventurer.' Watt had previously heard 'that Bourboulon is a partner in the Vitriol Works Javelle near Paris'.[5]

These manufacturers of *liqueur de Javelle* deservedly aroused indignant opposition from Berthollet, Watt, and others. They certainly were not the true and original inventors of the process which they brought over to England. But they were not without some merit. When James Rennie later said that he was 'not sure that De Bonneuil and his associates have been altogether justly treated by authors'—and even asserted that they were 'certainly the inventors of the liquid' which they popularized—he had some justification for doing so.[6] There is no doubt that, in emphasizing the advantages of absorbing the gas in an alkaline solution, rather than in water as advocated by Berthollet, they were right from a practical bleaching point of view.[7] Moreover, they undoubtedly possessed considerable chemical knowledge and experience, and were to prove of great importance in the dissemination of chlorine bleaching, especially in Lancashire. They also provide the earliest example of a tendency, which was to become more marked, towards integration of the acid, alkali, and bleach manufactures. Sulphuric acid, now widely used as a bleaching 'sour', was also required, together with salt, in producing hydrochloric acid,

---

[1] See above, pp. 264 and 271.   [2] See below, pp. 297 and 313.

[3] See below, p. 284. Dr. and Mrs. Clow, *op. cit.*, pp. 189–90, refer very briefly to Bourboulon de Boneuil & Co. (misspelt 'Bonneuil'), but do not realize that this firm was the same as that established in Liverpool by the *eau de Javelle* manufacturers from France.

[4] T. Henry to J. Watt, 12 Sept. 1788, Dol. This must have been the same 'Mr. Brown', then visiting England, whom Berthollet had recently recommended to Watt as a close associate. C. L. Berthollet to J. Watt, 28 July and 6 Aug. 1788, Dol.

[5] J. Watt to M. Boulton, 27 Feb. 1788, James Watt Box IV, A.O.L.B.

[6] *Op. cit.*, p. 29.   [7] See below, p. 313.

which tended to replace it for that purpose; this reaction between sulphuric acid and salt also formed the basis of most processes for making synthetic soda, another bleaching requirement; similarly, it was used in the production of chlorine, either directly together with salt and manganese, or indirectly in making hydrochloric acid. Sulphuric acid and salt thus formed the basis of the 'heavy' chemical industry now emerging.[1] There appears to be no evidence that Bourboulon & Co. manufactured sulphuric acid at Liverpool, though this had been their main activity at Javelle and they required it for their production of soda from salt. In the latter field, of course, they were not alone: many patents had been taken out in the previous ten years— by Shannon, Higgins, Fordyce, Collison, Gerard, and others—and there had also been much research by such men as Roebuck, Watt, Black, and Keir.[2] But they appear to have been the first firm in Britain to combine the manufactures of soda and chlorine bleach. Several factors may have attracted them to England: the prospect of getting a patent or Parliamentary monopoly, availability of raw materials such as salt, and the tremendous potential market for soda and chlorine in the textile trade.[3] Merseyside was obviously a very favourable base for their operations.

Their arrival knocked another huge hole in the wall of secrecy which Watt and McGrigor were endeavouring to maintain. At the same time, the earlier gaps were now beginning to widen rapidly. In the scientific world, knowledge of Berthollet's work was spreading, and we find Charles Blagden, secretary of the Royal Society, writing to Sir Joseph Banks that 'Berthollet is undoubtedly the inventor of the quick method of bleaching, by means of the dephlogisticated marine acid; it is applicable I understand, with most advantage to manufactures of hemp, but how far it will succeed economically seems not to be yet decided'.[4] Moreover, when Watt, in his efforts to thwart the Frenchmen, approached Messrs. Parker and Middleton, of London, to whom Argand had revealed Berthollet's process, he reported to his wife[5] that they were

resolved to teach the secret & have employed a rider for that purpose who is to go into Lancashire & Scotland & to teach it for 10 Guineas each Bleachfield, this will be 10 times worse than the patent, as it will lay the matter entirely

[1] See Clow, *op. cit.*, especially chaps. II, IV, VI, and IX.
[2] See below, chap. X and the various works there cited.
[3] In France there was apparently much more conservative opposition to the new 'chemical' bleaching.
[4] Blagden to Banks, 12 Sept. 1787, B.M. Add. MS. 33272. 42–3; Dawson, *The Banks Letters*, p. 71.
[5] J. Watt to Annie Watt, 16 Sept. 1787, M. II, B.R.L.

open as far as they know themselves, which luckily is no farther than the
preparing it by Spirit of Salt [hydrochloric acid] . . . the expence of which
cannot be afforded, neither Parker or him [Middleton] are Chemists & know
no notion of making any other than Glass vessels for doing it in so that I expect
your father will have much the heels of them, unluckily Mr. Middleton is going
to Paris & may obtain further knowledge from Berthollet but I believe he
[Berthollet] does not practice the making it with [sulphuric acid] himself & I
believe has no notions of other than glass vessels. This mischief I feared but
cannot prevent as I have no right to dictate to any of the parties in a matter
which was given to me freely & I would not for 10 times the worth of it
appear greedy or too interested. I think however your father ought by all means
to go on with it as he will have some advantages which they will not have &
would otherwise be left behind by those who practise it.[1]

Watt was now relying on the advantages which he and McGrigor
had acquired by their lead in development of the new process—firstly
the use of sulphuric acid and salt in place of hydrochloric acid in
making the bleach liquor, secondly the use of other than glass vessels,
and thirdly the practical experience gained in bleaching operations.
The apparatus which Watt sent to McGrigor was evidently of glass,
for his wife informed him that 'All the Glasses are whole which you
sent & he [McGrigor] is going to get a set of glass tubes made as soon
as possible'.[2] But in a subsequent letter, in which she reported on their
first trials with the new apparatus at her father's bleachworks, she told
of unfortunate experiences with cracked glass retorts.[3] Replacements
made in Glasgow proved very unsatisfactory, 'as none of the stoppers
were air tight by which means we lost a great quantity of gas & almost
suffocated ourselves. I wish you would but hurry to try the Lead retort.'[4]
It is not surprising that Watt, who was so closely associated with Roebuck and Garbett, pioneers in the development of the lead-chamber process for production of sulphuric acid at Birmingham and Prestonpans,[5]
should soon have explored the possibility of using a lead retort, instead
of a glass one, in which to generate the chlorine gas from sulphuric
acid, salt, and manganese. At the same time, although McGrigor had
not been very successful with a wooden receiver, in which to dissolve
the gas, Watt took up the idea again.[6]

Meanwhile, his wife and father-in-law had begun their large-scale

---

[1] In another letter to his wife (28 Sept. 1787, *ibid.*) he remarked that 'Argands
interfering in Berthollets affair was both foolish and unlucky but cannot be
helped, after all they [Parker and Middleton] may tell what they know but few
will succeed in it'.   [2] Annie Watt to J. Watt, 12 Sept. 1787, *ibid.*
[3] Ditto, 5 Oct. 1787, *ibid.*   [4] Ditto, 14 Oct. 1787, *ibid.*
[5] Clow, *op. cit.*, chap. VI; H. W. Dickinson, 'History of Vitriol Making in
England', *Newcomen Soc. Trans.*, Vol. XVIII (1937–8).
[6] J. Watt to Annie Watt, 6 Oct. 1787, M. II, B.R.L. See above, p. 267.

experiments with the glass apparatus sent from Birmingham. They were 'busy preparing to begin the process' towards the end of September 1787,[1] at the Clover bleachfield, and had successfully made 'tryals with the spirit of salt' [hydrochloric acid], as Watt had directed,[2] as well as with sulphuric acid and salt, when these had to be suspended owing to the accidents with the glass retorts.[3] McGrigor, however, began 'trying the whiting quality of the Liquor' they had made, on 5 October,[4] and on the 8th Annie Watt reported that he had 'succeeded wonderfully. I think it is much whiter than you ever made the Birmingham cloth but the process you sent him last as you had it from Mr. Ber[tholle]t that was to whiten in 6 hours has done the worst.'[5] He had so far experimented 'only on bits but he means as soon as we can make enough of the Liquor to try it on a larger scale'. McGrigor himself wrote to Watt on 14 October, giving more details:

The Apparatus you sent me came safe and with them and the very distinct drawings & instructions with Annies Assistance I have been able to make about 40 Bottles of the Whitning liquor with which I have made sundry experiments in the whitning small pieces of Linen Cambric yarn & flax. The Cambric with two operations was a tollerable good white but the third made the Cloth quite tender & thick fine Linen appears to be more injured by the Liquor than coarse or thin Linen. The flax with four opperations came to a good white. I had some of it Heckled by which the greatest part of it was reduced into fine white Tow very similar to Cotton wool & may be readily spun into yarn by a water machine.

Samples of the flax and yarn were sent to Watt.[6] The latter was 'not just as strong as I could wish but I hope to mend that in further trials'.

   McGrigor had been impelled to press forward with his trials for the following reason:[7]

Last Wednesday on going to town I heard that the person you had mentioned in your Letter to Annie had arrived from London[8] at Glasgow and had told

---

[1] Annie Watt to J. Watt, 22 Sept. 1787, *ibid.*
[2] See above, p. 275.   [3] Annie Watt to J. Watt, 5 Oct. 1787, *ibid.*
[4] Ditto, 8 Oct. 1787, *ibid.* (The letter is dated September, but this must have been a slip.)
[5] This was probably Berthollet's method of bleaching with a dilute solution of chlorine in water, without addition of potash. See above, pp. 270–1.
[6] McGrigor sent some of the flax, while Annie Watt sent 'a Hank of Yarn which my Father bleached in 48 hours from the time it was taken from the reel'. Annie Watt to J. Watt, 14 Oct. 1787, M. II, B.R.L.
[7] J. McGrigor to J. Watt, 14 Oct. 1787, Dol. Also Annie Watt to J. Watt, 14 Oct. 1787, M. II, B.R.L.
[8] Evidently the emissary of Parker and Middleton. See above, p. 277. In a later letter Annie Watt named him as Thomas Fielder, 'Ridder [Rider, i.e. travel-ler] to Messrs. Joseph Flight & Comy. in London'. Annie Watt to J. Watt, 23

sundry people that he had made a discovery in Bleaching by which he could Bleach in the one half the time & at the half of the expence of the common method but nobody could tell me what his proposals were . . .

Having some of the yarn & a few pieces of the Linen & the flax in my pocket I thought it a proper time to show them to a few of my friends & told them that I had got the Secret from you.

I then applyed to two Houses here in the manufacture of Muslin . . . if they would mark a small piece of Muslin so that they could know it again when white [and] each of them gave me a shawl or napkin & I got a pair of Cotton Stockings from Johnston Bannatine & Co. [and] these I brought [back] with me and put them to work on friday last & finished them yesterday & returned them to the owners this morning by Mr. Hamilton . . .

McGrigor also revealed that, as well as the parties in London, others in Scotland appeared to have acquired knowledge of the new process, and had, in fact, preceded him in demonstrating their achievement to Glasgow manufacturers.

There is a person settled about four miles from Edinburgh in the manufacture of Great Salt who I am lately told was in town some time last spring & give out that he had made a discovery in Bleaching & one night in Company with a number of the Manufacturers here he got several little bits of cambric from some of the people present went into another room and in a very short time returned to the Company with the Cambric brought to some degree of whiteness but far from being a good white. The same person bought at that time several pieces of Brown Cambric which he said he was to whiten but the person who sold him part of the Cambrics & told me the Storrie says he has never heard anything of him since.[1]

McGrigor could 'now make the liquor in sufficient quantity for experiments in the small line',[2] and would 'endeavour to find out the best method of using the liquor so as to preserve the goods from being tendered by it without having any regard to the time it may take to whiten them'. He reported that in all the experiments he had made 'the Cotton goods appears to be much less hurt by the liquor than the

Oct. 1787, M. II, B.R.L. The London firm referred to was probably that of the Joseph Flight who appears in London directories from 1782 at 2, Freeman's Court, Cornhill, variously described as 'merchant' and 'linen draper', and then in Lowndes's directory of 1791 as 'iron liquor manufacturer'. There are, however, others of the same name in the directories, one a tallow-chandler, and another a Worcester china manufacturer.

[1] There is no clue as to the identity of this person. The Prestonpans Vitriol Company was situated only a few miles east of Edinburgh, but does not appear to have acquired knowledge of the new bleaching process so early. See below, pp. 293–5.

[2] Annie Watt wrote that 'when the receivers are mixed the[y] produce about 17 quart bottles of the liquor so that today we have made 28 Quarts'. Annie Watt to J. Watt, 17 Oct. 1787, M. II, B.R.L.

Linen'. He would be glad to hear from Watt 'so soon as you have made trial of making the liquor upon a large Scale', i.e. using the lead retort.

Watt replied that he had 'not been able to try a single experiment' recently, owing to pressure of business, but urged his wife and Mc-Grigor to persevere with Berthollet's method:[1] '. . . you may assure him [McGrigor] that the process with the weak liquor & boiling in Alkali every hour or two succeeded extremely well with me, and with less original liquor than in the other way, but no alkali must be put in the weak liquor, & the Cloth well washed from the alkali after each boiling'. His wife reported on 23 October that, despite trouble with ill-fitting retort stoppers, they had 'contrived to make a good number of Gallons which performs the whole Act of bleaching in 24 or thirty hours'.[2] Since the last letter, her father had bleached '3 shawls 12 hanks of yarn & some other things', which were brought to 'a much finer white', yet kept stronger than previously: 'he has found it [the liquor] whitens perfectly well in open vessels & even in wooden ones'. Annie Watt therefore urged her 'dear Jamie' to come up to Glasgow to see 'if it can be made in the great and at a proper expence'.[3]

Watt consequently began to make preparations for such a trial. In November he sent McGrigor drawings and directions for a lead retort or still, and on the 27th wrote to say that he would set out for Glasgow early the following week, asking him to 'get as many of the utensils for the operation ready as possible'.[4] A month later he wrote to Matthew Boulton:[5]

The apparatus for trying the making the [dephlogisticated marine acid] in great was not compleated till yesterday, I mean to try it tomorrow though I am some what afraid to attack so fierce & strong a beast, there is no bearing the fumes of it—After all it does not appear that it will prove a *cheap* way of bleaching & it weakens the Goods rather more than could be wished—however it may do some good in the way of expedition.

We do not know how these trials went, but from letters written after Watt's return to Birmingham it is clear that difficulties were still being experienced, as he informed Gilbert Hamilton:[6]

The stink [gas] of [dephlogisticated marine acid] whitens, but I think neither better nor faster than the liquor. I have tryed some expts. but I do not yet find

---

[1] J. Watt to Annie Watt, 16 Oct. 1787, *ibid.*
[2] Annie Watt to J. Watt, 23 Oct. 1787, *ibid.*
[3] They had heard no more about the man from London.
[4] J. Watt to Annie Watt, 6, 16, and 27 Nov. 1787, *ibid.*
[5] J. Watt to M. Boulton, 30 Dec. 1787, James Watt Box IV, A.O.L.B.
[6] J. Watt to G. Hamilton, 31 Jan. 1788, Boulton & Watt Letter Book (Office), Mar. 1786–Feb. 1788, B.R.L.

T

any improvement on the method of using it.[1] [Potash] wholy takes away its virtue. I think it may be an improvement to steep or wash the cloth in weak [sulphuric acid] each time before [it is] put into the liquor, that the [alkali] which may remain in the Cloth may be entirely conquered at the least expence . . .

Gilbert Hamilton replied, however, that McGrigor, having now got the still and receivers working 'without any trouble or smell', was 'in much higher spirits about it than ever',[2] and McGrigor himself sent a detailed account of his procedure.[3] He had strengthened the liquor which Watt had left in the receivers by giving two more half charges to the still. Firstly, he mixed and bottled 10½ lb. of water with 15 lb. of sulphuric acid; he then mixed 5 lb. 3 oz. of manganese with 10 lb. 6 oz. of salt and spread it very thinly over the bottom of the still.[4] He then fixed down the head of the still and poured a pint of water down the feed-pipe, followed by the bottle of dilute acid and another pint of water, before finally putting in the stopper, 'all of which was done without the least appearance of any fumes whatever nor did the opperation begin to work for two or three minutes. After that it wrought very well from 10 to 15 minutes at which time it began to work slowly when I applyed the fire & the opperation went on very well for an hour & a half.' The following morning he had the still cleaned out and repeated the above procedure, using the same quantity of materials, except that he put in a quart (instead of a pint) of water before and after adding the diluted acid. Again, the operation went on very successfully, for an hour and thirty-eight minutes, though towards the end 'the lead conductor that joins the wooden pipe in the large receiver became so warm that I was obliged from time to time to steep clothes in cold water & lay them upon the knee [bend] of the lead pipe in order to prevent the heat that was there from distroying the Puttey & Bladders that cover the Joining'.

Lead and wooden pipes had evidently replaced the former glass ones, joining the lead retort to the receivers, of which there were two, a large and a small one.[5] There is no evidence as to the material of which

---

[1] McGrigor followed Watt in making experiments with the gas, and thought it might prove useful: 'The Stink though it may not do as well as the liquor itself yet may be made to be of use in bringing forward some part of the preparation.' G. Hamilton to J. Watt, 12 Feb. 1788, Dol.

[2] G. Hamilton to J. Watt, 5 Feb. 1788, Dol.

[3] J. McGrigor to J. Watt, 11 Feb. 1788, Dol.

[4] For comparison of these quantities with the proportions later recommended by Berthollet, see below, pp. 296, n. 4, and 305.

[5] The small one was probably intermediate between the still and the big receiver, for absorption of any muriatic (hydrochloric) acid vapour, which distilled over. See above, p. 263.

these receivers were made, but it was probably wood.[1] In order to improve absorption of the gas, Watt had sent McGrigor a drawing of an 'apparatus for turning the liquor in the big receiver', i.e. an agitator. Thus he was grappling with the problems of early chemical engineering, as he had previously coped with those of mechanical engineering in the development of his steam engine. Thanks to his knowledge, skill, and perseverance, large-scale bleaching could now be commenced, as McGrigor was able to inform him in this same letter.

As I hope the liquor in the large receiver is now near fit for use I have provided 1500 yards of light linen which I propose to whiten in the great in the course of next week or the week after . . . I know the quantity of goods mentioned will require more liquor than the Contents of the receiver but I mean to make more during the time the whitening Process is going on. What I now wish very much is that we could fall upon some method to assertain the exact strength of the liquor after it is made.

McGrigor's desire for such accuracy was probably increased by his experience in applying the liquor to 'clearing' printed goods.[2] Within the past few days, 'to oblige a principal Manufacturing House here', he had

whitened with liquor 24 patterns of fine Muslin many of them Stript & Spotted with Red & Gold some of the Reds stood & others went & some of the Gold not so bright as it usualy appears in goods of that kind when they come to market but that can[t] be helped [Even so] upon the whole the owners were very well pleased with the goods & have sent them to London as patterns for the ensuing years Trade.[3]

To achieve perfection in such operations, it was especially important to know the precise strength of the bleaching liquor, so that it would 'clear' the whites, yet not harm the printed colours. Watt, therefore, at once turned his thoughts to discovering some suitable chemical test and soon achieved success. By early April 1788 we find McGrigor mentioning that he 'had made use of our liquid Blue as a test in trying the strength of the liquor & as I had no Cochenelle [cochineal] in the Country [i.e. at Clover bleachfield] I continued to use it. But means to make up some with the Cochenelle . . .'[4] The 'liquid Blue' test was almost certainly one using a dilute solution of indigo in sulphuric acid, which was described by Berthollet and was to become common

---

[1] See above, p. 278. According to Baines, *op. cit.*, p. 249, Watt used 'air-tight vessels of wood lined with pitch'.   [2] See above, p. 269.
[3] He had also bleached some hanks of yarn for another firm, who were 'very pleased with them', and he hoped to improve on this performance as a result of further experiments.   [4] J. McGregor to J. Watt, 5 April 1788, Dol.

bleaching practice.[1] One part of indigo was dissolved in eight parts of concentrated sulphuric acid and then diluted with a thousand parts of water. A measure of this test solution was put into a graduated glass tube, and the bleaching liquor was gradually added to it, until the blue colour of the indigo was completely destroyed. A comparative test was then made with liquor of known strength, previously determined by direct experiment on cloth, thus providing a standard. Watt discovered, however, that this test did not work accurately with bleach liquor made by absorption in an alkaline solution (i.e. with 'oxy-muriates', or hypochlorites), but found that a solution of cochineal answered the purpose perfectly.[2] Thus he was able to achieve greater precision in bleaching operations, by scientific measurement of strengths or concentrations, as well as quantities, of materials employed.[3]

At this point, however, when McGrigor was about to commence large-scale bleaching, the Bourboulon affair blew up again, more seriously and with wider repercussions, which revealed the extent to which knowledge of the new process was now spreading. General concern was aroused by a petition to the House of Commons from Anthony Bourboulon de Boneuil on behalf of himself and his partner, Matthew Vallet,[4] in which they claimed that they had invented and perfected 'a liquor made from native Salt', whereby textiles could be 'thoroughly whitened in a very short Space of Time, and without any Risque', thus saving the very great expense of time, land, and money, and the risk of damage or theft of goods, involved by long exposure in bleachfields. They also claimed to have invented and perfected 'a Fossil Alkali from native Salt', as good as 'the best Sal Sodie extracted from Barilla, an Article which has hitherto been necessarily imported from Foreign Countries'. The petitioners had certificates from people who had witnessed the bleaching effects of their liquor and from 'an Eminent Chymist' in London who had analysed it. They claimed that these processes were 'entirely new in this Kingdom', and in view of this and the great expense in establishing their manufactory here,

---

[1] Berthollet attributed the discovery of this test to Décroisille: *Ann. Chim.*, Vol. II (1789), p. 177. He was followed by later writers, e.g. O'Reilly, *op. cit.*, pp. 100–1; Rees, *loc. cit.*; Parkes, *op. cit.*, Vol. IV, pp. 143–5. It is quite possible, however, that Watt discovered it independently. See C. L. Berthollet to J. Watt, 28 July 1788, Dol. (inserted in J. Watt Commonplace Book, 1782).

[2] He informed Berthollet of this discovery: *Ann. Chim.*, Vol. II (1789), pp. 177 and 188.

[3] He was probably familiar with the hydrometer, which was coming increasingly into industrial use, but in addition to measuring specific gravity, it was necessary to have some test of actual bleaching power (Parkes, *op. cit.*, Vol. IV, pp. 165–6).

[4] *Journals of the House of Commons*, Vol. 43 (1787–8), p. 202: 8 Feb. 1788.

they sought to 'be secured in the full and exclusive Possession of their Invention for a longer Term of Years than are usually granted by Royal Patents',[1] together with a remission of the duties paid on the raw materials used.

This petition[2] led immediately to the summons of a public meeting in Manchester, on Tuesday, 19 February 1788,[3] attended by many of the leading cotton manufacturers and bleachers of the area, together with a number of doctors and chemists interested in industrial-chemical matters, and it was decided to oppose the petition of Bourboulon de Boneuil & Co., since it would tend

to create a Monopoly injurious to the Interests of the Manufacturers of this Kingdom . . . we are informed that various Persons within these Realms are already in Possession of Modes of Bleaching equal to that said to be possessed by the Petitioners; and also that other Persons have already obtained Patents for the Preparation of Fossil Alkali, from Native Salt, and that several Processes for this Purpose are well known.

A committee was appointed[4] to carry out the meeting's resolutions, by securing the support of the county M.P.s, by publicity in the Manchester newspapers and in the *Morning Chronicle* and *Evening Post*,[5] and by such other measures as were thought expedient. It was pointed out that 'several Members of this Committee are acquainted with Persons already engaged in Processes of Bleaching, which will greatly abridge the common Methods, and answer all the Purposes assumed by the Petitioners'. The view was also expressed that, but for the duties on salt, the various existing processes for procuring fossil alkali from native salt would long ago have been successfully developed.

James Watt, meanwhile, having seen the announcement of this meeting, had written to friends in Manchester to inform them that he had no doubt that Bourboulon's process was the very same as one of

[1] According to Parkes, *op. cit.*, Vol. IV, pp. 61–2, they applied for an exclusive right to the invention for a term of 28 years.

[2] Which was ordered to lie on the table of the House, pending investigation.

[3] *Manchester Mercury*, 19 and 26 Feb. 1788.

[4] The membership included the following manufacturers: Peter and Richard Ainsworth, Joseph Baker, Thomas Cooper, George Lomas, John Nash, Michael Norton, John Partington, Robert and Lawrence Peel, Samuel Gratrix, James [Jonathan?] Hardman, John Hargreaves, John Horridge, Thomas Kershaw, Thomas Phillips, Thomas Ridgway, Charles Taylor, John [James?] Varley, and John Wilson. Several medical and chemical men were also included: Dr. Percival, Dr. Eason, Dr. Ferriar, Thomas Henry, and John Collison. John Wilson was Chairman. Many of these men, as we shall see, were among the pioneers of the new process in Lancashire.

[5] Reports of the meeting also appeared in other newspapers, e.g. in Leeds and Glasgow, where textile interests would be similarly opposed to the threatened monopoly.

which he had acquired knowledge and had 'used for more than a 12 months' and which he had communicated to James McGrigor, 'who is now whitening 1500 yds of linen by the process'.[1] Since it was their intention, 'as soon as we have well constituted the utility &c. to teach the same to as many as will give a reasonable compensation for the trouble & expence we have been at we cannot but deprecate any exclusive privilege being granted or national reward to any other than the inventor'. In addition, therefore, to supporting the Manchester meeting, Watt asked Matthew Boulton, who was then in London, 'to represent to Mr. Pitt what I have said, with whatever else may occur to you',[2] Boulton having originally witnessed Berthollet's experiments together with Watt.

Watt, however, adopted a very cautious attitude towards the Manchester resolutions. As he wrote to Matthew Boulton, 'I don't like the sentiment of opposing a Monopoly of an invention, in general, therefore shall scarcely join them'.[3] Obviously the Parliamentary extension of his own steam-engine patent precluded him from supporting such a resolution. He could not attend Parliament himself, 'nor will put myself to no expence in petitioning ... & do not wish you to labour over much, as it appears in no capital light'. Moreover, he did not want too much publicity to be given to the new bleaching process, 'as it may raise up more pretenders';[4] the original inventor should be referred to as 'an Eminent Chymist at Paris', without mentioning Berthollet's name, and Watt himself wished to keep in the background.[5] 'All I wish to be done is that the Minister & a few sensible members were apprized of the truth after which if parl[iamen]t will reward this man for being an impudent rascal let them.'[6] He suggested that the Glasgow member, Islay Campbell, the Lord Advocate, should be spoken to: 'tell him his neighbour Mr. McGrigor is engaged in the process by my means'.

Matthew Boulton was glad to assist. 'I shall never think it any trouble to frustrate the Machinations of Theves & Pirates, nor to assist in sup-

---

[1] J. Watt to M. Boulton, 25 Feb. 1788, James Watt Box IV, A.O.L.B. Watt's original letter, to Dr. Percival, dated 23 Feb. 1788, was quoted by William Henry in the *Annals of Philosophy*, Vol. VI (July–Dec. 1815), p. 423. See above, p. 256. Watt had asked that these facts should be reported to the Manchester meeting without mentioning his name, saying that the original inventor was 'an eminent chemist and philosopher at Paris'.

[2] *Ibid.*   [3] J. Watt to M. Boulton, 27 Feb. 1788, *ibid.*

[4] Annie Watt to M. Boulton, 1 March 1788, *ibid.*

[5] Watt's lukewarmness was also occasioned by the fact that though the new bleaching method was 'more speedy than the common process', he was 'afraid it will cost more', and because knowledge of it was now getting into so many more hands. *Ibid.*   [6] J. Watt to M. Boulton, 27 Feb. 1788, *ibid.*

porting the just Claims of ingenious Men.'[1] He at once wrote to Berthollet, informing him of Bourboulon de Boneuil & Co.'s petition and asking 'if it is with your consent or whether you wish me to oppose it'.[2] Berthollet, in his reply,[3] indignantly repudiated Bourboulon and his associates, referring to their audacity in pretending to have developed a new bleaching liquor under the name of *lessive de javelle*, which they represented as an improvement on his own process.[4] This, as we have previously noted, was formed by receiving the gas into a potash solution—to form 'oxygenated muriate of potash'—with which both Berthollet and Watt had previously experimented. Berthollet admitted that by this means a much greater quantity of the gas could be absorbed, and that much smaller receiving vessels were required, but at least half the gas was lost by its combination with the alkali, so that 'cette liqueur ne produit pas à prix égal la moitié de l'effet de la liqueur qui est préparée par mon procédé', i.e. by absorption in water. He therefore asked Boulton to oppose Bourboulon's petition: 'il ne mérite d'être traité que comme un ravisseur mal adroit'.

Others were also joining in the opposition. McGrigor was aided by Gilbert Hamilton, who got the Glasgow Chamber of Commerce to express its views to the Lord Advocate.[5] Richard Ainsworth, of Bolton, had also written to McGrigor, sending him the Manchester resolutions, and McGrigor had replied with a general account of his and Watt's achievements in the new bleaching process and an assurance of his opposition to Bourboulon's petition.[6] Other members of the Manchester committee had got into touch with Watt. Dr. Percival, to whom he had originally written, was ill when his letter arrived, so Thomas Henry replied.[7] Henry had feared 'that, if no Chemical person should attend the Meeting, the Crofters &c. might be tempted to countenance the application of the Petitioners', so he went to inform them 'that several persons in these Kingdoms were already in possession of Processes for the preparation of Fossil Alkali from [salt]', and also that it was very likely that a new and much quicker method of bleaching would soon be developed. Henry was thoroughly familiar with foreign (especially French) scientific publications,[8] and knew of the experiments

---

[1] M. Boulton to J. Watt, 28 Feb. 1788, Box 20, B.R.L.

[2] M. Boulton to C. L. Berthollet, ? March 1788, Matthew Boulton Letter Book, 1780-9, A.O.L.B.

[3] C. L. Berthollet to M. Boulton, 11 March 1788, Dol.

[4] See above, p. 275, for Berthollet's similar denunciation in the *Annales de Chimie*, Vol. II (1789).

[5] G. Hamilton to J. Watt, 28 Feb. 1788, Dol.

[6] *Ibid.* Ainsworth was an acquaintance of McGrigor, whom he had visited during the previous year.   [7] T. Henry to J. Watt, 25 Feb. 1788, Dol.

[8] See above, pp. 232, 236, 243, 244, and 259.

with the new 'chemical' mode of bleaching, but he admitted to Watt that, 'tho' I had seen Mr. Berthollet's experiments with [dephlogisticated marine acid] I did not know that it had been applied *in the large way* to the purposes of whitening Cotton &c.'[1] Henry himself was making experiments, for a few days later Dr. Percival informed Watt 'Mr. Henry has lately prepared some of the Liquor you mention, & has shewn me two or three samples of cloth, bleached by it without injury to the dye, in great perfection, & in a very short space of time'.[2] It seems certain that, if he had achieved such success on printed cotton goods, he must have been experimenting for some time, though only in a small way.

Henry had discovered, however, on attending the meeting, that he was by no means the only 'Chemical person' in Manchester who was interested in the new process.[3]

I found there besides Mr. Charles Taylor & Mr. [Thomas] Cooper, both of whom are good chemists, a Mr. [John] Collison who is lately taken into partnership in the House of Livesey, Hargreaves & Co. who appeared well informed on Chemical subjects; and it was declared that not only that house was in possession of a mode of bleaching equal to that of de Boneuil, but it was believed that you and some others also knew it.

Collison, a London chemist, had already, in 1782, taken out a patent for manufacturing soda from common salt, using sulphuric acid,[4] and had now evidently turned to development of chlorine bleaching. How far he had progressed, in collaboration with Livesey, Hargreaves & Co., one of the leading cotton-manufacturing firms in Manchester, we do not know. As a member of the Manchester committee, he had written to Watt, who was not at all pleased to find that 'Collison has somehow learnt something about the process probably from M[iddleto]n & P[arke]r. He is but a bungler & did not use J. Fry well, therefore I mean to have nothing to do with him but shall thank him for his communication.'[5]

Livesey, Hargreaves & Co. failed soon afterwards,[6] so it is probable that Collison's schemes came to grief, but he provides another interesting link between chemistry and industry in this period. Watt's guess

---

[1] See above, p. 259.   [2] T. Percival to J. Watt, 6 March, 1788, Dol.
[3] T. Henry to J. Watt, 25 Feb. 1788, Dol.
[4] See above, pp. 133 and 165, and below, pp. 361–2, 364–6.
[5] J. Watt to M. Boulton, 27 Feb. 1788, James Watt Box IV, A.O.L.B. The partnership between Collison, Fry, and Jones in the Battersea alkali works had broken up by the beginning of 1784. J. Fry to J. Watt, 31st, 1st Month 1784, B.R.L.
[6] L. H. Grindon, *Manchester Banks and Bankers* (Manchester, 1878), pp. 45 and 106–7.

as to his source of information on bleaching may have been correct, but he may well have read Berthollet's publications. Another possible source was Dr. Richard Kirwan, the eminent Irish chemist, who, according to Rees, obtained information from Scheele 'previous to any publication by Mr. Berthollet', and passed it on to Charles Taylor, of Manchester,[1] and possibly to others. Confirmation of this statement is provided by the evidence of Taylor himself, at the Tennant trial in 1802:[2] 'To the best of my remembrance, in conversation in the year 1787, with a gentleman in London, he told me Mr. *Schele* had discovered a method of whitening the leaves of vegetables, and he thought that hint might be usefully employed in the manufacture [of textiles].' Taylor, as we have seen, had links with scientists both in this country and overseas, eventually (in 1799) becoming secretary to the Society for the Encouragement of Arts, Commerce, and Manufactures.[3] In the 1770s and 1780s he was one of the outstanding industrial figures in Manchester, a pioneer in the application of scientific methods to dyeing and calico-printing,[4] who naturally seized upon the new bleaching process, together with other scientifically-minded manufacturers in the area. Their success was evidenced by the following publication in the Manchester press:[5]

As a proof of the improvements which even the present advanced State of the Manufactures of this Town will admit, a Piece (28 Yards) of Grey Callicoe was brought to Mr. Joseph Baker of Barton [near Eccles, Manchester], on the Afternoon of one Day, was bleached perfectly the same Evening, was printed on the next Day, and exposed for public Inspection on the Third.[6] This was

[1] See above, pp. 259–60. Kirwan, resident in Newman Street, London, from 1777 to 1787, was an expert linguist as well as chemist and had contacts with many of the *savants* of Europe. See above, pp. 122–3.

[2] *Proceedings in a Suit of Chancery and the Trial of a Cause instituted in the Court of King's Bench, by Messrs. Tennant, Mackintosh, Knox, Cooper & Dunlop, in the name of Mr. Charles Tennant, of Darnly, near Glasgow, Against Messrs. James Slater, James Varley and Joseph Slater, near Bolton, Lancashire, For infringing on a Patent obtained by Mr. Tennant, for substituting Calcareous Earths, instead of Alkalis, in preparing the Oxygenated Muriatic Acid used in Bleaching . . . Before Lord Ellenborough, Chief Justice of England, and a Special Jury, at Guild Hall, on Thursday, December 23, 1802* (Manchester, 1803), p. 114. Henceforward this source will be referred to as *Tennant Proceedings* (1803).

[3] See above, p. 97.   [4] See below, pp. 344–7.

[5] *Manchester Mercury*, 18 March 1788. See above, pp. 260–1, for references in Rees and Parkes, *op. cit.* William Henry, writing in *Annals of Philosophy*, Vol. VI (1815), p. 423, was incorrect in stating that this demonstration was made at the meeting on 19 Feb. 1788.

[6] The German traveller, P. A. Nemnich, *Beschreibung einer im Sommer 1799 von Hamburg nach und durch England geschehenen Reise* (1800), p. 279, stated (English translation): 'In the year 1788 Thomas Cooper, [Joseph] Baker and Charles Taylor brought into use the new bleaching method of Berthollet, and brought it to such perfection that a piece of calico, just as it came from the weaver, was

performed in the presence of Mr. C. Taylor, and T. Cooper, Esq. [who, together with Baker, were members of the recently appointed committee] . . . Half of the Piece thus printed was exhibited for the inspection of the Committee, together with a Specimen perfectly bleached i 1 a few Hours from the grey State, by Mr. Thomas Henry, and met with the full approbation of the Judges then present.[1]

This was soon followed by another advertisement,[2] publicising the achievements of Thomas Henry, junior, who was collaborating with his father:

As an additional Proof of the Efficacy of the new Chemical Mode of Bleaching, we can inform the Public, that Mr. Thomas Henry, jun. has, in the Course of the present Week, by means of a Liquor, prepared by him, Bleached several whole Pieces of Callicoes, Honeycombs, and Stockinets, so as to be fully ready for the different Processes of Printing or Dyeing. The whitening Operation was performed on these Wares, in various Space[s] of Time, from three Hours to five Minutes. The Experiments were made in the Presence of Messrs. Greatrex, Dyers and Printers: the Pieces are not in the least impaired in their Texture; and have met with the Approbation of the best Judges.

This advertisement, which also appeared in the Glasgow newspapers, caused James McGrigor to remark that 'if all be true thats thereon said Mr. Henry knows more than we do yet'.[3] As we shall see, however, there is no doubt that at this time the Henrys were behind Watt and McGrigor in their development of the new process.

It also appears that Baker, Cooper, and Taylor were less advanced, although their success in bleaching a whole piece of calico indicates that they must have been experimenting for some time. Indeed, in the Tennant trial of 1802 they were said to have been the first to introduce 'the short mode of bleaching'.[4] They certainly formed an impressive combination of chemical-textile interests. Taylor, as we have noticed, was one of the leading dyers and calico-printers in Manchester. Joseph Baker had recently established a large vitriol works near Eccles, producing this acid by the lead-chamber process. His uncle had been a partner in the firm of Kingscote and Walker, London druggists, who

---

printed with several colours, finished and ready for sale within the space of twenty-four hours . . .' This is an exaggeration, but indicates the impression which their achievement made.
    [1] William Henry claimed that, although his father had produced, not a half piece, but only half a yard, the latter 'was declared to be superior in whiteness to the larger one'. See above, p. 261.   [2] *Manchester Mercury*, 1 April 1788.
    [3] J. McGrigor to J. Watt, 5 April 1788, Dol. He wrongly supposed that young Henry had got the secret from Bourboulon de Boneuil.
    [4] Evidence by John Horridge, *Tennant Proceedings* (1803), p. 156.

were apparently the first to establish this process in the London area, erecting a large works at Battersea in 1772.[1] This firm, however, went bankrupt after a few years, and then Joseph Baker, so Parkes relates in his *Chemical Essays*,[2]

connected himself with two other gentlemen, and they, under the firm of Baker, Walker, and Singleton, some time about the year 1783 established a sulphuric acid manufactory of very considerable extent, at Pitsworth-Moor near Eccles in the county of Lancaster. This manufactory, which was the first establishment of the kind in that county, consisted of four [lead] vessels each twelve feet square, and four others which were forty-five feet by ten each.

Dr. and Mrs. Clow have briefly noted the establishment of this vitriol works,[3] but they did not realize that this Baker was the same 'Mr. Baker' to whom they vaguely refer later on as participating in chlorine-bleaching with Cooper and Taylor.[4] Baker must undoubtedly have been attracted to the Lancashire area in the early 1780s by the rapid expansion of the cotton industry following the spinning inventions of Hargreaves, Arkwright, and Crompton, and the consequently rising demand for vitriol 'sours'. Then, in 1788, as Parkes emphasized,[5] 'When the new method of bleaching by oxymuriatic acid was introduced, the sale of sulphuric acid was much increased . . . From this time the demand became enormous, and almost unlimited in extent . . .' Many more and bigger vitriol works were established, and shrewd manufacturers such as Baker—like the Javelle firm in France—soon appreciated the possibilities of integrating the production of this acid with that of the new bleach liquor.

Thomas Cooper, who was associated with Baker and Taylor in their development of chlorine bleaching, was—like many other early industrial chemists—a doctor by training, who had become one of the leading Manchester bleachers. We have already noted his combination of scientific and industrial interests, including his friendship with James Watt junior.[6] These three men made a formidable industrial-chemical combination, linking chemistry closely with the cotton industry.

Their efforts, allied with those of Henry, Collison, and Manchester

---

[1] Parkes, *op. cit.*, pp. 403–6.   [2] *Ibid.*, pp. 406–7.

[3] *Op. cit.*, p. 144. Parkes was obviously their source, though they do not give any reference.

[4] *Ibid.*, p. 189. They also state that 'it was probably through young Watt that Cooper got his knowledge of chlorine bleaching' (referring to Dickinson, *James Watt*, p. 151); but Watt junior did not arrive in Manchester until October 1788 (see above, p. 96, and below, p. 345), and, as we have seen, Baker, Cooper, and Taylor had acquired knowledge of chlorine bleaching considerably earlier and from another source.   [5] *Op. cit.*, p. 408.

[6] See above, pp. 92, 96–7. See also below, pp. 303–8, 347–8.

manufacturers generally, together with the opposition organized in Birmingham, Glasgow, and London, led to the defeat of Bourboulon & Co.'s petition in Parliament.[1] Messrs. Milne, of Aberdeen, had also helped to organize the Parliamentary opposition, revealing their own earlier success in chlorine bleaching.[2]

James McGrigor, meanwhile, continuing his experiments in large-scale bleaching, was now able to inform Watt of his success with light linens:[3]

> . . . I cut up a parcel of coarse long lawns into 42 pieces containing 1500 yds these I sent out [to the Clover bleachfield] and followed them on the 18th February & I have now the pleasure to inform you that after meeting with many difficulties & disapointments I at last Accomplished the business very much to my Satisfaction & to the Admiration of every person who have seen the goods.

His difficulties he attributed to the wooden receiving and bleaching vessels being new and unseasoned,[4] and also to the fact that 'in order to preserve the goods I made both my Lyes & liquor too weak at the beginning'. After experimenting unsuccessfully with various concentrations of lyes and bleach liquor, he eventually succeeded by boiling them in 24 lb. of white soap and then steeping them in 220 gallons of liquor diluted with 100 gallons of water. After three such operations the goods still had to be subjected to a thrice-repeated sequence of washing, boiling (bucking), washing, grassing (for 24 hours), and souring before being finally washed, blued, and starched. The new 'chemical' bleaching had saved greatly in time, capital, and labour, but it had by no means entirely eliminated the traditional operations, especially in bleaching linens. It took from 19 February to 10 March to complete these goods, though about six days had been lost in trial and error at the beginning. The following quantities of materials had been con-

---

[1] *Journals of the House of Commons*, Vol. 43 (1787–8), pp. 480–1, 19 May 1788, and p. 507, 26 May 1788. The debate was 'further adjourned till this Day Six Months', and there is no further reference to it. In a letter to Thomas Henry, dated 8 June 1788, quoted by William Henry in *Annals of Philosophy*, Vol. VI (1815), p. 423, Watt stated: 'Through the help of my friends, such parliamentary interest was made, as must in some degree have contributed to defeat the plagiary Bourbollon [*sic*], whose sole pretensions are founded on his impregnating caustic alkalies with gas, a process previously discovered and publicly mentioned by Berthollet, but laid aside as it destroys half the efficacy of the acid'. See also Brewster's *Edinburgh Encyclopaedia*, article on 'Bleaching'.

[2] Parkes, *op. cit.*, Vol. IV, p. 62; *Annals of Philosophy*, Vol. VII (1816), pp. 99–100.   [3] J. McGrigor to J. Watt, 5 April 1788, Dol.

[4] This opinion was later confirmed by Berthollet in a letter to Watt of 28 July 1788 (inserted in J. Watt Commonplace Book, 1782); alternatively Berthollet suggested that some of the gas might have escaped.

sumed: 6 lb. black soap, 111 lb. white soap, 68 lb. pearl ashes, 36 lb. vitriol, 6 lb. starch, 6 lb. small blue, and about 1,400 gallons of bleach liquor.

In the making of this liquor, McGrigor reported that 'every distillation has gone on as well as I could have wished & without being troubled with any of the Stink worth speaking of'. He had used a standard full charge for the still of 10 lb. 6 oz. manganese mixed with 20 lb. 12 oz. salt; 20 lb. 8 oz. vitriol diluted with 21 lb. water and allowed to cool before being poured down the feed-pipe; and finally 7 lb. water, half poured in before the diluted acid and half afterwards.[1] The distilling operation lasted from $2\frac{1}{4}$ to $2\frac{1}{2}$ hours; for the first half-hour or so without heat being applied. The gas so produced was passed into a potash solution. Despite Berthollet's opinion, Watt and McGrigor (like Bourboulon & Co.) eventually found this more practicable. 'I put 2 libs. of desolved fine Pearl Ashes into the receiver for every operation & the liquor in my opinion is better & freer of incumbrances than any we made in the Glass vessels.' From the rumbling that went on in the receiver, McGrigor considered that there was no need for an agitator to stir the liquor. But he thought 'another six-feet wide receiver' would be an advantage, 'as the liquor in the second with two operations would certainly be as good as that in the first with one', and this would reduce the cost of the liquor by a third. In general, however, McGrigor was thoroughly satisfied with their achievements: 'I am so much pleased with the apparatus you created at Clover & the easie way that I have now a working with them that I still think it will be some time before a better method can be falen upon to do it in the great.'

He was dubious about the reported achievements of young Thomas Henry, and evidently did not think that Henry or Bourboulon & Co. could have developed a better method; but he would wait and see. He was understandably disappointed, however, at the appearance of so many rivals: 'I am Affraid that after all the pains we have taken the secret has got into too many hands for us to make much by the discovery.' He went on to report that George McIntosh (or Macintosh), 'who is a great hunter after secrets', had recently visited him and told him

that his son [Charles Macintosh] who he says is a very good Chymist had been making experiments at home in order to find out the secret [and] was now at Manchester & that he had received a Letter from him of which he would read

---

[1] Comparing these quantities with the half-charges previously used (see above, p. 282), we can see that the proportions of manganese and salt remained the same, but the vitriol was more diluted.

me a part. I heard him & the description his son gives is not far from the first Process [using hydrochloric acid] & he expresses great fears that the Prestonpans Vitriol work might suffer greatly by the discovery & desires his Father to apply to Mr. James Gordon to write Mr. [Patrick] Downie to see if he could make the liquor from [sulphuric acid and salt].

McGrigor had 'laughed at the storie', but was obviously disturbed by the appearance of such a powerful Scottish competitor. George and Charles Macintosh were outstanding figures in the early development of the chemical industry,[1] and their connection with the famous Prestonpans Vitriol Company, pioneers in the establishment of the lead-chamber process in Scotland,[2] formed a formidable combination, like that of Baker, Cooper, and Taylor in the Manchester area. Here again, we find strong links between science and industry, especially in the achievements of Charles Macintosh, who, after studying under William Irvine at Glasgow University and then under Joseph Black at Edinburgh, 'entered upon his vocation equipped with the best training that the "pure" chemistry of his day afforded'.[3] In later years he continued to combine theoretical with practical interests, and extended his knowledge by continental tours on which he met eminent French and German scientists. His early industrial achievements, like those of his father, were mainly in textile chemicals, producing dyes and mordants. In the late 1770s George Macintosh, in partnership with James Gordon, John Glassford, and others, had built a factory at Dunchattan, near Glasgow, for manufacturing the dye cudbear, and this was followed in 1785 by the establishment of a Turkey-red works at Dalmarnock, apparently the first in Britain; sal-ammoniac manufacture was added shortly afterwards. Glassford, a wealthy Glasgow merchant, in whose counting-house young Charles Macintosh received his early commercial training, had also acquired control of the Prestonpans Vitriol Company (of which Patrick Downey, referred to in Macintosh's letter, was manager), and this company, too, found a large part of its market in the textile industry, for which it provided 'sours'.

To extend foreign sales of cudbear and vitriol, Charles Macintosh was sent to the continent in 1786 and 1788, spending a considerable time in France, Holland, and Germany. It is not unlikely that he may then have heard about the new chemical bleaching; but he had not

---

[1] G. Macintosh, *A Memoir of Charles Macintosh, F.R.S.* (Glasgow, 1847), D. W. F. Hardie, 'The Macintoshes and the Origins of the Chemical Industry', *Chemistry and Industry*, June 1952; Clow, *op. cit.*, especially pp. 191–3, 209–18; and chap. XII.

[2] Clow, *op. cit.*, pp. 135–43, 145–6, 211, 247.   [3] Hardie, *op. cit.*, p. 3.

proceeded far with his experiments at the time McGrigor met his father, and had gone to Manchester to find out more about it. The Prestonpans Vitriol Company, it was feared, might lose a valuable market if the new bleaching method eliminated vitriol sours; on the other hand, however, if the new bleaching agent could be made from vitriol and salt, the company would get a great boost. In this they soon succeeded, as McGrigor informed Watt three months later:[1] 'Soon after young Mr. McIntosh returned from Manchester the Prestonpans Company sent up Mr. Downie who returned some time ago & Mr. Gordon told me the other day that he had made as much of the whitening liquor from [sulphuric acid, manganese, and salt] as would fill 20 or 30 Vitriol Bottles . . .'

Thus competition in Scotland looked like becoming much keener, though McGrigor remarked, 'I don't hear that any of the Bleachers here who have been making experiments have succeeded even in the small way.' English bleachers, too, were meeting similar difficulties. Gordon told McGrigor 'he was informed the New mode of Bleaching was perfectly exploded at Manchester owing to the great expence of it & the goods being all damaged in the whitening'.

It was only to be expected, however, that ingenious chemists and manufacturers would eventually overcome these problems as Watt and McGrigor had done. In France, Berthollet was continuing his experiments, in collaboration with industrial bleachers, and was independently discovering and openly revealing improved methods, many of which Watt had previously developed and hoped to keep secret. In March 1788 he asked Boulton to inform Watt of an improved wooden receiver which he was now using, of a type similar to that subsequently described and illustrated in the *Annales de Chimie*,[2] consisting of a large wooden cask fitted with three or four 'soucoupes'— like inverted tubs—the gas being piped beneath the bottom one and then passing successively into those above, while its absorption in the liquor was facilitated by the rotating arms of an agitator. Watt replied to Berthollet that he, too, had substituted a cask for the original Woulfe apparatus, but he gave no details of its construction.[3] It was probably similar to Welter's, though, as we have seen, McGrigor did not think an agitator necessary. Watt, being engaged in the industrial application of the new process, was much less communicative than the non-commercially minded Berthollet.

---

[1] J. McGrigor to J. Watt, 12 July 1788, Dol.
[2] C. L. Berthollet to M. Boulton, 11 March 1788, Dol.; *Ann. Chim.*, Vol. II (1789), pp. 163-4 and 309-11, and the plate at the back of that volume. This receiver was devised by Welter, Berthollet's assistant.
[3] *Ann. Chim.*, Vol. II, p. 163.

Two months later, in a letter to Watt,[1] Berthollet stated that, some time previously, he had communicated to Dr. Charles Blagden, secretary of the Royal Society, certain improvements in his bleaching process, which he had asked to be passed on to Watt,[2] but which he now repeated with some additions. The most important was concerned with the production of the bleach liquor. Berthollet later stated, in the *Annales de Chimie*,[3] that quite early in his attempts to secure the industrial application of his process he had tried to reduce production costs by decomposing common salt (or muriate of soda) with sulphuric acid (thus producing muriatic or hydrochloric acid), for reaction with the manganese, instead of using muriatic acid ready prepared, but that his efforts were unsuccessful, either from the sulphuric acid being too concentrated or from the ingredients being used in the wrong proportions. It appears, however, that soon afterwards Décroisille, then engaged in manufacturing experiments at Rouen, succeeded in utilizing sulphuric acid, which encouraged Berthollet to resume his experiments. His young assistant, Welter, suggested diluting the acid and this proved successful.[4] He at once communicated this information to Bonjour, at Valenciennes, and also to Watt, who then informed him that he had employed the same means almost from the start.[5]

This information was, in fact, conveyed to Watt in Berthollet's last-quoted letter, of 5 May 1788, in which he stated,

je ne me sers plus d'acide muriatique pour faire la liqueur; mais je fais un mélange de muriate de Soude, d'acide sulfurique (vitriolique) et de manganèse; mais il faut ajouter de l'eau, si l'acide sulfurique est concentré, comme il l'est ordinairement dans le Commerce; autrement l'acide muriatique se dégage trôp promptement; voici les proportions que je suis: deux onces et demi de sel marin, deux onces d'acide sulfurique avec volume égal d'eau, une once de manganèse. Ce mélange distilé avec l'appareil convenable fournit huit pintes de paris d'une liqueur qui a la force convenable pour être employée. Vous voyez que le procédé devient par là beaucoup moins dispendieux qu'il n'était.

He also gave Watt some further tips on the bleaching process.[6] He told him, for example, how to get rid of a yellowish shade caused in some

[1] C. L. Berthollet to J. Watt, 5 May 1788, Dol.
[2] See below, p. 305, how this information was also passed to Manchester manufacturers, through Sir Joseph Banks, President of the Royal Society.
[3] *Op. cit.*, Vol. II, p. 162.
[4] *Ibid.*, pp. 162–3. Berthollet recommended the following proportions of ingredients: 6 oz. manganese dioxide; 16 oz. sea salt; 12 oz. conc. sulphuric acid diluted with 8 to 12 oz. water (*ibid.*, p. 165). Much, however, depended on the quality of the manganese. And the more materials were used, the more the acid should be diluted.   [5] *Ibid.*, p. 163.
[6] In the above letter and in another of 28 July 1788, Dol. (inserted in J. Watt Commonplace Book, 1782).

textiles by iron compounds: 'je passe la toile dans l'acide sulfurique étendu d'une grande quantité d'eau immédiatement après la première lessive'; but this was not necessary for cotton goods. He similarly removed a yellowish colour caused by a slight quantity of manganese (which was apt to pass over into the liquor during distillation), by immersing the cloth in dilute sulphurous acid. The bad odour of the gas could be diminished by mixing powdered chalk ('la craye en poudre') with the liquor; this did not appear to harm the bleaching. Berthollet also made further comments on the use of a potash solution[1] for absorbing the gas, producing 'lessive de javelle'. Despite its advantages, which he enumerated—less odour, a more concentrated liquor and so lower transport costs, and much quicker bleaching action—he was still against it, partly because of loss of gas to the alkali, but mainly because the cost of the alkali more than doubled that of the liquor. Finally, he recommended to Watt the Chevalier Landriani, a Swiss chemist who was about to visit England and would like to discuss bleaching and dyeing with him.

Despite Berthollet's scientific liberality, Watt was undoubtedly disappointed to find that his process for using sulphuric acid and salt had been discovered, and apparently tried, though vainly, to get Berthollet to keep it dark. Gilbert Hamilton ruefully commented that he was 'sorry Mr. Berthollet has found out your process but I hope what you have wrote him will prevent his publishing it'.[2] The technical lead of Watt and McGrigor was thus being narrowed, but they still had the advantage of their lead vessels for producing the bleach liquor. When the Chevalier Landriani visited McGrigor, the latter, acting on Watt's advice, did not let him learn too much. He showed Landriani 'everything practised in the present mode of Irish Bleaching', i.e. the traditional methods, but when the Swiss chemist asked him if he still continued 'the short mode of bleaching', McGrigor spoke of 'the great difficulty . . . in making a large quantity of the liquor'.

When he mentioned a wooden receiver I told him I thought it could not easily be made Air tight when he recommended laying the Vessel all over on the inside with Bees Wax . . . The use of Lead was never mentioned by either him or I. But . . . he said that the conducting pipes that went into the receiver should not be made of metal of any kind but all of white wood.[3]

It thus appears that Watt and McGrigor were now using lead not only for the retort or still but also for the connecting pipes and possibly for

[1] Four ounces of potash to a pound of water.
[2] G. Hamilton to J. Watt, 13 May 1788, Dol. As we have seen, however, Charles Macintosh and the Prestonpans Vitriol Company had also discovered it.
[3] J. McGrigor to J. Watt, 3 Oct. 1788, Dol.

the receiver, and were pleased to find that Landriani knew only of wooden apparatus.[1] The latter, however, promised to inform them of any new bleaching discoveries he made after his return to Switzerland. This promise he kept. In May of the following year, for example, we find him writing to Watt from Vienna: 'Le procédé pour le blanchissage de Mr. Born [the eminent German chemist and mineralogist] est le meme que celui de Mr. Berthollet', except for minor differences.[2] He also suggested to Watt that, in order to 'revivify' or brighten cottons dyed with Adrianople or Turkey red (a process in which Watt was also keenly interested at this time[3]), he should use 'une solution de soude combiné avec l'air marin dephlog[istiqu]é de Berthollet', instead of simply using a soda solution, as Watt had been doing. Thus, by these contacts with European scientists, Watt was able to keep himself informed of bleaching and other chemical developments on the continent.[4]

Watt himself continued his experiments. Thus we find McGrigor acknowledging his observations 'as to the proportions you now use in making the liquor as also the strength you reduce it to in whitening'.[5] And McGrigor continued to be successful with large-scale bleaching. During October 1788 he was treating 80 pieces of fine linen, containing 2,000 yards; two of these pieces he sent as a present to Berthollet, with an account of the process followed.[6] Berthollet thanked McGrigor warmly for this present—a testimony to the success of his process— and suggested to Watt[7] that the process might be made still more successful

si l'on pouvait se servir du résidu de la distilation pour en extraire la soude, et comme dans la plûpart des procédés qu'on employe pour cet objet, on commence par décomposer le sel marin par l'acide sulfurique, il y a grande apparence qu'on doit réussir, et couvrir par là la plus grande partie du prix de l'acide muriatique oxigéné.

The distillation process, as we have seen, left a residue of sodium sulphate or bisulphate in the retort,[8] from which Glauber's salt could be

---

[1] Before very long, however, French chemists found a more suitable substance. Décroisille, for example, had abandoned wood for his receivers by 1790, though he did not say what other material he was using. *Ann. Chim.*, Vol. VI (1790), p. 206.   [2] Chevalier Landriani to J. Watt, 24 May 1789, Dol.
[3] See below, pp. 345–6.
[4] He also read foreign scientific publications, assisted by his son, who, as we have seen, translated a number of them (see above, pp. 96–7).
[5] J. McGrigor to J. Watt, 12 July 1788, Dol.
[6] Ditto, 3 Oct. and 11 Nov. 1788, Dol.
[7] C. L. Berthollet to J. Watt, 28 Dec. 1788, Dol.
[8] See above, p. 254, n. 1.

extracted and sold for pharmaceutical purposes; but, as Berthollet pointed out, it would be much more profitable to convert it to the carbonate, commercial soda, which could be used for making alkaline lyes or could be sold.[1] These considerations may, indeed, have motivated Bourboulon de Boneuil & Co.; other firms certainly appreciated them in later years.

Meanwhile, the Lancashire rivals of Watt and McGrigor were being very active. Thomas Henry claimed in June 1788 that he had 'perfectly succeeded in the preparation of the liquor, and made some considerable improvements in it'.[2] But, as we shall see, he was over-sanguine. At first he experimented on a small scale, probably in his own works, but soon 'prepared to embark in a much larger establishment for the purpose',[3] in partnership with the most famous Manchester dyer of the day, John Wilson, who had achieved success largely through chemical knowledge and experiment.[4] Here was yet another impressive alliance of science and industry. It is not clear, however, exactly when this partnership was formed. In the middle of 1788 Henry was still engaged in experiments, though he was hoping to go into commercial production. His works, however, were 'for the preparation of the liquor only',[5] i.e. he was proposing to sell his liquor to bleachers, not to become a bleacher himself. By September he mentioned having 'disposed of some bottles' to customers.[6]

His efforts, however, were bedevilled by the same practical problems as Berthollet, Watt, and others had had to grapple with, and it is not surprising, therefore, that he turned to Watt for assistance in solving them.[7] 'I could have wished to have substituted some other Vessels to Glass retorts in procuring the Gas, but it is so extremely penetrating that no luting I have tried will stand [it] in that part of the apparatus which becomes hot.'[8] In thus producing the gas, Henry used sulphuric acid (9 lb.) and salt (18 lb.), with manganese (6 lb.), the acid being

---

[1] Berthollet stressed the importance of this in the *Annales de Chimie*, Vol. II, pp. 180–1, and Vol. VI, pp. 208–9, but was unable to reveal the extraction process, which had been developed by others, e.g. the French chemist de Morveau. He considered, indeed, that the new process could hardly be economically viable, except for cottons and fine linens, without such by-product recovery. At the same time (*Ann. Chim.*, Vol. II, pp. 166 and 181), he pointed out the great cost-reducing advantage of making the sulphuric acid in the same works; this would cut out not only the vitriol manufacturer's profit and transport costs, but also those of concentrating the acid for sale; instead it could be used dilute, as originally produced.   [2] T. Henry to J. Watt, 4 June 1788, Dol.

[3] W. Henry's 'Tribute' to his father, p. 229.   [4] See below, pp. 340–3.

[5] T. Henry to J. Watt, 12 June 1788, Dol.   [6] Ditto, 12 Sept. 1788, Dol.

[7] He also corresponded with Watt on dyeing and other matters. See above, p. 243.   [8] T. Henry to J. Watt, 12 June 1788, Dol.

diluted with twice its weight of water.[1] These materials were dissolved in two large retorts, contained in sand baths, 'or at least distilled at two different operations', and the gas was absorbed in a receiver containing six gallons of water, in which 6 lb. of potash had been dissolved; but he found that when heat was applied marine (hydrochloric) acid vapours were also driven over in such quantities as to pass through the intermediate 'adopter' and into the bleach liquor, with harmful effects, and asked Watt if he knew of any way of stopping this. Much more seriously, he was 'staggered' to find that when the alkaline solution was saturated with gas and the liquor was left standing in bottles, crystals of an unknown salt were formed, exhibiting many of the properties of nitre,[2] and the liquor tended to lose its strength. This seriously upset his plans, for he had been hoping to sell the bottled liquor in this concentrated form. He had never expected the formation of this salt, 'on account of the supposed loose attachment of the Gas to the [Alkali]'. But it now appeared that the bleaching power of the gas was not, as he had thought, increased by thus saturating an alkaline solution. 'I am the more led to this opinion of our mistake by finding that several bottles which I have sent to some distance, tho' closely corked and luted, have lost their bleaching properties; whence I suspect that the Gas has entered into combination with some of the Alkali . . .'

Here again, of course, Henry was unknowingly treading in Berthollet's footsteps. His experiments on this salt were much the same as those of the French philosopher, who had named it 'oxygenated muriate of potash'.[3] Indeed it was only just at this time that Berthollet published his paper on it; Watt loaned Henry a copy a few months later.[4] Meanwhile, Henry's earlier plans were dislocated and he again appealed to Watt for help, though he wished him 'to answer no questions that may [be] inexpedient for you to resolve'.

My attention having been mostly directed to this scheme of concentration by means of [alkali], and fearing to be disappointed in my views, I shall be greatly obliged, by your opinion, whether you think the liquor, made without [alkali], will be capable of bearing carriage, for a few miles, and of being preserved for any time, & whether the absorption of the Gas, by the Water, can be increased by any mechanical contrivance, or by any apparatus, different from what I use, for I find that if the pressure against the motion of the Gas be too much increased, no Luting will stand it. For this reason the receiver cannot be closely stopped. Will wooden tubs, lined with wax, serve as Receivers?

[1] Ditto, 12 and 23 Sept. 1788, Dol.
[2] For instance, it 'made Gunpowder & Pulvis Fulminans'.
[3] See above, p. 271.   [4] T. Henry to J. Watt, 7 Jan. 1789, Dol.

It seems doubtful whether Watt gave Henry much help on these problems, the solution of which had caused himself and McGrigor such pains, though he did tell him about his test liquor and gauge for measuring bleaching strengths.[1] Henry appears to have continued his experiments in dissolving the gas in a solution of potash or soda, finding that this was particularly suitable for treating printed goods: 'Indeed without it we should destroy the Turkey red in striped goods, the gold thread in Muslins & the blue in Calicos.'[2] Henry appears to have been one of the first to recommend this bleach liquor particularly for clearing the maddered grounds of printed goods;[3] but bleachers and calico-printers were slow to adopt it.[4] Henry also recommended its use for heightening the colours of cottons dyed Turkey red.[5] Berthollet's recognition of Henry's achievements suggests that the two had been in correspondence.

Henry was also concerned with the economic as well as the chemical problems involved. The cost of the materials used to produce the aforementioned six gallons of liquor he estimated to be 7s. 8½d.[6] The price of the liquor, he said, was 'not yet absolutely fixed', though they had sold some bottles; they were 'waiting for a conference with Mr. Baker[7] . . . 2s./6 per Gallon is the price I propose . . .' He also gave some figures of bleaching costs by the new method.[8]

Two pieces of Velveret, the average weight of which is 17 lb. each, require nearly 7 gallons of my strong liquor—perhaps if just made somewhat less might have been sufficient. I suspect the liquor may be made cheaper, by using less [alkali] & consequently less of the other Ingredients. In the present mode & at 2/6 per Gallon, therefore 34 lb. [2 pieces] of Cotton will cost 17s./6, the same quantity of linen with your liquor at 2s. per Gal. would cost £1. 2s. 8d. Both these would, I fear be too dear.[9]

---

[1] As acknowledged by Henry in his letter of 12 Sept. 1788.

[2] T. Henry to J. Watt, 12 June 1788, Dol.

[3] *Ann. Chim.*, Vol. II, p. 186, where Berthollet attributes this discovery to Henry and Décroisille. The goods were first printed with 'fixing' agents or mordants and then dyed with madder, whereby the printed areas acquired different colours, according to the mordants used. The maddered 'grounds'— the unprinted areas, to which the madder had not been 'fixed'—had then to be 'cleared', i.e. whitened by bleaching. This was previously done by treatment with cowdung and bran, followed by long exposure in bleachfields.

[4] Parkes, *op. cit.*, Vol. IV, p. 85, footnote. See below, p. 323, n. 2.

[5] *Ann. Chim.*, Vol. II, p. 188. So did Décroisille; Landriani had also recommended this to Watt (see above, p. 298).

[6] T. Henry to J. Watt, 12 Sept. 1788, Dol. He gave no figures of other costs, e.g. rent, interest on capital, and wages.

[7] Of Baker, Cooper & Co.

[8] T. Henry to J. Watt, 23 Sept. 1788, Dol.

[9] Cf. the much lower figures for bleaching by traditional methods, as stated by McGrigor in the previous year (above, pp. 268–9).

Henry was therefore driven to conclude: 'If I cannot contrive to make the liquor saleable here, I shall be obliged to commence bleacher myself.'[1]

A month later he was 'convinced that every attempt to combine the dep[hlogisticate]d m[arine acid] gas with alkali in large proportions will be in vain, except for immediate use'.[2] He had discovered that 'without the [alkali] it keeps well', and also with a small proportion of alkali added; but it appears that he had abandoned his plan for selling the liquor. 'I think our other scheme will answer our expectations.' He and Wilson were evidently going into bleaching.

This evidence concerning Thomas Henry's early problems in preparing the new bleach liquor is supported by a later letter from his son, Dr. William Henry, who stated that his father's first venture was

an establishment for preparing bleaching liquor and selling it to consumers. This liquor was prepared by receiving the oxymuriatic acid gas into a solution of caustic potash. It was soon found, however, that the liquor, though very good at first, lost its power by keeping, in consequence of the decomposition of the acid, on a principle since explained by Mr. Chenevix, and the formation of hyper-oxymuriate and common muriate of potash. It was by disappointment from this cause (an effete liquor having been sent to Mr. Hoyle and others) that my father was influenced to take up the business of bleaching . . .[3]

His bleaching partnership with John Wilson, however, proved very unhappy. James Watt junior, who had by that time arrived in Manchester to learn the dyeing trade, soon reported gloomily on their prospects:[4] 'It is generally supposed that Henry & Wilson will ruin themselves in this business, by having attempted it on too small a scale, whence not being able to supply their customers with sufficient expedition these will of course apply to others [such as Baker, Cooper & Co.] on whose punctuality they can have more reliance.' Nevertheless, Henry remained hopeful, when writing to Watt at the beginning of the following year.[5]

The mode of Bleaching answers well; & in the small way, we are in, we have prepared a considerable number of pieces for dying & printing. For the former we have bleached 12, & for the latter 8 or 10 pieces of heavy goods such as Jennets &c. which measure near 40 yards each & are very close & compact, &

---

[1] T. Henry to J. Watt. 23 Sept. 1788, Dol.   [2] Ditto, 23 Oct. 1788, Dol.
[3] Quoted in Rees, *op. cit.*, article on 'Oxymuriatic Acid'. For Chenevix's researches, see above, p. 271, n. 2. For the inadequacy of the bleach liquor supplied to Thomas Hoyle, the Manchester bleacher, see below, pp. 303 and 315.
[4] J. Watt jun. to J. Watt, 7 Dec. 1788, Dol.
[5] T. Henry to J. Watt, 7 Jan. 1789, Dol.

I believe weigh about 12 lb. each, at the expence, in materials, of about 3/6. In our present situation, we do not attempt to make such goods perfect whites, but have no doubt that with further conveniences we shall be able to do it.

Alas for Henry's hopes! A few months later his partnership with Wilson was broken up and the bleaching concern had to be abandoned. This, according to Henry's son, was 'from no defect . . . of the process carried on', but in consequence of the 'dishonourable conduct' of his partner.[1] Confirmation of this view is provided in a letter of James Watt junior:[2] 'The partnership between old Wilson & Henry is dissolved, the old fellow has used Mr. Henry very ill, after having got all the knowledge he could out of him to turn him adrift. I am informed Wilson intends beginning the business with a son in law of his . . . Mr. Henry will in all probability give up the business.'

Doubtless Henry was shabbily treated by Wilson, but there is equally no doubt that his bleaching process was by no means entirely satisfactory. Thomas Hoyle, one of the leading Manchester dyers and calico-printers, giving evidence in the Tennant case in 1802,[3] remembered Henry having introduced a bleach liquor, which he himself had tried, in 1788, but which proved unsatisfactory. 'It seemed defective merely for want of strength . . . For want of a sufficient quantity of gas combined with alkali.'[4] He had seen goods bleached with liquor prepared by Henry: 'They were . . . only partially bleached . . . They were unequally bleached . . . They might be unequal from lying together in the liquor, and one part being more exposed to the action of the liquor than another.'

Not surprisingly, therefore, Henry gave up the bleaching business and from then on 'contented himself with imparting the knowledge he had gained to several persons, who were already engaged in the practice of bleaching, by the then established methods'.[5] We shall see that, as a kind of industrial consultant, Henry made important contributions to the spread of the new process.[6]

Meanwhile, Messrs. Baker, Cooper & Co. were 'erecting a very large works for preparing the liquor, & [also for] bleaching', at Raikes, near

[1] *Annals of Philosophy*, Vol. VI (1815), p. 423; Rees, *op. cit.*, article on 'Oxymuriatic Acid'. See also T. Henry to J. Watt, 25 Jan. 1790, Dol.

[2] J. Watt jun. to J. Watt, 25 May 1789, Dol.

[3] *Tennant Proceedings* (1803), pp. 103–4.

[4] From Hoyle's evidence, it would appear that Henry and others had previously absorbed the gas in water, 'but it was very inconvenient to use on account of the smell'.

[5] W. Henry, 'Tribute', p. 229.

[6] See below, pp. 315–17.

Bolton.[1] There were four partners—Joseph Baker, Thomas Cooper, Kempe Brydges, and Christopher Teesdale[2]—in this concern, which was, indeed, as Parkes said, 'of very considerable magnitude', eventually utilizing both water and steam power. The latter was to be employed for pumping water and 'turning their Wash wheels Squeezers (two Cylinders between which the pieces are passed after being washed for the purpose of extracting the Moisture) Devils and other machines for washing and dressing of goods'.[3] (Efficient washing was essential to the success of bleaching, to rid the fabrics of chemicals after successive bucking, bleaching, and souring operations.) Eventually Baker, Cooper & Co. also went into dyeing and calico-printing, and were proposing in 1791 to use some of their steam power for working a spinning mill.[4] At the same time, as we have seen, they had their own vitriol supply, from Baker & Co.'s works at Barton, near Eccles. They also had their own coal mines.[5] They were making, in fact, a remarkably ambitious attempt at vertical integration.

They were also applying scientific methods. Cooper was friendly with Watt, Priestley, and other scientists and philosophically minded manufacturers, and his correspondence with Watt shows that he had read the works of Berthollet and other French chemists, and had carried out scientific experiments.[6] Dr. Priestley, for instance, having told him of Watt's improvements on Papin's 'digester', he asked to be sent one 'for the making of some experiments upon bleaching'.[7] He regretted, however, that owing to pressure of business, his time for reading and

[1] T. Henry to J. Watt, 12 June and 12 Sept. 1788, Dol. James Watt jun. said they had 'taken one of the largest bleaching fields in this country not far from Bolton'. J. Watt jun. to J. Watt, 17 April 1789, Dol. See also above, pp. 260–1.

[2] T. Cooper to J. Watt, 8 May 1790, Box 6B, B.R.L.; *Manchester Mercury,* 24 Sept. 1793. Rees, *op. cit.,* article on 'Bleaching', includes John Horridge among the partners (see above, p. 260), but he was apparently not. Horridge, already an established bleacher in the Bolton area, was, however, certainly associated with Baker, Cooper & Co., for he announced in the *Manchester Mercury,* 7 April 1789, that he had 'lately made an Engagement with Joseph Baker, and Co. for an establishment in the Chemical Mode of Bleaching', and had 'bleached Forty Pieces of Muslinetts', to demonstrate the efficacy of the new method.

[3] J. Watt jun. to M. Boulton, 17 Aug. 1789, Watt jun. Box 1, A.O.L.B., and J. Watt jun. to J. Watt, 5 Sept. 1789, Dol.

[4] The Boulton and Watt engine was not finally erected till 1791. See letters by P. Ewart to J. Southern, 15 Feb. 1791; to Boulton & Watt, 16 Feb. 1791; and to J. Watt, 5 March 1791, B.R.L.

[5] See T. Cooper to J. Watt jun., 15 Dec. 1792, B.R.L., where they are referred to as 'Colliers & bleachers & Printers'. This letter also reveals the fierce opposition which their business had aroused among local manufacturers.

[6] See, for example, T. Cooper to J. Watt, 14 Sept. 1790, Dol. See also below, pp. 347–8, for his general scientific-industrial interests.

[7] J. Watt jun. to J. Watt, 1 Feb. 1789, Dol. See below, p. 327, for Chaptal's similar experiments with a pressured vessel.

experiment was greatly reduced, and admitted that he really knew far less about dephlogisticated marine acid

than anyone wd. reasonably expect from the constant use I am obliged to make of it. The Truth is, that having settled early in the progress of our concern the mode of preparing it in a gross way for use upon a large scale, the constant increase of our business (which being practical Novices we had to learn in detail) has hitherto afforded me no time or opportunity for any material Variation, or any speculative Experiments upon the Subject.[1]

Yet, though business affairs might limit their opportunities of philosophical speculation and experiment, scientifically-minded industrialists such as Taylor and Cooper had access to the same sources of information as Watt and Henry. During the campaign against Bourboulon & Co., they inserted the following announcement in the *Manchester Mercury*:[2]

Having an Opportunity of enquiring while we were in London, of Sir Joseph Banks [President of the Royal Society], whether he had received any Information from abroad respecting the new Mode of Bleaching with the dephlogisticated Marine Acid, he informed us . . . that he had a few Days before, received from Mr. Berthollet, who has applied this Method of Bleaching in France, an Improvement in the Process, which he desired us to mention at Manchester.[3] Mr. Berthollet, having in vain attempted to procure the dephlogisticated Acid from a Mixture of Vitriolic Acid, common Salt and Manganese, apprehended that it was necessary to make Use of the Marine Acid in its common State, in Lieu of the Vitriolic Acid and Salt; but having in a subsequent Attempt, diluted the Vitriolic Acid with Water, the Process for procuring the dephlogisticated Acid succeeded perfectly. The Proportions used by Mr. Berthollet in his Experiments in the small Way, were, two Ounces of good Vitriolic Acid, an equal Bulk of Water, three Ounces of Marine Salt, and one ounce of Manganese, from which may be procured eight Quarts of Liquor of a suitable Strength for Bleaching.

Thus Berthollet's method was thrown completely open, although, of course, its industrial application involved serious difficulties. Berthollet's publications were also available. Moreover, Watt was not the only one who had direct access to Berthollet himself. Middleton, as we have seen, went to Paris with this object,[4] and Charles Taylor followed him in 1789.[5]

I pass'd an entire Day very satisfactorily with Mr. Berthollet. He informed me that he had a method of seperating the Marine Alkali after preparing the

---

[1] T. Cooper to J. Watt, 14 Sept. 1790, Dol.
[2] *Manchester Mercury*, 20 May 1788.
[3] See above, p. 296, for Berthollet's communication to Blagden, secretary of the Royal Society.　　　　　　　　　　　　　　　[4] See above, p. 278.
[5] C. Taylor to J. Watt, 21 June 1789, Dol.

dephlogisticated Marine Acid which would yield him sufficient profit on the whole process without placing any value upon the dephlogisticated Marine Acid but that this Mode of seperating the Alkali he made a Secret of at present.[1]

Nevertheless, even with all this information, there were serious practical problems to be overcome before the new process could be successfully applied on a large scale. It is clear that Baker, Cooper & Co.—like Watt and McGrigor, and Henry and Wilson—spent a long period in tackling these problems. Charles Taylor, who was associated with their early experiments, though not a partner in the Raikes concern, stated that 'oxygenated muriatic acid gas' was first used in the bleaching of cotton goods by being dissolved in water, but that it 'frequently discharged the colours of printed goods called ginghams', and that it was also extremely volatile and injurious to the workmen.[2]

The next application was its combination with alkali; that was looked upon as a considerable improvement, because the colours were not so sensibly affected as when [it was] used with water only . . . [Nevertheless] it still had an unpleasant smell . . . [though] not to so great a degree as before, and it had not the property of preserving the colours so clear as could be wished. There wanted some further improvement . . .[3]

John Horridge, who was also with Baker, Cooper & Co. during their early experiments, described how the liquor was made.[4]

By mixing a quantity of vitriol, red lead, and salt, put into the cistern in the form of a barrel twisted round, made without fire, without any distillation.[5] It was a very good bleaching-liquor for bleaching whites, but took away the marks of goods, and it *destroyed both the colours and the fabric* . . . We damaged all of them. Cooper burnt many, set them on fire, and sent them up the chimney, he would not let them be seen at market.

At that early period, said Horridge, 'we did not know the use of alkali', and even when it was adopted, the liquor was liable to injure both the colours and the texture of the cloth, mainly because 'we were at no certainty, the ashes were not always of one strength'.[6]

It is obvious from their delay in starting at Raikes, after their

[1] See above, pp. 298–9.   [2] *Tennant Proceedings* (1803), pp. 114–15.
[3] Taylor also pointed out the considerable expense on alkali.
[4] *Tennant Proceedings* (1803), pp. 156–7.
[5] Perhaps because, as Henry found, heat caused marine (hydrochloric) acid vapours to be driven over, while the luting could not withstand the pressure of the gas. Red oxide of lead could be used as a substitute for oxide of manganese. H. Davy, *Elements of Chemical Philosophy* (1840 edn.), p. 173.
[6] If there was too much oxymuriatic acid, it attacked the colours and even the cloth, but if there was too much alkali, the cloths were not completely bleached. See also the evidence of Luke Howard, manufacturing chemist, *Tennant Proceedings* (1803), pp. 129–30, and Rees, *op. cit.*, article on 'Bleaching'.

apparent success at the Manchester meeting in February 1788,[1] that Baker, Cooper & Co. experienced considerable difficulties in trying to solve these practical problems. Though they began erecting this bleachworks fairly soon after that meeting,[2] they had still 'not yet begun to bleach' by the end of that year: 'they do not mean to start till they have everything ready to go on in great'.[3] These preparations were practically completed by the following April:[4]

Mr. Cooper is now making very great preparations for beginning bleaching in great . . . his partner Mr. Baker has made considerable improvements upon Bertholets method, for which they are now getting a patent; I saw the other [day] about 40 pieces [of] white goods which they had bleached by way of trial & [they] were equal to any bleached in the common way.

Thomas Henry said that they had actually obtained a patent,[5] but we have been unable to discover any evidence of this.

In May 1789 James Watt junior announced that 'Baker, Cooper & Co. have at length begun & will send in 3000 Pieces this week, of a colour far superior to anything which has hitherto been done by the common bleachers'.[6] The latter, he thought, 'will doubtless form a combination against them & let fall their prices', but consumers would benefit from such competition. By the end of the year they finally decided that they would have a Boulton & Watt engine,[7] their business being 'continually upon the increase'.[8] They were annoyed, however, by extortions on the part of dyers, who would not undertake their work without being allowed 2s. 6d. per yard for 'scouring' the cloth, prior to bleaching, 'whether they have done the scouring or not'. Baker, Cooper & Co. therefore decided 'to begin dying at least blacks and dark colours', and James Watt junior thought they would 'launch out to a very considerable extent'.[9]

Briefly outlining to Watt their methods of producing the gas,[10] Cooper referred to Berthollet's process,[11] commenting that it was 'unnecessary to tell you that for this purpose Vitr. Acid, Common Salt & Manganese are distilled in glass or earthen retorts. The Gas is

[1] See above, pp. 260–1 and 289–90.   [2] See above, pp. 303–4.
[3] J. Watt jun. to J. Watt, 7 Dec. 1788, Dol.
[4] Ditto, 17 April 1789, Dol.
[5] T. Henry to J. Watt, 7 Jan. 1789, Dol.
[6] J. Watt jun. to J. Watt, 25 May 1789, Dol.
[7] T. Cooper to J. Watt, 7 and 17 Dec. 1789, B.R.L. They also asked for Watt's permission to use his patent smokeless fireplace, since smoke from their furnaces, used for boiling alkaline lyes, was damaging goods in their dry-houses.
[8] J. Watt jun. to J. Watt, 30 Dec. 1789, Watt jun. Box 1, A.O.L.B.
[9] Ditto, 11 July 1790, Dol.
[10] T. Cooper to J. Watt, 14 Sept. 1790, Dol.   [11] See below, p. 313.

received in a Solution of Pearl Ash & the liquor made of such a strength as to allow from 10 to 15 waters.' Cooper considered this liquor—'oxygenated muriate of potash'—to be more practicable than the gas. They had been mainly concerned—like Watt and McGrigor—with making Berthollet's laboratory process a practical and economic large-scale proposition: 'The Improvements we have made upon the process, go entirely to the *cheapness* of procuring the liquor, but we obtain it in consequence less pure . . . We make our own Vitr. Acid, we waste none of the Gaz, we save the [by-] products, & we use the liquor to the utmost. Hence we can use it where others do not.'

Nevertheless, it is clear from the evidence at the Tennant trial in 1802 that Baker, Cooper & Co. never overcame all the problems of chlorine bleaching. The 'oxymuriate of potash' liquor suffered from practical deficiencies, and it was expensive. Moreover, it would appear that, with all their efforts at integration, they attempted too much, and were unable to withstand the commercial crisis on outbreak of war with France. By September 1793 they were bankrupt;[1] Baker & Co.'s vitriol works at Barton, near Eccles, were also involved in the crash.[2] John Horridge, who, as we have seen, had at first been associated with Baker, Cooper & Co., but had apparently withdrawn later, was appointed one of the trustees and eventually took over the Raikes bleachworks.[3] He formed a partnership for this purpose with another Bolton bleacher, James Varley, in 1796, but Varley 'followed the same scheme as Baker and Cooper, and destroyed a great deal of goods', so the partnership was broken up at the end of 1797.[4] Horridge, however, managed to carry on the business alone.[5]

Baker, Cooper & Co. do not appear to have attempted the sale of the liquor, but used it entirely in their own bleaching operations. Bourboulon de Boneuil & Co., of Liverpool, however, like Henry, established their works primarily with the object of selling *eau de Javelle* to bleachers. Having been thwarted in their attempt to obtain a Parliamentary monopoly, they at once sought to secure patents for their bleaching and alkali processes. This, of course, again raised indignant opposition from those manufacturers who were already practising the new process, or were hoping to do so. Henry wrote to

---

[1] *Manchester Mercury*, 24 Sept. and 5 Nov. 1793. The latter advertisement refers to the sale of their bleaching and printing works, collieries, etc.

[2] *Ibid.*, 25 Feb. 1794.

[3] J. Watt jun. to J. Horridge, 25 June 1795, Parcel F. 'H.', B.R.L.; *Manchester Mercury*, 7 June 1796.

[4] *Tennant Proceedings* (1803), Horridge's evidence, pp. 157-8. See below, p. 324, for James Varley.   [5] See below, p. 323.

let Watt know of the new moves in Manchester;[1] he himself had entered a Caveat and was supported by Baker, Cooper & Co. Another meeting of Manchester manufacturers was held on Tuesday, 17 June 1788,[2] at which a letter from Sir Joseph Banks to Charles Taylor and Thomas Cooper was read, 'giving an account of an improvement of the known process of M. Berthollet, a full abstract of which had some time before been published in the Manchester Papers by those Gentlemen'.[3] McGrigor, of course, also remained anxious to prevent such a patent being granted in Scotland.[4] This time, however, the opposition proved ineffective, because Bourboulon merely claimed a patent for 'a particular Apparatus by which the liquid . . . is prepared and applied',[5] which he was able to demonstrate was different from those of other manufacturers. McGrigor, for example, was asked to supply the Lord Advocate, Islay Campbell, with details and drawings of his own apparatus, but was loath to do so, because, as he told Watt, 'you had wrote me formerly that the use of Lead vessels was only known to us'.[6] On assurance of secrecy, however, Watt gave this information to the Lord Advocate, but the latter pronounced that Bourboulon's apparatus was 'in several Particulars very different from yours'.[7] On these grounds, Bourboulon had secured an English patent (no. 1678, 25 March 1789), and this was now extended to Scotland, though it left McGrigor 'at full liberty to follow his own practice, & to combat Mr. Bonneuil if challenged by him'.[8]

It might have been expected that the granting of these patents would have created alarm and despondency among those who had so long and bitterly opposed him. But Thomas Henry, for one, was not at all perturbed:[9] 'Neither of his patents are of value. The one is for throwing

---

[1] T. Henry to J. Watt, 4 and 12 June 1788, Dol.; *Annals of Philosophy*, Vol. VI (1815), p. 424. Collison, who had returned to London, was Henry's informant about Bourboulon de Boneuil.   [2] *Manchester Mercury*, 17 June 1788.

[3] G. Barton, Borough-reeve of Manchester, to Sir J. Banks, 18 June 1788, Banks Correspondence, Library of the Royal Botanic Gardens, Kew, Vol. I, p. 304. The meeting passed a vote of thanks to Banks. See above, p. 305, for the earlier publication in the Manchester press by Taylor and Cooper.

[4] J. McGrigor to J. Watt, 3 Oct. and 11 Nov. 1788, Dol.

[5] Ditto, 14 April 1789, Dol.   [6] *Ibid.*

[7] Islay Campbell to J. Watt, 29 April 1789, Box 4C, B.R.L.

[8] *Ibid.* At the same time, Bourboulon was granted a patent (no. 1677, 11 March 1789) for his alkali-manufacturing process (see below, pp. 364–6). The materials were 'such as have been already used', namely 'charcoal and iron mixed and melted with Glauber salt', but instead of being placed 'in an open reverberatory furnace', they were heated 'in pots or crucibles made of iron', with closely-shut lids. *Repertory of Arts and Manufactures*, Vol. IV (1796), pp. 96–100; *Abridgments of Patent Specifications* (Acids, Alkalies, etc., 1869), pp. 36–7. The Glauber's salt of course, could be obtained from the residue of the chlorine-manufacturing process (see above, pp. 298–9).   [9] T. Henry to J. Watt, 25 Jan. 1790, Dol.

up the Gas into an inverted matras [flask]; the other for fusing the sea salt, charcoal & iron filings, in *iron* crucibles instead of a reverberatory furnace.' The main difference between the bleach-making apparatus of Bourboulon and Vallet and those of Berthollet and Watt was in their receivers, which were a series of inverted glass matrasses or flasks, filled with water and having long necks descending beneath the surface of water in small earthenware pots below.[1] Each receiver was fed with gas from a separate 'glass or earthen retort or bottle' containing marine acid mixed with manganese; the gas, passing along tubing, was bubbled upwards into the receivers, gradually saturating the liquor and also depressing it into the pots below, from which it flowed into receptacles. The main virtue of this apparatus, stressed in the specification, was that it required no luting, since there was no resistance to the gas in the inverted receivers. It was only one, however, of various types that were gradually developed, and Bourboulon & Co., despite their patent, were in no position to prevent other manufacturers from utilizing different methods. Nor, indeed, as we shall see, were they able to prevent copying of their own apparatus.

In addition to this apparatus for preparing bleach liquor, Bourboulon & Co. also included in their patent a gas-bleaching method, whereby the gas, similarly generated, passed into a 'lodge' containing the goods to be whitened. Chaptal had previously expressed himself in favour of gas-bleaching, in a memoir to the Académie Royale des Sciences, not only for paper-making but also for linen and cotton fabrics, which would have to be hung in a gas-tight chamber.[2] But Berthollet and Lavoisier were very critical of this method,[3] the former having found by experiments that it was inferior to bleach liquor made by dissolving the gas in water; there was a greater loss of gas and the fabrics were more likely to be damaged or bleached unevenly, although bleaching was certainly more rapid.

It seems very unlikely, therefore, that Bourboulon & Co. had much success with their gas-bleaching process, and it is not surprising to find them concentrating on selling bleach liquor. In fact, immediately after the granting of their letters patent, advertisements appeared in the newspapers announcing that

Messrs. Bonneuil, Vallet, & Co. of Liverpool, take the Liberty to inform the Public, that they are now come to a Resolution, to make and sell their Patent Whitening Liquor, by which Muslins and other fine Cotton Goods . . . may

---

[1] *Repertory of Arts and Manufactures*, Vol. IV (1796), pp. 155–64; *Abridgments of Patent Specifications* (Bleaching, Dyeing, etc., 1859), pp. 46–7.

[2] *Mémoires* of the Académie Royale, 1787 (published 1789), pp. 611 ff.

[3] *Ann. Chim.*, Vol. I (1789), pp. 69–72, Vol. II (1789), pp. 176–7.

be whitened in Six Hours Time, Muslinets, Calico's, Dimities, and heavy Hosiery Goods in Twelve Hours, or thereabouts, and so well, that they will require but little Preparation afterwards. They have fixed the Price of their Liquor as low as possible, in order that all Sorts of Cotton Goods may be whitened, upon an Average, as cheap as by the usual Mode of bleaching. Their liquor may likewise be used, with great advantage, for whitening fine Goods made of Hemp or Flax, and it has also the Property of totally discharging all vegetable Colours, in printed Cottons, or Linens, and even metallic Colours with the Aid of a little vitriolic Acid, and of rendering the Goods (without any Detriment or Injury whatever) fully as white as they were before printing.[1]

They offered to supply manufacturers with small quantities of liquor for trial, together with printed instructions on its use. They would be ready to supply orders from the beginning of July.

This advertisement shows that the Frenchmen were claiming success with all kinds of textiles, with hemp and flax as well as cotton, with both heavy and fine fabrics, and also apparently with printed as well as plain goods.[2] They could only claim, however, that their process was no dearer, on average, than the traditional method. Their *eau de Javelle* no doubt suffered from deficiencies similar to those experienced by others who absorbed the gas in a potash solution. In particular, as industrial chemists selling the liquor, and not themselves bleachers, they would be faced with the same problems as Thomas Henry. It is true that they were making their own soda—though with what success we do not know[3]—but their liquor was still not particularly cheap, and it could not be transported easily. As a result, 'from the article being bulky, and from its constant loss of strength by the action of light, or by exposure to the air, the obstacles to its consumption continued to augment, until in the end they were obliged to relinquish the establishment altogether'.[4] Bleachers, as we shall see, found it preferable to erect their own plant for making the gas or liquor, which many were able to do with the aid of chemists who left Boneuil & Co.[5] The works

[1] *Manchester Mercury*, 16 June 1789.
[2] They claimed that their liquor would preserve 'the red Ends and blue Selvages wove in some Goods'. The practice was developing of running a line of coloured thread (dark blue or Turkey-red) along the edges (selvages) of goods sent out for bleaching, as a safeguard against damage. The liquor, if of appropriate strength, would be strong enough to bleach the goods effectually, without damaging these colours. Parkes, *op. cit.*, Vol. IV, pp. 170–7.
[3] Most of those who tried to develop synthetic soda manufacture at this period were unsuccessful. Even after Leblanc invented his process in 1790, it was introduced very slowly into Britain.
[4] Parkes, *op. cit.*, Vol. IV, p. 63.
[5] See below, pp. 313–15, 324–5, 330–1.

—described as 'An Estate in Garston, five miles from Liverpool, called *Beauregard*, situate near the Shore of the River Mersey'—was put up for auction in Liverpool on 6 June 1792.[1]

By that time, information about the new 'chemical bleaching' had been widely disseminated. Berthollet published in the second volume of the *Annales de Chimie* (1789) a detailed account of all its processes, with descriptions and drawings of the equipment, instructions about quantities of materials and strengths of lyes, bleaching liquor, and sours, and explanations of all the possible difficulties to be faced. This was specially intended for the use of manufacturers, but he stressed the necessity for employment of a skilled chemist to direct operations; even then much trial and error would be necessary to achieve success.

This work was quickly translated into English by Dr. Robert Kerr, an Edinburgh surgeon.[2] The first edition was sold off in four months and another was printed in the following year.[3] Kerr had been disappointed to hear 'that the oxygenated muriatic acid had been tried in this country [Scotland], and abandoned, as not answering the purposes expected from it'.[4] He had, in particular, 'been informed, though only by hearsay', that even James McGrigor had 'been induced, by some inconveniences attending the use of the oxygenated muriatic acid, to give it up, and recur to the ordinary process of bleaching'.[5] This would account for the apparent cessation of correspondence between Watt and McGrigor in regard to the new process after spring 1789.

Kerr hoped, however, that other bleachers would be able to benefit

[1] *Manchester Mercury*, 29 May 1792. Bourboulon de Boneuil was willing to assign his interest in the patents to the purchaser of the works. The stock included pearl ashes, iron filings (for the alkali-manufacturing process), Glauber's salt, fossil alkali, bleaching liquor, and spirits of salt.

[2] *Essay on the New Method of Bleaching* (Edinburgh, 1790). It was dedicated to the Honourable Trustees for Fisheries, Manufactures, and Improvements in Scotland, who had offered a premium for the most successful application of the new process. A translation of Berthollet's work also appeared in the *Repertory of Arts and Manufactures*, Vol. I (1794), pp. 53–72, 121–30, 190–201, 260–73, and Vol. II (1795), pp. 354–60.

[3] Together with a translation of Chaptal's memoir on bleaching in the *Mémoires* of the Académie Royale for 1787 and also of Berthollet's *Observations and Experiments on the Art of Dying with Madder*. The translator stated in his preface to this second edition that already he had 'been applied to, by an eminent Manufacturer, to assist and advise in the erection of the necessary apparatus for giving a full trial to the New Bleaching process'. Kerr provides yet another example of a medical man 'employing his leisure for the advancement of the arts', i.e. manufactures.

[4] Preface, p. xvi.

[5] *Op. cit.*, 2nd edn., p. 77, footnote. No effort has been made in this chapter to trace the general development of chlorine bleaching in Scotland because Miss Enid Gauldie, of Dundee University, is working on this subject.

from Berthollet's work. Many, undoubtedly, were keen to try the new method, as Thomas Cooper observed in 1790:[1]

Almost every Bleacher in this Country applies the dephlogisticated liquor to the finer kind of goods: we apply it indiscriminately to the coarse as well as the fine. The common mode of preparing it is now well known, partly from the persons who worked with Bourboulon & Vallet, & partly perhaps from Bertholletts process which I published in the Manchester Paper sometime in May 1788;[2] but as they are not much given to reading, I conjecture the former is the principal source of their knowledge.

At this same time Charles Taylor wrote to Berthollet: 'Le blanchîment sur les principes que vous avez établis s'est fort étendu dans notre voisinage, & est exécuté par différentes personnes avec beaucoup de succès.'[3] Abraham Rees in his *Cyclopaedia* similarly testified to the importance of 'the French chemists, whose opinions were regarded as law by the common bleachers, and whose treatises on the subject of bleaching were almost the only accounts published'.[4]

In one respect, however, practical bleachers did not generally follow Berthollet's advice. He strongly recommended absorption of the gas in water, rather than in an alkaline solution,[5] but this was found defective in practice,

as the volatility of the gas occasioned its speedy separation from the aqueous solution; a decomposition even by light alone in glass vessels took place; a rapid loss in the strength of the liquor when exposed; and much danger to the health of the workmen from its suffocating quality; at the same time, that in extracting the natural colours of the cloth, it also tended to discharge the colours dyed in the yarn.[6]

Berthollet, however, continued to recommend his process, on account of the expense of the alkali and because the bleaching power of the 'oxygenated muriatic acid' was reduced by absorption in an alkaline solution. But practical bleachers mostly preferred to use the *eau de Javelle* type of liquor, despite the expense, because the process could be better controlled, and 'though the acid thus combined [with alkali] whitened with somewhat less rapidity, yet it was not eventually in an inferior extent; and [it had] the advantages of preserving the colours'.[7]

[1] T. Cooper to J. Watt, 14 Sept. 1790, Dol. [2] See above, p. 305.
[3] Taylor to Berthollet, 21 Aug. 1790, quoted in *Ann. Chim.*, Vol. VII (1790), p. 244.
[4] *Op. cit.*, article on 'Bleaching'. As we have suggested above, p. 260, the author of this article was probably Charles Taylor, who would be writing from first-hand knowledge. [5] See above, pp. 264, 271, 275–6, 297. [6] Rees, *op. cit.*
[7] *Ibid.* Watt and McGrigor had eventually reached the same conclusion (see above, p. 293). See also above, pp. 306–8, for the similar experiences of Taylor and Cooper. Several of the witnesses in the Tennant trial of 1802 gave similar

From 1789 onwards there is evidence of a rapidly increasing number of bleachers adopting the new process. Perhaps the most important agents of its dissemination, as Thomas Cooper said, were French chemists from Bourboulon de Boneuil & Co., especially Matthew Vallet and Hugh Foy. Foy, giving evidence at the Tennant trial in 1802, mentioned having erected bleaching plant not only in Lancashire, after leaving Boneuil & Co., but also in Scotland, at Glasgow and Perth.[1] According to Parkes,[2] Foy 'waited upon most of the principal manufacturers, and proposed, for a considerable premium, to erect for them the necessary apparatus, and to instruct them in the process of making the Liquor de Javelle at their own respective works. This proposal was acceded to by many, and the oxy-muriate of potash was generally employed . . .' Vallet was similarly active, especially in the Bolton area.[3] The new method of bleaching, it was stated in 1795, had lately been introduced into that neighbourhood 'by M. Vallette (an ingenious Frenchman)'.[4] He entered the employment of—or perhaps into partnership with—Peter Ainsworth, of Halliwell bleachworks, near Bolton, described by James Watt junior in 1790 as 'the greatest bleacher in this county'.[5] Richard Ainsworth, son of Peter, is said to have been particularly active in applying chlorine bleaching,[6] and John Horridge said

testimony (*Tennant Proceedings*, 1803, *passim*). See also Brewster's *Edinburgh Encyclopaedia*, article on 'Bleaching', where the practical bleaching considerations are likewise stated.

[1] *Tennant Proceedings* (1803), pp. 205–12. See below, pp. 324–5. J. Mactear, in an article on the development of vitriol manufacture, Glasgow Phil. Soc. *Proceedings*, Vol. XIII (1880–2), refers (p. 419) to 'a set of hydrometers invented or arranged [in 1797] by a Mr. Foy, a chemist, at that time engaged in bleaching operations', apparently in Glasgow. At the time of his death, at 42 years of age, on 18 Dec. 1811, Foy was living at Rivington, near Chorley: 'a man universally respected, and well known as a principal agitator in the present process of bleaching goods'. *Manchester Mercury*, 7 Jan. 1812.   [2] *Op. cit.*, Vol. IV, p. 63.

[3] *Tennant Proceedings* (1803), evidence of Belshaw, pp. 144–54; Horridge, pp. 155–62; and Bentley, pp. 165–9.

[4] J. Holt, *General View of the Agriculture of the County of Lancaster* (1795), p. 213.

[5] J. Watt jun. to J. Watt, 11 July 1790, Dol.

[6] Most references to his achievements are derived from an article by J. D. Greenhalgh in the *Bolton Chronicle*, 12 Jan. 1884, p. 7, based partly on information supplied by the late Henry Ashworth, Esq., of Turton. According to this article, Richard Ainsworth brought two Frenchmen to the Bolton neighbourhood, Tennant and Vallette, who introduced chlorine bleaching. This Tennant is alleged to have put Ainsworth to considerable trouble and expense by experimenting with paper tubes, for conveyance of the gas, instead of metal ones (fearing the latter would corrode). Later, according to this account, he falsely claimed to have invented chloride of lime—hence the famous Tennant patent case of 1802. Eventually, however, he is said to have returned to France, with a testimonial from Bolton manufacturers, and was visited there by Ainsworth in 1818. This account has been followed, in whole or in part, by later writers, e.g. J. H. Partington, MS. 'A History of Halliwell' (1906), in Chetham's Library,

that Ainsworth's were the first he knew of to use potash in bleach liquor; Bentley, a former employee, stated that Vallet also experimented at Ainsworth's with gas bleaching. Vallet also erected a bleaching apparatus for Rothwell's, another large Bolton firm, and probably for others in that area and elsewhere.[1]

Thomas Henry, however, appears to have played an equally important role. John Kennedy, the great Manchester cotton spinner, stated in 1815 that Henry was not merely 'one of the first who carried it [chlorine bleaching] into practice', but also 'gave to some of the principal bleachers in the country the first instruction they received respecting the new process'.[2] Thomas Ridgway, for example, of Wallsuches bleachworks, Horwich, near Bolton, applied to Henry after the Manchester meeting of February 1788, 'to be instructed in the new process',[3] and was eventually so successful that by the 1830s he had built up 'the most extensive bleachworks in Lancashire'.[4] Thomas Hoyle, who similarly created one of the biggest dyeing and calico-printing works in Manchester, at Mayfield, Ardwick, also acquired his first knowledge of the new bleaching process from Henry, in 1788.[5] After experimenting unsuccessfully with Henry's liquor, he began to make his own, absorbing the gas in a caustic potash solution, contained in a series of vessels, apparently similar to Woulfe's apparatus.[6] (He had, he said, made some study of chemistry for use in his business.[7])

Manchester; Sir A. J. Sykes, *Concerning the Bleaching Industry* (1925); *Bolton Evening News*, 20 June 1928; Higgins, *op. cit.*, p. 78; Clow, *op. cit.*, p. 190. Tennant becomes 'M. Tenant'. But the story is utterly confused. Charles Tennant, famous for his bleaching patents of 1798 and 1799 (see below, pp. 322–6), and founder of the great St. Rollox chemical works, near Glasgow, was not a Frenchman, but Scottish born and bred, and was never employed by Ainsworth's (though the latter did adopt his patent process: see below, p. 323). The story about the paper tubes also sounds dubious, but, if true, Vallet must have been the Frenchman concerned; and it may well be that he did eventually return to France, with a testimonial. His name is perpetuated in 'Vallets', a hamlet in the Bolton neighbourhood (*Bolton Evening News*, 20 June 1928).   [1] See below, pp. 324, 330–1.

[2] J. Kennedy, 'Observations on the Rise and Progress of the Cotton Trade in Great Britain', read in 1815 to the Manchester Lit. & Phil. Soc., printed in his *Miscellaneous Papers* (Manchester, 1849), p. 17, footnote. See also Brewster's *Edinburgh Encyclopaedia*, article on 'Bleaching'.

[3] *Annals of Philosophy*, Vol. VI (1815), p. 423; Horwich is misprinted 'Harwich'. See also Baines, *op. cit.*, p. 248, and Higgins, *op. cit.*, pp. 76–7.

[4] Baines, *op. cit.*, p. 253. See also Sykes, *op. cit.*, pp. 80–4.

[5] See above, p. 303.

[6] *Tennant Proceedings* (1803), pp. 102–10. He was absorbing the gas in a caustic potash solution as early as May 1788.

[7] Hoyle's knowledge of theoretical and applied chemistry is also evident from a paper which he read to the Manchester Lit. & Phil. Soc. on 10 Nov. 1797,

By June 1789 he was able to announce that, having 'for some time past . . . bleached Cotton Goods, by a new, safe, and expeditious Method, and dyed them fancy Colours', he was now intending 'to carry on that Branch in a more extensive Manner'.[1]

Another early pioneer, Robert Hall, of Basford, near Nottingham, also appears to have obtained information from Henry, who told Watt in January 1790 that 'My Method of bleaching by vapour [i.e. chlorine gas] succeeds admirably in the hands of a bleacher in the neighbourhood of Nottingham'.[2] This bleacher was almost certainly Hall, who eventually became well known for his practice of gas-bleaching. He is yet another example of a scientifically-minded industrialist: 'product of one of the "commercial academies" of the town [Nottingham], a student of Lavoisier, Scheele, Berthollet and Black'.[3] Such a man would quickly seize on the new 'chemical' mode of bleaching. To find a cheap substitute for potash, some bleachers began to experiment with lime, absorbing the gas in a solution of lime and water, or lime-water, and thus producing 'oxymuriate of lime' (calcium hypochlorite). Bourboulon de Boneuil experimented with this method at Liverpool, according to Hugh Foy's evidence.[4] Henry also developed it, though 'in a way which adapted it only to white goods'.[5] It appears that he combined it with gas-bleaching, and that he told Hall of this method.[6] The successive processes which Henry

---

'Experiments and Observations on the Preparation and some remarkable Properties of the Oxygenated Muriat of Potash', *Memoirs*, Vol. V, pt. i (1798), pp. 221–42; *Nicholson's Journal*, Vol. II (1798–9), pp. 290–7; *Repertory of Arts and Manufactures*, Vol. XI (1799), pp. 46–58, 105–13. This paper was particularly commended by Dr. Richard Chenevix, *Phil. Trans.* (1802), p. 127.

[1] *Manchester Mercury*, 23 June, 1789.

[2] T. Henry to J. Watt, 25 Jan. 1790, A.O.L.B.

[3] J. D. Chambers, *Nottinghamshire in the Eighteenth Century* (2nd edn., 1966), p. xiii. The firm of Hall & White is also mentioned in S. D. Chapman, 'The Transition to the Factory System in the Midlands Cotton-Spinning Industry', *Econ. Hist. Rev.*, Vol. XVIII, no. 3, Dec. 1965, p. 535. Professor Chambers refers to Hall as 'reputedly the first manufacturer in England to introduce the use of chlorine in bleaching', but, although he was one of the early pioneers, his claims to be the first are extremely doubtful. In his evidence at the Tennant trial, Hall produced a drawing of an apparatus which he had used since 1789, the year in which he appears to have adopted chlorine bleaching, apparently with Henry's assistance (*Tennant Proceedings*, p. 194).

[4] *Tennant Proceedings*, p. 210.

[5] *Annals of Philosophy*, Vol. VI (1815), p. 424; Baines, *op. cit.*, pp. 248–9; Higgins, *op. cit.*, p. 80.

[6] Rees, *op. cit.*, article on 'Oxymuriatic Acid', actually states that this method was 'invented by Mr. Henry of Manchester, and practised, under his instructions, by the bleachers of cotton hose at Nottingham'. See also the article on 'Bleaching', in which this process is described.

developed are, in fact, well summarized in Brewster's *Edinburgh Encyclopaedia*:[1]

[Henry's method] at first consisted, sometimes in immersing the goods in a watery solution of the gas, or in an alkaline ley impregnated with it, and sometimes in exposing the goods previously moistened with water, to the action of the gas itself. Soon afterwards, he made a further improvement in substituting lime for alkali, as a means of condensing the oxymuriatic acid gas. An air-tight cylinder was prepared, on the floor of which rested a stratum of lime and water, mixed together to the consistence of cream. Through this the goods were passed by means of a wince; and the chamber being filled with gas, the goods were alternately exposed to the lime liquor, and to the acid vapour. Thus an oxymuriate of lime was formed upon the cloth, which, after a sufficient continuance of the operation, was taken out, and exposed to the usual processes of washing, &c.

This was exactly the same as the apparatus which Hall used from 1789 onwards,[2] in which the gas from a retort was piped to an enclosed receiver containing the goods to be bleached, hanging on a frame; in the lower part of this vessel was a mixture of lime and water, into which the goods could be lowered from time to time, by means of a winch. The gas entered above the level of the lime-water, until in 1791 a modified apparatus was developed, whereby it was bubbled up through the liquid; at the same time, combination with the lime was increased by agitation or stirring with a rake.

This modification was doubtless adopted because of the dangerous effects of the gas on both goods and workmen. Thomas Cooper, hearing 'that at Nottingham the dephlogisticated Gaz is applied immediately as it comes over to the Goods in a moist State', thought it could 'hardly [be done] with safety to the texture or consistent with the requisite Oeconomy', i.e. much gas would inevitably escape.[3] For these reasons, in fact, gas-bleaching was never very successfully employed, and according to Rennie's manuscript of 1816 was 'now, I believe, wholly laid aside'.[4] Absorption of the gas in water was not much more successful, but because of the price of alkali and the loss of bleaching power occasioned by its use, some manufacturers continued to try this earlier method. Theophilus Rupp, for example, a chemically

[1] Article on 'Bleaching'. It is possible that William Henry supplied this information. In the 'Tribute' to his father, he referred particularly to this article, as well as to his own letter in the *Annals of Philosophy*, previously cited.
[2] Hall's evidence in *Tennant Proceedings* (1803), pp. 41–4, 189–94. Parkes, *op. cit.*, Vol. IV, pp. 59–60, says that it 'succeeded tolerably well in bleaching small parcels of linen-yarn, linen, and cotton hose, and other small goods'.
[3] T. Cooper to J. Watt, 14 Sept. 1790, Dol. Berthollet had reached similar conclusions (see above, p. 310).   [4] 1 *Op. cit.*, p. 129.

trained German manufacturer in Manchester, described in a paper to the Literary and Philosophical Society an apparatus which he had devised for bleaching with a solution made by absorbing the gas in water.[1] The gas, produced in a lead retort, standing in a water-bath,[2] was passed through water in 'a range of four, five, or six hogsheads, or rum-puncheons, connected with one another, in the manner of Woulfe's distilling apparatus', which Rupp preferred to either Berthollet's or Watt's apparatus. 'Agitators, on Mr. Berthollet's principle, may be applied.' But the chief feature of his apparatus was his method of immersing the goods in the liquor. Rupp first made an interesting survey of previous bleaching methods:

The volatility of this acid and its suffocating vapours [when absorbed in water] prevented its application in the way commonly used in dye-houses. Large cisterns were therefore constructed, in which the pieces of stuff were stratified; and the liquor being poured on them, the cisterns were closed with lids. But this method was soon found to be defective, as the liquor could not be equally diffused; the pieces were, therefore, only partially bleached . . . Various other contrivances were tried without success, till it was discovered that an addition of alkali to the liquor deprived it of its suffocating effects, without destroying its bleaching powers. The process then began to be carried on in open vessels, and has been continued in this manner to the present period. The bleacher is now able to work his pieces in the liquor, and to expose every part of them to its action, without inconvenience.

This advantage was 'unquestionably great', but it was seriously diminished by the heavy expense on potash and by the fact that the bleaching power of the oxygenated muriatic acid was reduced 'in proportion as it is neutralized by an alkali'; the latter fact Rupp had

[1] 'On the Process of Bleaching with the oxygenated muriatic Acid; and a Description of a new Apparatus for Bleaching Cloths with that Acid dissolved in Water, without the Addition of Alkali', read 9 Feb. 1798, *Manchester Lit. & Phil. Soc. Memoirs*, Vol. V, pt. i (1798), pp. 298–313; *Nicholson's Journal*, Vol. II (1798–9), pp. 268–73; *Philosophical Magazine*, Vol. II (1798–9), pp. 293–301; O'Reilly, *op. cit.*, pp. 99–105, 186–91, and plate 5.

[2] Other such references (e.g. in Pajot des Charmes, *op. cit.*, O'Reilly, *op. cit.*, the *Tennant Proceedings* (1803), and Rees, *op. cit.*) show that by this time lead retorts, and also lead pipes and receivers, were coming into general use. Some manufacturers, however, used stone or earthenware retorts and receivers, while wooden receivers, lined with pitch, were also employed. By standing the retort in a water-bath (an iron boiler containing water) above the furnace, the temperature of the process was kept down and less trouble was experienced with hydrochloric acid and manganese being driven over into the receivers. Earlier a sand-bath had been used, e.g. by Berthollet, *Ann. Chim.*, Vol. II, p. 167, and Henry (see above, p. 300); see also Rennie, *op. cit.*, pp. 48–9.

Rupp recommended the following proportions of materials: 3 parts manganese, 8 parts common salt, 6 parts vitriol, 12 parts water. He found that the strength of the liquor was increased by diluting the acid more than was usually done.

proved by a series of careful tests. He had therefore devised an apparatus for using water instead of an alkaline solution. It consisted of a large cistern, with an air-tight lid, and fitted with two drums to carry the cloth, which could thus be repeatedly wound from one to the other, so that every part of it was exposed to the liquor. Rupp estimated that this apparatus would save 40 per cent on the present method, but it does not appear to have overcome the previous difficulties. Nevertheless, absorption of chlorine in water was still practised and Rennie stated in 1816 that it was 'still followed in some old establishments', though it could 'now be considered as one of the nearly obsolete agents in Bleaching, although it is unquestionably more powerful than any of the oxymuriates'.[1]

Clearly there were considerable differences of opinion as to the best method of bleaching in the 1790s, and in the prevailing uncertainty it appears that many bleachers still clung to the traditional 'grassing', or reverted to it after experiencing difficulties and disappointments. This uncertainty is illustrated by a letter addressed to Watt in 1798 by Thomas Creaser, of Bath, a friend of Dr. Beddoes and probably himself a doctor, who was writing on behalf of a brother-in-law, who wished to apply 'the new mode of bleaching' to coarse flaxen goods.[2] He had found 'by reading' that there were 'three modes employ'd viz. the addition of Lime to the Oxyg[enate]d Liquor, the addition of Alkali to the same, or the Oxyg[enate]d Liquor simply'; but he did not know which of these was preferable. What reply he was given we do not know.

In Scotland, as we have seen, the Trustees for Fisheries, Manufactures, and Improvements had sought to arrive at the best method by offering a bounty.[3] In Ireland, the Trustees of the Linen and Hempen Manufactures (the Irish Linen Board) tried the same device, and the outcome was a most interesting report, published in 1791,[4] which shows how knowledge of chlorine bleaching was spreading in that country. Dr. Richard Kirwan was probably the first to introduce it, on returning to live there in 1787, after having heard about it from Scheele,[5] or after reading Berthollet's publications. He certainly emphasized the importance of chemistry to the Irish linen industry.[6]

---

[1] *Op. cit.*, pp. 129–31. Brewster's *Edinburgh Encyclopaedia* stated that this mode of bleaching had 'been adopted only by a few'.

[2] T. Creaser to J. Watt, 23 Oct. 1798, Dol.   [3] See above, p. 312, n. 2.

[4] *Report on Experiments made by order of . . . the Trustees of the Linen and Hempen Manufactures, to ascertain the Comparative Merits of Specimens of Oxygenated Muriatic Bleaching Liquids, Sent by different Persons in Claim of a Bounty offered by the Trustees, to the Person who should produce the best Liquids* (Dublin, 1791).

[5] See above, pp. 259–60 and 289.   [6] See above, p. 186.

It also appears that his fellow-Irishman and chemist, Dr. William Higgins, was early in this field, for according to Brewster's *Edinburgh Encyclopaedia* 'Mr. Higgins of Dublin and Mr. Berthollet had both combined the oxymuriatic acid with potash, so early as the year 1788'.[1] Berthollet's method of bleaching with 'oxygenated muriatic acid' was discussed by the Irish Linen Board Trustees on 23 March 1790, when Kirwan attended and informed them that the new bleaching chemical was composed of 'Spirit of Salt distilled on Manganese'. Early in 1791, therefore, the Board arranged to offer a premium of £400 to the best formula for bleaching liquor.[2] The entries received were carefully tested and the resultant report showed that a good deal of progress had been made with 'this newly discovered bleaching Liquid'. Most of the report was taken up with chemical and bleaching experiments on various liquors by Robert Roe, of Ring's End, who was a member of the Royal Irish Academy. He 'presumed that the Reader has already received general Instructions from Mr. *Berthollet's Treatise* on this Subject, which is in the hands of almost every one', and therefore his main concern was 'to add Precision for the Use of People in Practice'. Roe explained how he had at first used Berthollet's apparatus for preparing the liquor, then what was obviously a variant of Bourboulon's patent apparatus, and finally a more simple and practical set of 'tubulated Retorts with Vitriol Bottles for Receivers', which he used when distilling from muriatic (hydrochloric) acid and manganese; but when distilling from vitriol, salt, and manganese he used 'a Kind of Alembicks with Stone-ware Bodies and Glass Heads fitted thereto'. He found it preferable to use diluted rather than mild lye in the receiver, but that the strongest and cheapest liquor could be produced with caustic lye, made by adding quicklime; the lye should be stirred from time to time to facilitate absorption of the gas. He also experimented with water in the receiver, but found the operation 'insupportably offensive', due to escape of gas; he warned that such liquor 'must always be made on the Spot where it is to be used . . . and it must be used immediately, as it will not retain the Gas for any considerable Time'. Roe made several other practical points, previously noted by Berthollet, Watt, and Henry, emphasizing particularly the importance of the indigo test for ascertaining the strength of the liquor. The experiments proved the merit of the new method in bleaching 'every species of Cotton Goods' (fabrics), thread, and yarn, also for linen yarn, but showed 'that it cannot be so generally useful in Linens' (linen cloth). It might, however, be 'partially useful', in finishing goods

---

[1] Article on 'Bleaching'.
[2] Wheeler and Partington, *op. cit.*, pp. 12–13.

which would otherwise have to be taken in for winter, or for which there was a suddenly increased demand.

In view of this lukewarm report, it is not surprising to find William Higgins, the Irish chemist, stating in 1799 that 'the oxygenated muriate of potash . . . does not appear to have made any great progress in our bleach-greens'.[1] In the Lancashire cotton industry, on the other hand, where its practical application presented much less difficulty, it came more rapidly into use. But even in this industry the cost of potash made its superiority over traditional methods economically doubtful. Watt's opinion—that chlorine bleaching would not, after all, prove a cheap way of bleaching[2]—was borne out by the experiences of Henry, Bourboulon de Boneuil, and others. 'The new process', as practised by Vallet, it was stated in 1795, 'is somewhat more expensive than the old.'[3]

The solution to these financial and technical problems was eventually to be found in the use of lime. Early experiments were not very successful. According to Rees, some bleachers at first 'put into the receiver, filled with water, a quantity of pulverized lime, then the goods themselves, and the whole agitated during the admission of the gas'; but the goods, being unequally coated with the lime, were unequally bleached.[4] A simple solution of lime-water 'was not . . . found to answer so well as the alkaline solution', because

water can dissolve no more than $\frac{1}{700}$th part of its weight of lime; a quantity wholly insignificant in neutralizing the oxygenated muriatic acid for the purpose of the bleacher; nor could pulverized lime, merely thrown into the water in the receiver, serve a better purpose, since . . . all beyond the quantity in chemical solution subsided and remained nearly useless at the bottom of the receiver.[5]

[1] W. Higgins, *An Essay on the Theory and Practice of Bleaching* (1799), p. 49. He also pointed out the various inconveniences of absorbing 'oxymuriatic acid gas' in water. Higgins, who became professor of chemistry to the Dublin Society and also chemist to the Irish Linen Board in 1795 (see above, pp. 185–6), was keenly interested in the utilitarian applications of chemistry, declaring, indeed, that 'the improvement of the arts and manufactures . . . is the ultimate and grand object of the science' (*ibid.*, p. xii). He produced this treatise expressly 'for the use of bleachers', written in a 'simple and familiar' style, since 'the majority of them are not acquainted with the theory of chemistry' (*ibid.*, pp. vii–x). Some, however, were scientifically minded, including John Duffy, of Ball's bridge, near Dublin, 'who from his knowledge of chemistry is very well acquainted with the principles of bleaching', and who provided Higgins with facilities for large-scale experiments (*ibid.*, p. 64). See below, pp. 327 and 336, for some of the results of his researches.

[2] See above, pp. 281 and 286, n. 5.   [3] Holt, *op. cit.*, p. 214.

[4] Rees, *op. cit.*, article on 'Bleaching'. This was similar to the method developed by Thomas Henry and used by Robert Hall (see above, pp. 316–17).

[5] *Ibid.* See also Tennant's patent specification of 1798, reprinted in *Tennant Proceedings* (1803), pp. 12–17.

It was probably for these reasons that, although Watt, Henry, Cooper, and others had experimented with lime, 'it does not appear [to have been] brought by them into their general practice, as they seem to have preferred the simple water or the oxymuriate of potass'.[1] In 1798, however, Charles Tennant, a bleacher at Darnley, near Glasgow, took out a patent for using a suspension of lime, in place of a solution of potash.[2] Tennant had been joined in partnership by Charles Macintosh, James Knox, Dr. William Couper (or Cooper), and Alexander Dunlop —another notable combination of bleaching with industrial chemical experience. Macintosh, as we have previously noted, had taken an early interest in chlorine bleaching, making his own experiments and visiting Manchester for information, on behalf of the Prestonpans Vitriol Company.[3] His main activities during the early 1790s,[4] however, were in the cudbear and Turkey-red works established by his father, to which he added the manufacture of mordants such as lead and aluminium acetates; his sal-ammoniac works, in which Dr. William Couper, an eminent Glasgow surgeon, had been a partner, was closed in 1792, but in 1796–7 he established a large alum works at Hurlet, near Paisley, in partnership with James Knox and others. At the same time, he continued to take an interest in the new bleaching process[5] and appears to have assisted Charles Tennant in its application; by the time of the 1798 bleaching patent Tennant, Macintosh, Knox, and Couper were in partnership and soon afterwards established the later-famous St. Rollox chemical works, Glasgow, for its exploitation. Tennant's experience of bleaching operations combined with the scientific knowledge and textile-chemicals experience of Macintosh and the other partners placed them in a very strong position.

Tennant, in his specification, disclaimed having discovered that oxymuriatic acid gas could be absorbed to some extent in lime-water, and, for the reasons above mentioned, pointed out that this was ineffective. What he claimed to have done was to have invented a new process 'of keeping the Lime in a state of mechanical suspension' in the

[1] Rennie, *op. cit.*, p. 134. This statement, however, is probably not true of Henry.

[2] Patent No. 2209, 23 Jan. 1798. The patent applied to the use of 'calcareous earth' (lime, either carbonate or quicklime, preferably the latter) and also 'the earths strontites and barytes', but as the latter were relatively scarce and expensive, lime was used almost exclusively. *Repertory of Arts and Manufactures*, Vol. IX (1798), pp. 303–9; *Abridgments of Patent Specifications* (Bleaching, Dyeing, etc., 1859), p. 68; Rees, *loc. cit.*; Parkes, *op. cit.*, Vol. IV, pp. 64–7; Rennie, *op. cit.*, pp. 134–5. Hardie, *op. cit.*, p. 5, and Clow, *op. cit.*, pp. 191–2, refer to Tennant's process very briefly.   [3] See above, pp. 293–5.

[4] See Hardie, *op. cit.*, pp. 3–5, and Clow, *op. cit.*, pp. 237–9 and 246 ff.

[5] See Clow, *op. cit.*, p. 250, for his correspondence with the German, Baron d'Aescher, concerning difficulties in managing it.

water, by agitation with 'a wooden paddle or rake', whereby the lime reacted with the gas to form a soluble compound (in contemporary terminology, 'oxygenated muriate of lime', or 'oxymuriate of lime', or, as it was later called, chloride of lime; in modern chemical parlance, calcium hypochlorite).[1] This substitution of lime for potash was of great benefit, not only because of the relatively negligible cost of the former, but also because the bleach liquor thus obtained proved a good deal more reliable, especially in treating coloured goods.[2] But, being a liquid, this product still suffered from the same drawbacks to its transport and sale as *eau de Javelle*.[3] Tennant therefore undertook to grant licences for the use of his patent process, at £200 each,[4] and Charles Macintosh immediately went on a tour of Lancashire, where, after successful trials, 'a great many of the most considerable bleachers' contracted to use it.[5] These included Thomas Hoyle, John Horridge, the Ainsworths, Richard Bealey, and doubtless others. Horridge declared that Tennant's invention 'has saved me 1000 *l.* a year in ashes since I began; and since then I have had very few [goods] damaged, indeed scarcely any'.[6] At Ainsworth's, the saving was estimated at 'from twenty-four to twenty-six hundred pounds each year', and the goods were bleached 'at more certainty'.[7]

A number of other Lancashire bleachers, however, though they admitted not having previously used Tennant's process themselves, claimed that others had done so, and therefore began to practise it without licence. Tennant & Co. at once began proceedings in the High Court of Chancery, in 1798, to secure an injunction against the

---

[1] This reaction is expressed in the following formula:

$$2Ca(OH)_2 + Cl_2 \longrightarrow 2CaOCl + 2H_2O$$

To produce the gas, Tennant suggested a mixture of 30 lb. manganese with 30 lb. common salt, to which 30 lb. vitriol, diluted with its own bulk of water, should be added. For the receiver, he suggested 140 gallons of water and 60 lb. of finely powdered quicklime; he also suggested adding 30 lb. of common salt, but merely to increase the liquor's specific gravity and so help to keep the lime in suspension.

[2] All witnesses at the Tennant trial agreed on this. See also Brewster's *Edinburgh Encyclopaedia*, article on 'Bleaching'. But the new liquor proved unsuitable for maddered goods, for which oxymuriate of potash or soda, as originally recommended by Henry, was still to be preferred (see above, p. 301). The latter, however, was very slowly adopted; bleachers still clung to the traditional methods for these goods. Parkes described it in 1815 as 'one of the last improvements in calico-printing', especially important for the finest ginghams and chintz work, with their variety of delicate colours. By that time, however, it had 'almost entirely superseded the operation of crofting' (*op. cit.*, Vol. IV, pp. 84–6 and 141–143). See also Rennie, *op. cit.*, p. 132.

[3] Parkes, *op. cit.*, Vol. IV, p. 64.   [4] *Ibid.*, p. 65.

[5] This and the following statements are taken from *Tennant Proceedings* (1803).

[6] *Ibid.*, p. 160.   [7] Bentley's evidence, *ibid.*, pp. 167–8.

following bleachers, all in the Bolton area: James Slater, James Varley and Joseph Slater, of Little Bolton and Sharples;[1] John Glover, of Great Lever; Thomas Thweat, of Burnden; James Shaw, of Tong-with-Haulgh; Thomas Slater and John Sudren, of Turton; James Lomax, of Bradshaw-hall; John Thweat, of Harwood, and Jonathan Hardman, of the same place. But Tennant & Co. were opposed not only by these Lancashire bleachers, but also by Robert Hall, of Basford, and an injunction was refused. Tennant & Co. thereupon commenced an action in the Court of King's Bench, and the trial commenced before Lord Ellenborough, Chief Justice, and a special jury, on 23 December 1802. These legal proceedings, as we have already had occasion to notice, produced a mass of evidence, not only about the application of the wet-lime process, but also about earlier methods of chlorine bleaching. It was clearly demonstrated that both lime and agitators, either separately or together, had been used for some years previously, as we have already seen, notably by Thomas Henry, Robert Hall, and Bourboulon de Boneuil, but also by various others.[2] Robert Hall and Hugh Foy gave evidence at the trial. The former described the apparatus which he had used since 1789,[3] and the latter not only referred to the use of lime by Boneuil & Co., and his fitting of agitators in receivers for making potash bleach liquor for various manufacturers, but also stated that he had actually erected such an apparatus for Tennant himself in 1796 and had told him that lime might be used if kept in agitation.[4] Vallet had similarly supervised the fitting of agitators for Ainsworth's and Rothwell's in 1796-7.[5] James Morris, of Bolton,[6] and Samuel Gratrix, of Disley,[7] had used the wet-lime process

[1] Slaters, of the Dunscar bleachworks, Bolton, apparently established *c.* 1750, were leading opponents of Tennant. A valuation of their stock in 1792 (the firm then being Holme & Slater) included manganese, spirits of salt, and a 'Chimicall house', so they were certainly using chlorine bleaching by that date; but evidence in the Tennant case shows that they did not adopt the wet-lime process till 1798. (The author is grateful to the Bleachers' Association for access to these and other records of early bleaching firms.)
A letter from Isabella Banks, the novelist, grand-daughter of James Varley, who was in partnership with the Slaters at the time of the Tennant case, claims that a dismissed workman had carried the chloride of lime process to Tennant. Varley was an intellectual type of manufacturer, having a library of books in many languages, of which he was apparently able to understand fourteen. *Manchester City News, Notes and Queries*, 27 Oct. 1883. Varley had previously been in partnership with John Horridge in the Raikes bleachworks formerly belonging to Baker, Cooper & Co. (see above, p. 308).   [2] See above, pp. 316-17.
[3] *Tennant Proceedings* (1803), pp. 41-4 and 189-94.
[4] *Ibid.*, pp. 205-12. The agitators were powered by water-wheels.
[5] *Ibid.*, pp. 144-54, evidence of William Belshaw.
[6] *Ibid.*, pp. 45-8 and 213-16, affidavit and evidence of Alexander Atherton.
[7] *Ibid.*, pp. 195-201, evidence of Peter Pilkington, and *ibid.*, pp. 202-5, evidence of Samuel Gratrix.

and agitators in their bleaching operations since 1795–6, in each case with an apparatus similar to that patented by Bourboulon de Boneuil.

A very interesting feature of this case is the evidence of the important role played by chemists in all these developments. We have already referred to the activities of Vallet and Foy; the latter was described at the trial as 'a chemist, [now] at Glasgow', and there are references to several others, including, of course, Thomas Henry. Slater, Varley and Slater were 'instructed by Robert Raper, now of Little Bolton, chymist, in the use of lime as a substitute for pot or pearl ashes, in preparing the oxygenated muriatic acid gas liquor', in September 1798, and Raper stated that 'he had practised it at the bleaching works of the Honourable Fitz Morris, at Llewery, in Wales, eight or nine years ago'.[1] Luke Howard, 'a manufacturing chemist', of Lombard Street, London, and Plaistow, Essex, but formerly of a town 'within seven miles of Manchester', had 'made experiments at several times on the preparation of bleaching-liquor, in consequence of having a demand from neighbouring manufacturers. I have made a bleaching-liquor in small quantities several years in consequence of such demands.'[2] After experimenting unsuccessfully with absorption of the gas in water, he had used a solution of potash or pearl-ash; he had also tried to neutralize the acid with chalk and marble, though not with lime. Ainsworth's, in addition to employing Vallet, also utilized the services of another chemist called Lloyd.[3] An Edinburgh chemist named Davy had 'invented, and afterwards sold to the bleachers in his neighbourhood', a liquor precisely the same as Tennant's, prior to the latter's patent;

---

[1] *Ibid.*, p. 54. The works referred to must have belonged to the Hon. Thomas Fitzmaurice (1742–93), of Llewenny Hall, co. Denbigh, brother of Lord Shelburne. He had attended Dr. Black's lectures in 1760–1 (J. Black to J. Watt, 15 March 1780, Dol.), and was evidently interested in the industrial applications of chemistry. His brother, Lord Shelburne, of course, was a well-known patron of scientists such as Priestley, Magellan, Warltire, etc. Evidence about the bleaching interests of the Hon. Thomas Fitzmaurice is provided by the following contemporary reference: 'During our stay we made an excursion to Llewanny, the seat of the Honourable Mr. Fitzmaurice, brother to Lord Landsdowne. The gentleman has an estate of 17,000l. a year, and only one son; and yet he has plunged himself into a business which might make even a tradesman tremble. He is a bleacher of linen. The buildings which he has erected, and the machines and apparatus which he has placed in them, are really astonishing. He has a shop in Chester, at which he sells his linen when it is bleached.' R[ichard] T[wining] to the Rev. T[homas] T[wining], 20 July 1785, Richard Twining (ed.), *Selections from Papers of the Twining Family* (1887).

[2] *Ibid.*, pp. 127–43. Howard had been apprenticed in 1787 to Ollive Sims, a Quaker apothecary at Stockport (*Camden Miscellany*, Vol. XXII (1964), p. 117). For Howard's scientific and industrial interests, and his partnership with William Allen, see above, pp. 137–8.

[3] *Ibid.*, p. 169. They also did business with Howard (*ibid.*, p. 170).

'he manufactured it as an article for sale'.[1] Another Edinburgh chemist, William Kinnaird, had similarly, since 1792–3, 'been in the practice of making a bleaching liquor with lime and oxymuriatic acid', using mechanical agitation, and had sold this liquor to neighbouring bleachers; he declared that such a liquor 'has been used in the regular way of business for bleaching, by manufacturers in and near Edinburgh since the year 1793, and that it has been still more extensively prepared for trade in the north of England since the year 1791'.[2]

In view of all this evidence, it is not surprising that Lord Chief Justice Ellenborough intervened to close the case (with the jury's agreement), describing Tennant's patent as 'a scandalous patent . . . as unfounded in all sorts of merit, as it is in law'. Tennant was the Arkwright of the chemical industry, exploiting other men's ideas. In 1799, the year after the wet-lime patent, he took out another,[3] for producing oxymuriate of lime, etc. 'in a dry, undissolved, or powdery form, without suspension or solution in water', i.e. for absorbing chlorine in dry hydrate of lime—lime very slightly slaked—to produce bleaching powder, which solved the transport problem and therefore made it possible for St. Rollox to supply bleachers with comparative ease. This, however, was not his own invention, but that of Charles Macintosh,[4] who in the following year was squeezed out of control of the business, although he appears not to have finally severed relations until 1814.[5] Tennant then proceeded, by exploitation of this process, to achieve 'world fame and great fortune'; linked with the manufacture of sulphuric and other acids and with that of Leblanc soda, production of bleaching powder became one of the bases of the nineteenth-century chemicals industry.

Bleaching powder, however, did not create an immediate revolution in the industry. Production at St. Rollox grew very slowly for some years, from 52 tons in 1799–1800 to 147 in 1805, 333 in 1820, and 910 in 1825; by that time it was beginning to increase more rapidly, to reach 9,251 tons in 1870.[6] Bleaching powder did not, as is usually

---

[1] *Tennant Proceedings* (1803), p. 183.

[2] *Ibid.*, Appendix. At the same time as the legal proceedings on the Tennant patent were going on in England, a similar action, between Tennant and Kinnaird, was taking place in the Court of Session in Edinburgh. Kinnaird's 'Memorial' is reproduced as an Appendix in the *Tennant Proceedings*. In view, however, of Miss Gauldie's researches on Scottish bleaching, we are not dealing with this case here.

[3] No. 2312, 30 April 1799; *Repertory of Arts and Manufactures*, Vol. XIII (1800), pp. 1–5; *Abridgments of Patent Specifications* (Bleaching, Dyeing, etc., 1859), p. 71.

[4] Hardie, *op. cit.*, p. 5; Clow, *op. cit.*, p. 192.

[5] Hardie, *loc. cit.*

[6] Higgins, *History of Bleaching*, p. 90; Clow, *op. cit.*, p. 193.

suggested, prove universally successful,[1] while most bleachers still preferred to make their own 'chemic', using the wet-lime process. The latter, as we have seen, was quickly adopted in Lancashire, while in northern Ireland, according to William Higgins, there were thirty such plants in operation by 1799.[2] Higgins decidedly preferred it to the oxymuriate of potash, because it was 'cheaper, and less liable to injure the texture of the cloth'.[3] But it still did not dispense with 'grassing' for linens: several weeks' exposure in the bleachfields was still necessary, in addition to repeated boilings in potash and immersions in the oxymuriate of lime.[4] Higgins therefore tried to reduce costs further by substituting sulphuret of lime—made from sulphur (much cheaper than potash) and lime—in place of potash in the 'bucking' process.[5] Despite his strong advocacy, however, it does not appear to have come into general use.[6]

At about the same time, the French chemist, Chaptal, published an account of a bleaching process, particularly for yarn, by means of vapour from boiling alkaline liquor contained in a pressured vessel similar to a large Papin 'digester'.[7] According to the *Journal de Physique* this method was rapidly adopted in England. De la Métherie reproduced a letter from a 'savant' in London referring to Chaptal as 'très-estimé parmi nos fabricans', and stating that his work had been translated in several English journals.[8] This correspondent reported that difficulties were at first experienced owing to uneven bleaching, where the

[1] See above, p. 323, n. 2, for the unsuitability of oxymuriate of lime in bleaching maddered goods, and below, p. 331, for John Marshall's failure with it in bleaching linen yarn.
[2] Higgins, *Essay on the Theory and Practice of Bleaching*, pp. 42–3. Higgins provided (pp. 37–47) a detailed description of the plant used: a large conical lead still, placed in a water-bath (a copper or iron boiler), with lead pipes for conveying the gas, via an intermediate lead vessel (for dissolving any escaping 'muriatic acid gas'), to a large wooden receiver, capable of holding 800 gallons of water and 80 lb. of well-powdered quicklime, which could be stirred with an 'agitator'. Higgins suggested a charge for the still of 60 lb. oxide of manganese, 60 lb. common salt, and 50 lb. sulphuric acid diluted with its own bulk of water. The oxymuriate of lime thus produced was of sufficient strength to be thrice diluted. Higgins' description was reproduced in O'Reilly, *op. cit.*, pp. 87, 90–1, 119–21, 178–81, and plate 3.   [3] *Ibid.*, p. 50.   [4] *Ibid.*, p. 51.
[5] *Ibid.*, pp. 53–71; *Nicholson's Journal*, Vol. III (1799–1800), pp. 253–7; O Reilly, *op. cit.*, pp. 122–32.   [6] Rennie, *op. cit.*, pp. 123–4.
[7] *Journal de Physique*, Vol. LI (1800), pp. 305–9; O'Reilly, *op. cit.*, pp. 132 ff.
[8] See *Nicholson's Journal*, Vol. IV (1800–1), pp. 469–71, Vol. V (1801), pp. 233–5; *Philosophical Magazine*, Vol. V (1799–1800), pp. 351–3; *Repertory of Arts and Manufactures*, Vol. XII (1800), pp. 197–201; Parkes, *op. cit.*, Vol. IV, pp. 148–53; Rennie, *op. cit.*, pp. 180–2. This method of bleaching had apparently been practised much earlier in the Levant, and it had been suggested to John Wilson, the Manchester dyer, many years previously by Peter Ottersen (Rees, *op. cit.*, article on 'Bleaching'). Chaptal introduced more scientific control.

materials rested on the wooden frames or where they were folded, but that this was overcome by winding and unwinding them continuously on a winch during the bleaching process, utilizing Rupp's apparatus. Chaptal apparently considered that this caustic–alkali–vapour treatment, followed by washing and a few days' 'grassing', would be sufficient to bleach fully, but O'Reilly (1801), who mentioned this process being used in Ireland as well as in France, referred to it only as an effective substitute for traditional 'bucking' and stated that it was to be followed by washing and then treatment in a solution of oxymuriatic acid or oxymuriate of lime, and then by further washing and 'grassing'.

No one who reads the long detailed contemporary descriptions of bleaching operations by Berthollet, Chaptal, Pajot des Charmes,[1] O'Reilly,[2] Higgins, Rupp, etc. can possibly be under the impression that chlorine had reduced a complicated traditional process to a simple operation. All these accounts, and those of encyclopaedists such as Rees, Nicholson, Aikin, Parkes, and Brewster in the early nineteenth century, emphasize (as Berthollet had done originally) the necessity for scientific control of these difficult chemical operations, with particular emphasis on analysis of raw materials and testing of the strengths of lyes, bleach liquors, and sours. Nevertheless, by the early nineteenth century all the leading bleachers seem to have acquired sufficient knowledge and skilled personnel to manage the new process. They were encouraged to make their own bleach liquor by an Act of 1798,[3] allowing a drawback of the duty on salt used in making oxymuriatic acid, but only to those making it for their own consumption, not to those manufacturing it for sale.[4] This Act 'induced not only the bleachers but also

---

[1] *L'Art du Blanchiment des Toiles, Fils et Cotons de tout genre* (Paris, 1798), immediately translated into English by William Nicholson and published in 1799. Running to 351 pages, with many drawings of apparatus, it provides the most detailed contemporary account of the new bleaching methods.

[2] *Essai sur le Blanchiment* (Paris, 1801); a summarized English translation at once appeared in the *Philosophical Magazine*, Vol. X (1801), pp. 97–111, 247–64, and 299–317.　　　　　　　　　　　　　　　　　　　[3] 38 Geo. III, cap. 89.

[4] See below, pp. 360–8, for the effects of the salt duties on early chemical manufactures. Another burden was the excise duty of £30 per ton on sulphate of soda (Glauber's salt), which could be obtained from the residues of chlorine distillation (see above, p. 254, n. 1); until this was removed in 1815, there was little incentive to recover this by-product (Clow, *op. cit.*, p. 193). There was only a limited sale for the sulphate itself, but it could be converted into the carbonate (alkali), the economy of which Berthollet had emphasized (see above, pp. 298–299). The French chemist, de Morveau, had developed such a process, which was kept secret (see above, p. 299, n. 1), but Pajot des Charmes (*op. cit.*, English trans., pp. 254–5) and O'Reilly (*op. cit.*, pp. 94–7) later referred to several such processes, and so, too, did the British chemist, F. Accum, in *Nicholson's Journal* (new ser., Vol. II (Aug. 1802), pp. 241–4), from which it appears that a number of manu-

the large calico-printers to make their own bleaching liquors, especially since the legal determination respecting the validity of Mr. Tennant's patent; so that an apparatus for this purpose, is now considered to be a necessary adjunct to every printing establishment in the North of England.'[1]

The rapid increase in the number of bleachers and calico-printers adopting the new process makes it impossible to chart the expansion of chlorine bleaching in further detail. A few of the more interesting examples, however, are worth mentioning. Bleaching, even before the introduction of 'Chymical' methods, was a specialized branch of textile manufactures, but in some important instances it was practised by firms whose main concern was spinning yarn. Towards the end of 1788, for example, the Strutts were planning to start bleaching at Milford, and in January 1790 we find Samuel Oldknow advising William Strutt about 'the New bleaching Liquor' his brother had prepared.[2] Sir Richard Arkwright, as might be expected, had acquired knowledge of the secret and had got several bottles, but claimed that 'his method of making it is his own & he asserts that he is not beholden to any man for what he knows of the process. Mr. W. [Watt?] & he do not communicate nor visit in the Character of Chymists'. The Strutts were so successful that by June 1801 the capacity of the Milford bleachworks was nearly 500–600 bundles of cotton per week.

Professor Rimmer has produced extremely interesting evidence on the efforts of John Marshall, the great Leeds flax-spinner,[3] which again provide a fascinating example of the links between science and industry, and also illustrate the greater difficulties of applying chlorine bleaching in the linen industry as compared with cotton. Marshall, a a man of wide scientific interests, both theoretical and experimental, had in 1790, as we have previously seen, acquired knowledge of

---

facturers were now utilizing the waste sulphate, 'sold cheap by the bleachers'. But when William Henry sought to develop such a process, his efforts were frustrated by the Excise regulations (see above, p. 249).

[1] Parkes, *op. cit.*, Vol. IV, pp. 68–9. Baines, *op. cit.*, p. 250, stated in 1835 that, although Tennant's bleaching powder was coming into more widespread use, 'the great bleachers use liquid chloride of lime, which they make in leaden stills, steam being used to expel the gas from the materials, and the gas being received into a cream of lime'.

[2] R. S. Fitton and A. P. Wadsworth, *The Strutts and the Arkwrights 1758–1830* (1958), p. 295, n. 2. Thomas Oldknow, Samuel's brother and partner, managed their bleachworks at Heaton Mersey. G. Unwin, *Samuel Oldknow and the Arkwrights* (1924), pp. 106 and 151.

[3] W. G. Rimmer, *Marshall's of Leeds, Flax-Spinners 1788–1886* (1960), pp. 50–3. Detailed accounts of his experiments are preserved in Marshall MS. 55, Brotherton Library, University of Leeds.

bleaching with 'dephlogisticated marine acid' from James Booth's philosophical lectures, which he had supplemented with information from Scheele's *Chemical Essays*.[1] It was not, however, until some years later, in 1796, that he bought a bleach-yard and began to carry out a series of practical experiments. He began by making copious notes on 'Berthollet's method of bleaching by oxygenated muriatic acid, translated in the Repertory of Arts from the Annales de Chimie'.[2] Then he acquired knowledge, through William Wood, of practical bleaching operations in the Bolton area. The information obtained there throws very interesting light on cotton bleaching at that time.[3] For distillation, equal quantities of manganese, salt, vitriol, and water were used, in a 'pot or matrass' generally containing about eighteen gallons, while the receiver or 'tub' was about twenty times as large; no intermediate vessel was used, no safety-pipe, and no agitator in the closed receiver, so the apparatus was much simpler than that used by Berthollet or Watt. No use was made of the residue of distillation, 'but at Nottingham they have lately begun to apply it to some use, the process of which is yet secret'. The liquor so made was afterwards diluted with eighteen parts of water for bleaching cotton; they never tested the strength of the liquor, but relied on the proportions of materials used. This bleach, after use, was strengthened with fresh liquor and used repeatedly until too foul. It was the general custom at Bolton to add potash to the water in the receiver, though this was 'not universal'. They had 'never had a piece of cloth damaged these seven years', though 'at Nottingham where they bleach their stockings chiefly in gas they have frequently very heavy losses by damages.'

Their usual course is first a work with lime, then after it is well washed & without lying down at all in the croft a work with ashes [potash]. They then lay it out in the croft about 2 days on each side, then take it up & steep it 24 hours in the liquor, next give it a light work with ashes, then steep it 24 hours again in the liquor & last of all a light work with ashes & the color is a perfect white.[4]

---

[1] See above, pp. 78 and 153–4.

[2] Marshall MS. 55, pp. 2–11; *Repertory of Arts and Manufactures*, Vol. I (1794), pp. 53–72, 121–30, 190–201, and 260–73; Vol. II (1795), pp. 354–60.

[3] *Ibid.*, pp. 14–20. This information was probably obtained mostly from Ainsworth and Vallet, of Halliwell, who are, in fact, specifically mentioned on p. 20 and also crop up later. Wood was perhaps the scientific Nonconformist minister (see above, pp. 157–8).

[4] Ainsworth & Vallet recommended a different sequence of operations, involving bowking, grassing (for three or four days), souring, washing, and then another bowking and washing, before immersion in the bleach liquor, diluted with 14 to 10 parts of water, for 12 to 20 hours, followed by washing—this whole sequence being then repeated.

Marshall began bleaching experiments in March 1798, 'with the bleaching liquor sent by Ainsworth & Vallet'.[1] Its cost was somewhat below that prepared by Henry[2]—2*d*. per lb., or 3*d*. per pint (2*s*. per gallon), including carriage and bottles—but he found that it had to be used in a much more concentrated state for linens than for cottons; it had little effect when diluted with more than four parts of water.[3] The cost of bleaching with it was therefore much too high.[4] So Marshall decided to make his own bleach, but first obtained more information, from Manchester,[5] and himself visited Bolton, where he found that by using lime in the receiver (which was lead-lined) they were able to make the liquor at less than ½*d*. per lb.[6] Meanwhile he also improved his knowledge by reading Lavoisier's *Elements of Chemistry*,[7] the relevant parts of the *Encyclopédie Méthodique*,[8] Rupp's article in the Manchester Literary and Philosophical Society's *Memoirs*,[9] Pajot des Charmes' *Art of Bleaching*,[10] and other works.[11] He also obtained more information from Vallet[12] and eventually got him over to Leeds to supervise more experiments in September 1799,[13] absorbing the gas in water only. These, however, proved unsuccessful: the yarn was bleached very unevenly, owing to its lying heavily together. Marshall therefore reverted to absorption of the gas in a lime solution, and also tried Rupp's system of moving the yarn on rollers;[14] the bleaching was good, but the costs were still too high. He also tried Tennant's new patent bleaching powder, but it proved seven times dearer than using his own liquor, and also ineffective.[15] Chaptal's method of steam-alkaline bleaching was also a failure.[16] Marshall therefore reverted to

[1] Marshall MS. 55, pp. 21–2, 26–30. [2] See above, p. 301.

[3] Higgins recommended dilution with three parts of water for linens (see above, p. 327, n. 2).

[4] Even at 1*d*. per pint, and if one steeping were sufficient, it would still be 'not quite as cheap, as the common mode of bleaching'.

[5] Marshall MS. 55, p. 30. At 'Barker's bleach-yard near Manchester', they used 40 lb. manganese, 40 lb. vitriol, 20 lb. salt, and no water, and this would impregnate 500 gallons of water, producing a stronger liquor than Ainsworth and Vallet's at no more than 1*s*. per gallon. [6] *Ibid*., p. 31.

[7] *Ibid*., p. 32. [8] *Ibid*., pp. 11 and 19.

[9] *Ibid*., pp. 33–4. [10] *Ibid*., p. 40.

[11] Such as Home's *Experiments on Bleaching* (1756) and Curry's *Elements of Bleaching* (Dublin, 1774). He was particularly interested in their experiments on potash. Marshall MS. 55, pp. 23–4 and 35–6. [12] *Ibid*., p. 44.

[13] *Ibid*., pp. 45–50. They used 60 lb. each of salt, manganese, vitriol, and water, and 300 gallons of water only in the receiver, which was fitted with an agitator.

[14] *Ibid*., pp. 54–7.

[15] *Ibid*., p. 57. At Glasgow, on the other hand, cotton bleachers found it satisfactory and 'as cheap as the liquor they make themselves'. It is interesting to note that Marshall was using 'the Hydrometer & the blue test' to ascertain the strength of the liquor (*ibid*., p. 57). [16] *Ibid*., pp. 63–71.

bleach liquor, still not very successfully,[1] and in the following years had to use a combination of chemical bleaching with traditional 'grassing'. Even the most up-to-date applied science and the most thorough practical experimentation had not yet made the new method fully successful in flax and linen bleaching.[2]

Combinations of bleaching with spinning—either of cotton (as by Arkwright, Strutt, and Oldknow), or of flax and linen (as by Marshall) —appear to have been unusual. Bleaching was, for the most part, a separate, specialized trade, or was combined with dyeing and calico-printing, which had similar chemical characteristics, distinct from the mechanical operations of spinning and weaving. Bleaching also tended, in some notable cases, to become combined with general chemical manufacturing of acids, alkalies, etc., along the lines envisaged by Berthollet. The examples of Bourboulon de Boneuil & Co. and Baker, Cooper & Co. were not very propitious, but others proved more successful. Tennant, as we have seen, proved remarkably successful in Scotland. In Lancashire, Bealey's of Radcliffe provide another success story.[3] An old-established bleaching firm, they had become a large concern by the early 1790s, with eight water-wheels and eleven dash-wheels in their works, which were extended in 1791 to include the manufacture of vitriol, principally for their own use; by 1799 they had six lead chambers '12 ft. by 10 ft. by 10 ft., roofed like a cottage', in which the charges of sulphur and nitre were burnt to produce the acid.[4]

Not all bleaching firms, however, were large integrated concerns like Bealey's. There were a great many small 'crofters' in the late eighteenth century, some of whom adapted themselves to the new methods, though in general they were less willing and able to do so. Richard Pilkington, father of the founder of the famous St. Helens glass firm, provides an interesting example of a small but progressive crofter.[5]

[1] He had, however, succeeded in reducing costs of production substantially. At first, using Ainsworth and Vallet's liquor, the cost of bleaching a bundle of linen yarn was 3s. for the liquor only; by the beginning of 1803 it was 9d. (*ibid.*, pp. 21–2, 26 and 73).

[2] Charles Bage, of Benyon's, another Leeds flax-spinning firm, was still experimenting unsuccessfully with absorption of chlorine gas in water in 1815 (Rimmer, *op. cit.*, p. 51, n. 1).

[3] Sykes, *op. cit.*, pp. 70–3; T. Swindells, *Manchester Streets and Manchester Men* (Third series, Manchester, 1907), pp. 125–6; J. Mactear, *Glasgow Phil. Soc. Proc.*, Vol. XIII (1880–2), pp. 420–2; O. Guttman, 'The Early Manufacture of Sulphuric and Nitric Acid', *Journ. of Soc. of Chem. Ind.*, 31 Jan. 1901, pp. 5–8, statements by J. Forbes Carpenter.

[4] The contemporary account of their vitriol manufacture, quoted by Mactear, contains some interesting references to Kirwan's chemical analyses.

[5] T. C. Barker, *Pilkington Brothers and the Glass Industry* (1960), pp. 22–4.

Starting as a bleacher at Horwich in 1785, he, too, visited Bolton, in November 1790, 'bought a Bottle of Chimical licquor'—probably from Ainsworth and Vallet—and by the following April 'began to make Chimical Licquor' for himself. Two years later he was 'fixing to make Chimical a New way'; but in February 1796 he sold out for £300 to Messrs. Ridgway & Son, of Wallsuches bleachworks, Horwich, who were on their way to becoming one of the biggest bleaching firms in Lancashire, after assistance from Thomas Henry.[1]

As Professor Barker observes, the greater technical knowledge and more expensive equipment required by the new process tended to squeeze out the smaller, less progressive firms. (On the other hand, of course, it should be said that chlorine bleaching effected considerable capital savings in land and goods in process.) Dr. Green has pointed out that 'the hazards of the long years of war between 1792 and 1815', as well as the technical changes in bleaching, 'worked against the smaller businesses' in the linen trade of northern Ireland.[2] Joshua Gilpin, visiting the Linen Hall in Dublin in 1796, found that bleachers were generally 'men of Fortune',[3] and Edward Baines later referred to the 'large capital' invested in Lancashire bleachworks, in power-operated machinery, etc.[4] Some large concerns, however, were to be found even before the chemical revolution in bleaching. A capital outlay of £3,000 was estimated for a well-equipped Irish bleachworks in the 1750s, including buck-houses and keirs, water-mill, washing-stocks, calender, beetling engine, dry-house, etc.[5] Slater's bleachworks at Dunscar, Bolton, was valued in 1792 at £2,081 (including buildings and stock), while their annual turnover, which had been only £505 in 1785, rose to an average of between £10,000 and £14,000 between 1799 and 1809.[6] Peter Ainsworth (d. 1807), of Bolton, was said to have been 'worth more than £10,000'.[7]

On the other hand, even bigger and technically advanced firms could fail, as was strikingly demonstrated by Bourboulon de Boneuil & Co. and Baker, Cooper & Co. They, however, like Henry and Pilkington, were relative newcomers to the bleaching trade, whereas most of those who emerged successfully from this technical revolution were old-established firms, like Ainsworth's, Rothwell's, Slater's, Ridgway's, and Bealey's. As we have seen, success depended not merely on scientific know-how, but also on practical bleaching experience, and involved an enormous amount of trial-and-error.

[1] See above, p. 315.
[2] E. R. R. Green, *The Industrial Archaeology of County Down* (Belfast, 1963), p. 5.
[3] 'Journal', Vol. XXI, 2 Sept. 1796.   [4] Baines, *op. cit.*, p. 253.
[5] C. Gill, *The Rise of the Irish Linen Industry* (Oxford, 1925), p. 246.
[6] Sykes, *op. cit.*, pp. 74–9 and 114–15.   [7] *Ibid.*, p. 68.

Moreover, chlorine affected only one part of the bleaching process: bucking, washing, souring, drying, etc. still had to be done, and keirs, dash-wheels, squeezers, dry-houses, etc. were still among a bleacher's basic equipment. Established commercial connections were also important; there was much distrust of 'chymical' methods—not unjustified in view of the many defects of early chlorine bleaching—and newcomers to the trade would find this a much greater obstacle to overcome than the established, reputable firms. Parkes could still refer, in 1815, to 'the prejudice which is generally entertained against what is called chemical bleaching'.[1] It was many years before chlorine bleaching became universally successful and before 'grassing' or 'crofting' disappeared. Well into the nineteenth century, one comes across references to, and illustrations of, the traditional methods still being used, especially for bleaching flax and linen, though often in combination with the new process.[2] Even in the cotton industry, the triumph of chlorine bleaching was nothing like so swift as is often imagined.

The bleaching of rags for paper-making was, of course, closely related to textiles bleaching, but space does not permit its treatment here.[3] It is, however, particularly interesting to note the findings of the American paper-manufacturer, Joshua Gilpin, whose manuscript 'Journal'[4] contains an immense amount of information about paper-making practices throughout Great Britain and Ireland in the mid-1790s.[5] Particularly significant are his references to the influence of

---

[1] *Op. cit.*, Vol. IV, p. 75. There was similar opposition in France, as Berthollet deplored in the *Annales de Chimie*, not only from conservative bleachers, but also, no doubt, from customers who found their goods damaged.

[2] Rennie, *op. cit.*, p. 166, writing in 1816, primarily about linens, pointed out that chlorine bleaching had not superseded crofting, but that it had certainly 'very much diminished the period which crofting was wont to require'. A drawing of Slater's works at Dunscar, Bolton, in 1837, shows that this firm, though among the earliest to adopt chlorine bleaching, were still 'grassing' some of their goods at that date (Sykes, *op. cit.*, pp. 78 and 103). See also the drawing of Messrs. Monteith's bleachfield, near Glasgow, in the *Penny Magazine*, 27 July 1844, p. 289, reproduced in Chaloner and Musson, *Industry and Technology*, plate 197.

[3] Chaptal was apparently the first to demonstrate the effectiveness of chlorine bleaching in paper-making in 1787: *Mémoires* of the Académie Royale for 1787 (Paris, 1789), pp. 611 ff.; *Repertory of Arts and Manufactures*, Vol. I (1794), pp. 355–60 and 427–32. Patents were soon taken out in England, e.g. by Clement and George Taylor, 1792 (no. 1872), Hector Campbell, 1792 (no. 1922), and John Bigg, 1795 (no. 2040).

[4] See above, p. 123, n. 5.

[5] He was also interested in textiles bleaching and referred, for example, to the use of 'chemical liquor' at various Glasgow bleachfields, including those of James Robertson & Co. and John Robinson ('Journal', Vol. IV, 20, 21, and 23 Oct. 1795). In both paper and textiles bleaching he refers generally to preparation of the bleaching agent from manganese, salt and vitriol, in lead or earthenware retorts.

Berthollet's works,[1] and his visits to chemists such as J. Parkinson, Fordyce, and Bancroft in London,[2] and Dr. Percival, Professor of Chemistry at Trinity College, Dublin.[3] There is no doubt about the widespread importance of chemical literature and personnel in paper as in textiles bleaching, though here again practical industrial experience was also requisite.

Research and development in bleaching processes continued into the nineteenth century, and the links between science and industry remained very close. Davy, Dalton, Gay-Lussac, and other scientists took up where Berthollet, Henry, and the rest had left off. Davy directed his great abilities in electro-chemical analysis and scientific theorizing towards an accurate chemical identification of 'chlorine', as a separate element, and towards a better theoretical explanation of its bleaching action.[4] He showed 'that the pure gas is incapable of altering vegetable colours, and that its operation in bleaching depends entirely upon its property of decomposing water, and liberating its oxygene',[5] or, as he explained in his *Elements of Chemical Philosophy*, bleaching is due to oxidation, and the oxygen is derived from water, 'which is decomposed by double affinity; that of hydrogen for chlorine, and of the colouring matters for oxygen'.[6] The 'oxymuriates' (hypochlorites), he demonstrated, are actually 'compounds of metallic bodies [e.g. potassium and sodium] with chlorine and oxygen; and the oxygen is held in them by a very weak attraction, and therefore is easily given off to colouring . . . matters'.[7] He also produced practical suggestions for industrial bleachers, such as the use of oxymuriate (hypochlorite) of magnesia (rather than of lime or potash) for clearing the whites of printed calicoes.[8] He showed that although a chlorine solution in water bleaches quickly, it weakens the cloth through formation of muriatic (hydrochloric) acid. Oxymuriates (hypochlorites), formed by dissolving chlorine in alkaline lixivia or limewater, were more satisfactory, but Davy showed that even the residual muriate of lime was liable to damage cloth. He therefore suggested oxymuriate of magnesia, since it acts more slowly and the muriate is not harmful to cloth. This suggestion was tried out successfully in the

[1] For example, at Simpson's paper mill in Edinburgh (*ibid.*, Vol. IV, 13 Oct. 1795) and at Apsley Mills, near St. Albans (*ibid.*, Vol. IX, 24 March 1796).

[2] *Ibid.*, Vol. IX, 27, 29, and 31 March 1796.

[3] *Ibid.*, Vol. XXI, 8 Sept. 1796.

[4] Sir H. Davy, *Philosophical Transactions*, 1809–12, *passim*, and in *Elements of Chemical Philosophy* (1812).   [5] *Phil. Trans.* (1811), p. 29.

[6] *Op. cit.*, pp. 242–3. See above p. 254.

[7] *Ibid.* They 'seem to owe their bleaching powers entirely to their loosely combined oxygene' (*Phil. Trans.*, 1811, p. 30).   [8] *Ibid.*

works of John Duffy, of Ball's Bridge, near Dublin, who had earlier provided William Higgins with similar facilities for large-scale tests;[1] when applied to 'whitening [the grounds of] printed calicoes . . . it does not destroy even reds or yellows fixed by mordants'.

John Dalton similarly carried out researches on oxymuriate of lime and discovered a test of its strength, using green sulphate of iron, which bleachers could use in preference to the earlier indigo or cochineal tests.[2] He also carried out water analyses for bleachers.[3] Thus chlorine bleaching continued to provide fascinating examples of the interactions between science and industry, which gradually brought about a revolution in this process of textiles manufacture. Theory and practice, research and development, were closely related. Scientists did not merely experiment in their laboratories and spin theories in philosophical transactions; they also participated in industrial-bleaching operations. At the same time, 'the scientific bleacher', to use Parkes's term, familiarized himself with philosophical publications and experiments, and utilized hydrometric and chemical tests to secure exact control of bleaching procedures.[4]

It is a commonly held view that only in comparatively recent times have research and development come together, and that during the Industrial Revolution there was a wide gap between them.[5] This view is obviously debatable. In bleaching, and in various other industrial operations of the late eighteenth and early nineteenth centuries, there is abundant evidence of close links between men of science and of industry, between laboratory and workshop—so close, indeed, that research and development were often almost merged together and there was hardly any gap at all. This is particularly evident in the intro-

[1] It was apparently through Higgins that Davy was invited to Ireland in 1810–11 to deliver lectures on bleaching and other chemical-industrial subjects (Wheeler and Partington, *op. cit.*, pp. 15 and 20–1).

[2] J. Dalton 'On the Oxymuriate of Lime', *Annals of Philosophy*, Vol. I (1813), pp. 15–23, a paper read to the Manchester Lit. & Phil. Soc. on 2 Oct. 1812. At that time, however, Dalton still clung to the old theory that 'oxymuriatic acid' (chlorine) was a compound of muriatic acid and oxygen. See also his *New System of Chemical Philosophy* (1810), Part II, pp. 297–313. He was severely criticized, therefore, by W. Henderson, 'On Sir H. Davy's Theory of Chlorine, and its Compounds', *Annals of Philosophy*, Vol. II (1813), pp. 13–18 and 122–33.

[3] See above, p. 79.

[4] *Op. cit.*, Vol. IV, pp. 72–3. Baines, *op. cit.*, p. 253, pointed out that in large bleachworks 'the managers are men of science, who are eager to adopt every chemical and mechanical improvement'.

[5] This view was expressed, for instance, by Professor Postan, in his paper on 'The Historic and Economic Problems of Technological Change', at the annual conference of the Economic History Society in Manchester, April 1966. See also M. M. Postan, *An Economic History of Western Europe 1945–1964* (1967), pp. 153–5.

duction of chlorine bleaching. As Berthollet himself said in 1790, 'c'est peut-être la première fois qu'une expérience a pu, dans quatre ans, produire de grandes manufactures en activité'.[1]

[1] Memoir by Berthollet to the Bureau du Commerce, 2 Feb. 1790, quoted by Ballot, *op. cit.*, pp. 531–2.

IX   Chemical Developments in Dyeing[1]

There is room for much more research into the applications of chemistry in the textile trades during this period. In addition, therefore, to the preceding chapters on the Henrys and on bleaching, we have drawn together various other scattered references to chemical developments in dyeing. Dr. and Mrs. Clow have already revealed some of these, mainly in Scotland.[2] Our evidence, chiefly from the Manchester area, supplements their findings.

There is no doubt, as we have previously noted, that traditional empirical methods were still predominant in this process. The remarks of Thomas Henry and Theophilus Rupp on the lack of applied science in dyeing[3] can be backed by other contemporary evidence. Thus Edward Hussey Delaval, F.R.S., in 'An Experimental Inquiry into the Cause of the Permanent Colours of Opake Bodies', communicated to the Manchester Literary and Philosophical Society by Charles Taylor in 1784,[4] made similar comments:

As the practice of Dying in its present state is not regulated by any scientific rules, it is seldom improved by the introduction of new processes: and the methods of varying the uses of the materials, which are already known, are rarely ascertained without repeated trials.

All the operations of the art, excepting only a few which have arisen from accidental discoveries, owe their origin to remote ages.[5]

[1] A previously unpublished contribution by A. E. Musson. A number of references from the Boulton and Watt papers have been provided by Eric Robinson.

[2] Especially the achievements of George and Charles Macintosh in Glasgow (*op. cit.*, chap. X). For similar textile-chemical developments in Leeds, by manufacturers such as Benjamin Gott and John Marshall, see *ibid.*, pp. 220–3, and Rimmer, *op. cit.*, pp. 14, 50–3, and 103. See also above, pp. 153–5.

[3] See above, pp. 82 and 241.

[4] *Manchester Lit. & Phil. Soc. Memoirs*, Vol. II (1785), pp. 131–256.

[5] *Ibid.*, p. 253.

Charles Taylor, though one of the most chemically knowledgeable dyers of his day, had to make the following admission to James Watt: 'When I attempted in a former Letter to give you some Theory of the Turkey Red Dye I know my account must appear to you very unsatisfactory, but tho I have made some Thousands of Experim[ts] to ascertain it I confess myself still in the dark about the true Theory.'[1]

It is clear from the letters of James Watt junior, whilst apprenticed to Maxwell, Taylor & Co., of Manchester,[2] that dyeing processes were still largely traditional, empirical, jealously guarded trade secrets of the workmen. He found that 'the dyers and printers here . . . have formed an association together not to let their employers know anything about their business or processes; they look upon these as so many *Arcana* which are their peculiar property and by means of which the body at large can always have a check upon their masters'.[3] He had to resort to bribery in order to extract their secrets. Such workmen, as Thomas Henry remarked, were scornful of chemical theory,[4] and most of their employers probably shared this 'practical' outlook.

Nevertheless, there is plenty of evidence to indicate that these attitudes were gradually changing and that more attention was being paid to applied chemistry, by some of the most important dyeing and calico-printing firms. Here again, as in bleaching, the French led the way,[5] and the published works of French chemists such as Dufay, Hellot, Macquer, d'Apligny, Berthollet, and Chaptal were frequently referred to in Britain, while skilled French dyers, as well as bleachers, were attracted to this country. From the middle of the eighteenth century, however, it is clear that chemical investigations in Britain were considerably assisting these branches of the textile trades, as exemplified in

[1] C. Taylor to J. Watt, 18 Jan. 1789, Dol. The 'former Letter' was that of 22 Nov. 1788, Dol. See below, p. 345, for Taylor's process for Turkey-red dyeing.
[2] See above, p. 96, and below, p. 345.'
[3] J. Watt jun. to J. Watt, 25 Oct. 1789, Dol. See also his letters of 14 July 1789 and 3 Nov. 1790, Dol.   [4] See above, p. 241.
[5] See, for example, H. Wescher, 'Great Masters of Dyeing in 18th Century France', *Ciba Review*, no. 18 (Feb. 1939); J. J. Beer, 'Eighteenth-Century Theories on the Process of Dyeing', *Isis*, Vol. 51, pt. i (March 1960); and H. Guerlac, 'Some French Antecedents of the Chemical Revolution', *Chymia*, Vol. V (1959), pp. 73–112.
    After the investigations of Sir William Petty and Robert Boyle in England in the 1660s, as Edward Bancroft observed, 'it does not appear that anything considerable was done, for nearly the space of a century, by men of science in this kingdom, towards improving the arts of dyeing and calico printing; they being, probably, discouraged by the difficulties which from the very imperfect state of chemical knowledge, must have occurred, in every attempt to improve upon what the dyers were able to perform without any principle or theory'. E. Bancroft, *Experimental Researches concerning the Philosophy of Permanent Colours* (2nd edn., 1813), Vol. I, p. xlviii.

the researches and writings of Dr. Francis Home, *Experiments in Bleaching* (1754), John Wilson, *An Essay on Light and Colours and what Colouring Matters are that Dye Cotton and Linen* (1786), and Dr. Edward Bancroft, *Experimental Researches concerning the Philosophy of Permanent Colours* (1st edn., Vol. I, 1794; 2nd edn., 2 vols., 1813), as well as in the works of Thomas Henry and others.

John Wilson, bleacher, dyer, and printer, of Manchester and Ainsworth, near Bolton, is a particularly interesting figure, 'to whom the velvet industry owed more than to anyone else on its dyeing and finishing sides'.[1] As Thomas Henry declared in 1786,

great improvements have been made [in dyeing], within these few years; improvements principally owing to the ingenuity and public spirit of Mr. Wilson, of this [Literary and Philosophical] society; who by the application of chemical principles, and by a diligent investigation of the nature of colouring substances, laid the foundations on which the present fabric is erected.[2]

Wilson, it is clear, had familiarized himself with the work of French chemists, and, as he himself emphasized, it was largely due to applied chemistry that the Lancashire dyeing industry began rapidly to overhaul the French. In reply to Delaval's criticism of the lack of applied science in textiles, he asserted,

These Observations would have been very proper forty Years since, when the dying of Cotton was not so well understood as it is now. And they may be still proper to some Persons who are ignorant . . .

But I can assure Mr. *Delaval*, that the Art of dying Cotton has for some Years past, fallen into such Hands here, as have spared no Pains to bring it to great Perfection. The Persons I mean are well versed in Natural and Experimental Philosophy, in Chymistry and Optics, so far as relates to their own Business: And Dying is now as well understood by those Persons, as it is in any Part of *Europe*.[3]

Wilson was speaking from his own knowledge, having 'commenced Dyer about forty Years ago', when he found that his manufactured goods, cotton velvets, 'could not be dyed and finished, so as to be acceptable to either Merchant or Shopkeeper', so that he had to learn how to dye and finish them himself, and to train his workmen. Not being 'brought up to the Business', he had to acquire 'a Knowledge in Chymistry, Optics, and other Branches of Philosophy, so much as

[1] A. P. Wadsworth and J. de L. Mann, *The Cotton Trade and Industrial Lancashire, 1600–1780* (Manchester, 1931), p. 180. See also Aikin, *op. cit.*, pp. 163–6, and J. Butterworth, *The Antiquities of the Town, and a Complete History of the Trade of Manchester* (Manchester, 1822), pp. 71–4.

[2] T. Henry's paper on 'the Art of Dying', *Manchester Lit. & Phil. Soc. Memoirs*, Vol. III (1790), pp. 348–9.   [3] Wilson, *op. cit.*, p. 6.

necessary for my purpose', and, at the same time, go to 'great pains, in trying Experiments on such colouring Matters as were likely to answer'.[1]

His interest in optics was demonstrated by his references to the works of Newton, Boyle, Bradley, Derham, and Priestley, in explaining light and colours. He followed Newtonian theory in asserting that colour is 'an inherent Property in Light', and that light consists of 'inconceivably small Particles of Matter, of different Magnitudes, which are emitted from a luminous Body in right Lines, and in all Directions, with an unparalleled Velocity'. There are different 'Species of Rays' of light, producing different colours, as revealed in the spectrum by glass prisms. Different substances contain different 'Colouring Particles', which 'form Colours by Attraction and Reflection from Rays of Light', e.g., a substance which appears red absorbs all but the red rays, which are reflected to our eyes.

His knowledge of chemistry was likewise demonstrated by his brief descriptions of dyeing processes, using various dyes, such as logwoods, weld, madder, annotta, safflower, Turkey red, indigo blue, etc. He also referred to the action of mordants or fixing agents:

Colouring Particles do not unite with Cotton without the Intervention of certain Acid, Neutral or Metallic Salts, as Alum, Alum neutralized with Ashes or Chalk, Cream of Tartar, Vegetable Acids, or Solutions of Metals in Vegetable Acids, Copperas, Verdigris, Sugar of Lead, Corrosive Sublimate, &c.

These have an Affinity to the Cotton or Linen, and also with the colouring Matters, and will unite with them, and so enter the Substance of the Cotton or Linen, and with them adhere to it. And these Salts will alter the Colours of the Particles, sometimes by their different Unions and different Properties.[2]

It is true that these mordants and dyes had long been used, that many dyers' 'receipts' or recipes were empirical, that the chemical explanations were very vague, and that contemporary knowledge of the chemical constitution of such agents as 'Cream of Tartar' or 'Sugar of Lead', or of traditional additives such as animal dung and blood, was very deficient. But it is also clear that Wilson had derived an immense amount of knowledge from the experimental chemistry of his day, and that he was conducting his operations along rational scientific lines, with careful regard to known chemical reactions, and to quantities and strengths of the materials used. As Thomas Henry pointed out, 'Though long experience may establish a number of facts, yet, if the rationale of the manner by which they are produced be not understood, misapplications are liable to be made; similar practices are pursued, where

[1] *Ibid.*, p. 7. [2] *Ibid.*, p. 13.

the cases differ essentially; and improvements are attempted at hazard, and often on false principles.'[1]

Chemical theory and experiment were now helping in the rationalization and improvement of dyeing processes. A modern chemist, looking back at dyeing in this period, may regard it as still in the 'eotechnic' era;[2] but what most struck Wilson, as 'a practical dyer' in the 1780s, was the fact that forty years earlier 'the dying of Cotton was little understood, compared to what it is now'—compared 'to what Perfection the Art is [now] arrived in Manchester'—thanks largely to increased chemical knowledge and experiment.

Wilson's own evidence is supported not only by Thomas Henry, but also by other contemporaries such as John Aikin, who emphasized his important contributions to the dressing and dyeing of cotton velvets, velverets, and velveteens. 'He [Wilson] was originally a manufacturer in the fustian branch in Manchester, and early engaged in the making of cotton velvets, which by unwearied efforts he brought to the utmost degree of perfection.'[3] The methods of finishing cotton velvets and thicksets were imperfect 'till the present mode of dressing was invented and brought to perfection'. The new dressing process—'firing off the pile', first with hot hand-irons,[4] later with red-hot cylinders—led to improvements in bleaching and dyeing, pioneered largely by Wilson.

Before this time, the lighter drabs and fancy colours might be said rather to hang on the surface, than to be fixed in the substance of the cotton goods. But the necessity of passing through the ordeal of dressing over glowing hot iron, caused them to employ more fixed drugs and astringents, with more powerful menstruums, in order to discharge the rustiness contracted by the fire; in all which attempts they kept improving till the dressing in the grey took place, and goods were brought to considerable perfection by alternate dressings and bleachings before they were dyed.

The dyers and printers then began to develop a much wider variety of patterns, so that 'the art of printing here [in Lancashire] came to rival that of London and that branch has in great measure been transferred from thence to the town and neighbourhood of Manchester'. John Wilson played the leading role in these developments:

Mr. Wilson having a turn for chymical inquiries, investigated the different known processes for dying . . . Resolving to give full scope to his improve-

---

[1] *Manchester Lit. & Phil. Soc. Memoirs*, Vol. III (1790), pp. 389–90.
[2] Clow, *op. cit.*, p. 223.   [3] Aikin, *op. cit.*, p. 164.
[4] *Ibid.*, pp. 163–5. This process, 'first employed . . . [by] Mr. Whitlow, the governor of the house of correction' in Manchester, was quickly adopted by Wilson and others, replacing earlier methods such as the use of razors and 'spirits of wine', to secure a smooth pile.

ments, he took a house and grounds at Ainsworth near Cockey-moor,[1] and commenced a capital dresser, bleacher, and dyer, first and principally of his own goods, which he brought to such high perfection, as to acquire the highest character both at London and in foreign markets.

He trained articled apprentices, because 'none of the workmen previously employed in dressing, bleaching, or dying, would suit his purpose, on account of their attachments to old methods'. After succeeding with black colours, he tried others and finally 'procured from the Greek dyers of Smyrna the secret of dying Turkey red,[2] which has been described at large in two essays read before the *Philosophical and Literary Society of Manchester*, which he printed and distributed among his friends after he had retired from business.'[3]

He was not very successful with this process, except with machine-spun yarn,[4] but his general improvements in dressing and dyeing 'induced the manufacturers to get most of the rich colours dyed by him'. Among other dyed velvets and velverets, 'china blues' were his invention.[5]

Wilson was not alone in the early application of chemistry to dyeing. Another Manchester manufacturer, Peter Meadowcroft, provides a similar, though less notable, example. 'Having a turn to chymistry, he made experiments till he produced fast colours [in silk handkerchiefs] in different shades of chocolate, and a colour approaching to scarlet, which he long kept to himself, and established the article to his own deserved emolument.'[6]

[1] He appears in Elizabeth Raffald's *Manchester Directory* (1772) as 'Wilson, John, Printer, Dyer, and Manufacturer, St Mary's Church-yard, and Cockymore'.

[2] Wilson (*op. cit.*, pp. 20-1) himself described how, at the cost of 'several Hundred Pounds', he had in 1753 'sent a young Man to Turkey, on Purpose to learn to dye it'. This young man, who had previously lived with a merchant in Smyrna and had learnt the language of the Greek dyers, 'got Admittance into their Dye-houses, and was instructed'; he also brought back many bales of the best madder root.

[3] The first paper, read to the Literary and Philosophical Society on 20 March 1782, does not appear to have survived. Some references to it are to be found at the end of the second paper, previously quoted, which was published in 1786. Wilson then stated that he had 'quitted the Business twelve Years ago'.

[4] He was also unsuccessful, about 1762, in experiments to dye cotton velvets red with cochineal. Yet later, in 1779, James Berkenhout, of London, received £5,000 from the government for what Wilson called 'this useless Colour on Cotton'. Wilson, *op. cit.*, pp. 25-6. See also above, p. 155.

[5] According to the German traveller, P. A. Nemnich (*op. cit.*, pp. 272-3), Wilson 'first introduced the blue, black, red and several other colours for the dyeing of fustians. He subsequently purchased the secret of dyeing Turkey red.' For the latter process he was awarded a premium of £50 by the Society of Arts in 1761. W. Bailey, *The Advancement of Arts, Manufactures, and Commerce* (1772), p. 331.

[6] Aikin, *op. cit.*, p. 161; Butterworth, *op. cit.*, p. 69. Meadowcroft does not

It is evident, however, that in order to improve dyeing, especially the Turkey-red process, many foreign dyers had to be attracted to this country. In 1786, for instance, Louis Borelle, a Frenchman, and his brother were awarded a government grant of £2,500 for having communicated the secrets of the Turkey-red process to the Manchester Committee of Trade two years previously.[1] Another Frenchman, Pierre J. Papillon (or Cigale), made similar overtures to the Manchester manufacturers in 1785, before going to Glasgow,[2] while yet another, Angel Delaunay, also played an important role in the development of this process in the Manchester area.[3] Indeed, it was reported in 1786 that twenty French dyers had come to Manchester.[4] The French themselves, however, had acquired their knowledge of Turkey-red dyeing from the Near East, just as John Wilson had previously sent a young man to Turkey to discover the secrets of this process, while another leading Manchester dyer and calico-printer, Charles Taylor, actually brought over a Greek.[5]

No doubt these processes were mainly empirical, but manufacturers such as Wilson and Taylor were also bringing about improvements by means of applied chemistry. Taylor, like Wilson, is another outstanding figure in the cotton industry who has been largely neglected in historical studies. He preceded Thomas Bell in the development of roller-

---

figure in Raffald's *Manchester Directory* (1772), but in Scholes's of 1794 he appears as a 'soap-boiler, calico, silk and cotton, and silk-handkerchief dyer, 32, Hilton-St. Soap-house, 29, Long-mill-gate'.

[1] *Journals of the House of Commons*, Vol. XLI, pp. 289, 467–8, and 882; Wadsworth and Mann, *op. cit.*, p. 181; Clow, *op. cit.*, p. 217. He had actually arrived in England in 1781. His name is variously spelt 'Borelle', 'Borell', and 'Borel'. He not merely revealed his process to Manchester manufacturers, but also set up his own dye-works at Blakeley (Blackley), near Manchester. In the *Manchester Mercury*, 28 Oct. 1788, he announced that he had 'now entered upon a New Plan for the Dying of that useful Colour' (Turkey-red). He retired from business in 1804, his works then being put up for sale (*ibid.*, 31 July 1804).

[2] *Journals of the House of Commons*, Vol. XLI, p. 468; Wadsworth and Mann, *op. cit.*, p. 181; Clow, *op. cit.*, p. 217. See also below, p. 345, n. 4.

[3] He established a works for Turkey-red dyeing, etc., at first in partnership with another Frenchman, Charles Payant, at Crumpsall, in the Manchester area (*Manchester Mercury*, 24 June 1788). The partnership was soon broken up, however (*ibid.*, 13 April 1790), and he later moved to Blakeley (*ibid.*, 25 March 1794). He went bankrupt in 1808 and his works were sold (*ibid.*, 12 and 26 April 1808). He died in 1811 (*ibid.*, 22 Jan. 1811), but the business was carried on by his sons. His name is perpetuated to this day in Delaunay's Road, between Crumpsall and Blackley.

[4] H. Sée, 'The Normandy Chamber of Commerce and the Commercial Treaty of 1786', *Econ. Hist. Rev.*, Vol. II (1929–30), p. 312.

[5] See the reference to 'our Greek' by Taylor in a letter to James Watt, 22 Nov. 1788, Dol., and by James Watt jun., while apprenticed to Maxwell, Taylor & Co., in a letter to his father, 3 Nov. 1790, Dol.

printing,[1] and he was also, as we have seen, among the first Lancashire manufacturers to adopt chlorine bleaching.[2] 'In the year 1785 he established in Manchester a dye-house for dyeing Turkey or Adrianople red, using the same method as was practised in Rouen.'[3] Wadsworth and Mann suggest that 'this was probably on the Borelle method', but it is clear from a letter of James Watt junior that by 1788, if not earlier, Taylor had improved considerably on this.

I forgot to ask Mr. Henry the last time I saw him whether he had sent you Borelle's process; if he has, you will find a considerable difference between it and ours, which is very much to our advantage. The Dyers of this town [Manchester] are exceeding jealous of one another, and as Mr. T. has now the lead in the Turkey red they are all of them upon the watch to learn what they can of his method.[4]

Thomas Henry had been largely instrumental in getting young James Watt apprenticed to Maxwell, Taylor & Co. He particularly recommended this firm because 'Mr. T.'s chemical knowledge would fall in with Mr. J. Watt's disposition . . . he [J. Watt jun.] would become better acquainted with the processes of printing and dyeing.'[5] James Watt junior's entry into this trade was no doubt a result of his scientific education[6] and of his own and his father's considerable interest in dyeing, and probably also because of the bleaching and dyeing concerns of his grandfather, James McGrigor. James Watt corresponded extensively with Berthollet, Henry, and Taylor on dyeing as well as on bleaching matters, again displaying his great knowledge of and

[1] Nemnich states (*op. cit.*, pp. 273–4) that in 1770 'a new process was discovered of printing cotton goods and linen cloth. It was done by means of engraved wooden rollers which were coated with dyes, and so printed the cloth . . . This was the invention of Charles Taylor and Thomas Walker of Manchester, who also secured a fourteen years' patent for it from the Government. The two men were then in partnership, but when the partnership was dissolved various people took advantage of this and made use of this method of printing even before the expiry of the patent.' The patent (no. 1,007, 14 March 1772) was granted to Taylor, Walker, and Joseph Adkin, father and son, turners. B. Woodcroft, *Chronological Index of Patents of Invention* (1854). Bell's later patents, of 1783 and 1785, were for printing from engraved copper cylinders, which Taylor subsequently improved (Nemnich, *op. cit.*, p. 279).
[2] See above, pp. 259–60, 288–91, 305–6, and also Nemnich, *op. cit.*, p. 279.
[3] *Ibid.*, pp. 278–9. See also Wadsworth and Mann, *op. cit.*, p. 181, n. 2, and Clow, *op. cit.*, p. 217.
[4] J. Watt jun. to J. Watt, 27 Dec. 1788, Dol. Taylor's Turkey-red process is described in great detail in a letter from J. Watt jun. to J. Watt, 29 Dec. 1788, Dol. It may be compared with that of John Wilson (*op. cit.*, pp. 21–2), that of Thomas Henry (*Manchester Lit. & Phil. Soc. Memoirs*, Vol. III (1790), pp. 380–9), and that of P. J. Papillon (Bancroft, *op. cit.* (2nd edn. 1813), Vol. II, pp. 247–61).
[5] T. Henry to J. Watt, 23 Oct. 1788, Dol.   [6] See above, chap. V.

z

expertise in applied chemistry.[1] Berthollet, who acknowledged Watt's valuable help, sent him a copy of his *Éléments de l'Art de Teinture* (1791), another for Priestley, and also one 'for the Manchester Society', via Watt and Henry.[2] And when Charles Taylor visited Paris in 1789, he carried a letter of introduction from Watt to Berthollet and others.[3]

James Watt junior shared with Taylor his interests in the works of foreign chemists. Thus we find him reading 'd'Appligny's treatise on dying, which contains much interesting knowledge and some very probable theories, though Mr. Taylor says it is full of errors'.[4] Taylor wrote to his son at Hamburg to obtain 'some chemical books',[5] and he, together with Thomas Henry and other 'chemical men' in the Manchester area, subscribed for a copy of Crell's *Annals*, translated from the German by the French chemists Lavoisier, Berthollet and Morveau.[6]

Taylor was obviously very interested in both British and foreign books on chemistry, and he also tested theory against practice. This is illustrated by the series of experiments which he carried out with James Watt junior, preparatory to writing articles on dyeing and calico-printing for James Keir's *Dictionary of Chemistry*.[7] But they soon ran into difficulties in regard to chemical theory.

For this some time back Mr. Taylor and myself have been consulting and extracting from different authors on the art of dying, in order if possible to arrive at some plausible Theory of it, for the article which he is writing for Mr. Keir's Chemical Dictionary; but I am afraid we shall not make much of the theoretical, whatever we may do of the practical part, as it appears to me that we neither of us have a sufficient knowledge of modern chemistry for the purpose.[8]

Their difficulties are not surprising, in view of the dubious nature of contemporary theory,[9] but it is clear that interest in theory—in trying

[1] This correspondence is of such extent that it cannot be dealt with here. It merits, indeed, separate attention, as for Watt's alkali interests, and Mr. Robinson intends to publish such a paper shortly.

[2] F. Swediaur (Paris) to J. Watt, 11 April 1791, and J. Watt to J. Watt jun., 22 April 1791, B.R.L.   [3] See above, pp. 305–6.

[4] J. Watt jun. to J. Watt, 27 Nov. 1788, Dol.

[5] J. Watt jun. to J. Watt, 7 Dec. 1788, Dol.

[6] J. Watt jun. to J. Watt, 1 Feb. 1789, Dol.

[7] C. Taylor to J. Watt, 13 Jan. 1790, and J. Watt jun. to J. Watt, 14 Jan. 1790, Dol. James Watt jun. also helped Keir with the article on bleaching. J. Keir to J. Watt jun., 10 Feb. 1790, A.O.L.B. Unfortunately, Keir abandoned this project after producing the first part of Volume I of the *Dictionary*, so that these contributions on bleaching and dyeing were never published and do not appear to have survived.   [8] J. Watt jun. to J. Watt, 11 July 1790, Dol.

[9] See, for example, Beer, *op. cit.*,

to find a rational explanation of observed processes—was linked with a great deal of fruitful experiment by such men as Wilson, Taylor, Henry, and others.

Thomas Cooper, another Manchester pioneer, along with Taylor, in the early development of chlorine bleaching,[1] was also interested in the chemical processes of dyeing, and later, after emigrating to the U.S.A., wrote *A Practical Treatise on Dyeing and Callicoe Printing* (Philadelphia, 1815), which has been referred to as 'probably the most important single influence which guided the infant dyeing industry in the United States along scientific and modern lines'.[2] In this book, Cooper frankly acknowledged his ignorance of many chemical matters, but, like Henry, emphasized the necessity of chemistry for improving knowledge of dyeing processes: 'The art of dyeing is yet in its infancy. No one but a good chemist, who is at the same time a good dyer, can form any judgment of the very many unascertained points that yet remain in this art.'

Like Henry, Taylor, and others, Cooper drew very heavily on the works of French and German chemists, but it is clear that he had also carried out much chemical experimenting of his own, though unfortunately there appears to be little surviving evidence of his activities in this field. He was a man of wide intellectual interests: graduating in law at University College, Oxford, in 1779, he extended his studies to medicine and natural philosophy. His application to the Royal Society at the end of 1789, though unsuccessful, was signed by Priestley and Watt.[3] He became a member of the Manchester Literary and Philosophical Society and shared the chemical interests of Henry, James Watt junior, and others.[4] We find him approaching James Watt, on Priestley's advice, to ask for one of Watt's improved 'digesters', with which to make bleaching experiments;[5] and he also asked James Watt junior if he could recommend anyone 'to keep his library in order and to make experiments under his direction'.[6] Here, obviously, was no

---

[1] See above, pp. 290–1, 303–8. He was among the Manchester 'chemical men' referred to by James Watt jun.

[2] S. M. Edelstein, 'The Contributions of Thomas Cooper', *American Dyestuff Reporter*, Vol. 43, no. 6 (15 March 1954), pp. 181–2. At the time when he wrote this book, Cooper was Professor of Chemistry in Dickenson College, Carlisle, Pennsylvania. In the following year he was appointed Professor of Chemistry and mineralogy in the University of Pennsylvania; a few years later he moved to South Carolina College. Obviously he was a man of considerable scientific distinction, as well as having a practical knowledge of bleaching, dyeing, and calico-printing. See *D.N.B.* and *Dictionary of American Biography*.

[3] See above, p. 92.

[4] See above, p. 96–7.

[5] J. Watt jun. to J. Watt, 12 April 1789, Dol.

[6] J. Watt jun. to J. Watt, 25 May 1789, Dol.

mere practical bleacher and dyer, but a thoroughly scientific manu-facturer.

The growing scientific interest in dyeing among Manchester manu-facturers and 'philosophers' is also evident from the *Memoirs* of the Literary and Philosophical Society. As Wadsworth and Mann have observed,

The papers which Wilson, Charles Taylor and Henry gave to the Literary and Philosophical Society mark the beginning of the application of scientific method in the dyeing and finishing trades, just as the lectures on dyeing, bleach-ing and calico printing held under the auspices of the Society [in the College of Arts and Sciences] may be said to be the beginning of technical education in Manchester.[1]

We have previously noted the utilitarian objects of these papers and lectures.[2] Thomas Henry particularly aimed at linking theory with practice. Other philosophers who addressed the Society were of the same practical mind. Edward Delaval, F.R.S., for instance, to whose paper on colours we have previously referred,[3] was led to his experi-mental inquiry

from a persuasion of its utility, to those interesting and elegant Arts whose object is the preparation, and use, of colouring substances . . . It should be the office of experimental philosophy, to examine the powers and properties of all the materials requisite to technical uses. Nor should its views be confined to the theories, which result from those researches, but directed to the practical application of them.[4]

Delaval himself was especially interested in optics and chemistry, and wished to apply scientific principles 'to the practical uses of several Arts: particularly to those of Dying, Painting, and such others as depend upon the knowledge, and management, of Colouring Materials'.[5] Like Thomas Henry, he deplored the lack of science in dyeing, being con-vinced that 'the perfection of the arts of dying, and bleaching, and of preparing and preserving painters' colours, essentially depend [*sic*] on the knowledge of this subject'.[6] His lengthy paper is packed with information—much of it drawn from scientific experiments—upon the colouring properties of a great many vegetable, animal, and mineral substances, with references to the work of other scientists, notably of the French chemist, Hellot, on dyeing. He found that

the transition from physical experiments, to practical operations of dying, is easy and obvious. For, the experiments which I have made . . . have guided

---

[1] *Op. cit..* p. 182.   [2] See above, pp. 89–93 and 241–3.   [3] See above, p. 338.
[4] *Manchester Lit. & Phil. Soc. Memoirs*, Vol. II (1785), p. 131.
[5] *Ibid.*, pp. 141–2.   [6] *Ibid.*, pp. 177–8.

me to the discovery of several bright and permanent dyes, in the execution of which I have, principally, used cheap and common ingredients, that have not before been applied to such purposes.[1]

Dr. Edward Bancroft, F.R.S., a more famous textile chemist, whose work on the *Philosophy of Permanent Colours* we have previously mentioned,[2] had similarly practical interests. Thus in 1775 he was granted a patent for his newly discovered yellow dyestuff 'quercitron', obtained from the bark of the American black oak,[3] and he gave valuable advice to practical dyers. James Watt junior, for example, referred to his having met Bancroft in Manchester, when he obtained from him certain information about a rare green dyestuff imported from the East Indies.[4] Doubtless other dyers were similarly able to benefit from the learned doctor's knowledge.

Wilson, Meadowcroft, Henry, Taylor, Cooper, James Watt junior— the list of chemically minded Manchester manufacturers grows longer. To it can be added the German, Theophilus Rupp,[5] and also Thomas Hoyle, founder of one of the leading Manchester dyeing and calico-printing works, at Mayfield, Ardwick. Hoyle, a member of the Literary and Philosophical Society, read a paper on chlorine bleaching in 1797,[6] which clearly demonstrated his interests in both theoretical and applied chemistry, including extensive knowledge of the works of French chemists. That such interests were becoming increasingly wide-spread among Manchester manufacturers is also illustrated by the attendances at the chemical lectures given by Thomas and William Henry, Adam Walker, John Warltire, Thomas Garnett, and others,[7] and by the sale of the works of French chemists such as Lavoisier[8] and Hellot, Macquer, and d'Apligny.[9] Undoubtedly many Manchester industrialists were taking Thomas Henry's words to heart, and were finding practical benefit, as well as philosophical interest, in applied chemistry.

It is true that in dyeing, as distinct from bleaching, there were no really revolutionary developments during these years, while traditional skills and materials remained important. But it is misleading to contrast the 'palaeotechnic' period after William Perkin's production of the first dyestuff from coal-tar in 1856 with the preceding 'eotechnic'

---

[1] *Ibid.*, p. 255.　　　　　　　　　　[2] See above, pp. 133 and 340.
[3] Woodcroft, *op. cit.* The patent was no. 1,103, 23 Oct. 1775. Bancroft was described as 'of Downing Street, Westminster, Doctor in Physick'.
[4] J. Watt jun to J. Watt, 25 Oct. 1789, Dol.
[5] See above, pp. 82–3 and 317–19.
[6] *Manchester Lit. & Phil. Soc. Memoirs*, Vol. V, pt. i (1798), pp. 221–42. See above, pp. 315–16.　　　　　　　　[7] See above pp. 92–3, 101–11, 242.
[8] *Manchester Mercury*, 21 Sept. 1784.　　　[9] *Ibid.*, 30 June 1789.

period.[1] In this earlier period it is clear that rational theory was evolving and that applied chemistry was making important contributions to dyeing, by introducing more systematic scientific methods in place of blind empiricism. As Dr. Abraham Rees observed:[2] 'Though we have to record no brilliant discoveries or improvements in the practice of dyeing, within these few years, yet the art has continued progressively to improve, the different processes have been simplified and amended; and what some years ago was considered a matter of chance and uncertainty, is now reduced to fixed principles.' Samuel Parkes, too, emphasized that 'of all the arts [manufactures], none are more dependent on chemistry than those of Dyeing and Calico-printing', which could now hardly be carried on successfully without a good deal of chemical knowledge.[3]

This continued to be evidenced in the first half of the nineteenth century. We find the famous scientist John Dalton, for instance, analysing water supplies for bleaching firms[4] and experimenting on dyeing materials such as indigo;[5] he clearly had some practical acquaintance with commercial dyeing operations. Leading dyers and calico-printers such as James Thomson, of Primrose, near Clitheroe, and John Mercer, of Oakenshaw, near Blackburn, were also scientifically minded. The former, so Edward Baines tells us, combined 'in an eminent degree scientific with practical knowledge', as demonstrated by his patents of 1813 and 1816 for using acids or metallic oxides to print patterns on Turkey-red and other colours and then discharging them with chloride of lime.[6] John Mercer, likewise, though beginning life in very humble circumstances, as a handloom weaver, and teaching himself chemistry at first from such works as James Parkinson's *Chemical Pocket-Book* (1799 and later editions), was eventually to reveal remarkable ability in both theoretical and applied chemistry.[7] We can only refer very briefly here to his manganese-bronze colours, his famous 'mercerizing'

[1] Clow, *op. cit.*, p. 223. The Clows define the 'palaeotechnic' phase as that in which 'chemistry abandoned remnants of alchemical tradition and evolved a rational theory' (*ibid.*, p. xiv).

[2] A. Rees, *Cyclopaedia* (1819), Vol. XII, article on 'Dyeing'.

[3] S. Parkes, *Chemical Essays* (1815), Vol. I, pp. 38–46.

[4] See above, p. 70.

[5] J. Dalton, 'On the Nature and Properties of Indigo, with directions for the valuation of different Samples', *Manchester Lit. & Phil. Soc. Memoirs*, 2nd ser., Vol. IV (1824), pp. 427–40.

[6] E. Baines, *History of the Cotton Manufacture in Great Britain* (1835), pp. 277–9; J. Graham, 'On the History of Invention as applied to the Dyeing and Printing of Fabrics. Part 1st, Chemistry', *Manchester Lit. & Phil. Soc. Proceedings*, Vol. I (1860), pp. 216–18.

[7] E. A. Parnell, *Life and Labours of John Mercer* (1886); *Journ. Chem. Soc.* (1867), p. 395; *D.N.B.*

process, and many other textile-chemical developments; his researches on catalysis and on the ferro-cyanides; his friendship with Dr. Lyon Playfair (who was a chemist at Thomson's works in the early 1840s); his papers to the British Association, his membership of the Chemical Society, and Fellowship of the Royal Society. These men, and their forerunners in the late eighteenth and early nineteenth centuries, cannot sensibly be regarded as 'eotechnic'. Contemporaries certainly did not think of them in this way:

The arts of dyeing and calico printing have received great assistance from such men of science [as those referred to in this chapter] . . . and in most of the establishments in which these arts are wrought out, some one or more of the principals are, practically, men of eminent scientific talent; in fact, it is almost indispensable that they should be, in order that the trades be rendered profitable.[1]

---

[1] Love and Barton, *Manchester As It Is* (1839), pp. 220–1. This contemporary view has recently been supported by Mrs. Susan Fairlie, 'Dyestuffs in the Eighteenth Century', *Econ. Hist. Rev.*, Vol. XVII (1964–5), pp. 488–510. While her article is mainly concerned with the drysalting trade in traditional vegetable, animal, and mineral dyestuffs, Mrs. Fairlie points out the importance of the efforts by experimental chemists to develop new agents (alkalis, acids, bleaches, mordants, and dyes) and to achieve more scientific control over dyeing operations, even though they did not properly understand the chemical processes involved. By the end of the eighteenth century, 'virtually every dye drug had been analysed and experimented with', and there is little doubt that 'a chemically minded industrialist was better able to regulate quality, quantity, temperature, etc.' than earlier purely empirical dyers.

X James Watt and Early Experiments in Alkali Manufacture[1]

The discovery of chemical methods of producing soda from common salt was a basic prerequisite of the expansion of the chemical industry in the nineteenth century, but just as the steam revolution of that century had been preceded by a hundred years or more of constant experiment both in the laboratory and in industry, so the chemical revolution also had its long antecedents. In both the steam revolution and the chemical revolution, James Watt was a key figure, though it is usual to associate his name only with the former movement, and Watt's attempts to produce soda from common salt form the crux of this chapter. That Watt, Dr. Joseph Black, Dr. John Roebuck, and James Keir were somehow involved together in pioneering the production of synthetic soda has long been known, but the details of their relationships with each other in this enterprise have remained vague because of too many gaps in the evidence.[2] A good deal more information has come to light in the Doldowlod MSS., and elsewhere, which makes a more coherent account possible. Some of the letters used here will appear in full in a volume of Watt's scientific correspondence with Joseph Black and others, to be published shortly.

The attempt to replace domestic kelp and foreign barilla by a chemical method of alkali production, especially as the importation of the

[1] A previously unpublished chapter by Eric Robinson. Mr. Musson has made a number of valuable suggestions. Footnotes on chemical matters and Appendix II, 'The Lime-Salt Process for Making Soda', are by Drs. W. V. Farrar and K. R. Farrar. The author is also indebted to Dr. J. L. Moilliet for helpful discussion and advice.

[2] For different and varying accounts, cf. A. and N. L. Clow, 'Vitriol in the Industrial Revolution', *Econ. Hist. Rev.*, Vol. XV (1945); A. and N. L. Clow, *The Chemical Revolution* (1952), chap. IV; R. Padley, 'The Beginnings of the British Alkali Industry', *University of Birmingham Historical Journal*, Vol. III (1951), pp. 64–78; R. E. Schofield, *The Lunar Society of Birmingham* (Oxford, 1963; actual date of publication, 1964), pp. 66–7 and 76–9.

latter was always subject to interruption in the eighteenth century by war with France, attracted the inventiveness of numerous men of scientific leanings in the eighteenth century,[1] and it was natural that John Roebuck, the pioneer vitriol manufacturer and perhaps the leading industrial chemist of his day, and Joseph Black should share a common interest in this subject. This has led to the suggestion that it was the existence of this shared interest between Black and Roebuck which led Black to introduce Watt to his industrialist friend,[2] but this seems doubtful, and the first correspondence linking Roebuck with Watt in alkali matters appears to date from 1766. We have Watt's evidence, however, that it was Black who first invented the theory: 'As to the Alcali affair You know Dr. Black first invented the theory which he communicated to me I tryed experiments, & found it succeed After I had given it up, I went on with Experiments till I brought it to a probability of succeeding in practice Doctor R. was taken in to the scheme in the beginning of it . . .'[3] In the letter actually despatched, of which this is a draft, Watt corrects his statement about Roebuck to: 'Dr. R. was taken in the scheme soon after me . . .'[4]

By 1766, Watt was corresponding with Roebuck about his alkali experiments.[5] Thus on 19 February 1766 Watt wrote to Roebuck:

I give a Summary view of my Experiments on Lime & Salt

1st Lime was mixed with about its own bulk of Salt & sett in a cool Cellar it rarified in Del:[iquescence] the liquor being a brine of Salt. owthing [nothing] being expected from it it was neglected for 5 months it was then found Covered with Crystals of mild alcali

2d a flat plate was filled with Lime made into a paste with strong brine drops of caustic alcali appeared on its surface which soon shot into Crystals during its Crystalization a considerable quantity of brine ran from it having somewhat of an alcaline taste.

3d three flat plates were filled one with a mixture of Coal ashes & Lime equal parts moistened with Strong pickle the second with lime moistened with pickle of $\frac{1}{2}$ strength the third with Lime moistened with pickle of $\frac{1}{3}$ strength. in about 4 days the ashes & Lime produced drops of alcali on the surface which in about 2 weeks were Cristal: & a quantity of brine ran from it

[1] See Sir William Ramsay, *Life and Letters of Joseph Black, M.D.* (1918), pp. 64–5.

[2] Clow, *op. cit.*, p. 94. Following Clow, Schofield, *op. cit.*, p. 67, says: 'By 1765 he [Roebuck] and Joseph Black were experimenting on the decomposition of salt', although he also says, p. 77, quoting Watt to Small, 27 Oct. 1769, A.O.L.B., that Black and Watt first experimented together on Black's theory of alkali and Dr. Roebuck was taken into the scheme soon afterwards.

[3] Draft of a letter, J. Watt to William Small, 20 Oct. 1769, Dol.

[4] Watt to Small, 27 Oct. 1769, A.O.L.B. Cited by Schofield, *op. cit.*, p. 77.

[5] The chemistry of Watt's process is discussed in the Appendix to this chapter.

considerably alcaline the second also had produced a small Quantity of Cristals like wool.

the third produced scarce any.

The Lime (still wett) of the second experiment was mixed with the alcali produced by it the mixture lixiviated filtrated & evaporated the alcali was Lost & the Com[mo]n Salt reproduced My other Experiments not being yett brought to conclusion shall reserve them for another Letter. . . .[1]

Further experiments are described in letters from Watt to Roebuck dated 29 February 1766, 3 March 1766, 20 April 1766, 14 May 1766, 1 November 1766, and an undated letter of 1766 or 67.[2]

The amounts of alkali produced by these methods were probably very small, and Watt, realizing the effect of the air, was concerned to devise a method whereby the surface exposure could be increased with as little additional expense as possible.[3] This he proposed to do by taking equal parts of lime and small sand mixed with brine to form a stiff paste, which would be moulded into 'bricks' and these would then be built into a wall three feet thick, nine feet high and thirty feet long. The production of alkali from such a wall was calculated by Watt on this occasion to be about sixty pounds of alkali *per diem*. The cost of erecting three such walls, taking into account the lime, the sand and the work would be about £80. By November 1766 Watt was calculating that ten houses would clear about £2,000 *per annum*,[4] and by the end of 1766 the scheme was thought to be sufficiently practicable for Watt to be thinking about the setting up of racks rather than worrying further about the chemical methods to be employed.[5] Watt's part in all this seems to be that of the industrial chemist and practical man-of-affairs as against Black's role as the laboratory chemist, and Roebuck's as the entrepreneur and provider of capital.

In 1767 Watt was employed with Robert Mackell to make a survey for a canal uniting the Forth and the Clyde, and, since there was a debate about the line to be followed, he was obliged to attend a committee of the Commons in order to give evidence. On this trip he visited Birmingham and, though he had been there before, this was the occasion of his meeting Erasmus Darwin, and of his being shown round Matthew Boulton's Soho works by William Small and Boulton's partner, John Fothergill. He also met the Scots chemist, James Keir, at Dr. Small's house and was so brought into contact with the principal

[1] Dol.

[2] All these letters, Dol., Watt to Roebuck. See also James Watt's *Experiment Book* [?1775], Muirhead Box 1, B.R.L.

[3] Watt to Roebuck, 3 March 1766, Dol.

[4] Watt to Roebuck, 1 Nov. 1766, Dol. See Appendix I.

[5] Watt to Roebuck, no date but probably late 1766 or early 1767, Dol.

members of the Lunar Society of Birmingham. Watt was soon in correspondence with both Darwin and Small, but particularly with Small, for whom he seems to have conceived a considerable affection. The subject of alkali manufacture very soon appears in the correspondence. On the 7 January 1768 Small warned Watt: 'One Experiment was shewed to me, and two were described to me since you left me which will certainly soon lead the operators, altho they are not great cymists, to your alkali secret. This is what I alluded to, and I beg you will attend to it.'[1] In the light of later correspondence, it is natural to assume that the two chemists referred to were Keir and, perhaps, his later partner Alexander Blair, but this does not seem like Small to refer so slightingly to a close friend, and since Small was also an intimate of George Fordyce, the possibility arises that he was already familiar with Fordyce's experiments.

After Watt's meeting with Small, the latter was in a position of some difficulty. Watt had confided his alkali research to him, but Keir was also investigating the same problem, quite independently, and was also telling Small about his work. Small encouraged Watt to press on with his discoveries: 'I am glad the Alcali is attended to. I know nothing for which I would more willingly have quit[t]ed physic, if the invention had been mine.'[2] In July 1769 he let Watt know that a friend of his was at work on the subject and that therefore he was relieved that a patent had been obtained, as Roebuck had led him to believe.[3] Black had discussed this very matter with Watt in a letter of 23 January 1769:

As to the proposal of taking a Patent I have no objection to it provided you are satisfyed that the thing will do & that You as well as Dr. Roebuck will join me in it—perhaps it would be best that I should apply for it in my own name alone giving you two a private obligation to share the advantage of it with you & You going equal with me in the Expence—My reason is that I have no witnesses of my invention & of my right to the Patent but your selves two & you could not be admitted as witnesses if you were claimers of the Patent—or Parties.[4]

Watt, at this time, was feeling exhausted and did not reply promptly, so his letter to Black of 27 January 1769 was occasioned by a reminder from Black on 26 January 1769, in which Black said, 'I imagine despatch to be of great consequence in the Business which was the subject of my last & wonder you have delayed to answer me.'[5] In the same letter

---

[1] William Small to James Watt, 7 Jan. 1768, A.O.L.B.
[2] Small to Watt, 5 Feb. 1769, Dol.
[3] Small to Watt, 26 July 1769, Dol. Small had already repeated his advice of despatch by May. See Watt to Black, 29 May 1769, Dol.
[4] Black to Watt, 23 Jan. 1769, Dol.   [5] Black to Watt, 26 Jan. 1769, Dol.

Black said that he could find other partners if Dr. Roebuck did not choose to participate. Watt reassured Black about Roebuck's wanting to take a share, but was anxious that the experiments should not be conducted at Kinneil, Roebuck's house, lest 'it might be entrusted to improper people'. On the other hand he was anxious in his poor state of health for 'an Industrious partner'.[1] In the same letter Watt estimated that: 'in a house of 18 feet wide 36 long & 15 high 9 tuns of alcali might be made in a year that is taking things at the least some processes gave much more.' If his estimates were correct, the amount produced, though small, would compare well with a process employed by Birch, a Manchester textile manufacturer, described by Thomas Percival in 1780,[2] where 9 cwt. 1 qr. 12 lbs. was produced in about three weeks and where much more labour and fuel was required. Watt repeated further experiments to Black on the 28 January 1769.[3]

Dr. Roebuck was still not satisfied in May 1769, however, that the best process had been settled upon, and since it was he who was the industrialist among them and since it was the salt from his works at Bo'ness which was to be decomposed into alkali, his voice must have counted for a great deal.

During the whole of 1769 experiments continued. The work seems to have been shared between Black and Watt, but as Watt was also engaged in canal business and engine building, he was obliged to relax some of his efforts. In April he wrote to Small: 'As to the Alcaly Doctor Black has it at present under tuitun [*sic*] he proceeds slowly according to the way of Philosophers. Whether it be lost or not I can do no more than I do.'[4] Again in July, he continued his report on alkali progress:

both Dr. Black & I am busy with large experiments on the Alcali & Black is forthwith to sollicit a patent under the title of . . . the art of making Alcaline salt from C[ommo]^n: salt & lime. I mentioned to Dr. Roebuck (who is also concerned in it) that I had told you of our discovering the means of decomposing salt by lime. If he talks to you on the Subject you will seem to know just as much about it as he pleases to believe you do. & I will be much obliged to

[1] Watt to Black, 27 Jan. 1769, Dol.

[2] *Proc. Roy. Soc.*, Vol. LXX (1780), p. 345, quoted by Padley, *op. cit.*, p. 66. Birch's process consisted of the evaporation of liquid drained from dunghills, followed by calcination of the residue.

[3] Dol. Watt was still experimenting with the exposure of a mixture of moistened lime and salt to the atmosphere. It appeared to him that the formation of crystals of alkali on the surface was accelerated by frost. He decided, however, that more experiment was needed, and concluded: 'Now I must leave it to yourself whether it is worth while to secure it for it is certainly one of the few things that can be compleatly secured by a patent.'

[4] Watt to Small, 28 April 1769, A.O.L.B.

you to tell him what you properly can of any other experiments going on for that purpose. I mean of their forwardness.[1]

In October 1769 Small was again told by Roebuck that a patent had been secured. Small now felt himself free to tell Watt about Keir's experiments and to suggest that Keir be taken into the partnership with Watt, Black and Roebuck.[2] He discussed the matter in Birmingham with Roebuck, who was agreeable. Watt, in a characteristically self-deprecatory letter, suggested that Keir would be a much more useful partner than himself: 'I apprehend he has money & Abilities to push himself & keep his own with Mankind which I want.'[3] But he considered that the decision should be left to Black and Roebuck. Watt was a little worried that he might have let slip to Keir more than he should have done, but was at least partially reassured by Small:

As to what passed at my house in Mr. K's Company about the alcali I cannot now distinctly recollect it. You did mention decomposition by quicklime however I am certain but not your method except to me. Mr. K had made many expts. for a considerable time some years I believe with lime. I had assisted at a few, not that I believed lime would do the business, being a theorist, but rather to persuade my friend to take no further pains about an impossibility as I believed it to be. Soon after you went he made [a]lcali in some trials, & in othe[rs] failed [In a]ll he used different methods [from] yours, & in most added superfluous subs[tances]. By degrees he left thes[e] out.[4]

In any event, by January 1770 Keir was involved with the others, though no partnership agreement had been drawn up.[5]

One point in Small's letter to Watt of 10 October 1769 may have a particular importance to the later course of events. In it Small had mentioned a book published thirty-five years earlier in which Black's method of alkali production had been described. Watt, in his reply, had asked Small to identify the book, and in the letter of 5 November quoted above, Small wrote: 'The book you ask about was written by a Dr. Cohausen. I have found his method since I last saw you, described in Potts[6] works, w[hi]ch I had read before, but overlooked

---

[1] Watt to Small, 5 July 1769, A.O.L.B.

[2] Small to Watt, 10 Oct. 1769, A.O.L.B. See Appendix I.

[3] Watt to Small, 20 Oct. 1769, Dol. See Appendix I. Watt had already described Keir in a letter to James Lind, 25 Oct. 1768, Dol., as 'a mighty chemist before the Lord, and a very agreeable man'.

[4] Small to Watt, 5 Nov. 1769, A.O.L.B. See Appendix I.

[5] John Roebuck to James Watt, 17 Jan. 1770, Dol.; 'I also conversed with Mr. Keir about the Alkali and can readily put it on a proper plan with him provided our Experiments give encouragement. I have seen Dr. Black but shall not write to Mr. Keir till I have conversed with yourself.'

[6] Johann Heinrich Cohausen, *Europa arcana medica* (2 vols., Frankfurt, 1757–1760); J. H. Pott, *Observationum et Animadversionum Chymicarum, Centuria Prima*

that.' This may have shaken the confidence of the Kinneil group in the validity of any patent they might have taken out and therefore may have urged them to the normal eighteenth-century business-method of secrecy. In addition the experiments were not progressing well. On the 20 September 1769 Watt had told Small: 'As to the Alkali we have been trying some experiments which are not yet come to Issue & Indeed are not Operating so fast as we wish them [The] 2 Drs. therefore seem willing to defer Compleating the pat[en]t till they are more *au fait* & I cannot urge it.'[1] One of the problems was making the lime into cakes or bricks by some easy practical method. Black had reported in May that he was having no success in forming the lime into plates with some instruments he had shown Watt.[2] Six months later he seems to have made little progress, as John Roebuck's son, John, wrote: 'I was yeterday in Edin. & saw Dr. Black at his Laboratoire. I think he is going very lazily about the Cakes & seems rather to despair of them, as they have as yet produced very little.'[3] Under these circumstances the addition of Keir to the team must have promised new vitality, so that in February 1770 Roebuck invited Keir to Kinneil: 'Please acquaint Mr. Kier [*sic*] after presenting my Comp[limen]ts that Dr. Black is to be at Kinneil this day sen'night. And then we shall give Mr. Kier a line with a plan of Exper[imen]ts Dr. Black would have been at Kinneil to day if necessary Business had not call'd me here.'[4]

Such then was the situation of 1770 and for the next ten years the story becomes thin. Black had moved to Edinburgh in 1766, Watt became increasingly involved in canal and steam-engine business, while Roebuck's financial affairs grew more and more distressed until in 1773 he went bankrupt. Keir seems to have been the most persistent of the group in continuing with his experiments, and his progress, or lack of it, was reported by Small to Watt. Thus on 3 February 1771 Small wrote: 'Captn. Keir is in London. He has thoughts of coming to reside

---

(Berlin, 1739), *Centuria Secunda* (Berlin, 1741). In 1736 Duhamel de Monceau had also written a memoir on the preparation of soda, on which Dr. Gittins has very recently made the following comments: 'his process differed from Leblanc's only in the method of converting the sodium sulphide to sodium carbonate. He used vinegar to convert the sulphide into acetate which was then evaporated to dryness and strongly heated to yield soda. Duhamel's method of displacing the sulphur with acetic acid, although not an economic proposition, did confirm that salt and soda had the same chemical base, and it also suggested that sodium sulphate was a more likely substance than sodium chloride to be converted to sodium carbonate.' (L. Gittins, 'The Manufacture of Alkali in Britain, 1779–1789', *Annals of Science*, Vol. 22, no. 3 (Sept. 1966), pp. 175–89.)

[1] A.O.L.B.   [2] Black to Watt, 27 May 1769, Dol.
[3] J. Roebuck jun. to Watt, n.d. Nov. 1769, Dol.
[4] John Roebuck to Matthew Boulton, 10 Feb. 1770, Dol.

in Fife somewhere. He has made many exper[imen]ts about fossil
Alkali [Na_2CO_3], with w[hi]ch he is better satisfied than I am, for I
think the yield too small.'[1] But by October he was more sanguine: 'As
to friends Mr. Keir has turned glass maker at Stourbridge and has
married a beauty. I grow ashamed and am *mortellement ennuy[e]use . . .*
Your fossil alkali, according to his Exper[imen]ts will answer well I
think. So you see we want to converse with you on more than one
subject.'[2] Small also reported the activities of a London company
making alkali from salt, but unfortunately does not name the firm:
'Some company of chemists or other near London makes what is sold
for fossil alkali, & sells it for 37/per hundred. It contains about one
third of sea salt, & is very white. I have not been able to learn their
process, but have seen & examined their production.'[3]

Two letters from James Keir in 1771 and 1772 give us more information than has previously been available about his part in furthering the
alkali experiments. On 14 December 1771 he wrote to Watt:

Dr. Small informs me that you desire to know the result of my experiments on
our method of procuring fossil alkali. My experiments are not sufficiently long
continued, nor do I think that they were made with sufficiently large quantities
of materials, to ascertain the quantity of alkali producible. I can only acquaint
you that from a pound of dry salt I have obtained a quarter of a pound of dry
mild alkali, within a twelvemonth, and that the mass was not then exhausted,
for it still continued to throw out more and more of the alkaline efflorescence.
Some masses that had been kept 18 months were not exhausted. I make no
doubt but the quantity of alkali producible is sufficient to make the Scheme
profitable, if the excessive slowness of the process does not render the quantity
of buildings, labor, &c, too great. When I was last in London I heard it mentioned by some persons curious in chemistry and who especially had made many
attempts to discover the method of obtaining the alkali of common salt, that
Dr. Black had actually discovered that method. I was afraid lest some persons
might prevent our obtaining a patent, if we should think it necessary, by entering a caveat in general against all patents for obtaining alkali from sea-salt,
especially as I found upon Inquiry that you and Dr. Black had not taken that
precaution. I accordingly entered a caveat in my own name which caveat shall
not prevent you or your friends connected in the Scheme from taking out a
patent when you and they think proper. I was in hopes at that time to have
offered my Service to you all to execute the Scheme if you chose to join me,
as I had thoughts of renting a coal and salt work in Scotland. I am now fixed
in an inland County, and though I consume a great deal of alkali, yet I believe

[1] Small to Watt, 3 Feb. 1771, A.O.L.B. It is interesting to observe Keir contemplating a return to a traditional salt-making area which was to become a
centre for alkali manufacture.
[2] Small to Watt, 19 Oct. 1771, A.O.L.B.
[3] Small to Watt, 16 Dec. 1771, A.O.L.B.

I must only wish success to those who chuse to undertake the scheme, I should nevertheless be glad to know when you are determined to take out a patent . . .[1]

Two months later he told Watt:

The surface from which I obtained the quantity mentioned of alkali was about a hundred square inches. The mass was placed in a cellar, but in other trials I found any part of the house from the uninhabited Garrets to the cellars fit for the purpose; Nor can I say that I found any circumstance to be necessary but a certain moisture, which must be given either by a moist air as that of cellars, or by a frequent sprinkling with water. If the mass be kept too moist, or too dry, the separation of the alkali does not take place. I have more than one tryed the decomposition of vitriolated tartar of nitre by the same method, but without success. I never tried the Salt of Sylvius. The alkali I obtained[2] was mild, & as dry as could be without making it red hot.[3]

For the next year or two Small continued to press Watt to get on with his alkali manufacture, and eventually became suspicious of the patent position: 'It is unaccountable to me that you & your friends do not pursue the Fossil alkali. It is made near London bunglingly, & yet to advantage. Have you patent or caveat or what have you? I am perfectly satisfied of the practicability of your method.'[4] Throughout the seventies, however, the 'Lunar group', if we may so call them, made no effective progress with their plans, and left it to others to take the first steps in this important chemical field. Keir alone took action in so far as he entered a caveat at the Patent Office in 1771 against the granting of any patent for producing alkali from common salt.[5]

Nothing more is heard of the matter until 1780, when there was suddenly another great flurry, arising out of Alexander Fordyce's and George Fordyce's petition to the Commons for a drawback on salt duties, because George Fordyce alleged that he had discovered a method of manufacturing alkali and marine acid (i.e. hydrochloric acid) from common salt.[6] Keir informed Watt, who in turn immedi-

---

[1] Dol.   [2] An alternative word, 'mentioned', is written over 'obtained'.
[3] James Keir to Watt, 27 Feb. 1772, Dol.
[4] Small to Watt, 16 Nov. 1772, A.O.L.B. See also Small to Watt, 15 March 1773, A.O.L.B.
[5] *Commons Journals* (hereafter abbreviated to *C.J.*), Vol. XXXVII, p. 915. See also E. Hughes, *Studies in Administration and Finance* (Manchester, 1934), pp. 429–31.
[6] George Fordyce, nephew of Alexander Fordyce, was Professor of Chemistry at St. Thomas's Hospital. He was a friend of William Small and a member of the same Masonic Lodge. He was also a friend of James Keir and had communicated Keir's paper, 'On the Crystallisations observed in Glass', to the *Proc. Roy. Soc.*, Vol. LX (1776). Alexander Fordyce was already notorious in Birmingham as a result of the failure of his bank in 1772, when Boulton lost money and might have lost much more. Small had been much distressed by the disgrace of a member of the Fordyce family. See below, pp. 364–5, for Fordyce's alkali patent of 1781.

ately wrote to Black telling him that he and Keir were petitioning Parliament separately to be granted any privilege allowed to Fordyce.[1] A draft of Watt's petition still exists.[2] At the same time he sent a letter by Keir's hand to Lord Dundas, who was himself to be involved later in alkali manufacture, begging Dundas to exert his influence on behalf of all three of them. It is interesting to observe that the reasons assigned in this letter for abandoning their earlier ideas of procuring a patent were the duties on salt, 'and the disagreeable circumstance of being attended by excise Officers together with the moderate price of alkaline salts arising from the plentifull importation of American Potashes'.[3] In the letter Watt wrote to Keir he also promised to send a similar letter to William Adam, M.P., and this is of interest since James Adam was the witness before the Select Committee of the Commons on behalf of Peter Theodore de Bruges, another of the petitioners requesting a share in any drawback granted.[4]

The Commons Journals for 1780 refer to petitions requesting exemptions from the payments of salt duties made by Alexander Fordyce (22 May), James Keir (31 May), Peter Theodore de Bruges (31 May), James Watt (31 May), John Collison (1 June), Joseph Fry (19 June), Richard Shannon (19 June) and Isaac Cookson (21 June). Alexander Fordyce said that he had an alkali manufactory at South Shields; Keir claimed to have discovered his alkali process in 1771; de Bruges claimed to have built an extensive works; James Watt petitioned on behalf of himself and Joseph Black; John Collison and Joseph Fry set up their alkali works at Battersea in 1782 to make alkali for Fry's soap-making works at Bristol, but quarrelled in 1784; Richard Shannon claimed to have erected an alkali works costing 'some thousands of pounds' and was 'now manufacturing Barilla in great Quantities'; Isaac Cookson,

---

[1] Fordyce's method was probably a variant of that of Malherbe, in which sodium sulphate (from salt and sulphuric acid) is heated with iron and carbonaceous materials. The chemistry is complex but may be roughly expressed by the equation $Na_2SO_4 + Fe + 2C \rightarrow Na_2CO_3 + FeS + CO$.

[2] Watt to Black, 30 May 1780, Dol.

[3] Watt to Keir, 31 May 1780, Dol.

[4] *C.J.*, Vol. XXXVII, pp. 914–15. James Adam testified that de Bruges had acquainted him with his method of making alkali and that he, Adam, had employed Charles Gall to offer a sample of it for sale. Adam then gave a specimen of it to Samuel More, 'Secretary of the Society of Arts and Sciences', to be made into glass, and another sample to John Adam, soap-boiler near Chelsea, for making into soap. According to Adam, de Bruges had laid out more than £2,000 in works for making alkali and had collected over 700 tons of raw material. Samuel More gave evidence of his experiments in a glass-house, and Charles Gill said that a Mr. Impey, in Great St. Thomas the Apostle, told him that it was very suitable for both soap- and glass-making. Impey was said to be 'one of the principal Dealers and best Judge in that Article'.

a glass-maker of Newcastle upon Tyne, petitioned on behalf of himself and Edward Wilson of South Shields, saying that glass-makers were the best to prepare their own fluxes and salts because when they bought them from others they were often adulterated.[1] Watt commented to Black:

> But their processes are not formidable, part of them proceed on decomposing hepar sulphuris. Consequently [they] must first make Glauber salt by [oil of vitriol = sulphuric acid] and Fordyce is on the same plan only his process is more perfect than the rest/one man has a patent for doing it by [fixed air = carbon dioxide] Others only exchange the vegetable for the Fossil [alkali] by elective attraction.[2]

All these wished to share in the drawback, but Samuel Garbett also petitioned against the drawback being allowed to Fordyce.[3] The duties of £20 a ton on salt were more than this new industry could have hoped to have survived. At first sight it might have appeared at the time that these were independent petitions, but in fact some of the petitioners concerted action with each other. For example, in the parliamentary recess of November 1780, Watt was corresponding with Joseph Fry, who had set up an alkali factory at Battersea in partnership with John Collison and another Quaker, William Jones.[4] Watt wrote to Fry, 21 November 1780: 'I shall be glad to hear how you do, how you go on with your Alkali & if you mean to make any stir about the exemption this parliament with whatever else you think proper to communicate on that subject. As to myself though I can by no means promise my attendance in London yett I will be of any service in forwarding the scheme which lies in my power.'[5] Fry's lengthy reply was mainly directed to asking Watt to perfect a method of making waste salt unusable for culinary purposes so that it could be freed from duty for alkali manufacture.[6] Fry's alkali manufacture was connected with his soap production and he was not particularly interested in a very pure substance as some salt was needed for graining the soap. About this time Watt obtained knowledge of de Bruges' method, as he

---

[1] Gittins, *op. cit.*, pp. 176–7.

[2] Watt to Black, 9 June 1780, Dol. Watt's chemical symbols have been replaced with the names of the substances. These symbols will appear in the forthcoming edition of the Black–Watt correspondence. Hepar sulphuris = sodium sulphide, made by reducing sodium sulphate with iron and/or carbon. 'Exchanging the vegetable for the fossil alkali by elective attraction' means a double decomposition such as $K_2CO_3 + Na_2SO_4 \longrightarrow K_2SO_4 \downarrow + Na_2CO_3$. Potassium carbonate (vegetable alkali) is much inferior to sodium carbonate (fossil or mineral alkali) for many purposes, including soap and glass making.

[3] *C.J.*, Vol. XXXVII, pp. 891 and 913–16.   [4] Padley, *op. cit.*, p. 72.

[5] Watt to Fry, 21 Nov. 1780, Boulton and Watt Letter Book, beginning July 1780, B.R.L.   [6] Fry to Watt, 11 Month 26th 1780, B.R.L.

notes in his copybook for 20 June 1780: 'In the evening mett Mr. Drummond who showed me a paper by which I learned that Mr. Du Bruges process was decomposing hepar Sulph by [tartaric acid?] and I suppose burning away the [acid] mett Mr. Garbet who is determined to present a petition ag[ains]t the Bill on acc[oun]t of the marine [acid].'[1] At the same time Roebuck was nosing round the public offices and Samuel Garbett was anxious about the competition his sulphuric acid would meet with from the hydrochloric acid incidentally produced by some of these alkali-manufacturing schemes, particularly Fordyce's.

All the parliamentary negotiations in June 1780 were complicated by the Gordon Riots which were taking place at the time. Watt's letters to his wife at this period are full of the riots and there is not much information in them about his alkali affairs.[2] Watt arrived in London to join Keir and Boulton on Sunday, 4 June, and stayed until Friday, 9 June; he was back in London again on 18 June and stayed until 6 July.[3] It was a most inconvenient time also for the engine business, as Watt had a great amount of work on hand. The rounding up of necessary witnesses for the Commons committee was no light task, and it is little wonder that Watt wrote to his wife, in exasperation: 'The man that is to serve all Britain with Alkali is a lyar, an Irishman, a Bankrupt & a Blockhead.'[4]

As he complained to Boulton: 'My being so long at London has been an immense loss to our business Mr. Blair has been here these three days and Mr. Keir has been wholy occupied with him so that I have not seen K yet.'[5] As the select committee got down to work we find John Whitehurst testifying that he had seen Keir make alkali the day before;[6] Alexander Blair, Keir's future partner and fellow army-officer, disclosing that he had heard about Keir's discoveries from Dr. Irvine some five years earlier;[7] and Samuel More, James Watt and Mathew Boulton all giving evidence in Keir's behalf.[8] Boulton and James Black, Dr. Black's brother, gave evidence for Watt. Boulton was also asked by Watt to track down in London Dr. Cleghorn, 'to whom Dr. Black partly communicated our method of Alkalizing common salt', so that

[1] James Watt's Pocket Notebook, 1780–2, Dol. Nothing can be found in the literature about de Bruges' process.
[2] James Watt to Mrs. Ann Watt, 6 June, 7 June, 8 June, 18 June, 21 June, 24 June, 29 June 1780, B.R.L.
[3] Watt to L. Henderson, 13 June 1780, Boulton and Watt Letter Book (Office), B.R.L.
[4] James Watt to Mrs. Watt, 24 June 1780, B.R.L., not 28 June 1780 as in Padley, *op. cit.*, p. 71, n. 24.
[5] Watt to Boulton, 9 July 1780, mechanical copy, Dol.
[6] *C.J.*, Vol. XXXVII, p. 913.
[7] *Ibid.*, pp. 913–14.   [8] *Ibid.*

he could appear as a witness if required.[1] These efforts stopped Fordyce's attempt to secure an exclusive privilege, but they failed to secure a general drawback of the salt duties, which continued until 1823–5. It seems to have been fiscal considerations and the implacable opposition of the Salt Office which occasioned the defeat,[2] together perhaps with Fordyce's reluctance to share the privilege with others.[3]

Nevertheless a new attempt was made in the following year, 1781, to procure some alleviation of the salt duties when salt was used in alkali manufacture. Watt had written optimistically to his brother-in-law, Gilbert Hamilton, on 14 August 1780:

I was occupied the beggining of this summer attending parliament on a bill for granting an exemption from the duties on common salt employed in making fossil Alkali—We lost the Bill but are promised success next session—that Scheme is a good one—I know of several methods of doing it, but many experiments must be made to determine w[hi]ch is best before it can be made a trade of—In the mean time I will be obliged to you to keep your eye[s open] in search of any materials containing [sulphuric acid] in an active form that can be had cheap in quantity.[4]

In February 1781, Keir with de Bruges petitioned the Commissioners for Trade and Plantations, but their Lordships would not change their stand.[5] Then a bill was introduced in the Commons against which Garbett filed a petition,[6] but which passed the Commons on 1 May, only to be delayed in the Lords so that no further discussion could take place before the adjournment. In 1782, by 22 George III, cap. 39, s. 2, alkali-makers were eventually allowed to use salt free of duty; but because of alleged abuses the privilege was withdrawn in 1786 (26 George III, cap. 90, s. 1).

The years 1779–1789 saw six patents taken out for the manufacture of alkali—Richard Shannon's Patent 1223 (May 1779), Bryan Higgins's Patent 1302 (July 1781), Alexander Fordyce's Patent 1303 (August 1781), John Collison's Patent 1341 (November 1782), James Gerard's Patent 1369 (May 1783), and A. B. de Boneuil's Patent 1677 (March 1789). In addition in 1788:

J. C. de la Métherie proposed to calcine sodium sulphate with carbon in the belief that it would yield sulphurous acid and soda. Some sodium sulphide was formed, and he suggested that it should be treated with acetic acid and the resulting sodium acetate heated to yield soda. This process was practically the

---

[1] Watt to Boulton, 18 June 1780, mechanical copy, Dol.
[2] E. Hughes, *op. cit.*, p. 432.   [3] Black to Watt, 1 Sept. 1780, Dol.
[4] Watt to G. Hamilton, 14 Aug. 1780, Dol. mechanical copy.
[5] *Journal of the Commissioners for Trade and Plantations* (1780–2), p. 391 (28 Feb. 1781).   [6] *C.J.*, Vol. XXXVIII, p. 415 (24 March 1781).

same as that described by Duhamel fifty years earlier, and the first part of it was also the basis for Leblanc's investigations.[1]

Since de la Métherie and James Keir were in correspondence in 1789 and 1790[2], there may have been discussions on this topic between de la Métherie and members of the Lunar Society, to whom he sent his regards through Keir.

Richard Shannon's method like all the other methods began by converting salt to sodium sulphate, and in this was similar to the Leblanc process. The sodium sulphate was then converted to sodium sulphide by treating it with charcoal and lime. The crux of the problem was then the conversion of sodium sulphide to sodium carbonate, and the efficiency with which the patentees solved this problem distinguishes between them.

Shannon, Higgins, and Fordyce described how they proposed to remove the sulphur. Dr. Gittins believes that Shannon's understanding of the process was uncertain, but that the nature of his raw material—sodium sulphate, salt of steel, calcareous earth (limestone or chalk), phlogiston (charcoal) and other salts—suggests that he may have obtained a reasonable yield.[3] Fordyce used iron, or 'any calx of iron or any ore of iron, or anything containing iron, or any other metal capable of destroying sulphur'.[4] Higgins, having produced sodium sulphide, threw into the furnace twice its weight of lead and agitated them together. If Fordyce and Higgins had employed limestone instead of iron or lead they would have completely anticipated Leblanc, and as it was they were very close indeed to his method. Collison and Gerard both consulted Dr. Joseph Black. He was asked by John Collison and Co. of Southwark to analyse a sample of soda sent to him on the 7 February 1782, that is just before Collison took out his patent, and his report was: 'very strong and powerful. It contains more alkali than the best Alicant barilla in the proportion 68 to 44, and more than the best kelp in the proportion 68 to 10. There is no need to use lime in drawing the leys from it, as it is already in a caustic state.'[5] It was probably Black's knowledge of Collison's process which discouraged him from being more active in promoting the schemes of James Keir, or from developing the envisaged partnership with Keir and Watt.

---

[1] Gittins, *op. cit.*, p. 175.
[2] See Mrs. A. Moilliet, *Sketch of the Life of James Keir, Esq., F.R.S.* (1868), pp. 90–1 and 117.   [3] Gittins, *op. cit.*, p. 179.
[4] Patent 1303. Dr. Gittins points out that sodium hydroxide was probably first formed and this would then react with atmospheric carbon dioxide to form sodium carbonate. In 1777 Père Malherbe's process used scrap-iron.
[5] Clow, *op. cit.*, p. 98.

Three years later, on 15 January 1785, Black was consulted by James Gerard, surgeon, of Liverpool, about the establishment of works for the manufacture of 'fossil alkali' or soda.[1] Gerard said that he proposed to make sal ammoniac, sulphuric acid, Glauber's salt (sodium sulphate) and alkali. He also makes it clear that he is acquainted with the patents of Higgins, Fordyce, and Collison, and knows that Keir, Boulton and Watt may be making alkali. Black, however, had acted as a consultant to Collison and was involved with Keir and Watt, and so was professionally debarred from entering into any agreement with Gerard, even if he had wished to do so. Gerard's patent talks of 'calcining sulphates with coal and igniting sulphides with tar on iron plates', while de Boneuil's speaks of 'fusing sulphate, charcoal and iron in closed iron crucibles on flat surface'. The descriptions are too imprecise for us to know exactly what occurred. Only Gerard in a letter to Black, 7 June 1785, mentions that he gets sal ammoniac as a by-product.

Keir's method is described by Mrs. Amelia Moilliet:

They [Keir and Blair] established works at Tipton, near Dudley, for the manufacture of alkali, for the use of the soap-makers, from the sulphates of potash and soda. The method of extraction proceeded on a discovery of Mr. Keir's, contradicting a point in the doctrine of elective affinities held by chemists of the day. Their experiments seemed to show a stronger affinity of sulphuric acid for either of the two fixed alkalis than for lime. Mr. Keir found that—by presenting the salts in an exceedingly weak solution, and by calling in the aid of a chemical agent (for which he always professed the highest respect, and the functions of which in natural operations, were, he thought, greatly underrated) *Time* the rule of election was reversed. By passing the weak solution *slowly* through a thick body of lime, the sulphates were decomposed; the sulphuric acid uniting with the lime, and leaving the alkalis disengaged. The liberated alkali had then only to be brought into a concentrated form for sale.'[2]

Keir had originally considered that Ireland would be the best place for a factory because he would there have avoided the tax on salt.[3] Thomas Day, another member of the Lunar Society, in a book published in 1783, commented: 'I actually am acquainted with a gentleman who has just engaged in a considerable manufactory, which will save some hundred thousands a-year to this country, if it succeeds, who has refused the offer of having the duties taken off the principle [*sic*] article of his consumption, rather than expose himself to the influence of the

---

[1] Clow, *op. cit.*, p. 99, citing Ramsay, *op. cit.*, p. 69.
[2] Moilliet, *op. cit.*, p. 75. See Schofield, *op. cit.*, p. 157.
[3] *C.J.*, Vol. XXXVII, p. 914.

excise-laws.'[1] In fact, as it appears from Mrs. Moilliet's account, Keir had abandoned the salt process for alkali manufacture and was using the potassium and sodium sulphates which were waste products from vitriol and aquafortis manufacture. Since Keir had been busy at Soho with aquafortis manufacture, because the acid was used in stripping silver plate from its copper base, he was well informed on such processes.[2] Aqueous solutions of these sulphates, passed very slowly through a sludge of lime, produced an insoluble calcium sulphate and a weak solution of alkali.[3] Keir then used this alkali in his soap manufacture or concentrated and sold it. On 13 June 1781 he sold a small quantity, 'one pound of crystals of Mineral Alkali', to Boulton and Fothergill.[4] By 1801 the Tipton works was paying excise duty on soap produced at the rate of £10,000 annually. Keir does not appear to have sold much alkali to glass manufacturers, and it was not until the nineteenth century that synthetic alkali replaced kelp or barilla to any extent in that industry.[5] However, it is true to say that soapers' waste was used in the manufacture of bottle glass[6] and some of Keir's alkali may have been disposed of in this form.

Later references to the activities of the Birmingham group and their friends in the manufacture of alkali once more became fragmentary after this period. Boulton travelled to Scotland in the autumn of 1783 and met Black and John Robison in Edinburgh. He reported back that he had 'talked with Dr. Black and another chemical friend respecting my plan for saving alkali at such bleach-grounds as our fire-engines are used at instead of water-wheels: the Doctor did not start any objections,

[1] Thomas Day, *Reflexions upon the Present State of England and the Independence of America* (3rd edn. with additions, 1783), p. 125, quoted by Schofield, *op. cit.*, p. 157.

[2] Memorandum dated 1778, in Boulton's hand, Keir Box, A.O.L.B.

[3] For details of Keir's process, see J. L. Moilliet, 'Keir's Caustic Soda Process—an Attempted Reconstruction', *Chemistry and Industry* (5 March 1966), pp. 405–8, and B. M. D. Smith and J. L. Moilliet, 'James Keir of the Lunar Society', *Notes and Records of the Royal Society of London*, Vol. 22, nos 1 and 2 (Sept. 1967), p. 146. In the latter article it is suggested that 'other members of the Lunar Society may well have contributed' to Keir's process. It should be noted that, by using these cheap sulphate by-products, Keir 'avoided the high excise duty on common salt'.

[4] Richard Edwards to Boulton, 13 June 1781, Keir Box, A.O.L.B. This is the 'man Edwards' referred to by Boulton as assisting Keir in aquafortis manufacture at Soho and by Scale in a letter to Boulton, 21 Sept. 1780, A.O.L.B: 'Mr. Keir was consulted at the first going off but I have never seen him since—his Man Edwd[s]. says he is to leave Soho & go to Tipton in a month or 6 weeks for good.'

[5] See Padley, *op. cit.*, p. 77; A. E. Musson, *Enterprise in Soap and Chemicals* (1965), pp. 14, 23.

[6] Samuel Parkes, *Chemical Essays* (5 vols., 1815), Vol. III, p. 447–9; Dionysius Lardner (ed.), *The Cabinet Cyclopaedia* (1832), *sub* 'Glass', p. 187, quoted by T. C. Barker, R. Dickinson, and D. W. F. Hardie, 'The Origins of the Synthetic Alkali Industry in Britain', *Economica* (May 1956), pp. 158–69.

but, on the contrary, much approved it'.[1] Boulton was also consulted by Dr. Francis Swediaur, through a letter to Andrew Smith, about the methods of alkali manufacture from sea-salt used by Lord Dundonald.[2] Boulton, in his reply, said: 'Neither Dr. Withering nor I know anything of Lord Dundonald's process of making mineral alkali from sea-salt, nor do we believe it.'[3] This would seem to be, on the face of it, a typical example of Boulton's evasiveness when any new process of which he had any knowledge was under discussion, especially when, as with Swediaur, he had any reason to be suspicious of his correspondent. As Dr. Clow has shown in some detail,[4] the promotion by Lord Dundonald of the efforts of the two Newcastle manufacturers, William Losh and Thomas Doubleday, led to the manufacture of alkali by methods closely related to the Leblanc process, but this is not part of our story since the Birmingham group were not active participants.

Despite all these varied attempts to manufacture alkali in this country before Leblanc pioneered his process, the industry did not flourish.[5] For one thing the increase in vegetable alkalis continued to meet the rise in demand, although fluctuations in price must have been disconcerting to the manufacturer of gunpowder, soft soap and textiles. The increasing supply of potash from the American woodlands compensated to a large extent for the loss of imports from northern and eastern Europe where forests were being depleted. Moreover, the soap-makers needed the sodium chloride which existed in barilla and kelp alongside the alkali and potassium salts in order to separate out the soap. If the duty on salt had not been so high, the soap-maker would have found less advantage in the use of kelp. The maker of bottle-glass also employed the soaper's ashes and was not worried about the sandy impurities in the mixture, while they could also be used by the farmer as a fertilizer. Moreover the soap-makers discovered in the 1780s a method of extracting alkali from the refuse left behind in the soap pan, and so obtained an additional profit from the by-products of soap-making with vegetable alkalis. For these and other reasons, British industry was not under the same pressure as France to synthesize soda and in the 1780s there seemed to be no commercial future of any importance in Britain for synthetic processes. Even for the limited opportunities that existed, the efforts of Watt and his friends had not been the best calculated to gain success.

---

[1] S. Smiles, *Lives of Boulton and Watt* (1865), p. 329, quoted by Clow, *op. cit.*, p. 97.
[2] F. Swediaur to Andrew Smith, 24 March 1791, B.R.L., quoted by Clow, *op. cit.*, p. 100.　　　　　　　　　　[3] Boulton to Swediaur, 7 April 1791, B.R.L.
[4] Clow, *op. cit.*, pp. 100–7.
[5] The explanations given in this paragraph are based on Gittins, *op. cit.*

Though James Watt and his friends in Birmingham and Scotland were not to be the most successful exponents of alkali manufacture, their correspondence shows how once again they were alert to the new possibilities of applied science and ready to turn them to profit if they could. Their social relationships provide the springboard for their endeavours, and they were much assisted by the mutual trust and confidence which existed between them. In addition we see James Watt playing a leading part in parliamentary negotiations, seconded, but not led by, Boulton, and having no small influence in government circles on his own account. Watt's activities in alkali manufacture can be paralleled by his work on chlorine bleaching. Though the theoretical basis of his investigations into alkali manufacture were deficient, and the chemistry of chlorine bleaching derives primarily from Berthollet, it was because Watt had a keen interest in the chemical thinking of his time that he shared the friendship of both Berthollet and Black. On the other hand, Watt's own practical sense led him to undertake further experiments of his own into both processes, because he saw what material advantages might be obtained; and his close relationship with men of business sharpened his awareness of the value of new methods, especially if they could be patented. Indeed, it is Watt who continues to press for action while Black, perhaps more doubtful of the chemical feasibility of the process and made cautious by his confidential knowledge of other processes, ceased to take an active interest. Watt and Black's method proved a failure, but it should be remembered that Keir, by a different process from that of Leblanc, made a considerable fortune. Nor should Watt's enquiring mind be lightly dismissed. Unlike Erasmus Darwin, or Richard Lovell Edgworth, Watt cannot be described as a *dilettante*. He always investigated with a purpose and was suspicious of wild-cat schemes. At the same time, very little passed him by, and it is very fitting that his personal seal was an open eye with the word 'Observare' superimposed.

Appendix: the Lime–Salt Process for Making Soda[1]
James Watt was only one of a number of people who proposed to make soda (sodium carbonate) from lime (calcium hydroxide) and salt (sodium chloride); the prospect of making valuable alkali from the two cheapest inorganic chemicals then available was an attractive one. To the modern chemist, however, the process is puzzling. Double decompositions of the type:

(1) $2NaCl + Ca(OH)_2 = 2NaOH + CaCl_2$ (followed by absorption of carbon dioxide from the atmosphere)

[1] This appendix is contributed by Drs. W. V. Farrar and Kathleen R. Farrar.

do not occur unless one of the products is insoluble in water; in this case both are freely soluble. The reverse reaction is indeed more likely, since calcium hydroxide is sparingly soluble. From the experiments which Watt described in these letters, however, it would seem inescapable that, somehow or other, reaction (1) must work. This feeling is strengthened by the fact that Guyton de Morveau and Carny actually operated this process commercially in 1782;[1] the brief life of this enterprise ended mainly for fiscal reasons. These chemists may have based their process on a paper by Scheele,[2] in which the same reaction is described. A speculative explanation,[3] according to which the product is actually sodium bicarbonate, is not acceptable; the partial pressure of carbon dioxide is far too low for this. We have thought it worth while to study this problem experimentally.

A block made according to the directions of Watt, from salt, lime and sand, showed no alteration on keeping indefinitely in a moist atmosphere enriched with carbon dioxide. This confirms the chemist's instinct that reaction (1) will not 'go'. The greatest difficulty, however, in the experimental study of the history of chemistry is to reproduce the degree of *impurity* of the reagents originally used. In this case the salt used would probably come from sea-water, perhaps from Roebuck's salt-pans at Bo'ness. Now in sea-water the most abundant anion (after chloride) is sulphate; and a lime-sulphate process for soda is quite feasible, since calcium sulphate is insoluble in water:

(2) $Na_2SO_4 + Ca(OH)_2 = CaSO_4 + 2NaOH$ (followed by absorption of carbon dioxide).

This is indeed the equation for the process later used successfully by Keir.

We therefore evaporated some sea-water to about one-tenth of its original bulk, and again made a block with lime and sand. After about three weeks under the same conditions as before, the block began to show a satisfactory efflorescence of sodium carbonate. The source of Watt's soda, therefore (and of that of Guyton and Scheele), was the sulphate in his crude salt; his process would give variable results as his raw material fluctuated between the extremes of pure sodium chloride and concentrated sea-water. The addition of coal-ashes (which contain both carbonate and sulphate), as in some of his experiments, added a further variable.

[1] *Annales de Chimie*, Vol. 19 (1797), pp. 102–4.
[2] *Collected papers of C. W. Scheele* (1931), trans. by L. Dobbin, p. 200. (*Kongl. Vetenskaps Academiens Handlingar*, Vol. 10 (1779), p. 158.
[3] J. W. Mellor: *A Comprehensive Treatise on Inorganic and Theoretical Chemistry* (1922), Vol. 2, p. 715.

Watt, with great optimism, went straight from his meagre experimental results to ambitious plans for building walls of lime-salt-sand blocks. We found it difficult to make a good coherent block, and a drying stage is necessary. Presumably the efflorescence would be scraped off by hand; a tedious business, though standard practice in contemporary processes for making saltpetre.[1] However, if the results were as variable as we suppose, it is not surprising that Watt's process came to nothing. Guyton, who probably used concentrated sea-water (his factory was on the coast of Brittany), had a little more success.

The role of Black is enigmatic. He does not seem to have grasped the chemistry of the process, though it was well within his abilities to do so. He probably decided at an early stage that the process was worthless, but was reluctant to say so to Watt, and temporized.

[1] J. R. Partington: *A History of Greek Fire and Gunpowder* (Cambridge, 1960), pp. 314–23.

XI   The Profession of Civil Engineer in the Eighteenth Century[1]

A Portrait of Thomas Yeoman, F.R.S.,
1704(?)–1781

The history of the engineering profession in the eighteenth century is still only in sketchy form, and in all branches of the profession at that period there were men whose work deserves to be better known. Among canal and waterway engineers there were Jessop, Whitworth, the Grundy family, and Golbourne; among millwrights and mechanical engineers, men such as Thomas Hewes, Peter Ewart, and Thomas Yeoman; and the instrument-makers have had such short shrift from historians, that Ramsden, Nairne, Dollond, Short and others, are still little more than names to them. Yet the fact remains that, by the end of the eighteenth century, engineering was a profession, well established and respectable, in which specialist branches were beginning to develop. Thus James Watt, writing in 1781 to advise his cousin, Mrs. Marion Campbell, on the education of her son, could refer to civil engineering, even at that early date, as a recognized profession in its own right. The letter is worth quoting at some length, however, not only to confirm this point, nor even simply because it is a masterly resumé by a man who was both a mechanical engineer of genius and a civil engineer of distinction, but also because it shows the close links that existed between civil engineering and other mechanical occupations: 'The Businesses which require Mechanical genius and by which a modest man can get money are few—An Architect, an Engineer civil or military—a Ship Carpenter—a Jobbing Smith—a Surgeon—(Painting and engraving are poor businesses). . . .' The combination of professions referred to may chime oddly in the modern ear, but they were

---

[1] This chapter is based on an article by Eric Robinson in *Annals of Science*, Vol. 18, no. 4 (Dec. 1962), pp. 195–215.

closely related at this stage of technological growth. Watt then proceeds:

> An Architect requires taste and ingenuity—a modest assurance, a knowledge of mankind, and a stock of money or credit, with these it is a very lucrative business—a Civil Engineer requires invention, discriminating judgement in Mechnical matters, boldness of enterprize and perseverance, ability to explain his ideas clearly by words and drawing a good constitution to bear fatigue and vexation, a knowledge of, and ability to treat with one part and govern another part of mankind. With these qualifications joined to experience, a man may get a comfortable livelyhood without a stock but can scarcely without some uncommon good luck acquire a fortune but if he has a stock of money and is prudent in his undertakings he *may* get a fortune . . .'

At the same time, Watt gives some down-to-earth advice for the intending civil engineer: 'When he is 14 put him to a Cabinet Maker to learn to use his hands and to practise his Geometry—at the same time he should work in a smiths shop occasionaly to learn to forge and file. . . .' He also adds: 'before he attempts to make a theodolite let him be able to make a well joined Chair'.[1]

This civil engineer is clearly not divorced from the humbler mill-wright or from the instrument-maker. He is a man who both knows how to do the job and can do it himself. He is a practical man capable of using his hands, and will be the sort of engineer who is not confined to the drawing-board or dependent upon consultants in allied branches of his profession. Such a man, as we shall see, was Thomas Yeoman, one of the leading civil engineers of his day, who deserves to be ranked alongside Smeaton and Brindley for his contributions to his profession.

Yeoman, however, was an engineer of the wind-and-water age. On the whole, it is engineers of steam-power who have attracted the attention of historians. A Newcomen, a Savery, a Smeaton or a Watt have had much better chances of survival in the halls of fame than those engineers who harnessed by far the greater part of the power used in industry until at least 1830. But the concentration upon steam engineers is a distortion for the eighteenth century, and Yeoman's career provides an opportunity to correct some part of that distortion, and to appreciate more sympathetically the role of the engineer before the steam revolution was at its height.

Nothing is known of Yeoman before 1742 when he was manager of Cave's cotton-mill at Northampton.[2] His services there were both to

[1] James Watt to Mrs. Campbell, 15 Sept. 1781. Quoted by kind permission of Col. P. M. Thomas, D.S.O., T.D., M.A., from whose private collection this letter comes.

[2] J. H. Thornton, 'The Northampton Cotton Industry—an Eighteenth Century Episode', *J. Northants. Nat. Hist. Soc. & Field Club*, Vol. 33 (1959), pp. 242–59.

the cotton-mill and to many varied aspects of life in an eighteenth-century market-town. He acted principally as a surveyor and a mechanic for agricultural machinery, but also as a leading member of the Northampton Philosophical Society and as a lecturer on scientific subjects. Such a man played an equivalent part in market-town life to the country attorney, the doctor or the parson. He describes himself, however, quite modestly in advertisements of this period, as 'Thomas Yeoman, millwright'.

Tho. Yeoman, Millwright

At his house in Gold-street, Northampton

Makes and Sells all Sorts of Boulting Cloths for Size Mills, as
cheap as in London or elsewhere; and, by the Approbation of the
ingenious Dr. Hales, makes all his useful Machines, as
Ventilators for Hospitals, Goals, Graineries, or Ships,
Back-Heavers for expeditiously winnowing of Corn, Hollow
Sticks for ventilating Corn in Sacks. Specimens of each
Machine may be seen at his House aforesaid, Corn-Factors, Meal-men,
&c. may be furnished with his new-invented Machine for cleansing
Corn, described in the Gentleman's Magazine for December last,
one of them may be seen at Mr. Alderman Chapman's in Northampton;
likewise common Screens for Wheat, Barley, Malt, &c. all sorts
of Engines for raising Water are erected or repair'd,
particularly his improved Bucket Engine, upon Gervis's Principle;
also Bridge Weighing Engines for all sorts of Carriage. He makes
and sells Air Pumps, refracting and reflecting Telescopes, with
all Sorts of Mathematical and Philosophical Instruments, Electrical
Machines for the Studies of the Curious, &c. He also gives Plans
and Designs for erecting all Kinds of Mills now in publick use
(when the Distance is so great that he cannot personally attend)
especially those with modern and useful Improvements.
N.B. Gentlemen may have their Estates survey'd, and Plans
thereof neatly laid down, by the said Tho. Yeoman; who gives
Proposals, and takes in Subscriptions, for the ingenious
Mr. Neal's Mensurator.[1]

Here is evidence of Yeoman serving as a mechanical engineer to local industries—the cotton industry, the paper mills, butchers, corn merchants and farmers; as surveyor to local landowners; as weights-and-measures engineer for turnpikes, the municipality of Northampton, and all other persons or corporations requiring weigh-bridges; as water engineer for mills and farms etc.; as architect for mills; and as philo-

[1] *Northampton Mercury*, 27 April 1747.

sophical instrument-maker for surveyors and for gentlemen with scientific interests. The construction of ventilators had a multiplicity of uses in mines, gaols, workhouses, barracks and hospitals, as well as in granaries, ships, and many kinds of mill.[1] The list of activities in the advertisement has, however, a decidedly local tone and shows such a man to be a key person in a market-town.[2]

There is a suggestion that Yeoman came to Northampton from the Abbey Mills near Bromley, where he was employed as wheelwright and engineer,[3] but two letters at least survive to tell of his work in Northampton in 1742-3. The cotton-mill contained the famous spinning-machines invented by Lewis Paul and John Wyatt. In 1742 Yeoman reported in detail on the mill to its proprietor, Edward Cave.[4] It was quite a large concern with a minimum of twelve spinning-engines, employing fifty girls as well as the managers, who included Yeoman's deputy, a man called Harrison. The fees of the managers seem to have been heavy, amounting to £88 per week, against £150 per week for the fifty girls and £296 5s. 0d. for the cotton and waste (7920 lb. at 9d. per lb.).The mill was not a success. By 1749, if not earlier, it was on the decline and in 1756 the mill and its machinery were sold up.

It is possible that Yeoman himself had so many other irons in the fire that the mill was not properly supervised. An unsigned report on the mill, dated 8 October 1743, comments on 'the miserable decay of the Mill and great Wheels', the heavy expense incurred in repairing the cards, the mismanagement generally of the carding, the misuse of the water wheel and the general dirtiness of the whole concern.[5] There is also a note to the effect that the Superintendent, i.e. Harrison, 'seems a very indifferent manager'. The division of responsibility between Yeoman and Harrison did not work well. Harrison wanted Yeoman to make for the mill an engine with 20 spindles to go by hand, but Yeoman evaded the request because: 'he imagines I have got an estate

---

[1] See Rees's *Cyclopaedia* (1819), Articles on 'Ventilators', 'Gunpowder', and 'Granary'. See also Pneumatics Plate XVII. The subject of ventilators is taken up at greater length below, pp. 378-9, 384-7. See also above, pp. 35, 39, for Hales and Desaguliers.

[2] A comparison may be made with the activities of perhaps the most famous family of eighteenth-century surveyors, the Wyatts of Lichfield and Burton-on-Trent, who included in their number, John Wyatt, Boulton and Fothergill's London agent, James Wyatt, the famous architect, Samuel Wyatt and William Wyatt, carpenter, bailiff, and architect, and Charles Wyatt, wood-screw manufacturer. See Wyatt Box, Tew MSS., Assay Office Library, Birmingham, and the Boulton and Watt Collection, Birmingham Reference Library.

[3] Thornton, *loc. cit.*, p. 251.

[4] Thomas Yeoman (Northampton) to Mr. Cave, 14 Oct. 1742, Wyatt MSS., B.R.L.   [5] Wyatt MSS., Vol. I, B.R.L.

under him and so in all probability would not pay Me . . . [but] I am so far from getting an estate under him that this very Week my Wife changd the last guinea we have in the world.'¹

Yeoman had also begun to go on surveys and engineering business away from Northampton, since he mentions in the same letter of 1743 his 'journey into the west'. Nothing is known of that journey but some of his surveys in Northamptonshire during the year 1752 survive.² Clearly he was the sort of man to have been involved as a surveyor in eighteenth-century enclosures and he was certainly engaged a few years later, in 1757, in turnpike schemes, for there is a map by him accompanying 'An Act for Repairing and Widening the Road from Towcester, through Silverstone and Brackley, in the County of Northampton, and Ardley and Middleton Stoney, to Weston Gate, in the Parish of Weston on the Green, in the County of Oxford, 1757'.³ Turnpike work occurs at intervals later in his career when he had moved to London.

The combination of his interests in the Northampton cotton-mill and in turnpikes, placed him in a very strategic position for promoting the invention, by his friend John Wyatt, of a weigh-bridge. The cotton-mill not proving a success,⁴ Wyatt had turned to other schemes. In an undated letter from the Fleet Prison, where he had been committed for debt, he described to Yeoman, in detail, his new weighing-machine.⁵ The machine was also described and illustrated in a communication by Yeoman to the *Gentleman's Magazine* in March 1748.⁶ Before the weigh-

¹ Thomas Yeoman to John Wyatt, 26 Nov. 1743, *ibid.*, Vol. II.
² 'The Plan of the Inclosed Estates of Hitch Young Esqʳ. in the Parish of Great Houghton in the County of Northampton late the Estate of Bartholomew Clark Esqʳ. deceased. By Thos. Yeoman 1752'; 'Plan of the Inclosed Estates lying within the Town and Lordship of Hardingstone late Barth. Clark's Esqʳ. deceased by Thos. Yeoman 1752' (Bouverie Delapré 322); 'A Survey and terrier of the Arable Lands Ley and Meadow Ground late the Estate of Barth: Clark Esqʳ. lying in the Common Fields of Brafield and Little Houghton and now in the Occupation of John Coleman Surveyed and taken in the month of April 1752 by Thos. Yeoman' (Bouverie Delapré 326); 'A Survey of the Terrier of the Arable Land Ley and Meadow Ground lying in the East End Common Fields of Hardingstone, late the Estate of Bartholomew Clark Esquire Surveyed and taken in the Year 1752 by Thos Yeoman' (Bouverie Delapré 323). See also Bouverie Delapré 324a–f, *c.* 1752. My thanks are due to Mr. King of the Northamptonshire County Record Office.
³ Northamptonshire Record Office.
⁴ Robert Dossie said of Wyatt's spinning-machine in 1768 that it was 'wholly laid aside as unprofitable, after sixty or seventy thousand pounds had been spent in various attempts to establish its use'. *Memoirs of Agriculture*, Vol. I, p. 98, cited by D. Hudson and K. Luckhurst, *The Royal Society of Arts, 1754–1954*, (1954), p. 129.
⁵ Wyatt MSS., B.R.L.
⁶ Pp. 120–1, and Figs. i and ii on Miscellaneous Plate.

bridge was introduced, the immense machines used for this purpose were very clumsy:

the machine was erected in an open building, beneath which the road passed, so that a cart, waggon, or other carriage could be drawn under; strong chains were then passed beneath the body of the carriage, to attach it to the extremity of an immense steel-yard. The fulcrum of the steel-yard was suspended by a lever, or by pulleys and crane-work, from the top of the building; and when the carriage was properly secured, the steel-yard was hoisted up by the crane-work, so as to suspend the waggon, and it could then be weighed by applying the sliding-weight of the steel-yard to different parts of the divided bar.[1]

Quite apart from the danger of the wagon being damaged while it was being hoisted, there was also the labour of unharnessing and reharnessing the horses. When the number of turnpikes was increased, and special tolls were instituted for carriages above a certain weight in order to compensate for the damage done to the roads, a more expeditious method of weighing was required. It must also be remembered that the quantities of heavy raw materials and manufactured articles travelling the roads was increasing. It was also a great convenience to be able to weigh loads in any large town. Weigh-bridges of Wyatt's design were erected at Birmingham (Snow-hill), Shrewsbury, Chester, Liverpool, and Worcester.[2] Yeoman was responsible for erecting one at North-ampton, where it still stands.[3] In 1751, another inventor, Joseph Eayre, who had been 'employed in making almost all the weighing engines within 50 miles of London', claimed, with the support of Dr. Desagu-liers and others, that Wyatt had borrowed his idea, and so Eayre claimed to be the prior inventor.[4] However, in 1752, another corres-pondent criticized Eayre's engine as not being so accurate as was claimed, and so the charge of plagiarism seems to have been dropped, though it may have counted against Wyatt at a later date.[5]

Returning once again to our basic advertisement of 1747 we see that Yeoman was offering to make his improved bucket-engine upon

---

[1] Rees's *Cyclopaedia:* Article on 'Weighing-machine'. Several of these old machines were described by Leopold in his *Theatrum Staticum* (1724).

[2] *Gentleman's Magazine*, March 1748, p. 120. See W. A. Benton, 'John Wyatt and the Weighing of Heavy Loads', in *Newcomen Soc. Trans.*, Vol. IX (1928–9), pp. 60–77, and correspondence by H. W. Dickinson, pp. 73–6. Benton and Dickinson were unaware that Yeoman and Wyatt knew each other well, and that consequently Yeoman could not possibly have been mistaken about the nature of Wyatt's weigh-bridge. See also Miscellaneous Plate in *Gentleman's Magazine* for March 1748.

[3] In *Northampton Mercury*, 27 April 1747, Yeoman advertises himself, as making 'Bridge Weighing Engines for all Sorts of Carriages'.

[4] *Gentleman's Magazine*, Oct. 1751, p. 450, and Nov. 1751, p. 488.

[5] Against his claims to have invented roller-spinning (see below, p. 390).

Gervis's principle. The engine is described in the *Gentleman's Magazine* together with an illustration.[1] The original engine, on which Yeoman's improvement was modelled, had been made by George Gervis (or Gervas) at Sir John Chester's house near Newport Pagnell, and had been described by both Desaguliers and Switzer.[2] It seems to have been used for watering gentlemen's gardens, as was Wyatt's very early engine erected for Sir Harbord Harbord, and shows that the gardening taste of the eighteenth century made contributions to engineering as well as to architecture and to literary taste. The patronage of land-owners in their houses, their gardens, and their farms, was a most important factor in the early stages of technological advance. They also bought Yeoman's agricultural machines. Yeoman invented 'A machine for cutting away mole or rather ant-hills',[3] which was erected by the Duke of Cumberland for clearing away the ant-hills in Windsor Forest, and Yeoman also supplied his ventilators and winnowing machines to local gentry.[4]

The construction of ventilators and machines for cleaning corn seems to have been the first of Yeoman's activities to gain him a considerable amount of publicity. The ventilator was invented by the Rev. Dr. Stephen Hales, F.R.S., and presumably he looked around, as many inventors of his day, to find a craftsman capable of making his machine. How the meeting between the two came about, we do not know, but they were associated at least as early as December 1743.[5] Hales's ventilators, however, were made before this date and were recommended by Desaguliers in 1734.[6] Hales's *A Description of Ventilators* (London, 1743) does not mention Yeoman, though his later writings on ventilators are full of references to him, and so it is likely that the association between the two men began some time in 1743. Although Hales's ventilators had many other uses, their importance in helping grain-storage reflects a general European concern in this period on the same subject. We

[1] *Gentleman's Magazine*, Oct. 1747, pp. 459–60, and plate facing p. 461.

[2] J. T. Desaguliers, *A Course of Experimental Philosophy* (2 vols., 1734), Vol. II, p. 76, and plate 7, figs. 14 and 15; S. Switzer, *General System* (2 vols., 1729), Vol. II, p. 314 and plate 17, p. 316.

[3] *Gentleman's Magazine*, Dec. 1746, pp. 639–40, fig. i.

[4] *Ibid.*, fig. iii. His first agricultural ventilators were made for Northampton-shire gentlemen, 1745–52, and are listed in Hales, *A Treatise on Ventilators* (1758), p. 243.

[5] *Northampton Mercury*, 26 Dec. 1743: 'Among the many Excellent and Useful Machines invented by Our own Countrymen, the VENTILATOR, by the Ingenious Dr. Hales is not the least . . . this is to inform the Publick, that THO. YEOMAN, Operator for Mr. Cave's Cotton Engines at Northampton, will undertake to make any sort of them . . .'.

[6] Desaguliers, *op. cit.*, Vol. II p. 568. Desaguliers, in fact, should apparently share with Hales the credit for this invention. See above, p. 39.

know that English corn in this period was sometimes exported to Holland, gaining thereby the export subsidy, and was stored in Dutch granaries until it could be reimported in times of scarcity. In France, Duhamel de Monceau, F.R.S., member of the Royal Academy of Sciences, and Inspector of the Marine, was concerned with using ventilators for preserving and drying corn in granaries.[1] M. Pommyer, one of the French king's engineers, was also reported to have made an improvement upon Hales's design.[2] A similar development was reported from Silesia where a farming mill used for cleaning corn from tares was said to be easily 'converted into Dr. Desagulier's machine for throwing fresh air into rooms; and by a fine stove, set within the case . . . cover'd with tin or iron plates with holes, may give warmth to the wind, for making the corn dryer, if there be occasion for it'.[3] Such examples as these, together with the preoccupation of the Society for the Encouragement of Arts, Manufactures and Commerce with agricultural design, show how much technical advance was made in the agricultural industry before the better-known developments in the manufacture of textiles. Even the commoner sorts of implements must have required some sort of mechanical ingenuity—such as the 'common Screens for Wheat, Barley, Malt, &c.',[4] or the 'Corn and Flour Mills, French Stones, Sack Tackle &c., after the newest improv'd Method now used at London, Bristol, Bath, and other Places; Paper Mills, Felling Mills, Wind Mills, Horse and Hand Mills of all Sorts'.[5]

But besides making these contributions to the commercial life of the Northampton area, Yeoman, like other craftsmen of his age, had something important to give the social and intellectual life of his town, through his membership of a philosophical society. Such men as John Whitehurst of Derby, James Watt, when in Glasgow, Peter Clare of Manchester, and Yeoman in Northampton, are characteristic of a democratic spirit in the scientific life of England in the eighteenth century, unparalleled in France at the same period.[6] The Northampton Philosophical Society was one of many such societies that have not yet received their full due. This society met at Yeoman's house in Goldstreet, Northampton, and we know that from 13 December 1743 onwards experiments in electricity were conducted by Yeoman, Sir Thomas Samwell, Sir Joseph Jekyll, Mr. Collier, and Mr. Charlewood

---

[1] *Gentleman's Magazine*, April 1749, p. 160; April 1752, p. 181; April 1753, pp. 165–9. [2] *Ibid.*, Feb. 1757, p. 60.
[3] *Ibid.*, Sept. 1747, with illustration.
[4] *Northampton Mercury*, 27 April 1747.
[5] *Ibid.*, 26 Dec. 1743.
[6] M. Daumas, *Les Instruments Scientifiques aux XVIIe et XVIIIe siècles* (Paris, 1953), *passim*.

Lawton, all members of the society.[1] In April 1747, some experiments conducted by Yeoman to confirm a theory of Hales's were reported in the *Gentleman's Magazine*:[2] 'Mr. Yeoman having electrify'd a man, while a vein was open in his arm, the blood flow'd then much faster, and slower on ceasing to electrify.' Hales took this as evidence to confirm 'his suspicion, that electricity will not promote vegetation'.

At this period, electrical experiments were very popular in the philosophical societies. A letter from Henry Baker, F.R.S., to another member of the Northampton society, Dr. Philip Doddridge, is interesting not only in its testimony on this point but also because it shows how these societies were springing up all over Britain, and not least in the eastern counties:

> It gives me no little pleasure to observe, that societies for inquiring into the productions of nature, and the improvements of art, are forming in different parts of the King's dominions: there are such established at Edinburgh, Dublin, York, Bristol, Peterborough, Spalding, and several other places. When ingenious people meet to communicate their several observations, with a sincere desire to discover truth, great advances may be made in knowledge.
>
> No doubt the members of your society have made some experiments in Electricity, a subject which has lately engaged all the curious world, from the discoveries lately made. . . .[3]

Yeoman's certificate for the Royal Society, which was signed by Henry Baker, states that he 'was several times President of a Philosophical Society at Northampton'.[4] He was not, however, a founder-member when the Northampton society began in November 1743, but joined shortly afterwards, and was responsible for drawing-up the rules of the society.

The Northampton Philosophical Society undertook the same kind of things as most philosophical societies: lectures were read on magnetism, electricity, mechanics, hydrostatics, pneumatics, optics, meteorology and other subjects, and were illustrated by experiments conducted by members of the society. A collection of scientific instruments and apparatus was assembled for this purpose at Yeoman's house. Baker

---

[1] *Gentleman's Magazine*, Sept. 1746: 'A Report of the Proceedings of the Philosophical Society in Northampton, from its Institution, Nov. 11, 1743, to the general Meeting, Nov. 5, 1745'.   [2] *Ibid.*, April 1747, p. 200.

[3] J. D. Humphreys (ed.), *The Correspondence of Philip Doddridge, D.D.* (1831), Vol. V, p. 28. Henry Baker to Doddridge, 24 Nov. 1747. This correspondence should be collated with the manuscript correspondence of Henry Baker in the Rylands Library, Manchester.

[4] Dated 19 May 1763. Yeoman was elected 12 Jan. 1764 and admitted 19 Jan. 1764. I am grateful to Mr. I. Kaye, Librarian of the Royal Society, for this information.

and Doddridge corresponded about the effects of an earthquake: 'In the House of Mr Yeoman where our little Philosophical Society meets, it threw down a Board from the Tester of a Bed, yet Mr Yeoman himself did not feel it';[1] as well as about psychic phenomena and inoculation against cattle distemper.[2] William Shipley and Baker corresponded about a new barometer invented by Shipley;[3] and Charlewood Lawton sent Baker fossil specimens. Through this correspondence with Baker, the Northampton Philosophical Society was kept in touch with the investigations of other societies.

The members were very interested in invention and communicated some of their inventions to the *Gentleman's Magazine* for publication; a pump worked by turning a winch on which is a crank and fly,[4] a model of a machine for carrying timber without horses (improved from a hint in the *Acta Germanica*)[5] and a 'new-invented Universal Dial'.[6] In February 1747, Charlewood Lawton published in the *Gentleman's Magazine* 'A Table of the Specifick Gravity of several sorts of Wood'. He also made and presented to the society his own improvements of the hydrostatic balance, thermometer, and hygrometer,[7] and borrowed others from William Hanbury, F.R.S.,[8] for exhibition to the society. It may be that some of the statistics of Northampton's mortality communicated to the same journal were also supplied by the society or one of its members. Sir Joseph Jekyll of Darlington Hall was one of the first officers of the Northampton Infirmary where Yeoman installed one of his first ventilators in 1748[9] and would have been in a position to obtain these statistics.

Yeoman lectured to the society on Hales's ventilators and their application and 'read divers useful observations from letters which he had received from that very learned and worthy gentleman',[10] but his chief contribution was in his lectures on electricity. Here we see another eighteenth-century activity, important to the scientific life of the age, in which engineers took part—the lectures given by itinerant lecturers. On 29 December 1746, in the Birmingham newspaper, *Aris's Gazette*,

[1] Doddridge to Baker, 17 Oct. 1750, Baker MSS., Rylands Library, Manchester. See Baker to Doddridge, 6 Oct. 1750; and Baker to Doddridge, 8 Feb. 1750.

[2] *Ibid.*, Baker to Doddridge, 3 Jan. 1751.

[3] *Ibid.*, William Shipley to Baker, 10 May 1748. See also D. G. C. Allan, *William Shipley, Founder of the Royal Society of Arts* (1968).

[4] *Gentleman's Magazine*, Sept. 1746, immediately preceding the account of the society, p. 475. See also fig. ii.

[5] *Ibid.*, plate IV, fig. v.

[6] *Ibid.*, pp. 477–8. See also fig. iii, opp. p. 475.

[7] *Ibid.*, March 1748, p. 121.

[8] Hanbury also signed Yeoman's certificate of election to the Royal Society.

[9] Hales, *op. cit.*, pp. 21–2. He also installed one in Northampton Gaol in 1749 (*ibid.*, p. 62). [10] *Gentleman's Magazine*, Sept. 1746.

it was announced that Mr. Yeoman would give a course of lectures on electricity. Further information was given on 5 January 1747:

> This is to acquaint the curious in the Town of Birmingham and its Neighbourhood That they may now be agreeably entertained with a Variety of surprising Experiments in ELECTRICITY (that Branch of Philosophy which engrosses so much Conversation everywhere, and is the Subject of so many learned Debates). These Experiments are made with a new and compleat Apparatus, and exhibited after the most accurate Manner, and according to the latest Discovery made in London and elsewhere. . . . N.B. Besides the Common Experiments, those of Professor Muschenbrook and Mons. Mornier are exhibited, as also several others very lately invented by Mr. Watson and Mr. Neal;[1] and among the rest, those now known in London by the Name of the Lightning, the Ignis Fatuus, the Shooting Stars, and Aurora Borealis. The Muschenbrook Shock will be granted to those that chuse not to feel it who oin'd in the Same Company with those that do. . . .

These lectures had been given by Yeoman in Northampton a month earlier.[2] The advertisement tells us that Yeoman had recently been to London to inform himself on recent developments in the subject, and also that he made electrical apparatus for sale. Some of the principal experiments were described in an advertisement two months later:

1. To shew what Bodies are Electrical, and what are not.
2. To prove that these Wonderful Effluvia will not be conveyed by an Electric *Per-se.*
3. To shew it is the Property of these Effluvia to attract and repel alternately all light Bodies that are brought within the Sphere of its Activity.
4. To prove the Ingress and Egress of the Electrical Effluvia to and from the excited Globe.
5. To shew that Electricity differs from Magnetism, and that the one does not interfere with the other.
6. To shew the Action of Electricity upon the Nerves.
7. To prove that the Electrical Power accelerates the Motion of Fluids.
8. The Motions of the Planets and Comets shewn by an Electrical Orrery.
9. Spirits of Wine Fired by the Electrical Spark.
10. To shew that a Watch when electrified will go faster than before.[3]

The membership of the Northampton Society was not distinguished by comparison with such a society as the Lunar Society of Birmingham. The best-known member was Dr. Doddridge, principal of the Northampton Academy, but William Shipley, from whom came the idea of the Society for the Encouragement of Arts, Commerce and

---

[1] We saw above (p. 374) that Yeoman sold Neal's Mensurator.
[2] *Northampton Mercury*, 1 Dec. 1746.
[3] *Ibid.*, 16 Feb. 1747. Also 23 March 1747.

Manufactures,[1] was a man of no little distinction. Doddridge delivered two papers to the society: one on the doctrine of pendulums and the other on 'the laws of the communication of motion, as well in elastic as non-elastic bodies'. Baker also communicated some of his papers to the Royal Society. The significance of Doddridge's membership, however, lies in the connection between the dissenting academy and the local philosophical society. It was not unusual for a dissenting academy to have employed itinerant lecturers, and it may be that there were such connections between the Northampton Academy and Yeoman. One of the early members of the society was John Fergusson, who was a pupil of Doddridge and who thus pursued his scientific interests. Here then are such members of the society as we have been able to identify:

(1) Dr. Philip Doddridge (F.R.S.)
(2) Thomas Yeoman (F.R.S.)
(3) William Shipley
(4) Charlewood Lawton (F.R.S.)
(5) Sir Joseph Jekyll
(6) Mr. Collier
(7) Henry Lawton
(8) George Paxton*
(9) Samuel Paxton*
(10) Mr. Bartholomew Goodman*
(11) Mr. Golding
(12) Samuel Poole[2]
(13) William Hanbury[3] (F.R.S.)
(14) Sir Arthur Harsley
(15) John Fergusson
(16) Henry Woolley[4]*
(17) Sir Thomas Samwell[5] ⎱corresponding members
(18) Henry Baker (F.R.S.) ⎰
(19) Edward Cave (possible member)

★ = Founder Member.

[1] The Rev. Henry De Foe Baker to S. Bentley, 21 Sept. 1830: 'Shipley you know conceived, & H. H. gave Birth to the Soc^y of Arts', Baker MSS., Rylands Library, Manchester. See D. G. C. Allan, *op. cit.*

[2] *Northampton Mercury*, 5 June 1749: Samuel Poole advertises for the sale of Tobacco-Pipe Clay.

[3] Another F.R.S. and an unsuccessful candidate for Parliament in the county of Northampton in 1748.

[4] *Northampton Mercury*, 24 March 1746. Writing master 'at the House and Shop near All-Saints Church in the Drapery, late in the Possession of Mr. John Fowler'.

[5] *St. James's Chronicle*, 5 April 1766. 'Sir Thomas Samwell owns wine vaults in Northamptonshire. Formerly in partnership with John Hooper deceased—now in partnership with Capt. John Wye of Oporto'.

From this small but lively society, Yeoman migrated to London and thereafter played a significant part in some of the leading societies of the age. That transition must have been assisted by the friends who had moved to London before him—particularly Edward Cave and William Shipley. It is no accident that so many of the references already given come from the *Gentleman's Magazine*, since Cave, one of the proprietors of the Northampton cotton-mill, became the editor of this important journal. As early as 1743, Cave was proposing to act as publisher for Yeoman, who was said to be preparing for the press a treatise on mechanics, principally for the use of millwrights,[1] and Cave was taking in subscriptions for it at St. Johns Gate. Yeoman's activities gradually became more and more widespread from about 1749 onwards, and he presumably moved permanently to London sometime in the mid-fifties.

The manufacture of ventilators for ships, hospitals and gaols was one important side of his expanding business, and there is a fair amount of information about this given in the *Gentleman's Magazine*, in Hales's own publications, and in Admiralty and other records. After fitting ventilators in Northampton Hospital (1748) and Northampton Gaol (1749), he put them into the gaols at Shrewsbury, Maidstone, Bedford, and Aylesbury, and also into St. George's Hospital. The last four were worked by windmills.[2] He was also employed in the same year to do the same sort of work for the Navy, installing ventilators at Haslar Hospital, the prison at Porchester Castle and elsewhere.[3] For this he received the official thanks of the Navy: 'The Commissioners of the sick and wounded Seamen wrote me word, that they find most beneficial good Effects from these several methods, which were put in practice by the direction of Mr. *Yeoman*, in the Hospitals at *Portsmouth*, *Gosport* and *Plymouth*.'[4] We also know that he fitted ventilators in both Houses of Parliament and in Drury Lane Theatre,[5] but at what date is unknown, and it is probable that he was responsible for fixing them in Newgate Prison in 1746, in the county gaols of Winchester and Durham, as well as in the Savoy and in Sir John Oldcastle's small-

[1] *Gentleman's Magazine*, July 1743, pp. 354–5. I have been unable to trace this work.
[2] Hales, *op. cit.*, pp. 21–2, and 62. It should be noted that at this period Yeoman also studied wind-forces in Northamptonshire.
[3] Record Office, Admiralty Department Out Letters, 5, p. 238, cited by A. E. Clark-Kennedy, *Stephen Hales, D.D., F.R.S., An Eighteenth Century Biography* (Cambridge, 1929), p. 168.
[4] Record Office, Admiralty, Navy Board Out Letters, 2188, p. 452, cited by Clark-Kennedy, *op. cit.*, p. 168.
[5] T. Mortimer, *The Universal Director* (1763), p. 72. I owe this reference to the kindness of Mr. G. Buckland.

pox hospital (1753).[1] He also made proposals for ventilating Hertford Gaol, estimating the cost at £150 if it were done with a windmill such as he had used at Bedford, Aylesbury, and Maidstone Gaols and St. George's Hospital.[2] Sometimes the ventilators were designed to be worked by hand as exercise for the men. The armed forces showed particular interest in Hales's ventilators. Pringle recommended the use of them in army hospitals and even suggested the use of portable ventilators.[3] In the same year, 1753, Hales published 'Some Considerations about means to draw the foul air out of the sick Rooms of occasional Army-hospitals, in private Houses in Towns'.[4]

No invention of the eighteenth century seems to have been more widely advertised than Hales's ventilators and as late as 1819 they were still considered sufficiently important for Rees's *Cyclopaedia* to pay a good deal of attention to them. Rees mentions their uses in dispelling foul air from ships, mines, gaols, workhouses, barracks and hospitals, as well as for making salt, drying corn, hops, malt, and even gunpowder, and 'for drying linen hung in low, long, narrow galleries, especially in damp, rainy weather, and also in drying woollen cloths, after they were fulled or dyed, and in this case they might be worked by the fulling water-mill'.[5] Hales himself read a description of his ventilators to the Royal Society in May 1741, and then in 1743 the *Gentleman's Magazine* quoted from it[6] and continued a steady propaganda in favour of the ventilators for the next twenty years. Other inventors on the Continent were thinking along the same lines at roughly the same period. Martin Triewald, Captain of Mechanics to the King of Sweden, published an account of 'a ventilater, or Machine for furnishing Ships with Fresh Air, invented by him in the Spring of 1741, for which the King and Senate not only granted him a privilege in October following, but all their Ships of War have one of them'.[7] A model of Triewald's invention was sent to France for the benefit of the French Navy, but in this country it was considered to be less efficient than Hales's ventilator.[8] In Britain, Hales had a rival in S. Sutton, who published in 1749 an account of his invention for ventilating ships, bound up with Dr. Mead's *Discourse on the Scurvy*.[9]

[1] *Gentleman's Magazine*, Sept. 1746, p. 494, and March 1754, pp. 115–16.
[2] W. J. Hardy, *Hertford County Records*, Vol. II, p. 96.
[3] *Gentleman's Magazine*, April 1753, p. 172. See also D. W. Singer, 'Sir John Pringle and his Circle, Part II', *Annals of Science*, Vol. 6 (1950), p. 3.
[4] *Gentleman's Magazine*, April 1753, pp. 172–3.
[5] See article on 'Ventilators'.
[6] *Gentleman's Magazine*, July 1743, Register of Books.
[7] *Ibid.*, p. 448.   [8] *Ibid.*, p. 503.
[9] *Ibid.*, Register of Books, Feb. 1749, p. 96. See also Oct. 1747, p. 468, and May 1758, p. 210. Also Clark-Kennedy, *op. cit.*, p. 168.

From about 1749 onwards, Yeoman was kept busy fitting the ventilators to a great variety of ships, beginning with slavers and transports. In that year, the Earl of Halifax gave orders for ventilators to be fitted in five Nova Scotia slave-ships, and impressed by the improved health of the slaves on board these ships, the Lords of the Board of Trade and Plantations sent Yeoman across to Rotterdam, with orders to fix ventilators on board the other slave-ships in the Nova Scotia service.[1] When he had finished his work, Yeoman left a model in Rotterdam with instructions for its use. In June 1749 a further report claimed that the ventilators were to be fitted to 'each of the transport ships which are to carry 500 Germans to the British plantations: so that 'tis not questioned but this invention will be brought into general use in the navy'.[2] Lord Halifax's patronage of Yeoman proved to be his making, just as it was of service to William Shipley in Northampton by encouraging him to form a national society for the public good.[3]

On 17 August 1756 Yeoman was appointed to be in charge of the ventilation arrangements on board all H.M. ships.[4] Thus he became 'Chief Marine Superintendent of H.M. Navy', as he is described on his certificate for election to the Royal Society. The Admiralty has among its records a diagram of a windmill ventilator by Yeoman, sent to Portsmouth 29 March 1765, for guidance in fitting the system into the *Sandwich* and the *Arrogant*.[5] Such a position as Yeoman now held was clearly of importance and a considerable promotion from the days when he was Operator of the Cotton-Mill at Northampton, and reflects the new position that he had won for himself in his profession.

That Hales relied very largely on Yeoman's practical knowledge is shown by a letter from Hales to William Stukeley in which he says that,

[1] Record Office, Admiralty, Navy Board Out Letters, 2185, 2187, cited by A. E. Clark-Kennedy, p. 162. In April 1749, it was reported that 'the 84 transports destin'd for Nova Scotia are all order'd to have on board the ventilators invented by the Rev, Dr. Hales'. *Gentleman's Magazine*, April 1749, p. 185. Hales *op. cit.*, p. 84, gives 1751 as the date of Yeoman's visit to Rotterdam, but in view of the above notice this would seem to be too late for his first visit. Yeoman did, however, visit Rotterdam in May 1751. See *Gentleman's Magazine*, May 1751, p. 235, where Yeoman is still described as 'of Northampton'.

[2] *Gentleman's Magazine*, June 1748, p. 282. See also Sept. 1749, p. 408. Copy of a letter from one of the Settlers in Nova Scotia: 'our health and preservation has been in a great measure, under Almighty God, owing to the prudent measures taken by those, who had the direction of this good work in having ventilators and air pipes in all the ships, and furnishing rice and fresh provisions'.

[5] Thomas Mortimer, *A concise Account of the Rise, Progress, and Present State of the Society for the Encouragement of Arts, Manufactures and Commerce*, 1754 (1763), pp. 5–7, cited by D. G. C. Allan, *Journal of the Royal Society of Arts*, Vol. 110 (1962), pp. 855–9.

[4] Record Office, Admiralty, Navy Board Out Letters, 2185, 2187, cited by A. E. Clark-Kennedy, *op. cit.*, p. 162.   [5] *Ibid.*, p. 169.

since Yeoman had fitted most of the ventilators, it would be well to have his advice on any proposed fitting of them.[1] Yeoman was not, however, the only engineer consulted by Hales, since Keane Fitz-Gerald, F.R.S., reported on their use in mines and said that Hales had applied to him 'for assistance in contriving a machine to work the ventilator, by the help of the fire-engine, which is now generally used in all mines for drawing off the water'.[2] Erasmus King had advocated the use of steam-engines for working the ventilators in 1752.[3] Yeoman seems to have had nothing to do with steam-engines himself. He may, however, have advised on the fitting of ventilators to gunpowder mills, mentioned by Rees, Smeaton and Yeoman were associated in various projects during the 1760s,[4] and Smeaton was known to be an expert on the erection of such mills.

By 1758 Yeoman was installed in Little Peter Street, London. Mortimer's *The Universal Director*, 1763, has an entry about him, describing him as 'Surveyor and Civil Engineer', a very early use of the specialized title of 'civil engineer'. This reference draws attention to another important branch of Yeoman's career, when it points out that 'This ingenious Artist had distinguished himself by conducting the works of several inland navigations'. The eighteenth century is the great age of fen drainage and river improvement. With regard to the former, Sir John Rennie observed:

> During the past century, the drainage of the Bedford Level, as well as other districts, has been submitted to, and has occupied in succession the attention of the ablest engineers of the day; among whom may be mentioned the names of Perry, Elstobb, Grundy, Golborne, Armstrong, Kinderly, Smeaton, Jessop, Chapman, Page, Robert and William Mylne, Huddart, Rennie, Telford, Walker, G. and J. Rennie, Cubitt, Rendel, and others.[5]

He might well have included Yeoman's name among them. At different times he was associated with some of the greatest engineers of his age—Smeaton, Brindley, Golborne, and Robert Mylne among them—in the improvement of rivers and harbours, as well as in drainage. Yeoman's participation in this sort of work seems to have started in

---

[1] Stephen Hales to William Stukeley, 23 Dec. 1759, Bodleian Library, cited by A. E. Clark-Kennedy, *op. cit.*, p. 239.

[2] *Phil. Trans.*, Vol. L (1758), pp. 727–32, with Plates.

[3] *Gentleman's Magazine*, April 1752, p. 181. The reference occurs in an article describing Hales's ventilators as they were when installed that month at Newgate Gaol. See also plate facing p. 180.

[4] See below, p. 388.

[5] Sir John Rennie, F.R.S., *An Outline of the Progress of Civil Engineering in Great Britain since the time of Smeaton to the Present Day* (1846), p. 29.

the 1760s and not, as had been previously suggested, in 1735.[1] There is in the Essex Record Office a MS. plan of the River Chelmer from Chelmsford to Maldon dated 1765 and signed by Yeoman. The map was printed in the same year. Yeoman had surveyed the river both in 1762 and 1765, and on 18 April 1766 gave evidence before the Commons that the scheme proposed would cost £13,000.[2] An Act was passed but nothing was done under it as the money could not be raised, and the work had to wait until 1793 when a new Act was passed, and John Rennie carried out its provisions.[3] In 1765 Yeoman was surveying the River Stour, and was making plans for draining the marshes nearby and improving the port of Sandwich.[4] The next year he was named by the Trustees for the River Lea to assist Smeaton in making a survey. In 1767 Smeaton reported that he had 'with the assistance of Mr. YEO-MAN, Engineer, carefully viewed, measured, and calculated the quantity of water discharged at Sir WILLIAM WAKE'S turnpike in the driest season for passing the barges'.[5] The same year he also proposed improvements in the drainage of the Level of Ancholme,[6] and, with Nichols and Robert Mylne, he also gave in a joint report 'on Embankment of the Thames'.[7] He was consulted in 1767 along with Smeaton, Brindley and Mylne as one of 'four eminent surveyors' regarding a proposal to erect a water-wheel at the fifth arch of London

[1] Mrs. E. C. Wright, 'The Early Smeatonians', *Newcomen Soc. Trans.*, Vol. XVIII (1937-8), pp. 101-10, states that an Act was obtained for the improvement of the River Chelmer in 1735. She cites in support of her statement, Russell, *History of the County of Essex*, Vol. I, p. 292: 'The ingenious Mr. Yeamans is to have the direction of this work, which will, when completed, be of the greatest advantage to the county'; and Gough, *British Topography* (2 vols., 1780), Vol. I, p. 368: '"A plan for making the river Chelmer navigable, by —— Yoman" was published in 1735'. Professor A. W. Skempton, 'The Engineers of the English River Navigations, 1620-1760', *Newcomen Soc. Trans.*, Vol. XXIX (1953-5), pp. 25-54, asserts that in 1735 Yeoman published 'A plan for making the river Chelmer navigable'. Mr. P. J. Carne, however, who has written a thesis, 'A History of the Chelmer Navigation down to 1830', has discovered no evidence in the Rayleigh Archives at the Essex Record Office to support either of these statements. I am grateful to Mr. F. G. Emmison, County Archivist, Essex, for this information. A scheme for the improvement of the Chelmer was mooted in 1732 and in the following year J. Hoare made a survey. There is no suggestion anywhere that Thomas Yeoman was concerned with this abortive proposal.
[2] *Commons Journals*, Vol. xxx, p. 741, 18 April, 6 Geo. III. Professor T. S. Willan kindly supplied me with information about the Chelmer improvements.
[3] J. E. Marriage, 'Navigation on the Chelmer', *Essex Review*, Vol. 64 (1955), pp. 249-52.   [4] E. C. Wright, *op. cit.*, p. 101.
[5] *The Report of John Smeaton, Engineer, concerning the situation of the Mills and Bleach-Field at Waltham Abbey in respect of Water* (May 16, 1767).
[6] T. Yeoman, *A Report of the State of the Level of Ancholme and the methods proposed for the more effectual draining the same* (1767).
[7] A. E. Richardson, *Robert Mylne: Architect and Engineer, 1733-1811* (1955), p. 77.

Bridge.[1] In 1768 he reported with Brindley and Golbourne on the improvement of the Forth and Clyde Navigation.[2] A further edition seems to have been published in 1769.[3] Also in a collection entitled 'Cambridgeshire Roads and Fens', in the Bodleian Library, is a twelve-page report by Yeoman, dated 15 November 1769 at Castle Street, Leicester-fields, concerned with the 'draining of the North Level of the Fens and the outfall of the Wisbeach river'.[4] Yeoman was also called on for testimony in 1769 on the Grand Trunk and Birmingham canal schemes,[5] and completed the work on the River Stour begun in 1767.[6] He continued to be busy in the next decade, being called upon for plans and estimates for the navigation of the Stroudwater to the Severn in 1774.[7] The following year he was involved in controversy with Murdoch McKenzie, 'late Maritime Surveyor in His Majesty Service' over revived proposals for improving the port of Sandwich[8] and did a small survey of watercourses in Staplehurst, Kent.[9] In 1777 he surveyed with Robert Mylne all the streams, mills etc. of Cooks, Bissons, and others near Bow.[10] This evidence alone would show that Yeoman was at the head of his profession and associated with many of the leading members of it.

What confirms this impression is Yeoman's membership of three of the leading intellectual societies of eighteenth-century London. In 1764 he was elected a Fellow of the Royal Society, but does not seem to have played a very active part in its deliberations. It is noteworthy, however,

---

[1] *Gentleman's Magazine*, July and Aug. 1767, pp. 337–9, 407–8.
[2] *A review of several matters relative to the Forth and Clyde Navigation* . . . (1768). This report involved criticisms of Smeaton's plans for the canal.
[3] *St. James's Chronicle*, 1239, from Sat. Feb. 4 to Tues. Feb. 7, 1769: 'This day were published, Price 4s. with a Plan of the Canal, Reports by James Brindley, Engineer, Thomas Yeoman, Engineer, and F.R.S., and John Golborne, Engineer, relative to a navigable Communication betwixt the Firths of Forth and Clyde, with Observations.'
[4] See also Nene River Board Archives. This report is dated Brigg, 1769.
[5] Matthew Boulton to Samuel Garbett, 12 Feb. 1771: 'I have seen the testimony of his [Colmore's] two Engineers Smeaton & Yeoman & I value the opinion neither of them nor Brindley nor Simcox in this Case nor of the whole Tribe of Jobbing Ditchers, who are retained as evidences on any side which first applies for them . . .' (Matthew Boulton, Letter Book [D], 1768–73, A.O.L.B.)
[6] B.M. Add. MSS. 5489, ff. 108–37 for the Stour, cited by E. C. Wright, *op. cit.*
[7] J. Phillips, *A General History of the Inland Navigation* (1803), p. 211; *Gentleman's Magazine*, Jan. 1781, p. 47.
[8] British Museum, Add. MSS. 5489, ff. 108–37.
[9] In the archives of St. Bartholomew's Hospital, London, is a map of these watercourses by Yeoman. A report is attached of a view taken by the Treasurer of the Hospital, Yeoman and the Commissioners of the turnpike road from Cranbrook to Staplehurst and Maidstone. For the references to this map and to the one in the Essex Record Office I am grateful to Mr. F. W. Steer.
[10] Richardson, *op. cit.*, p. 109.

that six of the eight Fellows who signed his certificate of election were members of the Society for Arts, Manufactures and Commerce: the Earl of Shaftesbury, Lord Romney, Peter Wyche, William Hanbury, James Parsons and Henry Baker. Yeoman was elected a member of the latter society on 23 April 1760 when he was proposed by the man midwife, Sir Richard Manningham. His subscription continued to be paid until 2 August 1782,[1] though he died in January 1781. At all times he was an honoured and valued member of that society, but it is also possible that he made a special contribution to its very foundation. When Shipley came to London from Northampton to promote the idea of a national society, he numbered among his London acquaintances Henry Baker, F.R.S., but he also brought with him 'a recommendation to the Reverend Dr. Stephen Hales of Teddington'.[2] The archivist of the Royal Society of Arts has said that 'The introduction of William Shipley to Dr. Hales was of crucial importance in the history of the Society of Arts. Whoever gave Shipley his "recommendation" to the Doctor brought together the idea of the Society and the possibility of its realization. Perhaps it was Lord Halifax, who certainly knew Hales, or some other "ingenious and public spirited" Northamptonshire gentleman.'[3] No friend of Shipley could have been on closer terms with Hales than Yeoman, who might very well have effected the introduction.

However that may be, Yeoman himself did not immediately join the Society, probably because he was still resident in Northampton. On 16 March 1762 John Wyatt junior wrote to John Wyatt, the inventor, saying that the latter's invention of a spinning-machine had been refused a premium, because it had been invented before the date of the Society's advertisement.[4] Wyatt's unsuccessful supporters were Mr. Pinchbeck, Mr. Dossie and Mr. Yeoman. The records of the Society show Yeoman as a participator in a number of its committees from 1761 onwards, and deliberating on such topics as 'planks for preserving Ships Bottoms',[5] some mills sent by Mr. Daux,[6] a device for winnowing corn,[7] a model of a drill plough[8] or a tide-mill.[9] The number of subjects is too numerous to be enumerated in detail.[10] On

[1] R.S.A. MS. Subscription Books.   [2] Mortimer, *op. cit.*, p. 8.
[3] D. G. C. Allan, *Journal of the Royal Society of Arts*, loc. cit., p. 857.
[4] J. Wyatt (King St, London) to J. Wyatt (Birmingham), 16 March 1762, Wyatt MSS., B.R.L. It is possible that this correspondence adds weight to the idea that Wyatt was the inventor and not Lewis Paul.
[5] R.S.A. Min. Comm., 21 March 1761.
[6] R.S.A. Min. Comm., Committee of Manufactures, 23 Aug. 1760.
[7] *Ibid.*, Committee of Mechanics, 8 Jan. 1761.   [8] *Ibid.*, 2 April 1761.
[9] *Ibid.*, 9 April 1761.   [10] I hope to go into more detail in a separate paper.

23 November 1763, however, Yeoman was elected, along with Wyatt, one of the two Chairmen of the Committee of Mechanics.[1] Throughout the period 1764–78 his name regularly appears as one of the Chairmen of this committee, but he also on occasion chaired the Committee of Manufactures.[2] From time to time he also attended meetings of the Committee on Agriculture and Chemistry. During his membership of the Society he co-operated in different projects with Hales, Smeaton, Pinchbeck, Mylne, Edward Cave, and others. He was particularly concerned with all matters relating to agricultural machinery, ships and textiles. Once again we see that this Society was one of the most important meeting places in eighteenth-century London for men with scientific and engineering interests.

Finally we come to the culmination of Yeoman's career, his election to the Presidency of the Society of Civil Engineers (Smeatonians) on 15 March 1771.[3] Present at the first meeting were some of the best engineers of the time, including Smeaton, Grundy, Nichols, and Mylne. It is interesting to observe that they agreed to meet fortnightly while Parliament was in session, and when, presumably, their services in London would be most required. At the second meeting, on 29 March 1771, Robert Mylne was elected vice-president, and Golborne and Whitworth joined the small club. Other early members were William Jessop, Joseph Priestley, John Gott, Pinchbeck, Murdoch McKenzie. They were joined by Matthew Boulton, John Whitehurst, and others, but these two were elected when Yeoman no longer presided. On 28 April 1780, the meeting was adjourned until 12 May because the President was ill and not capable of attending. On 15 May 1781 Pinchbeck was elected President in Yeoman's place. Though this club seems never to have been heavily attended, and had not reached the status of a great society such as it is today, nevertheless its very existence proclaimed that a new profession had arisen and had become conscious of its identity.

Civil Engineers are a self-created set of men, whose profession owes its origin, not to power or influence; but to the best of all protection, the encouragement of a great and powerful nation; a nation become so, from the industry and steadiness of its manufacturing workmen, and their superior knowledge in practical chemistry, mechanics, natural philosophy, and other useful accomplishments.[4]

[1] R.S.A. Min. Soc., 23 Nov. 1763.
[2] E.g., R.S.A. Min. Comm., Manufactures, 20 Dec. 1768, 23 Feb. 1769, 19 April 1769 and 18 May 1769 and throughout 1764.
[3] Facsimile Copy of Minutes, Library of the Institution of Civil Engineers.
[4] S. B. Donkin, 'The Society of Civil Engineers (Smeatonians)', *Newcomen Soc. Trans.*, Vol. XVII (1936–7), p. 54, quoting from the Preface to the *Reports of the late John Smeaton* (1812).

One of those manufacturing workmen earned only the briefest of obituaries in the *Gentleman's Magazine*, for January 1781:[1] 'In Castle-Str. Mr. Tho. Yeoman F.R.S.', but the story of his profession is greatly enriched by his career. The variety of his talents as millwright, lecturer, ventilating engineer, river, harbour, and canal improver, agricultural machine-maker, cotton-mill manager and gentleman of science testifies to the importance of the engineer in the early industrial revolution. Moreover, from Yeoman's career, we can see that the transition from agricultural improvement to mechanical improvement was not a great one in the eighteenth century, and that, from the engineer's point of view, the Agricultural Revolution and the Industrial Revolution were intimately connected. Moreover, the vitality of market-town society was a sound basis on which to construct a community of scientific interest among engineers and gentlemen. That interchange was also developed in national societies such as the Royal Society and the Society for the Encouragement of Arts, Manufactures and Commerce. Yeoman epitomizes the union of practical and theoretical knowledge out of which a new profession grew, and though little known, he was one of that profession's finest adornments.

Smeaton and Brindley were accompanied and followed by a number of able men in rapid succession, amongst whom were Jessop, Whitworth, Mylne, Yeoman, Henshall, Golborne, Huddart, Rennie, Ralph Walker, Chapman, Telford and others, all stimulated to exertion by the magnificent career before them, each contributing, more or less according to their opportunities, great skill and invention of their own, in addition to that acquired from their predecessors.[2]

Most of these men have never been studied in detail. It is hoped that this chapter will do something to rectify the omission for Thomas Yeoman, F.R.S.

[1] His wife died in 1793 at Mile-end, aged 84: *Gentleman's Magazine*, Nov. 1793, p. 1060.
[2] Rennie, *op. cit.*, p. 8.

XII The Early Growth
of Steam Power[1]

This chapter, based principally on the various collections of Boulton
and Watt papers, but using also local sources, manuscript and printed,
wherever possible, seeks to answer some limited questions on the
production and use of steam engines in the early stages of the Industrial
Revolution.[2] The period covered does not extend beyond 1800, and the
material has been richest for the 1790s; the area under examination is
Lancashire, and more particularly, the Manchester district. Though the
period of time and the area chosen may appear small, by comparison
with the total extent of steam power in the Industrial Revolution, they
are of central importance to the wider story.

The following statement about the steam engine, made by J. Lord
in his *Capital and Steam Power* (1923), was still accepted by the majority
of historians before our original paper was published:

From 1775 to 1800 they [i.e. Boulton and Watt] possessed a monopoly of
steam-engine construction, and, therefore, their output comprises practically all
the engines that were erected in Great Britain before the year 1800. The records
of the firm contain the dates, power, and owners of all engines erected, therefore, it is comparatively easy to discover which trades employed steam-power,
and to what extent.'[3]

[1] This chapter is based on a jointly-produced article in the *Economic History
Review*, 2nd ser., Vol. XI, no. 3 (1959), but a considerable number of revisions
and additions have been made.
[2] The principal MS. collections used in this paper are the Boulton and Watt
Collection, Birmingham Reference Library (B.R.L.), the Matthew Boulton
papers in the Assay Office Library, Birmingham (A.O.L.B.), and the Doldowlod
MSS. (Dol.), the property of Major David Gibson Watt, M.C., M.P. We have
also consulted the Marshall papers in the Brotherton Library, University of
Leeds, and some papers relating to John Wilkinson and James Stockdale which
have recently been deposited in the Lancashire County Record Office, Preston
(L.C.R.O.).
[3] Lord, *op. cit.*, p. 147.

Interpreted strictly the statement is obviously untrue. Boulton and Watt held a monopoly of construction only in those steam engines covered by Watt's patents; they could not prevent the manufacture of Newcomen and Savery engines, and even their own patent rights were not defended without costly legal battles. This, of course, Lord knew quite well, but the slip in wording is interesting because it suggests the way in which his mind was working. Steam-engine construction for him was Boulton and Watt steam-engine construction. The weakness of his statement lies in the words, 'practically all the steam engines'. Is it possible to arrive at a clearer notion of what those words mean? This leads us to Lord's further statement that: '*It must be borne in mind that the output of Boulton & Watt represents the entire steam-engine building of the country, if not the whole world*; the pirated engines, that caused the firm so much anxiety and litigation, were not great in numbers, and the atmospheric engines, even as improved by Smeaton, were costly and inefficient, and ceased to be erected.'[1] Here both the pirate engines and the atmospheric engines are dismissed as being unworthy of serious consideration and the Savery engines are not even mentioned. It is in fact a common assumption among historians that the manufacture of Savery engines ceased early in the eighteenth century except for a few examples. Since Boulton and Watt engines were expensive to install,[2] Lord then deduces that the capital available in those industries, such as cotton manufacturing, that employed Boulton and Watt engines must have been large because of the amount of fixed capital invested in these engines. It is important therefore not only for the technological but also for the financial history of the Industrial Revolution that Lord's assumptions should be tested. Thus the questions we wish to attempt to answer are: first, how many engines not covered by Watt's patents were still being erected in Lancashire quite late in the century; and secondly, how many pirate engines were there in Lancashire, and how serious a menace were they to Boulton and Watt's potential market?

I

It is unfortunate that a list of the earliest steam engines in Manchester, left to the Literary and Philosophical Society by George Walker, one of the original members, was destroyed when the Society's library was burnt in an air-raid during the Second World War. Only one item

[1] Lord *op. cit.*, p. 166. Our italics.

[2] Robert Mylne, giving evidence before a Select Committee of the Commons on Watt's Engine Bill, 12 April 1775, declared that a Watt engine with the same diameter of cylinder as a Newcomen engine would cost twice as much but would do twice as much work. (Minutes, Dol.)

from it has been preserved, quoted in a newspaper account of the donation of Walker's MS.[1]

September 26, 1788: Fire engine or steam engine, Shudehill, to raise water on the old construction. Examined this day in company with the eldest son of Sir John Stanley. Cylinder in diameter 64 in. Two pumps of cast iron each 31 in. [bore]. Makes 11 strokes in one minute and each stroke is 7 ft. 9 in. depth in the pumps. Mr. Barton told us the quantity of water raised by each stroke was about 64 gallons. He reckoned the power of pressure on the piston equal to 8 lb. per inch square. Two boilers 14 ft. 6 in. and 14 ft. 4 in. in diameter . . . About five tons of coal consumed every 15 hours, or one day's work . . .

It seems very probable that this engine, which was of the atmospheric type, was that erected by Arkwright in the early 1780s to pump water for an overshot water-wheel, which drove the spinning frames in his factory on Shudehill, the first such factory built in Manchester. Apart from a few bald statements,[2] nothing has hitherto been known about this engine. But that Walker's account refers to Arkwright's engine is indicated by similar evidence which we have recently discovered in the surviving records of John Marshall, the famous Leeds flax-spinner,[3] who noted in 1790 that the engine of Simpson & Co. (Simpson being Arkwright's partner), at Manchester, had a cylinder of 64 in. diameter and worked two pumps of 30 in. bore; with a stroke of 7½ ft. and working at 12 strokes per minute, it raised 500 gallons of water per stroke to a height of 30 ft., to work a water-wheel of 30 ft. diameter, which drove 3,000 spindles; the engine is said to have consumed the fantastic quantity of 5 tons of coal per day.

A few more early engines are mentioned by other contemporary witnesses. James Ogden stated in 1783 that about the time when the Bridgewater canal was built from Worsley to Manchester, i.e. about 1760, 'the collieries nearest *Manchester* in *Newton-lane* were set to work . . . then first drained by a steam engine, and scarcely worth getting before'.[4] Other collieries in the Manchester area similarly adopted steam engines for pumping.[5] Ogden also informs us that water for the Manchester reservoirs was 'forced up by an engine near *Ancoats*'.[6] And he refers to Richard Arkwright's cotton mill just then being set up in

---

[1] *Manchester Guardian*, 15 April 1870.

[2] *Patents for Inventions. Abridgments of Specifications relating to the Steam Engine* (1871), Part I, A.D. 1618–1859, Vol. I, p. 77; A. Ure, *The Cotton Manufacture of Great Britain* (1836), Vol. I, p. 273.

[3] Marshall MS. 57, Brotherton Library, University of Leeds.

[4] [J. Ogden,] *A Description of Manchester . . . By a Native of the Town* (1783), pp. 6–7.

[5] Thus in 1774 we find John Porter of Bradford Colliery, Manchester, advertising for sale 'A Good Fire Engine, the Cylinder 32 Inches, with Beam, Pumps, &c.', *Manchester Mercury*, 29 March 1774. [6] Ogden, *op. cit.*, p. 13.

Miller's Lane, Shudehill, in which the carding and spinning machines were 'setting to work by a steam engine'.[1]

In addition to these items, we have a few general statements about early steam engines in the Manchester area, some of which seem at first sight to be contradictory. There is Peter Drinkwater's statement in 1789 that there were 'a great number of the common old smoaking engines in and about the town',[2] but against this is John Kennedy's that:

During the period of ten or fifteen years after Mr. Arkwright's first mill was built in 1771 at Cromford, all the principal works were erected on the falls of considerable rivers; no other power than water having then been found practically useful. There were a few exceptions where Newcomen's and Savary's steam engines were tried. But, the principles of these machines being defective and their construction bad, the expense in fuel was great, and the loss occasioned by frequent stoppages was ruinous.[3]

Drinkwater's statement is contemporary with what he describes, while Kennedy's was made some thirty years after the time to which he refers. The two statements are not mutually exclusive. By 1815 the number of steam engines in Manchester was much greater than in 1789, and, to a man living at the later date and prospering in the cotton trade, the provision of steam power in 1789 would by comparison appear scanty and inefficient. In another paper, moreover, Kennedy states that both 'Watt's and Newcomen's steam-engine supplied power for the mule spinner' in the 1790s, and that when he himself improved the mule for fine spinning in about 1793, 'the *power* in this instance was *Savary's*, which was used to raise water upon a water-wheel'.[4]

Some of the first engines in Manchester about which we have any details were on Savery's principles, and one engineer, Joshua Wrigley, seems to have been the best-known maker of this type of engine. Since he first appears in the directories as a pump-maker,[5] it would have been a natural progression for him to manufacture the pumping-engines as

[1] Ogden, *op. cit.*, p. 16.

[2] Peter Drinkwater to Boulton and Watt, 3 April 1789, B.R.L. Quoted by W. H. Chaloner in 'Robert Owen, Peter Drinkwater and the early Factory System in Manchester, 1788–1800', *Bulletin of the John Rylands Library*, Vol. XXXVII (1954–5), p. 87.

[3] J. Kennedy, 'Observations on the Rise and Progress of the Cotton Trade in Great Britain, particularly in Lancashire and the Adjoining Counties', read 1815, Manchester Literary and Philosophical Society, in his *Miscellaneous Papers* (Manchester, 1849), p. 15.

[4] J. Kennedy, 'A Brief Memoir of Samuel Crompton', read 1830, Manchester Lit. & Phil. Soc., in his *Miscellaneous Papers* (Manchester, 1849), pp. 71–2.

[5] Elizabeth Raffald's *Manchester Directory*, 1772: 'Wrigley Joshua, Pump-maker and Bell-hanger, Long Miln-gate'. The same entry appears in Raffald's *Manchester and Salford Directory* of 1781. The name sometimes is spelt 'Rigley'.

well. And these engines could be used to pump water for water-wheels, to drive textile or other machinery. Wrigley's engine, we are told,

usually raised the water about 16 to 20 feet high; and the water descending again gave motion to an overshot water-wheel. This is Mr. Savery's original project for working mills by his engine: but Mr. Rigley contrived his engines to work without an attendant; the motion of the water-wheel being made to open and shut the regulator, and injection-cock, at the proper intervals. They continued in use for some years, but were at length given up in favor of better engines.[1]

Wrigley's engines had a single cylindrical receiver and worked 'by suction only' i.e. by condensing steam in the receiver, so that atmospheric pressure forced up the water from below; and then, instead of being expelled from the receiver by steam pressure, as in Savery's original design, the water merely flowed out by gravity, into a cistern, controlled by a sluice, above the water-wheel; thus steam of little more than atmospheric pressure was required and the danger of bursting boilers was greatly diminished, while fuel consumption was also reduced, though still heavy by comparison with Watt's engine.[2]

Wrigley's was not the first effort to make the Savery engine a practical proposition. As Professor John Robison, of Edinburgh University and friend of Watt, pointed out at the end of the eighteenth century, the 'great durability' of the Savery engine, on account of its 'having few moving and rubbing parts', had 'occasioned much attention to be given to this first form of the [steam] engine, even long after it was supplanted by those of a much better construction'.[3] John Farey made the same point in regard to Wrigley's form of this engine: 'From the simplicity of its construction it is not liable to wear out for a very long time.'[4] It was also relatively cheap and easy to construct and maintain. Much earlier, in fact, Desaguliers had used similar arguments in favour of his own improved Savery engine.[5] Later, in 1766, William Blakey had taken out a patent for another modified version, fitted with self-acting valve gear and using oil on the surface of the water in the

---

[1] J. Farey, *A Treatise on the Steam Engine* (1827), p. 122.

[2] *Ibid.*, pp. 122–5.

[3] *Encyclopaedia Britannica* (3rd edn., 1797), article on 'Steam', repeated in *A System of Mechanical Philosophy* (4 vols., Edinburgh, 1822), Vol. II. Robison described 'a very ingenious attempt . . . made very lately to adapt this construction to the uses of the miners'. He also referred to Blakey's ingenious but unsuccessful efforts at improving it (as mentioned below).

[4] Farey, *op. cit.*, p. 124.

[5] See above, p. 39, for Desaguliers' improved 'fire-engine'. He stated that it cost only £80, compared with £300 for a Newcomen engine. *A Course of Experimental Philosophy*, Vol. II, p. 490.

receiver to reduce condensation of steam.[1] There were also repeated efforts at applying this engine to pumping water for a water-wheel, thus producing rotative motion for driving machinery, as had been shown in Leupold's *Theatrum Machinarum* (1727);[2] by the mid-eighteenth century this method of utilizing either Savery or Newcomen engines was becoming fairly widespread.[3] John Smeaton popularized this combination of 'fire-engine' and water-wheel in the 1770s and 1780s, especially for winding coal.[4] When Wrigley first adopted it we have been unable to discover.[5] Nor is it always clear whether engines built by him were of the Savery or Newcomen type, since he apparently made both.

Newcomen's engine, of course, was used throughout most of the eighteenth century simply as a pumping engine, to drain mines or supply waterworks. Its pumping action also came to be used, like that of the Savery engine, to drive machinery via a water-wheel. Various attempts were made to adapt it to rotative motion, whereby it could drive machinery directly, without a water-wheel, but it was not until the early 1780s, when Watt developed his rotative engine, that much success was achieved. From then on, however, though this is often overlooked, considerable numbers of Newcomen engines were made rotative, by means of a connecting-rod, crank and fly-wheel.[6] It is true

---

[1] *Abridgements of Specifications . . . Steam Engine* (1871), Part I, Vol. I, p. 55; Robison, *op. cit.*; Farey, *op. cit.*, p 121; J. J. Bootsgezel, 'William Blakey—A Rival to Newcomen', *Newcomen Soc. Trans.*, Vol. XVI (1935–6), pp. 97–110.

[2] *Op. cit.*, Vol. III, pp. 302 and 304.

[3] See *Abridgements of Specifications . . . Steam Engine* (1871), Part I, Vol. I: p. 41, 1724, Robert Bumpsted; p. 48, 1740, John Wise; p. 49, 1743, such engines used at Newcastle, etc.; p. 50, 1752, Champion, at Bristol; p. 55, 1766, Blakey. For its use by the Darbys at their Coalbrookdale furnaces, see Raistrick, *op. cit.*, pp. 65, 111–13, 143–4. Later, Wilkinson adopted the same method at Bersham. (*Abridgements*, p. 51, based on J. Aikin, *A Description of the Country from thirty to forty Miles round Manchester* (1795), p. 174). See also Farey *op. cit.*, pp. 296–7.

[4] Farey, *op. cit.*, pp. 296–306.

[5] In the *Abridgements* of steam-engine specifications, Wrigley's method of 'Raising water by a steam engine, and throwing it back into the mill-dam' is referred to under the date 1756, though (as is stated) Wrigley did not take out a patent. But this is obviously based on a statement by John Aikin in 1795 (see below, p. 401), in which there is no evidence to support so early a date for Wrigley's activities.

[6] Farey, *op. cit.*, pp. 406–22 and pp. 658 ff., has a wealth of information on Newcomen engines adapted to rotative motion. See also James Pickard's patent no. 1,263, dated 23 Aug. 1780, 'for a new method of applying steam-engines, commonly called fire-engines, to the turning of wheels; whereby a rotative motion, or motion round an axis, is performed, and the power of the engine is more immediately and fully applied . . . than by the intervention of a water wheel'. Watt alleged that this application of the crank was one of his own inventions, but that it was stolen by a workman at Soho (Dickinson and Jenkins, *op. cit.*, pp. 148–56). Watt, therefore, naturally averse from challenging patents,

that its motion was less regular than that of Watt's engine and the fuel consumption was greater, but in many operations, such as winding coal from pits, grinding corn, and crushing seeds, these drawbacks were not considered important.[1] It could also be applied to textile-spinning, though here it proved much less successful, especially by comparison with Watt's double-acting rotative engine.

The difficulties of its application in this field, and the persistence of the 'fire-engine' and water-wheel combination, is illustrated in the manuscript notebook of John Marshall, the Leeds flax-spinner, who clearly relied very heavily on Joshua Wrigley for information about water-wheels, steam engines, and millwork generally when setting up his spinning mill at Holbeck in 1780–1.[2] On the first page of this notebook Marshall refers to a Wrigley engine with a cylinder of 26 in. diameter, working a crank, instead of a water-wheel, evidently an adaptation of the Newcomen beam engine to produce rotative motion. There is also a reference to another Wrigley engine, of a 'new construction', driving a water-wheel; this engine had a 16 in. cylinder, 5 ft. stroke, 14 strokes per minute, raising 42 gallons of water 16 ft. at each stroke, and consuming only 6 cwt. of coal per day. As between these two types of motive power, Wrigley was definitely in favour of the latter: 'J. Wrigley says there is nothing gained by a Crank instead of a Water wheel because of the great weight they are obliged to use at the beam end. J[oshua] W[rigley] says the Boulton & Watt's Crank engines are the only ones that will produce a motion sufficiently regular for spinning.'[3]

Marshall did not yet, however, approach Boulton and Watt, but preferred to continue his reliance upon Wrigley, who apparently provided him with a steam engine and water-wheel in 1791 and also with much millwrighting advice. The engine appears to have been of the modified Savery type,[4] though this is not clear from Marshall's notes,

---

had to adopt the 'sun-and-planet' motion, but other engine-makers had less compunction in utilizing the crank, which had, in fact, been used for centuries in various types of machinery.   [1] Farey, *op. cit.*, p. 658.

[2] Marshall MS. 57, in the Brotherton Library, University of Leeds. For information generally about Marshall's activities, see Rimmer, *op. cit.;* Professor Rimmer does not, however, mention Marshall's association with Wrigley.

[3] *Ibid.*, p. 2. This view of Wrigley's was also that of John Smeaton, who had earlier, in 1781, expressed a similar opinion regarding the difficulties of applying the atmospheric engine directly to corn-milling, because of its irregular motion, and therefore preferred its combination with a water-wheel. Smeaton, *Reports* (2nd edn., 1837), Vol. II, pp. 97–8; Farey, *op. cit.*, p. 413, note (a).

[4] Farey, *op. cit.*, p. 444, note (a), states that 'an engine on Savery's principle was first tried'. J. Horner, *The Linen Trade of Europe during the Spinning Wheel Period* (Belfast, 1920), p. 254, has a reference to Marshall's mill: 'In 1791 a mill was built in Holbeck, Leeds, driven, at first, by one of Savory's engines, in

which mainly comprise information about water-wheels, shafting, gearing, etc.[1] 'J. Wrigley recommends the [water] wheel no higher than the fall—to lay the water on the top of the wheel [*i.e.* overshot type]. One 16 ft. Diam. 9 or 10 feet broad large enough for 2000 Spindles.'[2] But Marshall also collected opinions from other engineers, who differed somewhat from Wrigley in their recommendations, as to whether overshot or breast wheels were better, as to the depth of the 'buckets' and speed of rotation.[3] He also obtained information from various cotton-spinning mills, including those of Simpson & Co. (Simpson and Arkwright), Shudehill, Manchester, and Thackeray and Whitehead, Garratt, Manchester, as well as from several Yorkshire mills. But within a year he decided to abandon the Wrigley engine and water-wheel, and to replace them with a 28 horse-power Boulton and Watt rotative engine.[4]

Other manufacturers likewise were coming to appreciate the superiority of Watt's engine. In the previous decade, however, a considerable number of Wrigley's engines had been erected in Manchester and other parts of Lancashire, where they were used 'to impel the machinery of some of the earliest manufactories and cotton mills in that district'.[5] He may possibly have built Arkwright's engine on Shudehill in the early 1780s. In 1784 two of his Savery engines were erected at the cotton-spinning mill of Joseph Thackeray and John Whitehead, at Garrat (or Garret or Garratt), Chorlton-on-Medlock, a mill which till then had been worked solely by water-wheels on the River Medlock and which is said to have been the first cotton mill in Manchester.[6] The first of these Wrigley engines had an upright

combination with a water wheel . . .' Rimmer, *op. cit.*, p. 35, refers to it very briefly as an 'atmospheric engine'. [1] See below, pp. 443–5.

[2] Marshall MS. 57, p. 4. Marshall also notes that 'J. Wrigley recommends a water wheel to Benj. Kaye for 2000 Spindles to be 16 ft. high 9 ft. wide.' Wrigley was obviously seeking orders from other West Riding manufacturers.

[3] See below, p. 444.   [4] Rimmer, *loc. cit.*; Horner, *loc. cit.*; Farey, *loc. cit.*

[5] Farey, *op. cit.*, p. 122.

[6] Aikin, *op. cit.*, p. 174, writing in 1795, states that 'above thirty years since' a Mr. Gartside built a factory at 'Garret-hall', in which he attempted to work a number of swivel-looms (Dutch, inkle, or ribbon looms) 'with a capital water-wheel' on the River Medlock. The attempt, however, eventually proved unsuccessful and the mill was converted to cotton-spinning. According to Farey, it was 'the first cotton mill which was established in Manchester. The spinning machinery was on Sir Richard Arkwright's plan, and was worked by water wheels on the river Medlock; but the water power being found insufficient, two engines on Savery's system were put up by Mr. Joshua Rigley, in 1784, to return the water of the river for the water wheels . . .' *Op. cit.*, p. 662, note (a). On p. 125, where he describes these engines, Farey gives 1774 as the date of their installation, but the former date must be correct. This mill, as we shall see, was to witness further steam-engine innovations in the following years.

cylindrical receiver, 16 in. in diameter and about 22 ft. high; it raised the water 14 ft., making 12 strokes per minute and filling the receiver about 6 ft. high at each stroke, so that, according to Farey's calculation, it was of about $2\frac{2}{3}$ horse-power. The other engine was larger, having a cylindrical receiver 24 in. in diameter and 27 ft. high; it was of about 5 horse-power, raising the water 19 ft., making $7\frac{1}{4}$ strokes per minute and filling the receiver 6 ft. at each stroke. The first engine consumed 18 cwt. and the latter 32 cwt. of coal in 24 hours.[1]

The boilers were not well designed by later standards so that the fuel consumption was rather heavy. Wrigley also ran into trouble over smoke from his engines, but was unable to cure it without erecting a very tall chimney that still did not completely solve the problem.[2] Nevertheless he appears to have found a good market for his engines, and not only in Lancashire and Yorkshire. We are given a detailed description of one of his Savery engines 'which was erected at Mr. Kier's manufactory, St. Pancras, London, and which worked there for many years, to turn lathes, &c'.[3]

Farey's statements are supported by contemporary witnesses. Aikin, the historian of Manchester, confirms Wrigley's importance as late as 1795, when he refers to an engine of Wrigley's manufacture which had a 24 in. cylinder[4] and which may have been the larger engine at Garratt. Aikin declares that Wrigley, 'a common pumpmaker', 'never applied for a patent, but imparted freely what he invented to those who thought proper to employ him'.[5] His high opinion of Wrigley was not, however, shared by James Watt. The first reference to Wrigley's enterprises in Watt's correspondence seems to be in June 1781, when Watt wrote from Cornwall to Matthew Boulton, saying that he had accidentally seen a letter by one of the firm's engineers, Logan

---

[1] Farey's account, *op. cit.*, p. 125, is apparently based on a report by Smeaton, who made a trial of these engines, but they are not referred to in Smeaton's printed *Reports*. Farey gives 7 ft. as the height of the second engine's receiver, but this is evidently a misprint for 27 ft.

John Marshall, however, in his notes, gives different dimensions for Thackeray and Whitehead's surviving Savery engine in 1790 (by which time the other had been replaced, see below, p. 406): cylinder of 30 in. diameter, 14 strokes per minute, raising 300 galls. per stroke, to a height of 13 ft., to work a water-wheel of 12 ft. diameter, driving 2,000 spindles.

[2] James Watt to Matthew Boulton, 18 Aug. 1785, Boulton and Watt Letter Book (Office), June 1784 to March 1786, B.R.L.

[3] Farey, *op. cit.*, pp. 122–5. See also *Nicholson's Philosophical Journal*, Vol. I (1797–8), pp. 419–26, where it is stated that this Savery-type engine was erected at Mr. Peter Kier's 'manufactory of axle-trees near Pancras', Kentish Town, four years previously.

[4] Aikin, *op. cit.*, p. 174.

[5] This statement by Aikin is repeated, almost *verbatim*, in the *Abridgements* of steam-engine specifications, p. 51.

Henderson,[1] which mentioned 'some new engine to burn almost no coals and that some person at Manchester wanted him to engage in the business'.[2] Watt never got on with Henderson and suggested that he should be encouraged in this project, but Boulton, always anxious to keep in the firm anyone who could be of any use in engine erecting simply evaded Watt's point, and replied: 'As to the Manchester schemes I suppose we may add them to the same catalogue in wch are enrolled all Mr Hateleys, Mr Wasboroughs, Mr Matthews, Mr Pintos, Mr Jones's, the Spaniards and the Truro Mans & therefore I did not think it necessary to vex you wth nonsense or divert your attention one mom^t from the great Cornish object...'[3] This Manchester scheme was probably Joshua Wrigley's for driving cotton mills by his modified Savery engine. A little later, in November 1784, Watt was corresponding with Thomas Jordain of Manchester about the erection of a Boulton and Watt engine,[4] but Wrigley persuaded Jordain to let him build the engine instead of Boulton and Watt. At this date Wrigley was in partnership with the Rev. John Derbyshire. The latter had been a journeyman carpenter to John Wilkinson, but became a Methodist preacher, and then, after marrying a rich wife, set up a small foundry. He wasted his wife's fortune, however, and when he enquired of Boulton and Watt for an engine, they refused him because they did not trust his credit.[5] It may have been this which drove him into partnership with Wrigley. These two built some of the early cotton mills powered by Savery engines. Watt did not think very highly of them: he considered one of Wrigley's cotton mills 'most execrable' and reported that Derbyshire had 'again spoiled Parrs Mill at Liverpool'.[6]

A month after Wrigley had taken the Jordain order, Boulton and Watt agreed to erect an engine for the Clark Cotton Company,[7] near

---

[1] Lieut. Logan Henderson, formerly an officer in the Marines, joined Boulton and Watt in 1776. He worked mostly in Cornwall and eventually left the firm in 1783.

[2] James Watt to Matthew Boulton, 25 June 1781, Letter Book (Cornwall), June 1781–Jan. 1782, B.R.L. H. W. Dickinson and Rhys Jenkins, *James Watt and the Steam Engine* (1927), p. 299, give the year as 1780.

[3] Matthew Boulton to James Watt, 1 July 1781, Boulton and Watt Collection, Parcel D, B.R.L. Boulton himself dates this letter 1780 and this is followed by Dickinson and Jenkins, *op. cit.*, p. 299, but it seems to be a reply to Watt's letter of 25 June 1781. All the people referred to in this letter were rival engineers or infringers of Watt's patent. The list indicates the number of engine-makers active in 1781.

[4] James Watt to Thomas Jordain, 26 Nov. 1784, Boulton and Watt Letter Book (Office), June 1784–March 1786, B.R.L. Jordain does not appear in the Manchester directories.

[5] James Watt to Josiah Wedgwood, 18 Jan. 1785, same letter book, B.R.L.

[6] James Watt to Matthew Boulton, 18 Aug. 1785, same letter book, B.R.L.

[7] See below, pp. 411–12.

Lancaster, but Joseph Thackeray, for whom Wrigley had erected the two engines at Garratt, was a partner in the Cark concern and employed Wrigley as his engineer. This prompted Watt to write:

I learned at Manchester that Mr Thackeray employed one Joshua Wrigley as his Engineer, whom I object to having anything to do with; first because he is not a good workman if I may judge from a sample of his work I saw there. Secondly because he came here twice with people who wanted engines, and one of the times at least persuaded the person that he could put him up a more advantageous Engine on a principle of his own, which he is now about but it is the opinion of everybody that he will beshit himself though he has the Rev^d Mr John Derbyshire for his coadjutor, thirdly because having erected an Engine for the s^d Mr Thackeray & Comp^y the smoke of which poisons the whole neibourhood [*sic*], he came here pimping to find out how to cure the smoke, wch not succeeding, he has built a chimney 100 feet high. To a man of that stamp I will give no instructions.[1]

He was assured that neither Wrigley nor any other man unacceptable to him would be employed in erecting the engine, and the work therefore proceeded.[2]

Nothing more is heard of Wrigley in Watt's correspondence until June 1789 when the younger Watt wrote to his father from Manchester about Wrigley's engine at Garratt.[3] Nearly two years later, young James reported that Isaac Perrins, the Boulton and Watt engine erector,[4] had seen one of Wrigley's engines at the top of Deansgate, 'where the fire place was upon your smokeless construction'.[5] James Watt junior was anxious to give Wrigley a good trimming for infringing his father's patent; he inspected the fire-place himself and found it almost identical with his father's; but Watt himself felt that it was doubtful whether the courts would uphold his patent for a smokeless fire-place.[6]

When Watt junior, still in Manchester, wrote again about Wrigley in March 1791, Wrigley had 'orders for 13 Engines for this town & neighbourhood, all of them intended for working Cotton Machinery of one kind or other by the Medium of a Water wheel'.[7] This means that, apart from an unknown number of engines which Wrigley had previously erected, he had on his order books at this date one third of

[1] James Watt to John Wilkinson, 1 Sept. 1785, Boulton and Watt Letter book (Office), June 1784 to March 1786, B.R.L.
[2] John Wilkinson to James Watt, 12 Sept. 1785, Box 20, B.R.L.
[3] James Watt jun. to James Watt, 21 June 1789, Dol.
[4] See below, p. 419.
[5] James Watt jun. to James Watt, 13 March 1791, Dol.
[6] James Watt to James Watt jun., 17 March 1791, B.R.L. The patent, no. 1,485, was obtained in 1785 and the plan was followed in the Albion Mills in 1786.   [7] James Watt jun. to James Watt, 13 March 1791, Dol.

the number of engines erected by Boulton and Watt in the same area by 1800.[1] The sons of Boulton and Watt seem always to have been more alive than their fathers to these threats to the firm's business. To James Watt junior's warnings, his father merely replied: 'While Wrigley goes on with his water wheel engines he will be very harmless, he will hurt none but his customers'.[2] Then a new engineer arrived in Manchester and was said to have entered into partnership with Wrigley. He had designed a new boiler—which Watt junior described in some detail—and had taken out a patent for it.[3] Still his father remained unconcerned: 'The boiler you mention is neither new nor formidable, the effect from the same quantity of fire will be less than in the common boiler & it cannot last long . . .'[4] Nothing that his son could tell him seems to have disturbed his complacency, because he was supremely confident of his engine's superiority in mechanical efficiency over all others.

Meanwhile Wrigley seems to have done quite well at this period, though his good fortune could not last. James Watt was right in the long term. Joshua Wrigley died in 1810 and his estate was worth less than £40.[5] Among his unsuccessful ventures was the steam-powered grinding of logwoods, etc. for dyeing, as evidenced by an advertisement in the *Manchester Mercury* of 10 December 1799 announcing the sale by auction of 'Knot Mill', including 'the Machinery for grinding logwood, fustick, &c. provided with a powerful Steam Engine of a Cylinder of 24 inches diameter, and four French Burr Mill Stones, fit for grinding flour, &c. All which premises were lately occupied by Joshua Wrigley and Company . . .'[6] We last hear of him in the Boulton and Watt correspondence in March 1799 when he was to put up a 'Common Engine' with a 30 inch cylinder to pump water at the new Calender House of [John?] Horridge.[7] Whether this was one of his Savery engines or a Newcomen engine—the type more usually referred to as a 'common engine'—is not clear.[8] He probably carried on

[1] J. Lord, *op. cit.*, p. 179, n. 2.

[2] James Watt to James Watt jun., 17 March 1791, B.R.L.

[3] James Watt jun. to James Watt, 19 June 1791, Dol. The new engineer appears to have been John Roger Teschemacher, who took out Patent no. 1,808, 24 May 1791.   [4] James Watt to James Watt jun., 3 July 1791, B.R.L.

[5] *Lancs. and Chesh. Record Soc. Publications*, Vol. LXIII, *Index to Wills at Chester, 1801–1810*, Part II, M to Z.

[6] See also R. W. Procter, *Memorials of Manchester Streets* (Manchester, 1874), p. 109. He appears in Scholes's *Manchester and Salford Directory* of 1797 as 'millwright & logwood grinder, 75 Long-mill-gate, mill, Knot-mill'.

[7] James Lawson to Boulton and Watt, 7 March 1799, Parcel C, B.R.L.

[8] There is ambiguity because in 1789 we find James Watt jun. referring to one of Wrigley's Savery engines at Garratt as a 'common engine'. James Watt jun. to James Watt, 21 June 1789, Dol.

making engines in the following years; he continued to appear in Manchester directories as a millwright, at 75 Long Millgate, until the end of his life.[1] There is no doubt that he had a very long run in the Lancashire engine business.

To Joshua Wrigley's name, as a maker of Savery engines in Manchester, we must join that of Joseph Young, millwright.[2] In September 1790, Watt junior stated that Young:

has already constructed some engines upon the old plan, and . . . undertakes to erect engines of the power of 4 horses, at little more than £200, including the water wheel which they are to put in motion. These engines of his are upon the original Saverian construction, no piston, but merely a cylinder in which the water rises by means of a vacuum made at the top. He farther promises to engage that these engines shall not burn more coals than yours, and that they shall consume the smoke, which he affirms you have no patent for doing . . . The gross sum which your engines cost at first, startles all the lesser manufacturers here, and it is scarcely possible to make them comprehend the advantage to be derived from a regular motion, from a machine liable to a few repairs, and from an annual saving of fuel, when weighed against 2 or 300 £ more of ready cash.[3]

This attitude towards the cost of Boulton and Watt engines appears to have been very common among the smaller Lancashire mill-owners. Even when they were convinced of the greater efficiency of the Boulton and Watt engine, they would not meet the extra cost. They would take advantage of the engine by piracy if they could, but it seems to have been only the larger mills which were ready to invest so much capital in an engine. Thus in 1790 Watt junior failed to obtain an order for a Boulton and Watt engine from Messrs. Marshall and Reynolds, Manchester cotton manufacturers, because he was forestalled by Young with a cheaper Saverian engine and water-wheel.[4] It is not surprising to find Boulton and Watt acknowledging in June 1791 that 'Manchester has been backward in adopting our engines'.[5]

From this evidence about Young's and Wrigley's business, we can see that the transition from Savery engines was much slower than has been generally supposed. Savery's engine, we have very recently been told by such an authority as the late Dr. H. W. Dickinson, 'fell into

---

[1] Dean & Co.'s directory, 1804; Pigot's directory, 1811.

[2] Edmond Holme's *Manchester and Salford Directory*, 1788: 'Young, Joseph, millwright, Cupid's alley'; later, according to Scholes's *Manchester and Salford Directory*, 1797, of Young Street.

[3] James Watt jun. to James Watt, 30 Sept. 1790, Dol.

[4] James Watt jun. to James Watt, 30 Sept. and 3 Nov. 1790, Dol.

[5] Boulton and Watt to T. Cooper, 7 June 1791, Boulton and Watt Letter Book (Office), Dec. 1790–Aug. 1793, B.R.L.

disuse early in the eighteenth century'.[1] Yet as late as May 1796 we hear of a Manchester cotton spinning firm, Messrs. Nightingale, Harris & Co., still having 'A Savary's [engine] which now turns their works', but whose replacement by a Boulton and Watt they were considering.[2] Wrigley and Young between them evidently had a considerable output of engines, which may, in fact, have rivalled the number erected by Boulton and Watt in the Manchester area before 1800, though the latter were ousting them in the 1790s.[3]

II

During the early 1790s Watt's double-acting rotative engine began to be applied rapidly in various manufactures and especially in Lancashire cotton-spinning, for which its regular motion proved well adapted. Other engineers, unless they were prepared to 'pirate' Watt's patents, had to make do with variations of the older Savery or Newcomen engines. If they did not use a 'fire-engine' and water-wheel combination, they had to adapt the atmospheric (Newcomen) engine to rotative motion.[4] Many, in fact, did so and Farey informs us that, in addition to Watt engines, 'great numbers of atmospheric engines were also made for turning mills, particularly in the districts where coals were cheap'.[5] For some industrial purposes, as we have mentioned, where a regular motion was not a necessity, they 'answered the purposes to which they were applied, and were used for many years', some being 'still in use' in the 1820s.[6] In textile spinning, however, as Joshua Wrigley found, they could not compete successfully against Watt's double-acting rotative engine. In this process especially, 'a double or continuous action is of the utmost consequence to the regularity of the motion'.[7] It became necessary to devise a double-acting atmospheric engine, with two cylinders whose pistons acted alternately on the same crank and flywheel.[8] According to Farey, the two principal firms

---

[1] H. W. Dickinson, 'The Steam-Engine to 1830', in C. Singer, E. J. Holmyard, A. R. Hall, and T. I. Williams, *A History of Technology*, Vol. IV, *The Industrial Revolution c. 1750 to 1850* (1958), p. 173.

[2] M. R. Boulton to J. Southern, recd. 19 May 1796, B.R.L.

[3] Young had an even longer engineering career than Wrigley. Early in the nineteenth century he removed his business to Quay Street, where he remained until the 1840s (Pigot and Slater's directory, 1841).

[4] See above, pp. 398–9.   [5] Farey, *op. cit.*, p. 422.

[6] *Ibid.*, pp. 422 and 658.   [7] *Ibid.*, p. 427.

[8] *Ibid.*, pp. 422, 426–7, and 658–64. In Watt's patent double-acting engine, of course, only one cylinder was used and steam power was directly applied, both above and below the piston, whereas in the atmospheric engine steam was used merely to achieve a partial vacuum below the piston, so that atmospheric pressure forced down the piston; there was no power on the return stroke, since the piston

making such engines were James Bateman and William Sherratt of Manchester and Francis Thompson of Ashover in Derbyshire.[1] Not much is known of Francis Thompson's activities in Manchester,[2] but a good deal more about those of Bateman and Sherratt, who were Boulton and Watt's main competitors in this area. Aikin's contemporary account shows their importance:

A considerable iron foundry is established in Salford, in which are cast most of the articles wanted in Manchester and its neighbourhood, consisting chiefly of large cast wheels for the cotton machines; cylinders, boilers, and pipes for steam engines; cast ovens, and grates of all sizes. This work belongs to Bateman and Sharrard . . . Mr. Sharrard is a very ingenious and able engineer, who has improved upon and brought the steam engine to great perfection. *Most of those that are used and set up in and about Manchester are of their make and fitting up . . .* They are now used in cotton mills, and for every purpose of the water wheel, where a stream is not to be got, and for winding up coals from a great depth in the coal pits, which is performed with a quietness and ease not to be conceived.

---

was raised only by the 'counterweight' or pump-rods at the other end of the beam. Hence two cylinders were required to give power on both strokes and thus greater regularity of motion.

[1] *Ibid.,* p. 422. John Wilkinson, the famous ironmaster, also put up a few engines in Lancashire, not atmospheric but piracies of Watt's engine. Peter Ewart, of Boulton and Watt, whilst in the neighbourhood of St. Helens, saw a coal-winding engine of Wilkinson's make which he described as having 'a 14 in. Cylinder & a 3 feet stroke, worked double, with a rotative motion the same as those at the Dale only the working gear was very different'. (Peter Ewart, from Raikes, near Bolton, to John Southern, Soho, 15 Feb. 1791, B.R.L. For Coalbrookdale engines see Raistrick, *op. cit.,* chap. 9.) This was apparently the pirate engine built in 1790 for Nicholas Ashton's colliery near St. Helens. Wilkinson erected another pirate engine in the same area in 1791 for 'Michael Hughes Esq. Colliery at Ravenshead, copperwork (of the Parys Mine Co.) between Prescot and Wigan'. 'List of Engines made by John Wilkinson', drawn up by James Watt jun., Sept. 1795, B.R.L. We are indebted to Dr. W. H. Chaloner for a copy of this list. See also T. C. Barker and J. R. Harris, *A Merseyside Town in the Industrial Revolution. St. Helens 1750–1900* (Liverpool, 1954), pp. 85–6.

[2] Farey, *op. cit.,* p. 662, states: 'Two of the cotton mills in Manchester which are now on the largest scale, were commenced in 1794, with those engines, but being found insufficient for the increasing work, they were replaced by larger engines by Messrs. Boulton and Watt. Mr. Thompson's engines were found to be as difficult to keep in order as Mr. Watt's and very inferior in performance; hence, they were only adopted in a few instances, and have long since been laid aside.' See also Dickinson and Jenkins, *op. cit.,* pp. 312–13. Thompson also built large double-acting atmospheric engines for the worsted-spinning mill of Davison and Hawksley, at Arnold, near Nottingham (who had previously had a Savery engine) and also for a cotton mill in Macclesfield, as well as a few smaller ones. His main business, however, was in single-cylinder atmospheric engines, both pumping and rotative, for the lead mines, collieries, and other manufactures in Derbyshire and neighbouring areas. Farey, *op. cit.,* pp. 237–8, 422, 444, and 658–62; F. Nixon, 'The Early Steam-Engine in Derbyshire', *Newcomen Soc. Trans.,* Vol. XXXI (1957–9), and *Notes on the Engineering History of Derbyshire* (1956).

Some few are also erected in this neighbourhood by Messrs. Bolton and Watts [sic] of Birmingham, who have far excelled all others in their improvement of the steam engine . . .[1]

Even allowing for a little local patriotism, it is surely clear from this passage that in 1795 Bateman and Sherratt engines outnumbered Boulton and Watt engines in the Manchester area.

James Bateman first appears in the Manchester directory for 1773 as the partner of Walter Wilson, ironmonger, 24 Deansgate.[2] By 1781 he had started up as an ironmonger in Salford and seven years later his enterprises had ramified considerably: James Bateman and Company, ironmongers, Deansgate; iron founder, Water Street, Salford; iron forger, Collyhurst; iron founder and forger, Dukinfield.[3] In the 1792 and 1794 directories he appears in partnership with William Sherratt as an iron-founder at 7 Hardman Street, Salford.[4] He also appears later on as a coal merchant.[5] Bateman's gradual extension of his engineering activities enabled him to produce a wide range of castings and forged metal goods. In January 1782 he inserted a long advertisement in the *Manchester Mercury*, informing the public that he had lately erected an iron foundry (probably that in Water Street) and had begun to manufacture an astonishing variety of cast-iron goods, including stoves and grates, pots and pans, gates and railings, weights, iron wheels, spindles, cog-wheels, shafts, pinions, and rollers for carding and spinning machines, rollers for print-works, malt-works, and rolling and slitting mills, calender bowls, etc.[6] Soon afterwards he appears to have acquired another foundry, at Dukinfield. In 1782–3 Richard Crowder, of Dukinfield Lodge, was offering a furnace and foundry for sale,[7] and about the same time we find Bateman writing that he has at last agreed with Mr. Astley for his iron furnace and that Mr. John Taylor was to be concerned with him in it.[8] It is uncertain whether these are references to the same furnace, but Bateman was certainly at Dukinfield five years later.[9] Thus he provided himself with all the plant necessary for producing large engine castings.

[1] Aikin, *op. cit.*, pp. 176–8. Our italics.

[2] Raffald's *Directory*, 1773. Some details of Bateman's life are given in F. S. Stancliffe, *John Shaw's, 1793–1938* (1938), *passim*, and in a MS. article by the same author privately circulated to the members of John Shaw's Club, 'James Bateman, the Ironmonger'.   [3] Holme's *Directory*, 1788.

[4] Scholes's *Manchester and Salford Directory*, 1792 and 1794. The partnership appears to have been formed a few years earlier. See below, p. 411.

[5] Dean's *Manchester and Salford Directory*, 1808–9: 'Bateman and Sherratt, coal merchants, Mill-field, Gt. Ancoats-st'.

[6] *Manchester Mercury*, 1 Jan. 1782.   [7] *Ibid.*, 9 July 1782 and 29 April 1783.

[8] James Bateman to Jonathan Scale, 6 July 1783, A.O.L.B.

[9] Holme's *Directory*, 1788.

Up to this time, Bateman does not appear to have been making engines, since on 15 July 1782 Matthew Boulton wrote to him about an engine which Bateman wanted. This was for his own works, to be used in 'Boreing of small working barrells . . . also turning a Grinding Stone . . . also to turn two Strong Turning Lathes . . . & likewise to blow a dragon Furnice . . .[1] Bateman also showed a great deal of interest in the general commercial possibilities of the Boulton and Watt engine in the Lancashire cotton industry, supplying Boulton with information about cotton mills, and asking whether he might not get from Boulton a model of the engine with instructions for its working, so that he could introduce it in the right quarters.[2] In the light of later developments, this can hardly have been ingenuous. On 16 July 1782, Boulton supplied Bateman with complete details of a Boulton and Watt pumping engine,[3] and it was probably with this letter that he sent his draft of the advantages of Watt engines in working cotton mills, claiming for them greater regularity, more equable motion, economy of fuel consumption, freedom from interruption by floods or frosts, and general adaptability.[4] Bateman also sent his partner, John Taylor, to Soho to examine the Boulton and Watt engine, and promised information about Arkwright's mill.[5] Then, for some reason, the deal was not clinched, and Bateman did not buy the engine, though no doubt he had gained a good deal of useful information. At this stage, however, there does not seem to have been any rupture in friendly relations, for there is a note in James Watt's pocket diary that he called on Bateman on 13 August 1785 and saw round his works.[6]

As with Joshua Wrigley, there is a break in the information about Bateman until James Watt junior arrives in Manchester. In April 1789 Watt junior was negotiating with a leading Manchester cotton spinner, Peter Drinkwater, to sell him a Boulton and Watt engine.[7] James Bateman was also anxious to get the order and, making Drinkwater an uninvited visit, gave him an estimate for an engine of his own manufacture. Drinkwater would have nothing to do with Bateman and 'even told him that he would not have one of his engines

---

[1] Matthew Boulton to James Bateman, 15 July 1782, A.O.L.B. This letter clearly refers to the Salford foundry.

[2] James Bateman to Matthew Boulton, 11 July 1782, A.O.L.B.

[3] A.O.L.B.

[4] 'Description of a Free Mill to work a Cotten or Silk Mill', Matthew Boulton Letter Book, 1781–83, A.O.L.B., with a pencil note in Boulton's hand, 'probably sent to Bateman *c*. July 1782'.

[5] James Bateman to Jonathan Scale, 6 July 1783, A.O.L.B.

[6] Dol.   [7] Chaloner, *op. cit.*, pp. 87–8.

DD

if he would make him a present of it'.[1] Just a little earlier that year, in February, Watt junior visited a Bateman engine at the works of Moss, a dyer,[2] situated out of Manchester near St. John's Church. This engine was considered the best in the Manchester district at that time, though Watt junior did not think highly of it. 'It is a 34 Inch diam[r] 7 feet Stroke, goes generally about 16 Strokes a Minute ... It grinds the materials for dying, scrapes the logwood & turns the dressing and washing machines: the rotary motion, *as in all other engines here*, is applied upon M[r] Wilkinson's plan.'[3] Drinkwater's Boulton and Watt engine is generally said to have been the first direct application of rotative steam power to cotton spinning in Manchester; but it is evident from this letter that Bateman may have so applied Newcomen engines, without using a water-wheel, before this date. Young James Watt pointed out what a brisk active man Bateman was, 'perpetually on the look out', praising his own engines at the expense of Boulton and Watt's, and a dangerous competitor.[4] Bateman even had Soho men in his employ, such as John Varley, who had been got a job by James Lawson on John Lees's engine, but had gone instead to Drinkwater's, which he eventually left to work for Bateman.[5] This fellow proved to be most important to Bateman's business.

In June 1789 Watt junior reported that another Bateman engine had been erected by Messrs. Thackeray and Whitehead, cotton spinners, at Garratt, to replace one of the earlier engines of Wrigley's construction.[6] It must have been specially galling to Boulton and Watt to lose an order from a man who had already bought one of their engines, erected in 1786 for the Cark Cotton Company in which Thackeray was also a partner.[7] Yet still James Watt remained unmoved:

In respect to M[r] Bateman's calumny & efforts to extend his business at our expence we despise them & do not wish to procure business by sounding our

[1] James Watt jun. to James Watt, 4 April 1789, Dol. This is included among some letters at Doldowlod on this subject to which Dr. Chaloner did not have access at the time of writing his article.

[2] Probably of 'Moss and Rothwell, dyers, Water street', Lewis's *Manchester Directory* (1788).

[3] James Watt jun. to James Watt, 1 Feb. 1789, Dol. (our italics). He had also seen a cast-iron boiler made by Bateman for another steam engine, about 8 feet diameter and 5 feet deep, and in two parts intended to be screwed together. There is a sketch of its shape.   [4] James Watt jun. to James Watt, 25 May 1789, Dol.

[5] James Lawson to James Watt, jun., 3 March 1796, B.R.L.

[6] James Watt jun. to James Watt, 21 June 1789, Dol. If we can trust John Marshall's notes, this engine—'Thackwray's atmosph[eric] Engine', he calls it, as distinct from the earlier Savery type—was of fantastic size, with a cylinder of 120 inches diameter, 18 feet stroke, working 6 strokes per minute and raising 3,000 gallons of water per stroke to a height of 13 feet. It dwarfed all the other engines on Marshall's list.   [7] See below, pp. 411-12.

own praises if the merit of our Engines does not speak for us we shall be silent, and take such business as comes to us unsollicited/by the by you are guilty of an impropriety in speaking of Bateman & Wilson without putting *Mr* to their names, always speak of your opponents respectfully remember that '*Suaviter in modo fortiter in re*'.[1]

This might have been all very well in some circumstances, but Watt's opponents were learning fast and were unscrupulous. Renegade Soho workmen were coming to Manchester and taking employment as engine men and thus providing a pool of skilled labour. The new Bateman engine at Garratt was tended after September 1781 by Thomas Thorneycroft who had worked on Boulton and Watt engines,[2] and there were several other such men in and about Manchester.[3] By this time Bateman had entered into partnership with William Sherratt,[4] who, according to Aikin, was possessed of considerable engineering ability, particularly in the manufacture of steam engines. In 1791 their productive capacity was greatly increased by the purchase of land in Hardman Street, Salford, for the establishment of a new foundry.[5] They were left to develop and extend their engine business until 1796, when Boulton and Watt finally joined battle with them because they were infringing Watt's patents.

We must now retrace our steps a little before discussing the legal battle over piracies, in order to recount what happened at Cark, where a Boulton and Watt engine was built in 1786. The order was given to Boulton and Watt, against the advice of some of the partners who wanted a Newcomen engine, through the influence of James and Fletcher Stockdale, and their friend, John Wilkinson, the ironfounder.[6]

[1] James Watt to James Watt jun., 16 June 1789, B.R.L.

[2] Affidavit by Thomas Thorneycroft of Albury, 22 Feb. 1796, Boulton and Watt Collection, Packet C, B.R.L.

[3] In 1796, for example, we hear of several 'discarded Sohoites' being employed in Manchester. M. R. Boulton to J. Southern, 18 June 1796, B.R.L.

[4] They were associated together by 1788 when prospective lessees of a 'Water-Works' at Ashton Mills were told to apply for particulars to 'Mr. Bateman, Ironmonger, in Manchester, or Mr. Sherratt, at Duckenfield Furnace'. *Manchester Mercury*, 12 Feb. 1788.

[5] Deeds in the possession of Messrs. Mather and Platt, Ltd, whose present-day Salford Ironworks is in direct line of descent from the original Hardman Street foundry of Bateman and Sherratt. The firm also extended its ironfounding interests outside Lancashire, for we hear that 'Messrs. Sherratt, of Salford, had set on foot a Foundry, on the banks of the Trent and Mersey Canal, west of Shelton, which was carried on for some years in a spirited manner, but subsequently passed into other hands, and lately ended in bankruptcy.' J. Ward [and Simeon Shaw], *The Borough of Stoke-upon-Trent* (1843), p. 379. It is not clear when this foundry was established.

[6] James Stockdale to James Watt, 11 July 1785, M. IV. S, B.R.L.; John Wilkinson to James Watt, 24 Aug. 1785, Box 20, B.R.L. Since our article was originally published, Dr. W. H. Chaloner has produced an article on 'The Stockdale

A further decisive influence had been a copy of John Smeaton's report on the Soho engine at Pateley Bridge.[1] Watt gave a quotation for an engine costing between £1,000 and £1,100,[2] but was not anxious to do the job, as he did 'not above half approve of Manchester men'.[3] The engine did not actually begin to work until a year later, on 18 September 1786, mainly because Boulton and Watt did not have an engine erector free until August 1786. James and Fletcher Stockdale became increasingly disconcerted by this delay, which caused the firm considerable loss.[4] It is true that when the engine was set in motion the company were pleased, and some small trouble with it in July 1787 was soon put to rights, but the engine was expensive and the long delay in erecting it did no good to Boulton and Watt's reputation.

In 1789–90, therefore, when the Cark Company needed more power, they were doubtful about purchasing another Boulton and Watt engine. First they decided against it,[5] but then re-opened negotiations through John Wilkinson. Wilkinson, however, turned traitor on Boulton and Watt, as he had already done by his piracies of their engine.[6] He wrote to Stockdale saying that Boulton and Watt had shown so much indifference to the Cark proposal that he advised them to go elsewhere, and that he had a man called Rowe who was just then at Bersham erecting a new fire engine of his own invention, for which he had recently taken out a patent.[7] 'If it should be found to answer there will be no difficulty or delays on this Score—nor will there be any charge for the Patents that you will have any objection to.' Stockdale therefore ordered Rowe's engine for Cark, but had to wait for it as long as for the Boulton and Watt engine. By April 1791, the engine

---

Family, the Wilkinson Brothers and the Cotton Mills at Cark-in-Cartmell, c. 1782–1800', *Trans. Cumberland & Westmorland Antiq. & Arch. Soc.*, Vol. LXIV, n. s. (1964), pp. 356–72. This, however, has necessitated no revision of our account.

[1] James Stockdale to James Watt, 11 June 1785, M. IV. S, B.R.L.

[2] James Watt to James Stockdale, 11 July 1785, Boulton and Watt Letter Book (Office), July 1784 to March 1786, B.R.L.

[3] James Watt to John Wilkinson, 1 Sept. 1785, same letter book, B.R.L.

[4] Fletcher Stockdale to James Watt, 18 July 1786, M. IV. S, B.R.L.

[5] James Watt to James Watt jun., 16 June 1789, B.R.L.

[6] T. S. Ashton, *Iron and Steel in the Industrial Revolution* (2nd edn., 1951), pp. 79–80. See also above, p. 407, n. 1. The full extent of Wilkinson's piracies is shown in the 'List of Engines made by John Wilkinson', drawn up by James Watt jun. in Sept. 1795, B.R.L.

[7] John Wilkinson to Mr. Stockdale, 2 Aug. 1790, L.C.R.O. The 'Rowe' named must have been John Westaway Rowe, whose son, William Rowe, worked for Wilkinson as engineer and surveyor. See A. N. Palmer, 'John Wilkinson and the Old Bersham Iron Works', *Transactions of the Honourable Society of Cymmrodorion*, Session 1897–8, p. 37. See also J. W. Rowe's Patent, no. 1,749, dated 29 April 1790, for a semi-rotative engine.

parts had still not been despatched from Chester,[1] and even when the engine was erected it proved to be a complete failure and had to be replaced.[2] It was replaced, however, not by a Boulton and Watt engine but by a Bateman and Sherratt engine pirating Watt's principles.

III

For many years Boulton and Watt took few positive steps to defend themselves against piracies, confident of the superior efficiency of their own engine.[3] At last, however, when their Lancashire customers pointed to the infringers in the area and complained about paying premiums for what others enjoyed for nothing, Boulton and Watt were forced to take action.[4] Bateman and Sherratt, it appears, took to piracy because of the proved inferiority of their engines to those of Boulton and Watt. Their single-acting atmospheric engines were greatly inferior to Watt's double-acting engine for rotative motion. The old type of engine served well enough for pumping, winding coals, etc., but was not suitable for mule-spinning, so the makers were obliged to contrive a double-acting, two-cylinder atmospheric engine. At first, apparently, Bateman and Sherratt did not infringe Watt's patent, but effected condensation, as in the old engines, by injecting cold water into the cylinders.[5] This, of course, involved heavy costs in coal consumption, so they soon began deliberately to pirate Watt's separate condenser and his air pump for exhausting the cylinders.

The generals in charge of the battle against the Manchester pirates were James Watt junior and Matthew Robinson Boulton, who, at the beginning of 1796, had just succeeded in bringing John Sturges & Co. (Bowling Ironworks) and two other Leeds firms to heel for four atmospheric engines which infringed Watt's patent.[6] One of their informants in the Leeds battle was their old engine erector, Isaac Perrins, and it was probably he who set them on the track of the Lancashire pirates. Then Peter Ewart fell in with Thomas Thorneycroft, the former Boulton and Watt engine erector, who gave evidence of a Bateman and Sherratt engine which he had helped to erect in 1792 and which infringed Watt's patent by using an air pump and a

[1] John Wilkinson to James Stockdale, 6 April 1791, L.C.R.O. See also Wilkinson's letter from Brosely, 25 Feb. 1791, L.C.R.O.

[2] James Watt to James Watt jun., 2 June 1796, Boulton and Watt Letter Book (Office), Sept. 1795 to Aug. 1796, B.R.L. This letter book is hereafter referred to as 'Letter Book A'.

[3] See J. Muirhead, *Life of James Watt* (1858), pp. 401–2.

[4] M. R. Boulton to A. Weston, 23 Feb. 1796, Letter Book A, B.R.L.

[5] Farey, *op. cit.*, p. 662.

[6] Dickinson and Jenkins, *op. cit.*, p. 320.

condenser. This engine was built for Joseph Thackeray at Garratt.[1] It was actually put up by Richard Bradley (formerly coachman to Thomas Walker, cotton merchant, of Manchester), who appears to have become a partner in both the Garratt and Cark concerns. He was assisted in the construction of the engine by being shown the drawings of a Boulton and Watt engine (belonging to Barrow, Lees & Co.) at Werneth Colliery, and his informant was Thomas Livesey of that firm. The engine had two open-mouthed cylinders working a wheel placed between them, but instead of a condenser and air pump to each as in Maberley's pirated engine, Thackeray's and Bradley's engine had one of larger size for both cylinders.[2] Though Bateman and Sherratt were said by Thorneycroft not to have known fully about the air pump and condenser, they had inspected the Garratt engine as soon as it was set to work and must therefore have been aware of the infringement. And Thackeray, who later disclaimed any knowledge of Watt's patent, paid Thorneycroft's father a guinea a week to keep people out of the engine house.[3]

In face of this unscrupulous attitude, James Watt junior and Matthew Robinson Boulton organized a complete system of espionage,[4] in which James Lawson, disguised as a jockey, played a great part. From Thomas Thorneycroft, Thomas Livesey, Isaac Perrins, John Varley, and others, Boulton and Watt got to know all the pirate engines sold by Bateman and Sherratt.[5] While Boulton and Watt's solicitors in London prepared to ask the Court of Chancery for injunctions against the pirates, the team of spies in Lancashire collected affidavits and inspected all the pirate engines, keeping their whole investigation as close a secret as they could, 'until the Injunction like the exterminating angel, swoops down upon them, and Mercy they shall have none'.[6]

[1] M. R. Boulton to James Watt, 22 Feb. 1796, B.R.L. The Garratt mill is particularly interesting in the history of mechanical power in the cotton industry. We have already noticed (above, p. 400) that it was apparently the earliest Manchester cotton mill, powered by water, first for ribbon-weaving and then for spinning. Two Wrigley engines had been installed in 1784 to drive the water-wheels; further power had been obtained in 1789 by the erection of a Bateman engine (see above, p. 410), and then in 1792 the pirate engine was erected, a double-acting atmospheric engine, which, after being compounded for, as we shall see, in 1796, was still in use over thirty years later (Farey, *op. cit.*, p. 662).

[2] M. R. Boulton to A. Weston, 23 Feb. 1796, Letter Book A, B.R.L. For a detailed technical description of this engine, see Farey, *op. cit.*, pp. 662–5. For Maberley's engine, see Dickinson and Jenkins, *op. cit.*, p. 307.

[3] M. R. Boulton to A. Weston, 24 Feb. 1796, Letter Book A, B.R.L.

[4] James Watt jun. to A. Weston, 29 Feb. 1796, Letter Book A, B.R.L.

[5] The details of this process, though fascinating, are too involved for us to deal with here.

[6] James Watt jun. to James Lawson, 17 March 1796, Letter Book A, B.R.L.

Perrins gave information about the engines which he knew of and then brought over John Varley, who had been working on the air pumps and condensers for Bateman and Sherratt and was therefore able to supply an almost complete list.[1] This was supplemented by the skilful detective work of James Lawson, who travelled all over Lancashire to places where pirate engines were reported and succeeded in inspecting them personally.[2] These investigations produced copious information about the engines constructed by Bateman and Sherratt for the following firms:[3]

1. Joseph Thackeray and John Whitehead, cotton spinners, Garratt, Manchester. Rotative engine, erected by Richard Bradley, commenced working August 1792; two cylinders, 36 in. diam., 5 ft. stroke, 36 strokes per min.; estimated by Boulton and Watt to be of 46·5 h.p. at 21·4 strokes per min.[4]

2. Thackeray, Stockdale & Co., cotton spinners, Cark, near Lancaster. Double engine, with 36 in. cylinders, exactly like the Garratt one, and also erected by Richard Bradley; commenced working February 1794.

3. James Drury, Middleton Colliery, near Manchester. Pumping engine, single cylinder, 48 in. diam., 7 ft. stroke; commenced working January 1795. Described by Lawson as 'Mr Sherrats first', meaning apparently the first pirate engine entirely constructed and erected by Bateman and Sherratt.

4. Sir George Warren, colliery at Bullock's Smithy (modern Hazel Grove), near Poynton, Cheshire. Pumping engine, single cylinder, 41 in. diam., 7 ft. stroke; commenced working August 1795.[5]

5. Thomas Watson, of Watson, Fielding, Meyers & Co., calico printers, Preston. Two cylinders, 18 in. diam., $3\frac{1}{2}$ ft. stroke; commenced working September 1795. Bateman and Sherratt were also in process of erecting another pirate engine for this firm, in place of a 'common engine' previously erected by them.

---

[1] James Lawson to James Watt jun., 3, 15, 17 March 1796, B.R.L.

[2] See letters from James Lawson to James Watt jun., March 1796, B.R.L.

[3] Among the mass of correspondence on this subject, a letter from M. R. Boulton (Manchester) to James Watt, 6 June 1796, B.R.L., is particularly informative.

[4] Farey gives somewhat different details: 2 cylinders, 36 in. diam. 4 ft. stroke, 40 strokes per min., 70 h.p.; but at 30 strokes per min. 45 h.p. He is confused about the date, stating in one place (p. 427, note) 'about 1790', and in another (p. 662) 'about 1794'. This engine was still in use (1825) when Farey wrote.

[5] For further information on this engine, see W. H. Chaloner, 'The Cheshire Activities of Matthew Boulton and James Watt, of Soho, near Birmingham, 1776–1817', *Trans. of the Lancs. and Chesh. Antiq. Soc.*, Vol. LXI (1949), pp. 129–31.

6. Lord Derby, at Preston; purpose not stated. Two cylinders, 19 in. diam., $3\frac{1}{2}$ ft. stroke, said to be of 17 h.p.; commenced working February 1796.

7. Peel, Yates & Co., Burnley, probably for calico printing. Two cylinders, 21 in. diam., $3\frac{1}{2}$ ft. stroke; commenced working February 1796.

8. James Carlisle, Bolton; purpose not stated. Single cylinder, 24 in. diam., 4 ft. stroke; commenced working February 1796.

9. John Goodier, fustian calenderer, Bank Street, Manchester. One cylinder, 18 in. diam., 3 ft. stroke, said to be of 6 h.p.; commenced working March 1796.

10. Bateman & Sherratt, ironfounders, Salford. Single cylinder, 24 in. diam., 4 ft. stroke; commenced working March 1796.

11. Daniel Lees, of Barrow, Lees & Co., Werneth Colliery, near Oldham. Engine in process of erection by Thomas Livesey, two cylinders, 27 in. diam., 4 ft. stroke.

12. Daniel K. Burton, calico printer, Port Street, Manchester. Engine in process of erection, two cylinders, 16 in. diam., 3 ft. stroke.

13. James Mort, calenderer, 8 Tib Street, Manchester.[1] Engine in process of erection, said to be 'part of his own contrivance a little perfected by the assistance of T. Livesey', but parts supplied by Bateman & Sherratt; two cylinders, 16 in. diam.

14. Mr. Crossley, Manchester.[2] Lawson reported this engine 'about to be erected near Messrs. Owen & Co., with two Cylinders which I have seen lying on the spot'. No further reference; perhaps abandoned.

15. Mr. Rigby, Arding & Harwarding (Hawarden), between Holywell and Chester; for a boring mill. This engine, of unknown size, was said by Varley to have a 'close toped cylinder & every thing the same as B. & W.'s double Engines', but there is no other reference to it in the correspondence.

16. Lord Penrhyn's estate in the West Indies; Bateman and Sherratt pirate engine said by Varley to have been shipped from Liverpool, but no further information.

In the above list we have definite evidence regarding the construction by Bateman and Sherratt of sixteen engines (firms 1–14), though

---

[1] Mort had previously inquired about a Boulton and Watt engine, but on discovering the cost stated that 'he could not raise the money being a poor man', though he offered to pay by instalments. (J. Lawson to J. Watt jun., 19 March 1795, Box 38, B.R.L., and J. Watt jun. to Jas. Mort, 28 Sept. 1795, Letter Book A, B.R.L.) Here again we have evidence of a small firm's unwillingness or inability to meet the comparatively high cost of a Boulton and Watt engine.

[2] There are several Crossleys, cotton manufacturers, in the 1794 Manchester directory.

one or two may not have been completed. To these perhaps may be added two further engines (firms 15 and 16). All of these, both single and double cylinder engines, with the exception of Watson's first, a common engine, pirated Watt's separate condenser and air-pump. Varley also gave information that Bateman and Sherratt were proceeding with the construction of four new engines with air pumps, though he did not know where they were going. Later on, however, he reported that work on these engines had been stopped; perhaps they were converted to the ordinary atmospheric type, since Boulton and Watt refused to give licences for Bateman and Sherratt to manufacture any new engines utilizing the separate condenser and air-pump.[1]

In addition to all this evidence about piracies, we have come across information in this correspondence of a few ordinary atmospheric engines made by Bateman and Sherratt during these years. There was one built for a Mr. Norton, dyer, of Salford, and three for Messrs. J. & S. Horrocks, cotton spinners, Preston, one of which had been working 'some years', while the other two were still erecting (one pirating Watt's parallel motion). There is also other evidence of 'common' engines erected by Bateman and Sherratt about this time. One of their engines, 'which has never been used', was advertised for sale early in 1794 in a cotton-spinning mill 'at the top of Fleet-street, deansgate'[2] Richard Ainsworth, cotton spinner, Bolton, had an atmospheric engine built in about 1794,[3] and Mr. Cradock, of Fold's Colliery, Little Bolton, was in September 1795 offering the lease of a new Bateman and Sherratt engine, of 20 h.p.[4] What appears to have been the oldest surviving Bateman engine was found in 1885, by a great grandson of James Bateman, still working as a mine-pump at Talke-o'-th'-Hill in Staffordshire, bearing Bateman's name and the date 1791.[5] All together, we have here references to twenty-nine engines built or being built by Bateman and Sherratt between 1791–96, and there were others. Farey tells us, for example, that 'some' of their engines were sent to London, 'one or two' going to breweries.[6]

Injunctions were eventually obtained against Bateman and Sherratt, Thackeray and Bradley, and also Watson of Preston, on 2 May 1796, and though James Stockdale was not mentioned in the injunction, both the Garratt and the Cark engines were ordered to be stopped. There is not sufficient space here to deal with all the twistings and turnings adopted by Boulton and Watt's enemies. Suffice it to say the

---

[1] See below, pp. 420–1. [2] *Manchester Mercury*, 11 Feb. 1794.
[3] James Watt jun. to M. R. Boulton, 5 June 1798, B.R.L.
[4] *Manchester Mercury*, 15 Sept. 1795. [5] Stancliffe, *op. cit.*, p. 69.
[6] *Op. cit.*, pp. 422 and 444, note.

Salford, Garratt and Cark engines were all worked in defiance of the injunction until a writ of attachment was issued against Thackeray which brought him to his senses, with the help of a little encouragement from William Wilkinson and James Stockdale. Before they concluded peace Bateman and Sherratt had converted their engine back to a common engine. Thackeray did the same, but, in the words of James Watt, 'The Chief Justice in this affair was the Engine itself which refused to drive the machinery without the Condenser'.[1]

The terms of peace with Bateman and Sherratt were just but severe: (i) Bateman and Sherratt were to pay the premiums of their own engine with interest; (ii) they were to deposit their bond for £4,000 as a security against further infringements; (iii) they were to furnish an accurate list of the engines constructed upon Watt's patent principles, with the names of the purchasers, the dimensions of the engines, etc.; (iv) they were to pay Boulton and Watt's costs and expenses; and (v) Boulton and Watt were to have liberty to inspect their engines at all times.[2] On 3 June 1796 Thackeray capitulated also. By the middle of August 1796 all the pirates had come to terms except James Stockdale, who managed to hang out by delaying tactics until August 1797.[3] Middleton colliery was apparently wound up in consequence, and one or two firms converted their engines to ordinary atmospheric principles, but most agreed to pay up the premiums from the date their engines began working. From all these infringers in Lancashire Boulton and Watt recovered about £3,500.[4] But the most important result of all this conflict was not intended to be the money from the premiums— Boulton and Watt's legal costs amounted to between £5,000 and £6,000[5]—but the defence of the patent so that other pirates would be discouraged.

IV

In the process of defending their legal rights Boulton and Watt juniors learnt a great deal about the conditions of the engineering trade in each area they entered, and also observed the weaknesses in their own sales organization. In Lancashire it seemed to them that they would

[1] James Watt to A. Weston, 3 June 1796, Letter Book A, B.R.L.
[2] M. R. Boulton to James Watt jun., 25 May 1796, B.R.L.
[3] William Wilkinson's administration, 2 Aug. 1797, L.C.R.O.
[4] 'Memorandum for J. Pearson respecting payment of the premium & indemnities for the infringements at Manchester &c, 10 July 1796', Letter Book A, B.R.L. To the sums mentioned in this document we have added £550 for the Cark engine, awarded by William Wilkinson in his arbitration, see footnote 3.
[5] Muirhead, *op. cit.*, p. 402.

have an unchallenged command of the engine market once they had defeated Bateman and Sherratt's piracies:

> You will have been informed by Mr Watt of the submission of Bateman & Co whose defeat leaves us now without competitors in this quarter & every exertion should be made to supply the numerous Engines that are likely to be wanted in this part of the Kingdom. Bateman & Co were driving a roaring trade & had at the time they were served with Injunctions a great number of orders in hand. Indeed the field opened for us is very extensive . . .[1]

In an attempt to capture this market Boulton and Watt now made special efforts. In previous years they had been seriously handicapped by a shortage of engine-erectors.[2] In October 1790 they had sent Peter Ewart, 'a very ingenious and able Millwright', to erect engines and look after their interests in Lancashire,[3] but he also ran his own business and was constantly complaining about the lack of assistance.[4] After less than two years he gave up erecting for Boulton and Watt to go into partnership with Samuel Oldknow. He then advised them that they ought to have a full-time engineer in Manchester, for the erection and repair of their engines, and recommended James Lawson for the job.[5] Before Lawson was given the responsibility, Isaac Perrins acted for Boulton and Watt in Manchester. This man, a famous prize-fighter of his day, after erecting Drinkwater's engine and helping with others, moved permanently to Manchester about the end of 1793, took a public house, and continued to work for Boulton and Watt. He was an uncouth fellow, however, and insulted many of the customers, and he was eventually dismissed after threatening Lawson. Boulton and Watt either would not or could not remedy the engine erection delays in Lancashire, and did not seriously face the problem until the end of 1796.

Boulton and Watt juniors recognized that they needed in Lancashire 'a good workman of rather more intelligence than the common run of our Engineers, to be constantly resident' there, for the erection and repair of engines, because many customers would take repairs to Bateman rather than deal with Perrins.[6] This man was found in the

---

[1] M. R. Boulton to John Southern, 29 May 1796, Box 2, B.R.L.

[2] See above, p. 412, for their difficulties in regard to the erection of the Cark engine in 1786.

[3] J. Watt to J. Watt jun., 16 Oct. 1790, and P. Ewart to J. Watt, 10 Oct. 1790, B.R.L.

[4] See, for example, P. Ewart to Mr. Forman, 11 Feb. 1792, A.O.L.B., and P. Ewart to J. Southern, 25 March 1792, B.R.L.

[5] P. Ewart to Boulton and Watt, 4 Aug. 1792, and P. Ewart to J. Watt, 18 Aug. 1792, B.R.L

[6] M. R. Boulton to J. Southern, 29 May 1796, Box 2, B.R.L.

excellent James Lawson. They needed such a man also because customers wanted to discuss their problems with an intelligent person before they placed their orders.[1] Then they needed to settle their prices so that they could give an immediate quotation.[2] Perhaps most important of all they needed to reduce the delivery time, and in this connection M. R. Boulton said of Bateman and Sherratt: 'Their prices do not appear to be much if any lower than ours & the only inducement they offer is dispatch, this advantage is perhaps more pretended than real; it must however be our care not to leave [room for] this pretence . . .'[3] Boulton and Watt, through James Lawson's exertions, did tighten up their organization in Lancashire. With the establishment of the Soho Foundry in 1795–6, they also had more control over the delivery of engine parts than in the days when John Wilkinson supplied most of them.[4] Their previous dependence on Wilkinson and other ironfounders, and the trouble and delay which this caused to their customers, had given a strong competitive advantage to Bateman & Co., who had acquired their own foundry in the early 'eighties and so were able to make their own engine castings: thus Bateman & Co. could quote customers a price for an engine in the ordinary way, which was doubtless preferred to Watt's system of royalties, and were able to carry out the whole job, supplying all parts and erecting the engine.

Even after 1796 Bateman and Sherratt still had a great deal of engine business. A lull in their trade around Manchester was reported by James Lawson in November 1797,[5] but by February 1799 he was calling a different tune: '. . . B & S are busy with Engines in other places [and] they have more going on here that I know of'.[6] At this date they were copying Boulton and Watt engines exactly and leaving everything ready to put on the condenser as soon as Watt's patent expired. They were even making drawings which were copies of Boulton and Watt's engine drawings. At the height of the conflict with Boulton and Watt, they had asked if they could be granted licences to go on erecting engines on Watt's principles, but that permission was refused, since 'it would be to all intents and purposes making them partners in the Patent

---

[1] M. R. Boulton to M. Boulton, 23 May 1796, M. R. Boulton Box 2, A.O.L.B.

[2] M. R. Boulton to J. Watt, 31 May 1796, Parcel B, B.R.L.

[3] Enclosed in James Lawson to Boulton and Watt, 4 March 1797, B.R.L.

[4] Their difficulties had become acute in 1794–5, as a result of the quarrel between John and William Wilkinson, which completely disorganized the Bersham works, upon which Boulton and Watt depended for their cylinders. Transference of orders to the Dale Company did not solve the problem of delays and cancellations. See Ashton, *op. cit.*, pp. 75–9.

[5] James Lawson to M. R. Boulton, 15 Nov. 1797, B.R.L.

[6] James Lawson to Boulton and Watt, 23 Feb. 1799, Parcel C, B.R.L.

and whetting a knife to cut our own throats'.[1] They tried again in December 1799, when, as a special concession to Mr. Isaac Horrocks, a cousin of Messrs. J. and S. Horrocks, they were allowed to erect an engine with a condenser and air-pump for Messrs. Newsham and Horrocks, of Preston, but once again no general licence was granted them.[2]

Nevertheless, Bateman and Sherratt still continued to build atmospheric engines and we have evidence of several in the later nineties. Mr. Dunkerley, fustian manufacturer, of Oldham, ordered one of their engines in July 1796 and so did Mr. Hill of Pendleton.[3] At the auction in June 1797 of Messrs. Miles and Thomas Edwards, bankrupt cotton spinners of Kent Street, Southwark, London, one of Bateman and Sherratt's engines was put up for sale.[4] Mr. Wood, cotton manufacturer, of Carlisle, obtained a Bateman and Sherratt engine in 1798.[5] Messrs. Lees Cheetham & Co., cotton manufacturers, of Stalybridge, also ordered an engine from Boulton and Watt's rivals towards the end of 1800.[6] A steam engine in the cotton-spinning factory of Stephen Faulkner & Co., bankrupts, advertised for sale at Bolton-en-le-Moors in 1798,[7] and a pumping engine at Carr Colliery, near Prescot,[8] were also probably made by Bateman and Sherratt.

There is no doubt that Bateman and Sherratt were still finding customers for their atmospheric engines, especially in the rapidly developing spinning business. It is also interesting to discover that they made some of the earliest marine engines. One of these, an atmospheric engine, is said to have been fitted into a barge on the Bridgewater Canal in 1794.[9] Whether or not this was so, it is certain that in 1799 they

---

[1] James Watt jun. to M. R. Boulton, 23 May 1796, Box 38, B.R.L.

[2] James Lawson to Boulton and Watt, 20 Nov. 1799, Parcel C, B.R.L., and Boulton and Watt to Bateman and Sherratt, 11 Dec. 1799, Boulton and Watt Letter Book (Office), May 1799 to March 1800, B.R.L.

[3] James Lawson to Boulton and Watt, 22 July 1796, Parcel C, B.R.L. Two years later, however, we find Dunkerley replacing the Bateman and Sherratt engine by one from Boulton and Watt. James Lawson to M. R. Boulton, 7 July 1798, Parcel C, B.R.L.   [4] *Manchester Mercury*, 6 July 1797.

[5] James Watt jun. to M. R. Boulton, 6 Aug. 1798, B.R.L.

[6] James Lawson to James Watt jun., 8 Nov. 1800, B.R.L.

[7] *Manchester Mercury*, 21 Aug. 1798.   [8] *Ibid.*, 23 April 1799.

[9] Dickinson and Jenkins, *op. cit.*, p. 320, where it is stated that a tracing of this engine is in the Science Museum at South Kensington. The Museum, however, has informed us that no such tracing can be discovered. There is a possibility of confusion here with the steamboat built by John Smith of St. Helens, which is said to have steamed along the Sankey and Bridgewater canals to Manchester in June 1793. 'Among the visitors was Mr. Sherratt of the firm of Bateman and Sherratt' (*Liverpool Mercury*, 20 July 1832). Dr. J. R. Harris, however, is of the opinion that the correct date of this event is 1797. 'The Early Steam-Engine on Merseyside', *Trans. of the Hist. Soc. of Lancs. and Chesh.*, Vol. CVI (1954), pp. 115–16.

made a Newcomen engine for a boat, nicknamed *Bonaparte*, designed by the celebrated Robert Fulton, which was built in the Duke of Bridgewater's yard at Worsley, for hauling coal-barges on the canal to Manchester. The boat proved a failure, but the engine and boiler had a long and useful life, for various industrial purposes, until 1851.[1]

Meanwhile Boulton and Watt, though increasing their business in Lancashire, were still beset by some of the old problems of delay in fulfilment of orders and lack of engine erectors; they still had to rely very often on Perrins. Towards the end of 1800 James Lawson reported that they were experiencing difficulty in obtaining orders in Manchester, 'for tho' B. & W. stand high [in reputation] nothing will compensate for want of expedition in such a fluctuating trade as this'.[2] The slowness with which Watt's engine penetrated the Lancashire cotton industry, and the prolonged competition of Bateman and Sherratt's less efficient Newcomen engines were pointed out many years ago:

Notwithstanding the mechanical merits of the Soho Engine, and the great economy of its power, a large class interested in its adoption has been standing aloof and doggedly suffering enormous loss through waste of fuel, rather than submit to pay the 'oppressive tax' levied by Boulton and Watt of one-third of the saving in coals made by using their engine; which, in conjunction with the existence, in a flourishing state, of Bateman and Sherrard's manufactory of imperfect and most wasteful steam engines, may fairly be adduced as evidence of the avaricious ingratitude of communities to inventors, and of the obstinacy of the hostile prejudices generally arrayed against them.[3]

It is clear, however, from the contemporary evidence that the attitude of the Lancashire manufacturers was not determined merely by stupid conservatism and objection to Boulton and Watt's premiums. Bateman and Sherratt provided considerably speedier delivery and erection of engines whose initial capital cost was a good deal less than that of Boulton and Watt engines, and which were also simpler to keep in order.[4]

With the expiry of Watt's patent in 1800, Bateman and Sherratt were vigorously developing their business. They were reported to be planning the erection of 'a small Black Furnace in Manchester for

---

[1] W. H. J. Traice, 'Another Old Lancashire Engine at Worsley', *Manchester City News Notes and Queries*, Vol. IV (1881), pp. 1–2; B. Chapman, 'The World's First Steamboat', *Worsley Parish Magazine*, Feb. 1947, pp. 18–19; and J. A. Rogerson, 'Story of "Old Nancy" ', *A. E. I. News*, Vol. XIX, no. 6 (June 1949), pp. 8–9.

[2] James Lawson to James Watt jun., 14 Nov. 1800, B.R.L.

[3] *Patents for Inventions. Abridgments of Specifications relating to the Steam Engine* (1871), Part I, A.D. 1618–1859, Vol. I, p. 107.

[4] According to Farey, *op. cit.*, p. 671.

heavy Castings with Lancashire Ore',[1] and they were capturing orders from Boulton and Watt. The development of throstle spinning about this time necessitated more power, and several manufacturers were therefore enquiring about having their engines enlarged or getting new ones. Owing to delays by the Soho firm, some of them turned to Bateman and Sherratt: 'one (Holland & Bridge) have ordered a new Cylinder of Bateman and Sherratt, and I suppose the Chorlton [Twist] Co. will do the same'.[2] Farey tells us that when Watt's patent expired and steam-engine makers began to spring up rapidly in all the manufacturing districts, Bateman and Sherratt, with their long experience, 'had the greatest success in this new business' of building engines on Watt's principles, 'and as engines were very much wanted at that time, for the increasing manufactories in Lancashire, their trade became very extensive in a few years'.[3] They were also one of the first firms to manufacture Trevithick's high-pressure engines, again attempting, despite their experiences with Boulton and Watt, to do so in defiance of patent rights.[4] There is no doubt that these men prospered. At his death in 1824 James Bateman's personal estate was close on £80,000,[5] and when the Salford Foundry was sold in April 1838 it covered nearly 6,800 square yards, including pattern rooms, casting shops, smiths' shops, fitting-up shops, engine house, gas house and gasometer, crimper's shop, brass foundry and several other workshops.[6]

In the course of this paper we have listed references to forty-six engines built by Bateman and Sherratt by 1800, thirty-eight of them in Lancashire mostly in the Manchester area, and this has been done without the advantage of having the firm's own records. This figure is not far short of the total number of engines erected by Boulton and Watt in Lancashire by the same date (fifty-five).[7] And it is certain that Bateman and Sherratt produced many more engines of which no record has survived. In fact they may well have built more engines in Lancashire than Boulton and Watt had by the end of the century. Of their total horse-power, it is impossible to make an estimate, but figures of individual engines which have survived show that they were of comparable power to those of Boulton and Watt, though much less economical in coal consumption.

[1] George Lee to James Watt jun., 28 July 1800, M. IV, B.R.L.

[2] James Lawson to James Watt jun., 14 Nov. 1800, B.R.L.

[3] *Op. cit.*, p. 677.

[4] F. Trevithick, *Life of Richard Trevithick* (2 vols., 1872), Vol. II, p. 128. See also below, p. 464, n. 5.

[5] Will dated 21/6/1821 and Codicil dated 8/3/1824. Probate at Chester 14/5/1824.   [6] Advertisement, *Manchester Guardian*, 7 April 1838.

[7] According to Lord, *op. cit.*, p. 174.

V

The engines made by Wrigley, Young, and Bateman and Sherratt certainly do not exhaust the numbers of engines made and set to work in Lancashire by 1800. Ebenezer Smith & Co., of the Griffin Foundry, Chesterfield, who also owned the New Foundry opposite Shudehill Pits, Manchester, advertised in February 1787 that they made, among many other engineering products, 'Cylinders, Iron and Brass Working Pieces, Boilers, Chains, Regulators, Pipes and every other Article for FIRE ENGINES'.[1] This firm seems to have supplied a substantial proportion of the castings and forgings for machinery and millwork of all kinds in Manchester, including parts for Boulton and Watt engines.[2] They also made cylinders and other parts for Francis Thompson and soon began to produce their own engines. In 1790 they were receiving 'many applications for Fire Engine Work', i.e. for engines of 'the common sort'.[3] By the later nineties they appear as opponents of Soho, along with Bateman and Sherratt,[4] and there is evidence of at least one engine of their manufacture in Lancashire, somewhere on the Huddersfield Canal and described by James Lawson as 'one of the worst made I ever saw'.[5]

Another Manchester iron-founding firm was that of Thomas Rider (or Ryder) of Scotland-bridge, Manchester. This was an old-established concern, the only iron-foundry appearing in the first Manchester directory (1772). Thomas Rider was probably a relative of Nicholas Rider, of Marston Forge, near Northwich, Cheshire, who supplied boiler plates for a Watt engine in 1779.[6] After Thomas Ryder's death in 1779, the Scotland Bridge foundry—the 'Old Iron Foundry'—passed into the ownership of the Fletcher family,[7] about whose activities nothing is known, but in 1796 we find another Thomas Ryder, perhaps son of the former, at a 'Salford Foundery' and also, in partnership with George Nelson, at Marston Forge, where they were adver-

---

[1] *Manchester Mercury*, 27 Feb. 1787. Their foundry was at 7 Swan Street, near Shudehill pits, and was usually referred to as the Griffin or Newcross foundry (being in the Newcross district). See Manchester directories for 1788, 1794, and later years, and P. Robinson, *The Smiths of Chesterfield* (Chesterfield, 1957), *passim*.

[2] See, for example, Smith & Co. to Boulton and Watt, 26 June and 21 July 1790, Box 4 'S', B.R.L. [3] *Ibid.*

[4] James Lawson to Boulton and Watt, 3 Aug. 1796, Parcel C, B.R.L.

[5] James Lawson to M. R. Boulton, 5 July 1798, Parcel C, B.R.L.

[6] Boulton and Watt to Mr. Rider, 27 Jan., 12 Feb., 7 March, 19 April, and 16 Sept. 1779, Boulton and Watt Letter Book (Office), Jan. 1778–June 1780; Nicholas Rider to James Watt, 2 Feb., 30 March, and 29 May 1779, B.R.L.

[7] See below, p. 448.

tising their manufacture of engine boiler plates and also complete boilers, together with other cast and wrought-iron work.[1] We have discovered no evidence of their having produced steam engines, but they were certainly capable of doing so.

Alexander Brodie, a prosperous ironmaster and armaments manufacturer of Calcutts, Broseley, Shropshire, and Carey Street, Lincoln's Inn Fields, London, was also in partnership with two men named McNiven and Ormrod in a Manchester iron foundry.[2] In April 1798 Boulton and Watt discovered that Brodie was infringing their patent on engines which he had erected at Calcutts.[3] Brodie, in a most illiterate letter, said that he had no intention of infringing and offered to settle.[4] After some prevarication he signed in June 1798.[5] It appeared that he had erected a boring mill engine (April 1796), a pumping engine (October 1797), a blowing engine (April 1798), and a coal winding engine (January 1795), and for these he paid £602 in premiums.[6] It is interesting to notice, moreover, that by the early nineteenth century Brodie had a share in the Hazledine Foundry at Bridgnorth, Shropshire, at which the engineer John Rastrick was responsible for constructing the earliest high-pressure steam engines of Richard Trevithick.[7]

Thomas C. Hewes appears to have started steam-engine manufacture in Manchester in the late 'nineties,[8] though he is better known for his water-wheels. Probably there were other engine-makers whom we have not discovered. There were certainly several more iron-foundries and a host of machine-making firms in Manchester by the end of the century. And there were engine-makers in other Lancashire towns. Thus in 1790 we find that Robert Lindsay & Co. of the Haigh

[1] *Manchester Mercury*, 19 April and 4 Oct. 1796 (advts. re Marston Forge). See below, pp. 448-9.
[2] They first appear in Scholes's *Manchester and Salford Directory* (1794) at 'St. George's iron foundery, Knot-mill'. See W. H. Chaloner, 'John Galloway (1804-1894), Engineer of Manchester and his "Reminiscences"', *Trans. of the Lancs. and Chesh. Antiq. Soc.*, Vol. LXIV (1954), p. 104. Galloway's foundry was built on the foundations of Brodie's.
[3] Boulton and Watt to Brodie, 23 April 1798, Boulton and Watt Letter Book (Office), July 1797 to July 1798, B.R.L.
[4] Alexander Brodie to Boulton and Watt, 4 May 1798, Parcel F 'B', B.R.L. See also Alexander Brodie to Boulton and Watt, 10 May 1798.
[5] M. R. Boulton to James Watt jun., 17 June 1798, Parcel B, B.R.L.
[6] 'Calculations of Premiums upon the Engines erected by Mr B. Brodie', 3 Nov. 1798, Boulton and Watt Letter Book (Office), July 1798 to May 1799, B.R.L.
[7] S. Morley Tonkin, 'Trevithick, Rastrick and the Hazledine Foundry' *Newcomen Soc. Trans.*, Vol. XXVI (1947-9), pp. 173-4.
[8] See below, p. 446.

EE

Ironworks, near Wigan, were able 'to undertake the compleat construction of Fire Engines of every kind'.[1]

From the evidence presented it is clear that Boulton and Watt did not have a monopoly of engine-building in Lancashire during the term of their patent, 1775–1800. In fact it is probable that they built not more than a third of the engines in the county during that time. It is interesting to put this conclusion together with John Rowe's statement: 'By 1794, instead of monopolizing the use of steam power in Cornwall as they had done ten years earlier, Boulton and Watt had control and claims upon barely half the machines working in the county and their connections were with the less flourishing and losing mines.'[2] Thus in the two counties which made most use of steam power before 1800 Boulton and Watt had nothing like a monopoly: their engine had not succeeded in ousting the Newcomen engine, and piracies of their own were innumerable. In Lancashire, it seems that only the larger firms were able to afford Boulton and Watt engines. This must affect any estimate of the amount of fixed capital invested in steam power at this period, though that will be offset by the greater number of engines known to be at work.[3]

It is impossible not to be impressed by the great variety of functions which were being performed by Savery and Newcomen engines in Lancashire. At first they were used for pumping water from coal pits or for waterworks, and then for the motive power of overshot water-wheels. Engines of this last kind, and atmospheric engines adapted to a crank and fly-wheel, were used, along with pirated Boulton and Watt engines, for many industrial purposes: for winding coal, blowing furnaces, driving boring engines and lathes, milling corn, driving carding and spinning machines, grinding materials in dye-works, turning rollers in calico-printing and calendering works and squeezers and washers on bleachfields. Steam-powered mechanization was proceeding more rapidly in Lancashire in the late eighteenth century than has hitherto been supposed.

[1] *Manchester Mercury*, 14 Sept. 1790. For steam engines erected in the Prescot and St. Helens area, see J. R. Harris, 'The Early Steam-Engine on Merseyside', *Trans. of the Hist. Soc. of Lancs. and Chesh.*, Vol. CVI (1954), pp. 109–16.

[2] J. Rowe, *Cornwall in the Age of the Industrial Revolution* (Liverpool, 1953), p. 101.

[3] Since our article was originally published, Dr. J. R. Harris has made an estimate—using our own and other investigations—of the total number of steam engines produced in the eighteenth century: 'The Employment of Steam Power in the Eighteenth Century', *History*, Vol. LII (1967), pp. 133–48.

XIII The Origins of Engineering in Lancashire[1]

It is extraordinary how little is known about the engineers who produced the water-wheels, steam engines, textile and other machinery of the early Industrial Revolution. In most economic histories of this period there are merely a few brief and vague references to smiths, carpenters, and millwrights, based on Smiles or Fairbairn, with no contemporary evidence whatever. Most accounts of the development of mechanical engineering normally begin with Bramah and Maudslay, from about 1800, and carry on with such renowned nineteenth-century names as Fairbairn, Roberts, Whitworth, and Nasmyth. Before the nineteenth century, we are usually led to believe, mechanical engineering hardly existed. This belief is largely based on nineteenth-century evidence. William Fairbairn, for example, stated that when he first came to Manchester, in 1814, 'the whole of the machinery was executed by hand. There were neither planing, slotting, nor shaping machines; and, with the exception of very imperfect lathes, and a few drills, the preparatory operations of construction were effected entirely by the hands of the workmen.'[2]

There is no doubt that the most revolutionary developments in mechanical engineering did occur in the nineteenth century. But the origins are to be found in the previous century. From the well-known facts about the invention and application of machinery in the later eighteenth century, it seems evident that specialized machine-making must have been evolving quite rapidly for at least a quarter-century before 1800, and the beginnings must go back a good deal earlier. This is especially true of Lancashire, where the most rapid and striking developments occurred in the early Industrial Revolution. The tremendous growth of the Lancashire cotton industry, from about 1770

[1] This chapter is based on a jointly produced article in the *Journal of Economic History*, June 1960, but some very substantial additions have been made.
[2] Presidential address to the British Association at Manchester, 1861.

onwards, based on the mechanical inventions of Hargreaves, Arkwright, Crompton, and Cartwright, powered by water-wheels and steam engines, gave rise to an equally rapid development of mechanical engineering. Lancashire soon came to manufacture not only cotton, but also cotton machinery, steam engines, boilers, machine-tools, and, later on, railway locomotives, iron bridges, gas-works plant, and a vast range of other engineering products.[1] Some of the nineteenth-century Lancashire engineers, like those mentioned above, are famous, but the eighteenth-century beginnings of Lancashire engineering have never been revealed. It is our purpose to try to throw some light on these early developments, particularly in Manchester. Our main aim will be to show the various trades from which the new mechanical engineers were recruited and how they developed the manufacture of iron machinery, using machine-tools, the basis of advancing technology.

Before examining the contemporary Lancashire evidence, let us sketch the general background.[2] Machines or 'engines' of various sorts had been used in industry long before the Industrial Revolution: power-driven machinery was not a new thing in the eighteenth century. There had long been wind, water and horse mills for grinding corn, fulling cloth, working blast furnaces, hammers, and rollers in ironworks, and driving drainage engines in mines or fenland districts. There had long been specialized metal workers: iron-founders, brass-founders, blacksmiths and whitesmiths, locksmiths, clock-makers, instrument-makers, etc. There must also have been specialization in the making of such primitive textile machines as spinning wheels, hand-looms, and knitting frames. Much of this early machinery, however, was made of wood and leather, and carpenters made a good deal of it. Almost anyone, in fact, who was used to working in wood or metal might be employed to make machinery or set up millwork. We even hear of a 'Mechanical Priest in Lancashire', at the beginning of the eighteenth century, who was apparently an expert in the construction of windmills for draining mines.[3]

---

[1] See the *Victoria County History, Lancaster*, Vol. II, pp. 367–74, for the nine-teenth-century developments. See also below, chaps. XIV and XV.

[2] See C. Singer, E. J. Holmyard, A. R. Hall, and T. I. Williams, eds., *A History of Technology*, Vol. IV: *The Industrial Revolution c. 1750–c. 1850* (Oxford, 1958), chaps. XIII and XIV. A great deal of interesting material is also to be found in the Newcomen Society's *Transactions*. Much older works, but still very useful, are C. Holtzapffel, *Turning and Mechanical Manipulation* (3 vols., 1843), and R. Willis, 'Machines and Tools, for Working in Metal, Wood, and other Materials', *Lectures on the Results of the Great Exhibition delivered before the Society of Arts, Manufactures and Commerce* (1852), pp. 291–320.

[3] R. Bald, *A General View of the Coal Trade of Scotland* (Edinburgh, 1812), p. 7. See also Farey, *op. cit.*, p. 227.

At a fairly early date, however, a special class of millwrights emerged, who were of particular importance in setting up the early mills and factories, and from whom sprang many of the early engineers. James Brindley and John Rennie were perhaps the two most famous mill-wright-engineers in the eighteenth century, and well into the nine-teenth century well-known engineers began in the same way. The eighteenth-century millwright, says Fairbairn (himself originally a millwright), was 'a kind of jack-of-all-trades, who could with equal facility work at the lathe, the anvil, or the carpenter's bench. . . . He could handle the axe, the hammer, and the plane with equal skill and precision; he could turn, bore, or forge . . .' He generally had a good knowledge of arithmetic, geometry, and theoretical as well as practical mechanics.[1] It appears, in fact, that these millwright-engineers were not—as is often suggested—rough, empirical, illiterate workmen, but had usually acquired somehow a fairly good education or training.

Since wood was extensively used in the making of early machinery, there was a close link between the development of wood-working and metal-working machines such as lathes, planes, and drills.[2] The con-nection is well known in the case of Bramah and Maudslay at the end of the eighteenth century, but it is a good deal older. Iron, however, was to become the basic material of the Industrial Revolution. There was, as Professor Ashton and others have shown, a close relationship between the Midland iron industry and the manufacture of the steam engine—the supreme achievement of early mechanical engineering. Unfortunately, we do not know much about the actual making of the early Savery and Newcomen engines, but we do know a considerable amount about the manufacture of Watt's engines.[3] Watt experienced great difficulties at first in getting sufficiently accurate workmanship, but these were largely overcome by the skilled metal workers in Boulton's Soho factory and by John Wilkinson's cylinder-boring machine. The latter has been described as 'probably the first metal-working tool capable of doing large heavy work with anything like present-day accuracy',[4] but there were earlier boring and drilling machines in existence; Wilkinson, like Smeaton some years earlier, merely improved existing techniques. The Darbys of Coalbrookdale and other iron-founders also made parts for Watt engines.

Following the Darby's revolutionary achievements in iron-founding,

[1] W. Fairbairn, *A Treatise on Mills and Millwork* (2 vols., 1861–3), Vol. I, pp. v–vi. See above, p. 73. [2] Willis, *op. cit.*
[3] S. Smiles, *Lives of the Engineers*, Vol. IV: *The Steam Engine: Boulton and Watt* (rev. edn., 1878); H. W. Dickinson and R. Jenkins, *James Watt and the Steam Engine* (Oxford, 1927).
[4] J. W. Roe, *English and American Tool Builders* (New Haven, 1916), p. 11.

cast iron came into widespread use for machine parts. Cort's process provided in the later eighteenth century ample quantities of wrought iron; but production of accurate, standardized machinery in quantity necessitated the development of mechanical methods—machine-tools—instead of the laborious, costly, and insufficiently precise hand processes. Workmen, however, in the early stages of the Industrial Revolution were not, as is often suggested, dependent solely on hammer, chisel, and file. There were, as Smiles admits, some 'ill-constructed lathes with some drills and boring machines of a rude sort' in the later eighteenth century,[1] and Fairbairn, as we have seen, similarly refers, though disparagingly, to lathes and drills then in use. We know, in fact, that such tools have a long history. In their development, clock- and instrument-makers appear to have been particularly important. Professor Willis pointed out in his Great Exhibition lecture over a hundred years ago that the early engineering machine-tools probably evolved from the lathes and wheel-cutting engines of the clock- and watch-making trade.[2] Several eminent eighteenth-century engineers, such as Watt and Smeaton, were instrument-makers, and it has recently been shown that clock- and instrument-making was of particular interest and of considerable practical importance to Boulton and Watt in their development of engineering technique for accurate manufacture of steam-engine parts.[3] Other metal-working crafts likewise contributed to mechanical engineering. In the Birmingham trades such as button-making, for example, various machines for stamping, pressing, and cutting metal were developed. Many of these were included in the long list of tools and engines whose export was prohibited in 1785–1786, along with lathes, drills, and boring engines.[4]

Lancashire, especially the south-western area, also had numerous metal-working trades, which contributed to the growth of mechanical engineering in the vicinity.[5] The Lancashire and Cheshire Record Society's *Index to Wills* provides useful evidence on the emergence of engineers. Smiths and carpenters, of course, had carried on their trades for centuries, and the gradual growth in their number is not particularly significant. More interesting is the fact that millwrights and iron-founders rarely appear before the eighteenth century, but become

[1] S. Smiles, *Industrial Biography* (1863), p. 180.

[2] Willis, *op. cit.*, pp. 306–7. See also above, chap. I, *passim*.

[3] E. Robinson, 'The Lunar Society and the Improvement of Scientific Instruments', *Annals of Science*, Vol. XII (Dec. 1956), pp. 296–304.

[4] *Statutes*, 25 Geo. III, c. 67 (1785), and 26 Geo. III, c. 89 (1786).

[5] For a general account of these trades, see *V.C.H. Lancaster*, Vol. II, pp. 360–7, and G. H. Tupling, 'The Early Metal Trades and the Beginnings of Engineering in Lancashire', *Transactions of the Lancashire and Cheshire Antiquarian Society*, Vol. LXI (1949), pp. 1–34.

increasingly numerous with the mechanization of the cotton industry. We also discover more specialized manufacturers of textile machinery, such as hand-loom, shuttle, and reed makers. The introduction of the Dutch or 'engine loom' into the Manchester smallwares trade in the latter half of the seventeenth century, followed by that of the improved Dutch 'swivel loom' in the mid-eighteenth, was facilitated by the immigration of 'ingenious mechanics' from Holland, who were 'invited over to construct engines at great expence'.[1] What appears to have been the earliest attempt at a power-driven weaving mill in Lancashire was made in about 1760, when a Mr. Gartside built a factory at Garratt Hall, in which he attempted to work a number of swivel looms 'with a capital water-wheel' on the River Medlock.[2] Millwrights and machine makers must certainly have been required for such a venture.

Elizabeth Raffald's directories of Manchester and Salford, dated 1772, 1773, and 1781, give us our first comprehensive view of the various trades in the district at the beginning of the mechanical revolution in cotton manufacture. In those of the early 'seventies the term 'engineer' does not appear and only one millwright and one founder—apparently an iron-founder—are listed. But there are seven ironmongers, six braziers, or brass-founders, about a score of smiths (blacksmiths and whitesmiths), a dozen clock- and watch-makers, and about as many tin-plate and wire-workers and pin-makers—all metalworking trades—while the number of woodworkers (carpenters, joiners, turners, and cabinet-makers) was considerably greater.[3] Any of these firms might have turned to the manufacture of textile machinery. By this date, of course, the cotton industry was already of considerable importance in Lancashire,[4] and the directories show that specialization was developing in the making not only of hand-looms and spinning wheels but also of their parts. Those of 1772–1773 include two loommakers, two shuttle-makers, and a dozen reed-makers.[5]

---

[1] [J .Ogden,] *A Description of Manchester* (Manchester, 1783), p. 82; A. P. Wadsworth and J. de L. Mann, *The Cotton Trade and Industrial Lancashire, 1600–1780* (Manchester, 1931), pp. 103 and 284–8.

[2] J. Aikin, *A Description of the Country from thirty to forty Miles around Manchester* (1795), p. 174.

[3] The names listed in the directories are those of *firms*; there is no indication of the number of workers employed. It must also be remembered that these early directories may not have been complete.

[4] See E. Baines, *History of the Cotton Manufacture in Great Britain* (1835); G. W. Daniels, *The Early English Cotton Industry* (1920), and Wadsworth and Mann, *op. cit.*

[5] The reed is the appliance used in weaving for separating the threads of the warp and for beating the weft up to the web.

The 1781 directory shows a slow development of these various trades in Manchester and Salford during the intervening decade. The number of such firms grows, though not very appreciably. The most significant developments are the appearance of one or two more iron-founding firms and also of two undoubted textile engineers—Adam Harrison, 'cotton engine maker', and Samuel Smith, 'engine maker', both of Wright's Court, Salford, and perhaps, therefore, partners in one firm.

Later directories, in 1788, 1794, 1797, and 1800,[1] reveal a most striking growth of engineering during the last two decades of the century. Cotton-spinning machines, following the inventions of Hargreaves, Arkwright, and Crompton, were now being produced in considerable quantity to fill the mills which were springing up so rapidly, and which required water-wheels and steam engines to power them. Weaving, on the other hand, still remained for many years a process for the hand-loom, Cartwright's power-loom being developed comparatively slowly; and of course, more and more hand-looms were required for converting the increasing quantities of yarn into cloth. We find, therefore, a growth in the number of loom-makers, shuttle, reed, and beam makers. No doubt most of these were former carpenters and joiners, hand-looms being made mainly of wood: this evolution is, in fact, occasionally visible in the directories, where, for instance, we find tradesmen described as 'joiner and loom-maker', 'turner and loom-maker', 'joiner and shuttle-maker', or 'turner and shuttle-maker'. Some woodworkers also made spinning machinery, for we occasionally find a 'joiner and machine-maker', and there were also several 'turners in wood and metal'.

Edmond Holme's 1788 directory lists nine firms variously described as 'iron founders', 'iron forgers', 'machine makers', and 'engine makers', with a solitary millwright. By 1794 the number of such firms had grown to twenty-eight: the specialized producers of textile machinery were usually, it seems, described as 'machine-makers', while the iron-founders and forgers appear to have conducted a more general business. In his revised 1797 edition, Scholes added over two dozen more such firms, an indication of how rapidly engineering was developing in Manchester and Salford. In addition to the numerous machine-makers and iron-founders, millwrights become somewhat more numerous; one firm (Sutcliffe and Parkinson) is described as 'mill-wrights, machine-makers, and turners in wood and metal', another instance of engineering evolution. For the first time, moreover, in 1797 we come

---

[1] Edmond Holme's (1788), John Scholes's (1794 and 1797), and Gerard Bancks's (1800).

across the term 'engineer' in the Manchester and Salford directories:[1] Thomas Knight, 'engineer' of 2 Back-lane, and Isaac Perrins, 'victualler and engineer' of the 'Fire Engine' public house, 24 Leigh Street.[2] Moreover, this directory reveals the commencement of specialization in the manufacture of parts of power-driven spinning machines, similar to the earlier specialization in the production of shuttles, reeds, and beams for hand-looms. Thus James Burrows, of 1 Ancoats Street, is described as a 'water-spindle-maker, and turner'; Abraham Butler, of 32 Bank-top, as a 'cutter of machine wheels'; and Hodgin and Foster, of Trafford Street, are listed as 'rowler-makers'. It is noticeable that along with the rise of iron-founding and machine-making, there had also been a remarkable increase in the number of brass-founders, smiths, tin-plate workers, clock-makers, etc., listed in the directories, many of whom may have participated in the evolution of engineering.

The development of engineering in Manchester from the various older trades attracted the attention of Aikin in 1795:[3]

The prodigious extension of the several branches of the Manchester [cotton] manufactures had likewise increased the business of several trades and manufactures connected with or dependent upon them. . . . To the ironmongers shops, which are greatly increased of late, are generally annexed smithies, where many articles are made, even to nails. [Aikin then mentions various ironfounding firms in Manchester and Salford, making steam engines, boilers, etc., which we shall deal with later[4]] . . . The tin-plate works have found additional employment in furnishing many articles for spinning machines; as have also the braziers in casting wheels for the motion-work of the rollers used in them; and the clock-makers in cutting them. Harness-makers have been much employed in making bands for carding engines, and large wheels for the first operation of drawing out the cardings, whereby the consumption of strong curried leather has been much increased.

An equally interesting account of the early growth of textile machine-making is provided in a trade-union address issued by a Bolton mechanic in 1831.[5] The writer said he was giving facts 'of my own recollection', and was describing developments in the late eighteenth

[1] But see below, p. 436, for its earlier appearance in local newspapers.

[2] Perrins first came to Manchester in 1789 as an engine erector for Boulton and Watt. See above, p. 419, and below, p. 447.

[3] Aikin, *op. cit.*, pp. 176–8.

[4] See below, pp. 447–55.

[5] Preserved among material deposited in the Economics Library of Manchester University by Messrs. Dobson and Barlow, Ltd, textile engineers, of Bolton, and quoted by G. W. Daniels in 'A "Turn-Out" of Bolton Machine-Makers in 1831', *Economic History*, Vol. I (Jan. 1929), pp. 592–5.

century, after 'the [spinning] trade was laid open' by the quashing of Arkwright's patents.[1] His account is well worth quoting at length:

When a company intended to erect such a [spinning] mill, a spot was selected which promised sufficient [water] power, and was also convenient for working people. . . . An engineer was sent to superintend the works; he generally brought with him such workmen as he deemed necessary for the undertaking; but, if he could not accomplish that, he was under the necessity of supplying the defect by engaging such as the locality afforded. It is almost needless to state, that those artizans were a medley of trades. . . . Amongst them, the mill-wright at that time claimed the pre-eminence; the rest were composed of carpenters, joiners, smiths, clockmakers, who left their original trades for better wages; moulders and turners were then little known, the work being chiefly composed of wood, brass, malleable iron, and steel. Professional turners being then very rare, each artizan was under the necessity of learning to turn his own part. The moulding part being brass, had to be sent for some distance, and at great expence. The work was all done in places where it had to remain.

The writer goes on to state that when such a mill was completed, many of the engineering workmen would be dismissed, but that 'it was still necessary to retain a part of them on the premises, there being, at that time, no machine shops, and the apparatus must be kept in order'. It was not long, however, before machine-making became a specialized trade:

After the mills had been a few years established, some thrifty part of the men who had been employed in them began to study their own interest. They found the business lucrative and resolved to have a share in it. Some small shops were established for the construction of such machinery as could be constructed out of the mills, and I believe carding engines were the first things of the kind so undertaken; afterwards other parts of the apparatus began to appear in the shops, such as the jenny . . . and afterwards the mule. . . . Machinery began at this time to be improved, workshops spread themselves over the country . . .

John Kennedy, one of the leading cotton machine-makers and spinners in Manchester in the late eighteenth and early nineteenth century, gives similar evidence.[2] He confirms that carding engines were the first practically successful textile machines, in use from the

---

[1] Aikin states: 'It was about the year 1784 that the expiration of Sir Richard Arkwright's patent caused the erection of water machines [water frames] for the spinning of warps in all parts of the country, with which the hand engines [jennies] for the spinning of weft kept proportion.' *Op. cit.*, p. 179. See also Baines, *op. cit.*, p. 214.

[2] J. Kennedy, 'Observations on the Rise and Progress of the Cotton Trade in Great Britain, particularly in Lancashire and the adjoining counties', read before the Manchester Literary and Philosophical Society in 1815 and published in his *Miscellaneous Papers* (Manchester, 1849), pp. 5–25.

1750s. The inventors of these and of the later spinning machines, he says, 'soon found that if they could readily get a blacksmith's or a carpenter's assistance, they would be able to get their little apparatus more substantially made'; they were also enabled to enlarge them and eventually to apply non-human power, first of horses and then of water-wheels. The growth of the cotton trade, says Kennedy,

> created a new demand for artificers in various branches. . . . The artificers . . .
> became very useful in the construction of machinery. . . . By degrees, a higher
> class of mechanics such as watch and clock-makers, white-smiths, and mathe-
> matical instrument-makers, began to be wanted; and in a short time a wide
> field was opened for the application of their more accurate and scientific
> mechanism. Those workmen were first chiefly employed in constructing the
> valuable machines [water-frames, carding and roving engines] invented by Mr.
> Arkwright. . . . At that period mill-wrights, as well as the superior workmen
> above mentioned, were more generally employed in the establishments for
> spinning cotton than formerly . . .

In another paper[1] Kennedy points out that some of the early inventors, such as Crompton, were not themselves mechanics and their early
machines were very rudimentary, made with 'simple tools' and mostly
of wood. Arkwright's water-frame, however, had metal rollers and
'clockwork' (the term is interesting) and iron fixtures. These improvements were incorporated in the mule by Henry Stones, 'an ingenious
mechanic' of Horwich. The consequent need for metal rollers, spindles,
and other parts led to a great 'want of experienced workmen of every
kind', with the result that many craftsmen, such as smiths, joiners,
shoemakers, etc., came into the cotton industry from other trades,
attracted by the higher wages, and contributed many small technical
improvements. Thus in 1771 Arkwright and Strutt were advertising
for machine-making workers at their Cromford spinning mill:[2]
'Wanted immediately, two Journeymen Clock-Makers, or others that
understands [*sic*] Tooth and Pinion well: Also a Smith that can forge
and file—Likewise two Wood Turners that have been accustomed to
Wheel-making, Spole [Spool]-turning, &c.' By the 'nineties, Kennedy
states that specialized textile machine-making firms had emerged, such
as those of McConnel and Kennedy, Adam and George Murray, and
William Wright (a former apprentice and workman of Arkwright),

---

[1] J. Kennedy, 'A brief Memoir of Samuel Crompton', read before the Manchester Literary and Philosophical Society in 1830, in *Miscellaneous Papers*, pp.
50–83.
[2] *Derby Mercury*, 13 Dec. 1771, quoted in R. S. Fitton and A. P. Wadsworth,
The Strutts and the Arkwrights, 1758–1830 (Manchester, 1958), p. 65. See also
Derby Mercury, 20 Sept. 1781, where they were advertising for 'Forging & Filing
Smiths, Joiners and Carpenters' for the second Cromford mill.

all of Manchester, who enormously improved the mule. By that time, as we have seen, specialized makers of textile machine parts such as spindles and rollers were developing, and by the 1820s, if not earlier, there were even 'two or three classes of spindle-makers, separate and distinct trades',[1] together with bobbin-makers, flyer-makers, and others.[2]

Similar evidence comes from other cotton machine-makers. Thomas C. Hewes, a millwright and engineer, who set up in Manchester in the early 1790s, stated that he originally recruited many of his hands 'from cabinet-makers and clock-makers and things of that kind'.[3] Turners, joiners, smiths, and clock-makers appear on the wages books of Samuel Greg's cotton-spinning firm at Styal, Cheshire, in the 1790s,[4] where they were doubtless employed to make or repair the machinery.

Local newspapers are the richest sources of information on the recruitment of these early engineering workers. The *Manchester Mercury*, for example, from the 1770s onwards contains innumerable advertisements by both machine-making and cotton-spinning firms in various Lancashire towns for smiths, joiners, turners, filers, clock-makers, etc.[5] Sales advertisements also provide information in regard to lathes (some powered), wheel-cutting and fluting engines, and other equipment used in machine-making; smiths' and clock-makers' tools are very often mentioned. References to millwrights are much less frequent: already, it appears, the general skill of the millwright was being replaced by more specialized crafts.[6] The term 'engineer' is rarely used, 'engine-maker' or 'machine-maker' being more common. Robert Kay, of Bury, inventor of the hand-loom 'drop-box', is described as an 'Engineer' (4 August 1772); Richard Melling, deceased, of Wigan, as a 'Mason and Engineer' (30 June 1778); and Joshua Wrigley, Long Milngate, Manchester, as an 'Engineer' (8 June 1784).[7]

It is obvious from these selections that textile engineering was developing rapidly in Manchester and other Lancashire towns. And it

---

[1] *Select Committee on Artizans and Machinery, Fourth Report,* in *Parl. Papers* (1824), Vol. V, p. 253, evidence by Peter Ewart.

[2] *Ibid.*, p. 545, evidence of John Bradbury.

[3] *Ibid.*, p. 347. On p. 348 he refers similarly to 'joiners and wheelwrights, and cabinet-makers'. His name is wrongly printed 'Herves'. For his engineering achievements, see above, pp. 69–71, 98, and below, pp. 445–7.

[4] F. Collier, 'An Early Factory Community', *Economic History*, Vol. II (Jan. 1930), p. 119.   [5] See Appendix, pp. 455–8, for a selection from these.

[6] Cf. Fairbairn, *Mills and Millwork*, Vol. I, pp. vi–vii: 'The introduction of the steam engine, and the rapidity with which it created new trades, proved a heavy blow to the distinctive position of the millwright, by bringing into the field a new class of competitors in the shape of turners, fitters, machine-makers, and mechanical engineers.'

[7] For Wrigley, see above, pp. 396–405, and below, pp. 443–5.

is also evident that the long-established metal-working trades of Lancashire, especially clock-making, contributed greatly to this development. There is evidence that Lancashire craftsmen in these metal trades had a widespread reputation. Josiah Wedgwood, for example, appears to have acquired 'engine lathes' and tools from Liverpool, where he tried to get hold of a tool-maker from the clock-making trade.[1] He had been advised by Matthew Boulton to get 'a good Lancashire workman'.[2] Coming from Birmingham and the Soho factory, this was remarkable testimony to Lancashire skill. Wedgwood was very pleased eventually to secure such a craftsman, named Brown, 'a mathematical instrument maker, a wooden-leg maker, a caster of printers' types and in short a Jack of all trades', who could 'forge iron and file extremely well and cast in various metals', and whom Wedgwood proposed to employ in 'making and repairing Engine lathes, punches and tools of various sorts'.[3] One of those who supplied Wedgwood with tools and lathes was the clock-maker, John Wyke (1729-1787), of Prescot and Liverpool, who invented a wheel-cutting engine and became well known as a maker not only of watches and clocks but also of tools for this and other trades.[4] He was supplying James Watt with tools of various kinds as early as 1760.[5]

The manufacture of watch movements, watch tools, files, etc., was widespread in and around Prescot and the neighbouring villages towards Liverpool. It was, so Aikin tells us, 'much extended by improvements for first cutting teeth in wheels, and afterwards for finishing them with exactness and expedition'; their files, too, were said to be 'the best in the world'.[6] Workmen in these trades were naturally in great demand in the growing engineering industry, both in Lancashire and elsewhere. In the 1790s, for example, Boulton and Watt were asking Peter Ewart in Manchester to try to get them skilled metal-workers. But Ewart reported that it was almost impossible to get good millwrights, turners, and filers; he himself was having to 'make shift with Joiners and Carpenters. . . . The very few general good filers and turners that are here, are all engaged for a term of years in the different Cotton Mills.' He mentioned, however, two watch-

[1] J. Wedgwood to T. Bentley, 24 Aug. and 24 Oct. 1767, 22 Feb. 1768, Wedgwood Papers, Rylands Library, Manchester.
[2] M. Boulton to J. Wedgwood, 16 July 1767, B.R.L.
[3] J. Wedgwood to T. Bentley, 14 July 1768.
[4] J. Hoult, 'Prescot Watch-Making in the XVIII Century', *Transactions of the Historic Society of Lancashire and Cheshire*, Vol. LXXVII (1925), pp. 39-53; W. J. Roberts and H. C. Pidgeon, 'Biographical Sketch of Mr. John Wyke', *ibid.*, Vol. VI (1853-4), pp. 66-75.
[5] James Watt's Journal (Doldowlod), 20 March, 14 April and 17 Aug. 1760.
[6] Aikin, *op. cit.*, p. 311.

tool makers in London whom they might be able to get.[1] The wages of engineering workers in Manchester were 'almost ½ higher' than in Birmingham:[2] 'there is not a hand that is good for anything can be had here for less than 17 or 18 Shill^s pr. week'.[3] Lancashire therefore tended to attract skilled craftsmen from other areas. Thus in 1791 we find John Rennie, the famous engineer, deploring the fact to Matthew Boulton that 'in respect to workmen the Cotton Trade has deprived this place [London] of many of the best Clock Makers and Mathematical Instrument Makers so much so that they can scarcely be had to do the ordinary business'.[4] We also know that a number of workmen from the Soho works of Boulton and Watt migrated to Manchester.[5]

Clock-makers were evidently very important in the development of mechanical engineering. The lathes and wheel-cutting engines for making large clocks could easily be applied to manufacturing the moving parts of textile machinery. It is very interesting to find a reference, in an advertisement for the sale of a cotton-spinning factory in 1788, to 'the Toothed Wheels being finished in an Engine, and the Spindles, which are made of Cast-steel, hardened and ground in the same Manner as Cylinders for Horizontal Watches'. The constituent parts of the spinning machines in this mill were made with 'unerring Guages and other Tools'.[6] The tools and skills of clock-making were obviously important in the development of precision engineering. The term 'clockwork' was frequently used in reference to the mechanism of spinning machines. The working parts of Paul's roller-spinning machine (patented 1738) were said to be of a 'delicacy equal almost to that of clocks'[7] and Aikin states that the water-frame and mule were 'moved by clockwork', and that they were 'an aggregate of clock-maker's work and machinery most wonderful to behold'.[8] Several of the inventors of spinning machines were clock-makers. James Taylor, of Ashton-under-Lyne, inventor of such a machine in 1755, was a clock-maker,[9] and so too was John Kay of Warrington, who assisted Arkwright in the development of the water-frame;[10] indeed Arkwright himself claimed, very dubiously but significantly, to be a clock-maker.[11]

[1] P. Ewart to M. Boulton, 12 Dec. 1791, A.O.L.B.
[2] P. Ewart to Mr. Forman, 11 Feb. 1792, A.O.L.B.
[3] P. Ewart to J. Southern, 31 Jan. 1792, B.R.L. Dickinson and Jenkins, *op. cit.*, p. 267, wrongly quote 57 or 58 shillings.
[4] J. Rennie to M. Boulton, 19 Nov. 1791, Rennie Box, A.O.L.B.
[5] See above, pp. 410–11. [6] *Manchester Mercury*, 14 Nov. 1788.
[7] R. Dossie, *Memoirs of Agriculture, and other Oeconomical Arts* (1768), Vol. I, p 197.
[8] Aikin, *op. cit.*, pp. 172–3. [9] Wadsworth and Mann, *op. cit.*, p. 473.
[10] Baines, *op. cit.*, pp. 143 and 148–9. [11] *Ibid.*, p. 153.

On the other hand Peter Ewart, a well-known Manchester engineer and cotton spinner,[1] stated that the watch-tool and movement makers around Prescot and Warrington, though very highly skilled craftsmen, using 'the same sort of tools that the cotton machine makers use', were so specialized that 'when those men come to be employed in making cotton machines, we find that they have almost as much to learn as if they had never learnt any working in metal at all. . . . We have found them quite insufficient to do any ordinary filing and turning . . .'[2] James Lawson, Boulton and Watt's northern agent in the 1790s also considered the Prescot watch movement makers unsuitable for engineering, since 'they had only been used to small work—such as making watch hands and little watch works', while watch-tool makers could not be got 'without great wages—and none that I saw had been used to any such work as you [Boulton and Watt] wanted'.[3] The weight of the other evidence which we have discovered, however, indicates that clock-, if not watch-, makers, and above all clock-tool makers, were in very great demand for textile machine-making and contributed materially to the early growth of engineering.

The making of textile machinery must obviously have been the most important branch of early engineering in Lancashire, and the earliest specialized. Already, in the late eighteenth century, we have found references to gauges and precision engineering, which foreshadow the standardization lauded by Andrew Ure in the 1830s.[4] A great deal of the early textile machinery, however, was made under the direction of the mill-owners themselves, who frequently combined cotton-spinning with machine-making. The two activities, in fact, remained closely associated far into the nineteenth century.[5] On the other hand, there are several examples of craftsmen setting up first of all as machine-makers and then going over to cotton-spinning. It is about some such firms that we have most information. Peter Atherton, for example, who appears to have been originally an instrument-maker at Warrington,[6] and who established for himself a very high engineering reputation, also built cotton mills in both Liverpool and Manchester. When at Warrington, he was approached by Arkwright and Kay, the Warrington clock-maker, to help make the first water-frame, and 'agreed to lend Kay a smith and watch-tool maker, to make the heavier part of the engine, and Kay undertook to make the clock-maker's part

---

[1] See above, p. 99, and below, p. 441–2.
[2] *S.C. on Artizans and Machinery, Fourth Report*, in *Parl. Papers* (1824), Vol. V, p. 251.   [3] J. Lawson to M. Boulton, 28 Nov. 1797, A.O.L.B.
[4] See below, p. 478.
[5] D. A. Farnie, 'The English Cotton Industry, 1850–96' (unpublished M.A. thesis, Manchester University, 1953), chap. IV.   [6] Baines, *op. cit.*, p. 150.

of it . . .'[1] Atherton continued till his death in 1799 to manufacture textile machinery as well as cotton. In the advertisement of the sale of his Liverpool warehouses, it is very interesting to find, in addition to a large and varied assortment of carding, drawing, roving, and spinning machines,

2 capital Clock-maker's Cutting Engines, for Mill use; a Fluting Engine for Fluting Iron Rollers; . . a very valuable Cap Engine, for Watch Caps, &c. a great variety of excellent and valuable Tools for Smiths' and Mill use, consisting of Dies and Taps, Screw Stocks, &c. . . . . a great variety of Tools, for making Patent Jack Boxes, Guages, &c. a Fluting Engine, for fluting Wood Blocks for Jack Boxes; Wood and Iron Lathes, and Hand Lathes; and . . . several Wheels for turning Lathes; and an assortment of Smiths' Tools, Bellows, and Grinding Stones . . .; a large quantity of Turners' Tools, Joiners' Tools, Benches, &c. A variety of capital Tools for Clock-makers; an assortment of Files . . .[2]

James McConnel and John Kennedy, Adam and George Murray, Robert Owen, Peter Ewart, and other great Manchester cotton-spinners of the late eighteenth and early nineteenth century were also originally machine-makers and engineers. The first four of these were all from the same district of Galloway in Scotland and migrated to Lancashire in the early 1780s, to be apprenticed to the craft of machine-making with McConnel's uncle, William Cannan, of the firm of Cannan and Smith at Chowbent, near Wigan.[3] William Cannan, from the same Scottish parish, was originally a carpenter, who, after settling in Chowbent, drew there several other young Scots of the same trade; one of them, James Smith, he eventually took into partnership. McConnel and Kennedy, whose work, incidentally, included clock-making, left Chowbent in the late 'eighties and soon entered textile engineering in Manchester.[4] Almost from the beginning they combined this with cotton-spinning.[5] Adam and George Murray similarly, after their apprenticeship to machine-making with Cannan and Smith

[1] Baines, *op. cit.*, quoting J. Aikin and W. Enfield, *General Biography* (1799), Vol. I, p. 391. See also D. P. Davies, *A New Historical and Descriptive View of Derbyshire* (Belper, 1811), p. 489.   [2] *Manchester Mercury*, 17 Sept. 1799.

[3] J. Kennedy, 'A brief notice of my early recollections', in *Miscellaneous Papers*, pp. 1–18; D. C. McConnel, *Facts and Traditions collected for a Family Record* (Manchester, privately printed, 1861).

[4] They formed a partnership in 1791, together with Benjamin and William Sandford, two well-to-do fustian warehousemen, who provided most of the capital. McConnel, *Facts and Traditions*, pp. 137–8; Kennedy, *Miscellaneous Papers*, p. 17. They appeared in the 1794 Manchester directory as 'cotton-spinners and machine-makers'. See also above, p. 100, for Kennedy's scientific and technological interests.

[5] For a brief history of McConnel and Kennedy, see *A Century of Fine Cotton Spinning, 1790–1906* (Manchester, 1906), issued by the firm.

at Chowbent, also began as machine-makers in Manchester before becoming cotton-spinners.[1]

Robert Owen has left us an interesting account of how, about the same time, he first set up in business in Manchester, in partnership with a man named Jones, making cotton-spinning machinery.[2] Owen himself had no previous experience of textile engineering, having been brought up in the drapery business, but Jones was a small wire manufacturer who had acquired some knowledge of spinning machines.[3] They used 'wood, iron, and brass for their construction', and in 1792 Jones (who had by then broken with Owen) was advertising for 'a good Joiner or two accustomed to fit up Mules and Water Machinery', and for 'a good Turner, Iron Filer, and a Smith'.[4]

Peter Ewart, another of the leading Manchester cotton-spinners, brother of William Ewart, John Gladstone's partner in the Liverpool trade, was trained as a millwright under John Rennie and later secured employment with Boulton and Watt.[5] Towards the end of 1790 he was appointed Boulton and Watt's agent in Manchester, where he also set up as a millwright on his own account.[6] He erected some of the first Boulton and Watt steam engines in Lancashire mills, from which he himself secured orders for the millwork (shafting, gearing, etc.). In September 1792, however, he went into partnership with Samuel Oldknow, the famous cotton-spinner, in a bleaching and calico-printing business, which he considered more profitable than mill-wrighting;[7] but the partnership was dissolved after a year and Ewart returned to engineering. In 1798 he went into cotton-spinning as

[1] Kennedy, *Miscellaneous Papers*, p. 9; McConnel, *Facts and Traditions*, p. 133. In the 1794 Manchester directory Adam Murray is described as a machine-maker, in that of 1797 as a cotton-spinner.

[2] *The Life of Robert Owen, Written by Himself* (1857), Vol. I, pp. 22–3; W. H. Chaloner, 'Robert Owen, Peter Drinkwater and the Early Factory System in Manchester, 1788–1800', *Bulletin of the John Rylands Library*, Vol. XXXVII (Sept. 1954), pp. 78–102.

[3] Their partnership was formed in late 1790 or early 1791. At the beginning of 1791 they were informing the public that they had 'opened a Warehouse near the New Bridge, Dolefield, for making *Water Preparation* and *Mule Machines*' (*Manchester Mercury*, 18 Jan. 1791).

[4] *Manchester Mercury*, 21 Feb. 1792.

[5] W. C. Henry, 'A Biographical Notice of the late Peter Ewart, Esq.', *Memoirs of the Manchester Literary and Philosophical Society*, Vol. VII (1846), pp. 113–36; Dickinson and Jenkins, *op. cit.*, pp. 288–9; Ewart's evidence before the *S.C. on Artizans and Machinery, Fourth Report*, in *Parl. Papers* (1824), Vol. V, p. 250; see also above, p. 99, for Ewart's scientific and technological interests.

[6] J. Watt to J. Watt jun., 16 Oct. 1790; P. Ewart to J. Watt, 10 Oct. 1790, B.R.L.

[7] P. Ewart to J. Watt, 17 Jan. 1792; P. Ewart to Messrs. Boulton and Watt, 4 Aug. 1792, B.R.L.

FF

partner of Samuel Greg, later establishing his own business, but he continued his engineering interests, as is apparent from his evidence before the Select Committee on Artizans and Machinery in 1824.[1] Finally, in 1835, he became Chief Engineer and Inspector of Machinery in His Majesty's dockyards.

By the early nineteenth century, textile machine-making was becoming a large, rapidly expanding and specialized business. The directories of that time contain ever-growing lists of firms variously described as machine-makers, engineers and machinists, etc., as well as a considerable number of more specialized roller-makers, roller- and spindle-makers, spindle- and fly-makers, reed-makers, and shuttle-makers. Among them one or two well-known names eventually begin to appear, such as Sharp, Roberts & Co.,[2] but at the time when Richard Roberts came to Manchester, in 1816, there were already some reputable machine-makers there and in Salford. Amongst these, we may mention Adam Parkinson, of Sutcliffe and Parkinson, mill-wrights, machine-makers, and turners in the late 1790s, at 9 Water Street, Salford.[3] In Bancks' directory of 1800, however, Parkinson appears on his own as a machine-maker, at 28 Balloon Street; in Dean's directory of 1808–9 he is described as a 'millwright and [calico] printing machine maker, Riga-st.', while in Pigot's of 1811 he appears simply as a 'printing machine maker', at the same address, where he was now well established. Though in later directories he is again referred to as millwright as well as a machine-maker, he clearly provides a good example of engineering specialization. The importance of Parkinson's works is attested to by William Fairbairn, who was employed there as a journeyman during his first two years in Manchester, in 1813–15;[4] Fairbairn mentioned erecting printing machinery for Parkinson,[5] and there is an earlier reference in the *Manchester Mercury* to a cylinder-printing machine 'made by Parkinson';[6] he also made machinery such as squeezers for bleachworks.[7] Another firm, Jenkinson and Bow (or Bowe), of Blackfriars Bridge, Salford, also starting up at the end of the century,[8] were ranked 'very high as general machine-makers' by the early 1820s,[9] and William Jenkinson was to give some of the most

---

[1] *Parl. Papers* (1824), Vol. V, pp. 250–61.
[2] See below, pp. 478–9.   [3] See above, p. 432.
[4] W. Pole, *Life of Sir William Fairbairn*, pp. 101 and 103.
[5] *Ibid.*, p. 111.   [6] *Manchester Mercury*, 19 March 1799.
[7] D. Brewster, *Edinburgh Encyclopaedia* (1808–30), plates of bleaching machinery.
[8] John Jenkinson & Co., machine-makers, King Street, Salford, first appeared in Dean & Co.'s directory of 1804.
[9] J. Butterworth, *The Antiquities of the Town, and a Complete History of the Trade of Manchester: with a Description of Manchester and Salford* (Manchester, 1822), p. 291.

interesting evidence on textile-engineering before the Select Committee on Exportation of Machinery in 1841.[1] Their rise was paralleled by that of William and John Crighton, who appear to have been one of the leading machine-makers in Manchester in the early 1820s, with an 'extensive machine shop' in Water Street,[2] where young Joseph Whitworth started to learn the trade.[3] But these are only a few of the growing number of such firms, which also included, as we shall see, some important iron-founding and general engineering concerns. Almost all their names have long since been forgotten, and little historical evidence of their activities has survived, but they played a most important role in manufacturing the early textile machines of the Industrial Revolution in Lancashire.

Some of this early machinery was powered by horse-gins. In 1788, for instance, at Salvin's cotton factory, Garrat Lane, there was for sale 'a Good New Horse Wheel, 18 Feet Diameter, with lying Shafts, Drums, and Cog Wheels, which will turn six Carding Engines'; it was made by Thomas Leeming, machine-maker, of Bury Street, Salford.[4] Similarly in 1794 there is a reference to 'an excellent Horse Wheel', with upright shaft, gearing, etc., for driving a cotton-spinning factory in Lever Street, then being sold, which had been occupied by the firm of Wright and White, who had combined cotton-spinning with machine-making.[5] But horse-gins were soon superseded by water-wheels and steam engines, in the manufacture of which, together with gearing, shafting and pulleys, millwrights played an important role. To one of the most interesting of these, Joshua Wrigley, originally a pump-maker, we have already referred, as a maker of steam engines and water-wheels for cotton mills, etc.[6] We have seen how he supplied John Marshall, the famous Leeds flax-spinner, with a modified Savery engine and water-wheel for his Holbeck mill in 1790–1; he also appears to have carried out most of the millwork.[7] Such evidence[8] about early

[1] See below, pp. 477–8, 479–80.

[2] Butterworth, *op. cit.*, p. 246. W. & J. Crighton are listed in Pigot's directory of 1816–17 in Back Mill Street, later in Lower Mosley Street.

[3] *Manchester City News*, 25 Nov. 1865; A. E. Musson, 'Sir Joseph Whitworth: Toolmaker and Manufacturer', *Chartered Mechanical Engineer*, March 1963, p. 189. [4] *Manchester Mercury*, 15 Jan. 1788; Scholes's directory, 1794.

[5] *Ibid.*, 25 Feb. 1794. There was 'a Smith's shop' in the cellar. See also *ibid.*, 9 July 1793, when this firm was advertising for machine-makers in the manufacture of mules, carding engines, etc. Scholes's directory of 1794 lists William Wright (who was a former apprentice and workman of Arkwright: see above, p. 435) as a machine-maker and cotton-spinner at 57 Lever Street; in the following year he was in the same business at 34 Fleet Street, (*ibid.*, 24 March 1795; see also Scholes's directory, 1797). [6] See above, pp. 396–405.

[7] See above, pp. 399–400.

[8] In Marshall MS. 57, Brotherton Library, University of Leeds.

textile-mill construction is so very rare that it is worth considering in more detail. Wrigley was not the only millwright from the Manchester area whom Marshall consulted, for the latter's note-book contains a list of 'Names of Mechanicks &c.',[1] which, though headed by 'Joshua Wrigley erector of steam Eng[ine]s & Cott[on] Machinery —Man[cheste]r',[2] also includes 'Joseph Taylor maker of Cott[on] Machinery', and another such maker named Appleton, 'late servant of Wrigley's', both of Manchester,[3] together with 'Wake [Wyke] & Green—Cutting engines &c.—Liverpool', the eminent clock- and tool-making firm,[4] Samuel Lees, 'fluted Roller maker', and John Ogden, 'Spindle maker', both of Ashton-under-Lyne,[5] as well as a number of millwrights, spindle-makers, etc., in other towns. Another millwright, J. Sutcliffe (probably of Sutcliffe and Parkinson, of Manchester), was competing with Wrigley for the supply of water-wheels, and offering somewhat different opinions.[6] Wrigley, for example, considered that 'the wheel cannot be too broad nor the buckets too shallow' and considered that the shrouds should be 'not more than 8 in. because a greater bucket is so much loss of lever', whereas Sutcliffe said that 'the shrouds should not be less than 14 in. because the bucket will hold the water better to the bottom of the wheel'; Wrigley also considered that the overshot wheel should move at from 3 ft. 6 in. to 4 ft. of the circumference per second, whereas Sutcliffe favoured a faster rotation of 5 ft. or more per second.

[1] Marshall MS. 57, p. 6.

[2] That Wrigley made cotton machinery is also suggested by his earlier advertising two calender bowls for sale (*Manchester Mercury*, 12 March 1782).

[3] Peter Appleton, machine-maker, of 17 Thomas' Street, appears in Scholes's directories, 1794 and 1797, but Joseph Taylor is not listed.

[4] See above, p. 437.

[5] Samuel Lees (1759-1804) was the founder, in the 1780s, of the modern Park Bridge Iron Works of Hannah Lees & Sons Ltd. at Ashton-under-Lyne, long famous for its manufacture of 'fluted and special rollers for textile and other machinery', and more recently for 'hot rolled and bright steel bars'. See J. Beckett's articles on this firm in the *Oldham Chronicle*, 20 and 27 Jan. 1962, where it is also stated that when Samuel Lees, a whitesmith (or tinsmith) by trade, moved to Park Bridge, 'there already existed a small textile roller spindle concern carried on by two brothers named Ogden'.

Lees and Ogdens crop up repeatedly in the early history of the cotton trade and textile engineering. It was another Samuel Lees who at this same time, founded the later-famous firm of Asa Lees & Co. Ltd.; started at Holts Mill, Lees, near Oldham, the business was transferred in 1816 to the Soho Iron Works, Oldham, where rollers and spindles were also the chief products. By 1822 this works was employing '120 or 130 hands . . . in the roller-making business', with a warehouse in Manchester. J. Butterworth, *op. cit.*, pp. 244-5; E. Butterworth, *Historical Sketches of Oldham* (1856), p. 184. Jonathan Ogden was apparently 'the first individual who established a machine-making workshop in the village of Oldham'. E. Butterworth, *op. cit.*, p. 127.   [6] Marshall MS. 57, p. 4.

Wrigley's water-wheels and millwork were still, for the most part, massively built of wood. It was not till some years later that a revolution occurred in this field with the introduction of slender wrought-iron shafting, etc.[1] Thus Wrigley proposed an axle '16 ft. long and 2 foot Square' for a water-wheel 16 ft. in diameter and 9 ft. wide.[2] For 'pullies' on driving shafts, 'J. Wrigley says the best size & now most generally used is 14 in diam[ete]r',[3] while for the main 'upright shaft', providing the drive for the machinery on the various floors of the mill, 'J. Wrigley recommends Oak 18 in. diam[ete]r for two Stories, Iron 5½ square upwards'; but this was later amended, Wrigley deciding that timber 'is better than iron because [gear] wheels are fixed on firmer' and so recommending 'Deal 18 in. first floor 16 [in.] —2nd 14 [in.]—3rd'.[4] Wrigley also gave advice on gearing[5]—though it is not clear how far he went along with Marshall in his efforts to determine geometrically the correct pitch and depth of the teeth[6]— and generally on the transmission of power from the water-wheel throughout the mill, including the speed of spinning operations.[7] In addition to supplying the water-wheel and steam engine, moreover, he advised on the boiler for the latter:[8] 'J. Wrigley says it should be 4 times Diam[ete]r of Cyl[inde]r', compared with Boulton and Watt's 3½ times. Marshall mentions (with a sketch) a new type of boiler now being made by Wrigley,[9] but again he obtained information from other manufacturers.

Wrigley was only one of a rapidly increasing number of millwrights and general engineers in Manchester at this time. One of the most out-standing was Thomas C. Hewes, to whom we have referred several times previously.[10] Notable for his knowledge of mechanics, hydraulics, etc., he was very active in the 1790s and early 1800s in constructing water-wheels, millwork, and spinning machinery not only in Lancashire but also in many other parts of the country; he introduced improvements in water-wheel design and construction (especially the 'suspension wheel', built mainly of cast and wrought iron), which were further developed later in the nineteenth century by his one-time draughtsman, William Fairbairn,[11] who came to him after working

[1] See below, pp. 446, 462–3, 481.   [2] Marshall MS. 57, p. 6.
[3] *Ibid.*, p. 9.   [4] *Ibid.*, p. 18.
[5] *Ibid.*, pp. 24–5.   [6] See above, p. 154.
[7] Marshall MS. 57, p. 17.   [8] *Ibid.*, p. 23.
[9] See above, p. 404.   [10] See above, pp. 69–71, 98, 436.
[11] Fairbairn was employed by Hewes in 1816–17. He left after some friction caused by his competing with his employer in designs for the new Blackfriars bridge over the River Irwell between Manchester and Salford. Pole, *op. cit.*, pp. 104, 105–6, and 111. For Fairbairn's later achievements, see below, pp. 480 ff.

for Adam Parkinson. Hewes' reputation seems to have been based mainly on his water-powered mills, but he appears also to have erected some steam-powered ones.[1]

The scale of Hewes' activities can be gathered from his evidence before the Select Committee on Artizans and Machinery in 1824.[2] He then employed about 150 men, of whom about 40 were 'employed on heavy mill work, and about 110 on machinery', and he erected entire fire-proof mills, including the frame of the building, water-wheels, gearing, shafting, etc., and also the spinning machinery, though he obtained rollers and spindles from specialized 'roller makers and spindle makers'. He referred to mills built and machinery installed not only in Lancashire and Yorkshire, but also at Tiverton in Devonshire, in Gloucestershire, at Aberdeen in Scotland, and at Belfast, Bandon (near Cork), and other parts of Ireland; he had even exported a water-wheel to America.

In the construction of every description of mill-geering, as well as of spinning and other machines, he evinced abilities of the highest class, and in the planning and erecting of mills of various kinds, as well as other works of magnitude, he displayed a skill and a soundness of judgement that would alone have been sufficient to stamp his name in the remembrance of practical men.[3]

He was particularly interested in 'the introduction of iron, under divers modifications and arrangements, as a substitute for wood, and his exertions to that end were remarkably bold and successful'.[4] It seems highly probable that, in building textile mills, Hewes used not only cast-iron framing to render them 'fire-proof', but also wrought-iron shafting, in place of the earlier massive timber shafts, just as he replaced timber with cast- and wrought-iron in his water-wheels, and as he also used iron in bridges. In all these innovations he appears to have preceded Fairbairn, to whom most of the credit has generally been given. Andrew Ure's admiring comments in 1835 on Fairbairn's capacity to erect and power mills and fill them with machinery[5] might well have been applied many years earlier to Hewes. Ure did, indeed, admit that the revolution in millwork 'had commenced . . . before

---

[1] *Manchester Mercury*, 5 Feb. 1799: 'To be Let . . . A Spinning Factory and Steam Engine lately erected and very little used, with Upright and Tumbling Shafts, Drums &c . . . For further Particulars enquire of Mr. Thomas C. Hewes, Machine-maker, at his Factory Top of Portland-street . . .'

[2] *Parl. Papers* (1824), Vol. V, pp. 340–50.

[3] *Manchester Guardian*, 11 Feb. 1832.

[4] *Ibid.*

[5] A. Ure, *Philosophy of Manufactures* (1835), pp. 32–7; see below, p. 481–2.

their [Fairbairn & Lillie's] time',[1] and Fairbairn himself acknowledged his former employer's achievements.[2]

Another interesting, though less important, figure in Manchester in Hewes' time was Isaac Perrins, like Ewart a millwright and former employee of Boulton and Watt[3] (and also a well-known prizefighter in his day). He erected the first Boulton and Watt rotative engine in Manchester, in Drinkwater's mill in 1789, and after assisting with others moved permanently to Manchester about the end of 1793, took a public house,[4] and carried on erecting for Boulton and Watt. He was eventually dismissed for drunkenness and other irregularities, but continued to do general millwork as part of his own business.[5]

In addition to these millwrighting and 'machine-making' or textile-engineering firms, there were in Manchester and Salford a number of iron-founding and forging firms which carried on a more general engineering business.[6] The smelting and working of iron had been carried on in Lancashire for centuries, especially in the Furness peninsula, one of the chief centres of the charcoal-iron industry in the eighteenth century, producing not only pig and bar iron but also a great variety of forged and cast-iron goods, from kitchenware to cannon.[7] South Lancashire, however, produced a comparatively small quantity of iron. The Haigh ironworks was probably the most important centre of production.[8] Lesser known, but of some importance for Manchester, were Dukinfield furnace and foundry. Richard Crowder, of Dukinfield Lodge, was advertising this ironworks for sale in 1782–

---

[1] *Ibid.*, p. 36. See above, pp. 24, 69, 73, 76, for the contributions of Smeaton and Rennie. The shafting and gearing in Rennie's Albion Mill, built in the 1780s, had been of wrought and cast iron.

[2] See above, p. 70. In the latter part of his career, Hewes formed a partnership with Henry Wren, the firm becoming Hewes and Wren. After Hewes's death in 1832, the business was carried on by Wren, with a new partner, as Wren and Bennett. It was from this firm that James Nasmyth rented premises when he first set up as an engineer in Dale Street, Manchester, in 1834. J. Nasmyth, *Autobiography* (ed. S. Smiles, 1883), pp. 185–9.

[3] See above, pp. 101, 419.

[4] See above, p. 433.

[5] There are innumerable references in the Boulton and Watt papers to Perrins's work in Manchester. He also became chief of the town's fire brigade, in which his knowledge of steam pumping engines would be of considerable advantage; he lost his life in a fire accident in 1801. A. Redford, *The History of Local Government in Manchester* (1939), Vol. I, pp. 218–19.

[6] Several of these were important in the manufacture of steam engines. See above, chap. XII.

[7] Tupling, *op. cit.*, pp. 1–13; B. G. Awty, 'Charcoal Ironmasters of Cheshire and Lancashire, 1600–1785', *Transactions of the Historic Society of Lancashire and Cheshire*, Vol. CIX (1957), pp. 71–124.

[8] A. Birch, 'The Haigh Ironworks, 1789–1856', *Bulletin of the John Rylands Library*, Vol. XXXV (March 1953), pp. 316–33.

1783.[1] Both ironstone and coal were available, from which pig-iron and castings were produced; the latter included bored pipes, furnace bars, bearers, fire doors, and frames, and 'all Sorts of large Iron Work'. Several hundred tons of pig-iron were to be disposed of at that time. This works was shortly afterwards taken over by the Manchester and Salford iron-founding firm of Bateman & Co.[2] The bulk of the pig-iron supplies for the Manchester foundries, however, was brought in from outside, chiefly from the Midland iron-producing areas. Aikin (1795) tells us: 'The quantity of pig iron used at the different foundries in Manchester within these few years, has been very great, and is mostly brought (by canal carriage) from Boatfield [Botfield] and Co's iron furnace, Old Park, near Coalbrook Dale; and Mr. Brodie's furnace, near the Iron Bridge, both in Shropshire.'[3]

In Manchester and Salford, only one iron-founding firm appears in the first directory of 1772, that of Thomas Ryder (or Rider), Scotland Bridge, not far from the junction of the River Irk with the Irwell. This firm was apparently associated with Marston Forge, near Northwich, Cheshire, from which Nicholas Rider was supplying boiler plates for a Boulton and Watt engine in 1779.[4] Thomas Ryder died, however, in 1779[5] and the Scotland Bridge foundry passed into the hands of John Fletcher.[6] But he, too, soon died, in 1785,[7] and later, in the 1780s and 1790s, the 'Old Iron-Foundery', as it came to be called, at Scotland Bridge or Red Bank, is shown in the possession of Phoebe Fletcher.[8] The name of Ryder disappears from the Manchester scene until 1796, when Marston Forge was offered for sale and interested parties were asked to apply for particulars 'to Mr. Ryder or Mr. Nelson, on the Premises', or to 'Mr. Thomas Ryder, Salford Foundery, Manchester'.[9] They appear to have been unable to find a buyer, however, for a few months later we find that Thomas Ryder and George Nelson, 'having engaged the Iron Forge, at Marston, near Northwich', were now

[1] *Manchester Mercury*, 9 July 1782 and 29 April 1783.
[2] See below, p. 449.
[3] Aikin, *op. cit.*, pp. 177–8. For Brodie, see below, pp. 451–2. Iron was also imported from Cheshire and Yorkshire. See Awty, *op. cit.*
[4] See above, p. 424.
[5] Lancs. & Chesh. Rec. Soc., *Index to Wills*, Vol. XXXVIII.
[6] Raffald's directory, 1781: 'Fletcher John, iron founder, Scotland bridge'.
[7] Lancs. & Chesh. Rec. Soc., *Index to Wills*, Vol. XXXVIII.
[8] Holme's directory, 1788: 'Fletcher Phoebe & Co. iron forgers and founders, Red bank'. Scholes's directory, 1794: 'Fletcher Phoebe, Old Iron-foundery, 2 Foundery-lane, Red-bank'. Aikin, *op. cit.*, p. 177, lists 'Fletcher Phoebe, Old Iron-foundry, 2 Foundery-lane, Red bank', in his list of foundries operating in Manchester and Salford in 1795. The foundry is also clearly marked on Green's map of Manchester and Salford in 1794.
[9] *Manchester Mercury*, 19 April 1796.

offering for sale bar-iron, cast-iron goods, salt-pans and salt-pan plates, steam-engine boilers and boiler plates, other wrought-iron pans and boilers, and 'Smiths' work of any description'.[1] Where Ryder's 'Salford Foundery' was we cannot definitely say, but in 1804 John Ryder, iron-founder, appears at 6 Ravald Street, Salford, and later, in 1808–9, at Islington, Salford,[2] while in 1816–17 Thomas Ryder again comes on the scene.[3]

Meanwhile the 'Old Iron-Foundry' at Scotland Bridge passed into new hands. In 1797, the directory entry changes to 'Fletcher & Silcock';[4] the latter, Jacob Silcock, previously described as 'agent to the Griffin iron-foundery',[5] now appears as 'iron-founder', with a house near to the 'Old Foundry' on Red Bank.[6] Soon, however, he seems to have withdrawn, after Phoebe Fletcher's son, Thomas, entered the business.[7] But by 1808–9 the foundry had apparently been sold to Radfords and Waddington,[8] and Fletcher disappears from the scene.

The most important iron-founding and general engineering concern in Manchester and Salford in the later eighteenth century was that of Bateman & Co., later Bateman and Sherratt.[9] James Bateman (1749–1824) came of a Westmorland landed family, at Tolston Hall, near Kendal, but gave up his birthright to set up in the ironmongery business in Salford.[10] In the 1773 Manchester and Salford directory he is found in partnership with Walter Wilson, ironmonger, 24 Deansgate. Later directories in the 'eighties show him greatly expanding his business interests, to become also an iron-founder, Water Street, Salford; an iron-forger, Collyhurst; and an iron-founder and forger, Dukinfield. He appears to have entered into partnership with William Sherratt in the late 'eighties and in 1791 they built a new foundry in Hardman Street, Salford. In later directories they were also described as coal merchants. By the early 'eighties this firm was producing a wide

[1] *Ibid.*, 4 Oct. 1796.   [2] Dean & Co.'s directories, 1804 and 1808–9.
[3] In Pigot's directories, 1816–17, he appears as an iron-founder, at St. George's dock, near Oxford Road, with a house in Lower Mosley Street. See also the directories for the next few years.
[4] Scholes's directory, 1797.
[5] *Ibid.*, 1794. See below, p. 451, for the Griffin Foundry, established by Ebenezer Smith & Co. of Chesterfield.
[6] *Ibid.*, 1797. See also *Manchester Mercury*, 22 Jan. 1799, for an advertisement by 'Jacob Silcock, Old Iron Foundry, Scotland Bridge, in Manchester'.
[7] Bancks's directory, 1800, shows 'Fletcher Phoebe and Son' at the 'Old Iron Foundry', but in Dean & Co.'s, 1804, Thomas Fletcher only appears.
[8] See below, p. 454.
[9] See above, pp. 407–23, for a detailed account of this firm's very extensive steam-engine manufacture.
[10] F. S. Stancliffe, *John Shaw's, 1738–1938* (Manchester, 1939), pp. 45–6. Bateman was president for many years of John Shaw's Club in Manchester.

range of cast and forged iron goods, including machine parts for textile and other mills:[1]

Etna Foundery, Manchester, January 1, 1782. James Bateman, Iron-Monger, Deansgate, Manchester, Begs Leave to acquaint his Friends and the Public, That he has lately erected an Iron-Foundery, and begun to Manufacture a great Variety of elegant Pantheon, Bath, and other Stove Grates; Cast Iron Kitchen Grates, Cylander, Octagon, and the new invented Hob or Side Oven, Hot Hearths, Ironing Plates, Laundry Stoves, Pots and Pans of all Sizes in Sand or Loam, for Chymists, Soap Boilers, Crofters, Dyers, &c. Stoves and Pipes of all Kinds for Warehouses, Printers, &c. Furnace Doors and Bars, Velvet Irons, Clock and Sash Weights, Weights of all Sizes adjusted by an exact Standard, Box, Sad and Hatters Irons and Heaters, Waggon, Cart and Chaise Bushes, Iron Wheels for Coal Waggons, Gins and Barrows, Gudgeons and Spindles for Mills, Cogg Wheels of any Size, all Kinds of Shafts, Wheels, Pinions, and Rollers, Paper Screws, Boxes and Rolls for Copper and Slitting Mills, Garden Rollers, Calender Bowls, Iron Gates and Railing, Iron Doors and Chests, Brass Steps for Mills, and all other Castings to Patterns or Dimensions, with many Articles quite new in the Cast Iron Way.

N.B. The best Price given for Scrap Iron and old Cast Metal.

This advertisement was almost certainly for Bateman's foundry in Water Street, Salford. Shortly afterwards he negotiated with Boulton and Watt for a steam engine to work a boring engine, a small tilt-hammer, a grinding stone, and two turning lathes, and also to blow a dragon furnace.[2] The negotiations fell through, but there is little doubt that Bateman very soon installed a steam engine, probably of his own manufacture, for working his engineering equipment. After acquiring Dukinfield furnace and foundry and building the new foundry in Hardman Street, Salford, Bateman and Sherratt had a very considerable productive capacity. This, as we have previously described, was used principally for the manufacture of steam engines, both atmospheric and 'pirated' Watt engines. Indeed, Bateman and Sherratt almost certainly built more steam engines for Lancashire firms than did Boulton and Watt.[3] Aikin was obviously very impressed by their 'considerable iron foundry . . ., in Salford, in which are cast most of the articles wanted in Manchester and its neighbourhood, consisting chiefly of large cast-iron wheels for the cotton machines; cylinders, boilers and pipes for steam engines; cast ovens, and grates of all sizes'. Most of the steam engines in the Manchester area, says Aikin, were of their make.[4] They may also have made water-wheels.[5] Their applica-

---

[1] *Manchester Mercury*, 1 Jan. 1782.   [2] See above, p. 409.
[3] See above, p. 423.   [4] Aikin, *op. cit.*, pp. 176-7.
[5] *Manchester Mercury*, 12 Feb. 1788, where they were advertising a water-mill to let, at Ashton Mills, near Ashton-under-Lyne.

tion of iron to new structural purposes is also illustrated by their making a large 'circular iron weir, near ninety feet span', cast at Dukinfield furnace, for John Arden, Esq., near Stockport, and said to be 'the first of the kind ever made in this kingdom . . . it will be found more durable and cheaper than either wood or stone'.[1]

They were not the only steam-engine makers in Manchester. From the early 1780s, as we have seen, Joshua Wrigley and also Joseph Young, millwright, were producing engines, mostly of the old Savery type, to pump water for driving water-wheels in textile mills.[2] The expansion of the Lancashire engineering market also caused firms in other areas to establish branch works in Manchester. Ebenezer Smith & Co., of the Griffin Foundry, Chesterfield, established a 'New Foundry' in Swan Street, near the Shudehill pits, in 1789 and were advertising their pig-iron and a wide range of products 'in Forged or Cast Iron'.[3] In addition to a long list of household goods, such as pans, stoves, ovens, grates, etc., they were also selling

Cylinders, Iron and Brass Working Pieces, Boilers, Chains, Regulators, Pipes, and every other Article for Fire Engines, &c. Anvils, Hammers, Rollers, &c. for Forges and Rolling Mills. Large Screws, with all kinds of wrought and cast iron Work for Cotton, Paper, Rasping & Oil Mills. Cogg and Water Wheels . . . Crofters and Dyers Pans and Bottoms . . . Calendar Bowls and Heaters . . . Press Screws, . . .

This firm, as we have seen, was closely associated with Francis Thompson in the manufacture of atmospheric steam engines, and also made parts for Boulton and Watt engines besides fitting up millwork. They carried on business in Manchester well into the nineteenth century.

Another firm which entered the Manchester market was that of the Scotsman, Alexander Brodie (1732–1811), ironmaster and armaments manufacturer of Calcutts, Broseley, Shropshire, and Carey Street, Lincoln's Inn Fields, London.[4] His first appearance in the Manchester directories is in 1794, when he owned 'St. George's iron foundery, Knot-mill', in partnership with two others, McNiven and Ormrod.[5]

---

[1] *Ibid.*, 11 Aug. 1789. [2] See above, pp. 396–406.
[3] *Manchester Mercury*, 27 Feb. 1787. See above, p. 424. See P. Robinson, *The Smiths of Chesterfield* (Chesterfield, privately printed, 1957), for a short history of this firm, which mined iron ore and coal and operated several blast-furnaces, a forge, and a boring mill in the Chesterfield area. They were the biggest producers of pig-iron in Derbyshire in the early nineteenth century, with an annual output of 2,600 tons. J. Farey, *General View of the Agriculture and Minerals of Derbyshire* (3 vols., 1811), Vol. I. p. 397. [4] See above, p. 425.
[5] Scholes's *Manchester and Salford Directory* (1794). 'Messrs. Brodie & Co.'s Iron Foundery' is clearly marked on Green's map of Manchester and Salford in 1794, at Knot Mill, beside the River Medlock, facing the Duke of Bridgewater's warehouse at Castle Quay.

As we have seen, he was one of the main suppliers of pig iron to Manchester foundries.[1] Aikin states that he was 'well known for his very extensive manufactory of grates and stoves, as well . . . for kitchens and dining rooms, as ships'.[2] Brodie is said to have made 'many thousands' of pounds out of his 'patent Stove', though he was not the inventor.[3] He also appears to have carried out general engineering work, including piracies of Watt's engine at his Calcutts works, if not in Manchester.[4]

Ormrod, Brodie's partner in the St. George's iron-foundry, Knot-mill, was most probably Oliver Ormrod, who in the late 1780s was a plumber and glazier, and then, in the following decade, a brazier and brass-founder at 83 Market-street-lane. By the early years of the nineteenth century he had transferred the business to 49–50 High Street, where he was in partnership with his son, Richard, still as 'braziers, brass-founders, &c.'. By about 1810 he appears to have retired and his son then established the St. George's iron and brass foundry in Minshull Street, near London Road, which soon became one of the biggest engineering firms in Manchester.[5] Its size and the variety of its manufactures greatly impressed the Swiss steel manufacturer, J. C. Fischer, in 1825.[6] The works, built on both sides of the Rochdale Canal, was then turning out textile and milling machinery, steam engines, boilers, stoves, wheels, coupling boxes, weighing machines for bleachers and paper-makers, rails for collieries, ornamental railings, gas-generating plant, and steam-heating plant. Ormrod proudly showed Fischer his collection of wooden patterns (for castings) and declared: 'I challenge

---

[1] See above, p. 448.   [2] Aikin, *op. cit.*, p. 177.

[3] Colonel T. Johnes to M. Boulton, 29 June 1794, A.O.L.B. See *Repertory of Arts and Manufactures*, Vol. VII (1797), pp. 22–5, for the specification of the patent, dated 8 Dec. 1780, granted to 'Mr. Alexander Brodie, of Carey-street, Chancery-Lane, in the County of Middlesex, Whitesmith; for his invention of a Ship's Stove, Kitchen, or Hearth, with a Smoke-Jack, and Iron Boilers'. John Galloway, an eminent nineteenth-century Manchester engineer, whose foundry was built on the foundations of Brodie's, states that Brodie was 'the maker of a new stove for ships, and had a large connection, especially with Government'. Galloway's MS. 'Reminiscences' (Manchester Central Reference Library), p. 20. According to Brodie's obituary notice in the *Gentleman's Magazine* (1811), Part I, p. 89, he 'possessed an immense property'.

[4] See above, p. 425. As we have previously noticed, moreover, he had a share in the Hazledine Foundry at Bridgnorth, which was manufacturing Trevithick's high-pressure engines in the early nineteenth century. The St. George's Foundry may well have been one of those producing such engines in Manchester at that time. See below, p. 464.

[5] Manchester and Salford Directories for 1788 (Lewis's), 1794 and 1797 (Scholes's), 1802, 1808–9, and 1811 (Dean & Co.'s).

[6] W. O. Henderson, *J. C. Fischer and his Diary of Industrial England 1814–1851* (1966), pp. 62–3.

the whole world to show me anything as good as this. These patterns have cost me £10,000, and, as you can see, I keep them in a completely fireproof building.' The size of the firm's contingent in the Coronation procession of 1821—with separate sections of brass-workers, smiths, turners, fitters, filers, engineers, millwrights, pattern-makers, moulders, and boiler-makers, each section walking three abreast—indicates that they had a very considerable labour force.[1]

Aikin states that in 1795 there were two more iron-foundries in Manchester and Salford, in addition to those of Phoebe Fletcher, Bateman and Sherratt, Smith & Co., and Brodie & Co., making a total of six. These others were 'Bassett and Smith, Shooter's-brook iron-foundry, Ancoats-lane' and 'Smith William, iron-foundry, 23 Lee-street'.[2] These two firms are also listed in the local directories of the 1790s,[3] but do not appear to have been very important or to have lasted long. The Shooter's Brook foundry was eventually, in the 1820s, to become the site of Fairbairn and Lillie's great engineering concern, in Canal Street, off Ancoats Lane.[4]

In addition to these iron-founders, there were also many iron-mongers, brass-founders, coppersmiths, and other metal-working firms in Manchester and Salford, which doubtless supplied machine parts. A good deal of mill-work, however, was imported from other areas. It was reported in the 'nineties that 'many of the castings are done at [Smith and Co.'s works at] Chesterfield, some at Low Moor [Ironworks, Bradford] and some at [Samuel Walker & Co.'s] Rother-ham'.[5] Similarly, Peter Ewart's letters while agent for Boulton and Watt in Manchester show that most of the boilers for their engines were made by non-Lancashire firms, usually by John Wilkinson at Bersham, though a few were made in Manchester.

By the end of the eighteenth century, however, several more iron-foundries had been established in Manchester. Two of these, Peel and Williams and Galloway, Bowman and Glasgow, were soon to be among the biggest in the town. The latter's importance has already been revealed by Dr. Chaloner,[6] and we shall deal with the former in the next chapter.[7] Another, Radfords & Co., has hitherto remained in

---

[1] *Authentic Particulars of the Processions in Manchester and Salford, on Thursday the 19th of July, 1821, in Celebration of the Coronation of . . . King George the IVth* (Chetham's Library, Manchester); *also Manchester Guardian*, 21 July 1821.
[2] Aikin, *op. cit.*, p. 177.
[3] Scholes's 1794 directory also includes 'Gent John, iron-founder, house, Great Ancoats-street', but he may possibly have been a foundry manager.
[4] See below, p. 483.
[5] J. Lawson to M. R. Boulton, 15 Nov. 1797, B.R.L.
[6] See below, p. 459, n. 4.   [7] See below, chap. XIV.

oblivion, though it was of comparable size and of earlier establishment. As early as 1770 we find Thomas Radford, 'Brazier and Brass Founder, in Market-street-lane, Manchester', advertising for a journeyman clock-maker and also an apprentice to the watch- and clock-making business.[1] It is possible that he was already expanding from brass-founding into machine-making, though he continued to appear as a brazier and brass-founder, at the same address, in later directories.[2] He appears to have died in 1790, when his brass-founding business was taken over by Oliver Ormrod and George Stirrup,[3] but his wife, Elizabeth, and his sons carried on his other business activities, for they appear during the 'nineties as iron-dealers and iron-mongers in Hanging Ditch,[4] and then, in the early 1800s, formed a new partnership, Radfords and Waddington, which, in addition to continuing the ironmongery trade at that address, soon acquired the 'Old Iron-Foundry' at Scotland Bridge or Red Bank.[5] Their business grew so considerably that, soon after 1817, they established a new foundry—the 'Waterloo Foundry'— at the side of the Rochdale Canal, in Waterloo Street, David Street, off Portland Street, and disposed of the old foundry at Red Bank.[6] Rating assessments of 1820[7] show that their Waterloo Foundry was of about the same size as Ormrod's St. George's Foundry and Peel and Williams' Phoenix Foundry, and considerably larger than the works of Hewes and Wren, in addition to which they still had a large workshop for brass-founding, etc., in Hanging Ditch. Later they appear as engineers, machine-makers (with a special line in weighing machines), 'paper bowl manufacturers', and millwrights, as well as iron-founders.

Thus by about 1825 Manchester had a string of great iron-foundries and general engineering works, from Shudehill to Great Bridgewater Street, mostly along the line of the Ashton and Rochdale canals, including the works of Peel and Williams, Ebenezer Smith & Co., Fairbairn's, Hewes and Wren, Ormrod's, Radfords and Waddington, Galloway & Co., soon to be joined by Sharp, Roberts & Co. and

---

[1] *Manchester Mercury*, 29 May 1770.
[2] Raffald's directories, 1772, 1773, 1781; Holme's, 1788.
[3] *Manchester Mercury*, 26 Oct. 1790.
[4] Scholes's directories, 1794 and 1797; Bancks's, 1800.
[5] This partnership first appears in Dean & Co.'s directory of 1804, at which time the Scotland Bridge foundry was still owned by Thomas Fletcher; but Dean & Co.'s directory of 1808–9 shows the foundry taken over by Radfords and Waddington. The Radfords' partner was David Waddington (Pigot's directory, 1811).
[6] Directory and newspaper evidence shows them still at Red Bank in 1817. The 'Old Iron-Foundry' then appears to have been converted into a dye-works, being shown as such on Bancks & Co.'s map of 1831.
[7] See below, p. 466.

Whitworth's; while in Salford there were also large foundries, such as those of the Sherratts, the Hattons and others. And along with these great firms were dozens of other iron-and brass-founders, machine-makers, millwrights, and more specialized engineering concerns. Manchester and Salford, in fact, were the centre not only of the world's cotton industry, but also of the world's engineering.

These developments were not confined, however, to these two towns. One need only mention such outstanding firms as Fawcett's of Liverpool; Asa Lees and Hibbert and Platt's of Oldham; Dobson and Barlow and Hick, Hargreaves & Co., of Bolton; the Haigh Ironworks at Wigan—all originating in the late eighteenth and early nineteenth centuries—to show that engineering was growing with similar vigour in all the chief Lancashire towns. Evidence from newspapers and wills, as well as local histories such as Butterworth's, of Oldham, also indicate that the growth of these well-known firms was similarly accompanied, or preceded, in these towns by the rise and fall of a multitude of obscure machine-makers, engine-makers, millwrights, etc.

This survey of early engineering in Lancashire, particularly in Manchester and Salford, has demonstrated that by the early nineteenth century a very remarkable development had occurred in this area. Wood- and metal-workers of all kinds, using lathes, drills, wheel-cutting engines, and other tools, had been recruited into iron-founding and machine-making firms, which were producing textile machinery, water-wheels, steam engines, and other engineering goods. In textile machine-making especially, owing to the rapidly growing size of the market, a high degree of specialization was developing. This striking development of Lancashire engineering coincided with the accelerated growth of the cotton industry from the early 1780s onwards. No doubt the roots of this technical development lie deeper in history, in the skills and tools evolved by clock-makers, millwrights, smiths, and woodworkers in earlier years, but mechanical engineering as a distinct industry seems to have emerged in the last quarter of the eighteenth century. Thus Lancashire established the necessary technological basis for what Professor Rostow has called 'take-off' into sustained industrial growth.

Appendix: selections from the 'Manchester Mercury', 1770–1800, illustrative of the origins of early engineering workers in Lancashire:
26 September 1775: Aaron Ogden, whitesmith, Ashton-under-Lyne, advertised for journeymen whitesmiths 'accustomed to Jobbing, Kitchen Furniture, or Engine Work', and for an apprentice to these branches of the trade and to 'Jenny Spindle Making'.

9 February 1779: Lewis and Furnevall, 'Joiners and Cabinet-Makers', of Hill-gate, Stockport, 'Make and sell, all sorts of Spinning Machines ... from Sixty spindles to fifteen score each. Likewise Machines for Winding Cotton Warp, of an entire new Plan.'

15 February 1780: 'The Stock in Trade of a Turner, consisting of Two Wheel Lathes, and Four Foot Lathes, and other Utensils, proper for Turning Wood, Iron, or Brass ... For further Particulars, enquire of Benjamin Ashton, in Cold-house, Manchester.'

23 July 1782: 'Employment for Clock-Makers ... Robinson and Walmesley, Pall Mall, Manchester. N.B. Would be more agreeable if they have been before employed in Cotton Works.'

22 October 1782: 'Wanted, Wood Turners, Iron Turners, and Filers ... Messrs. Wm. Douglas and Co. [cotton spinners] at the Old Hall, Pendleton.' Later Douglas and Co. advertised for joiners 'to be employed constructing Cotton Mill Machinery' (7 June 1785).

29 March 1785: Dissolution of the partnership of Berry and Johnson, 'Joiners and Cotton Machine-makers', Chowbent.

27 September 1785: Advertisement by the owners of a cotton mill near Settle, Yorks., regarding the flight of a hired servant, 'by trade a Clock smith', lately employed as a 'Filer and Turner', now believed to be in the Manchester neighbourhood.

11 March 1788: 'To be Sold, A large Wheel Leath [Lathe], Suitable for Turning large Calender Rollers of Wood or Metal. One Ditto smaller, for turning Bengal Rollers, &c. One Ditto Foot Leath. One Ditto Pole Leath ... Abraham Clegg, Timber-merchant, Shude-hill, Manchester.'

1 July 1788: Sale of cotton factory at Birkacre, Chorley, including 'One large Laith, [driven] by Water; One Foot Ditto; ... Smith and Clock-makers Tools', including a 'Cutting and Fluting Engine'.

15 July 1788: Bankruptcy of Joseph Taylor and John Harker, 'Joiners and Engine-makers', Dole Field, Manchester.

24 February 1789: Advertisement by Thomas Porthouse, of Darlington, for 'Whitesmiths, Wood Turners, and Men that have been used to fitting up Cotton Machinery'.

4 August 1789: 'Wanted, Two or Three Joiners, and Two Filers ... Peter Whitaker, Cotton, Worsted, and Flax Machine Maker, Toad-lane, Manchester.'

8 September 1789: Advertisement by T. Grocott, Calender-street, Manchester, regarding his 'Turning Business ... both in Wood and Metal of all Kinds ... He also makes and repairs Calenders, Calender Bowls, and Printing Rollers of Wood and Metal.'

15 December 1789: 'Wanted Clockmakers and Turners, Enquire of Philip Chell, Machine-maker, Top of Deansgate, Manchester.'

4 January 1791: 'Wanted . . . a Smith, who has been accustomed to Forge every Article belonging to Water Machinery.

25 January 1791: 'Wanted, A Man who perfectly understands making of Mule Frames . . . and also Turning of Brass . . . Parker and Holland, Half-moon street, Manchester.'

17 January 1792: 'Wanted, Two or Three Spindle and Fly Forgers . . .'

31 January 1792: 'Wanted, Journeymen Joiners, and one or two that are perfect in fitting up water machinery and mules. Likewise, several Clock Makers, to fitt up the clock work for mules and water machines. A good Iron and Wood Turner—and a capital Smith and Filer, for spinning frames and mules. Also a Carder for mule preparation, and a person that can make even Rovings, on a stretching frame . . . Apply at Ford and Almond's Brass Foundry, Old Church Yard, Manchester.' John Ford was originally a tin-plate worker, Bottom Smithy-door (1772 *Manchester Directory*); began making 'Tin Rollers for Spinning Jennies' (9 May 1780); also became a brazier and brass founder (25 Aug. 1789, 16 February 1790); and was now not only able to make complete spinning machines, but was also apparently engaged in the manufacture of mule rovings. He died in October 1798, but the business was carried on by his widow and son (16 and 23 October 1798).

17 February 1792: 'Wanted . . . Two Journeymen Engine and Jenny Makers . . . James Hargreaves, Engine and Jenny Maker, Blackburn.'

30 October 1792: 'Wanted. Clock-makers and Joiners . . . Holywell Cotton Twist Company's Warehouse, Sussex Street, Manchester.'

17 September 1793: Sale of effects of John Howard, 'Whitesmith and Machine-maker', Little Hayfield, including roller-making machine, fluting engines, and several carding machines, drawing frames, mules, cotton roving, and yarn.

19 November 1793: 'Wanted for a Cotton Factory. A capital Joiner and Clock Maker, who has been accustomed to fit up Spinning Frames, Mules, and Preparation Machines . . . James Sheldon and Co., Old Quay, Manchester.' This firm later, as machine makers, advertised mules, carding engines, etc. (24 February 1795).

26 August 1794: Sale of Holt Town Cotton Mills, belonging to David Holt, including 'a large Quantity of excellent Materials and Implements, for the Purpose of Machine Making, and which consist of a large Assortment of dry Timber . . . Iron and Brass, Smiths, Joiners, and Clockmakers Tools'.

11 November 1794: 'Wanted, A Brass Founder. One who has been accustomed to Casting for Machinery . . . Robert Atherton, Machine Maker, near Mottram in Longdendale, Cheshire.'

13 October 1795: Samuel Haslam, 31 Market-street-lane, Manchester, inserted advertisement, 'To Cotton Factories and Journeymen Machine Makers', offering employment to 'Turners, Clockmakers, Filers, Joiners and Smiths, that are practised in Machinery'.

1 March 1796: 'Journeymen Machine-Makers. Wanted, Mule Fitters-up, Turners, and Clockmakers . . . A Brown, No. 11, Atherton-street, near St. John's Church', Manchester. Previously, on 26 Jan. and 2 Feb., he had also advertised for smiths and filers.

17 January 1797: Bankruptcy of William Platt, of Stockport, 'Cotton Machine Maker, Cotton Spinner, and Clockmaker'.

21 Feb. 1797: Sale of the cotton factory of Alexander Hunt (bankrupt) of Stockport, including several lathes, apparently worked by steam power, and also 'a Joiner's, Clockmaker's and Smith's Shop'.

XIV An Early Engineering Firm

Peel, Williams & Co., of Manchester[1]

There are many famous names in the history of Manchester engineering in the first half of the nineteenth century. Roberts, Fairbairn, Nasmyth, and Whitworth spring readily to mind. But these great engineers rose to fame in a city where engineering had already been developing rapidly for half a century.[2] They belonged to the second generation of Manchester engineers. The lustre of their names, however, has thrown into the shadows the achievements of their predecessors—of those who laid the foundations of Lancashire engineering in the early Industrial Revolution.

'The most noted engineers of the day,' says J. T. Slugg in his *Reminiscences of Manchester Fifty Years Ago* (1881), 'were Peel, Williams, and Peel, of the Soho Foundry, Ancoats, and Galloway, Bowman, and Glasgow, of Great Bridgewater Street'.[3] There is no doubt that Slugg was correct in his recollections of the importance of these two early nineteenth-century engineering firms. Dr. Chaloner has already revealed a good deal of the history of Galloway, Bowman, and Glasgow, who were clearly among the leading engineers of that period.[4] It is the purpose of this chapter to rescue from oblivion the early history of the other firm—Peel, Williams & Co.—which was actually the biggest engineering business in Manchester during the first two or three decades of the nineteenth century.[5]

[1] This chapter is based on an article by A. E. Musson in *Business History*, Vol. III, no. 1 (Dec. 1960). For contemporary engravings and plans of their works, see also A. E. Musson, 'Peel, Williams & Co., Ironfounders and Engineers of Manchester, c. 1800–1887. A Communication in Industrial Archaeology', *Lancs. & Chesh. Antiq. Soc. Transactions*, Vol. LXIX (1959), pp. 133–5.

[2] See above, chaps. XII and XIII.   [3] *Op. cit.*, p. 102.

[4] W. H. Chaloner, 'John Galloway (1804–94), Engineer of Manchester and his "Reminiscences"', *Lancs. & Chesh. Antiq. Soc. Transactions*, Vol. LXIV (1954), pp. 93–116.

[5] There are very brief references to the firm in the *Manchester Guardian*, 31 May 1887; Institution of Civil Engineers, *Minutes of Proceedings*, Vol. XC (1887),

This firm was founded at the end of the eighteenth century by George Peel and William Ward Williams. George Peel was one of the numerous members of the enterprising Peel family, descended from Robert Peele (d.1733) of Peele Fold, Oswaldtwistle, near Blackburn.[1] His father, Joseph Peel (d. 1820, aged 84), was the youngest brother of Robert Peel (1723–95), or 'Parsley' Peel, the founder of the Peel fortunes, so that George Peel was a cousin of the first baronet, Sir Robert Peel (1750–1830). His father was involved in the far-spread and complex cotton-spinning, manufacturing, and calico-printing businesses of the Peels and their various partners.[2] George Peel, born 14 May 1774, like his father a fourth and youngest son, probably had to make his way in the world. His family's cotton-manufacturing interests must have impelled him to seek his fortune in Manchester, the centre of Lancashire's rapid industrial expansion, where his uncles Robert and Laurence Peel were then living. He first appears in the Manchester directory of 1797, as 'Peel George, *reed-maker*, 3, Halliwell-street'.[3] However, it is by no means certain from the contemporary evidence that (as later writers have stated) he was responsible for the original establishment of the ironfounding business which developed in the early years of the nineteenth century as Peel, Williams & Co. Reed-making was certainly part of early textile engineering, and Peel may possibly have combined ironfounding with it in his early business in Halliwell-street.[4] But the strongest probability is that the original founder was William Williams, who first appears in the late 1790s in partnership with James Marshall in an ironfounding concern, which was dissolved, however, in 1797 after lasting less than a year.[5] He immediately formed another partnership with Thomas Knight—

---

pp. 435–6; T. Swindells, *Manchester Streets and Manchester Men* (1908), Vol. V, pp. 133–4; F. S. Stancliffe, *John Shaw's 1738–1938* (1938), p. 148.

[1] See J. Davies, *Pedigree of the Right Hon. Sir Robert Peel, and the Peels of Lancashire, from the year 1600 to 1846* (Manchester, 1846); J. Foster, *Pedigrees of the County Families of England*, Vol. I, *Lancashire* (1873); *Burke's Landed Gentry*.

[2] He is described in George Peel's will as 'of Faisley in the County of Stafford, Calico Printer' (Peel's will, dated 9 Jan. 1804, is in the Lancashire Record Office, Preston). According to the obituaries of his grandson (d. 1887), he appears to have migrated from the family cotton-spinning business at Bury to Fazeley, near Tamworth, Staffs., where the firm of Peels, Dickenson & Co. were established as cotton manufacturers and calico-printers in the early nineteenth century.

[3] Scholes's *Manchester and Salford Directory*, 1797.

[4] He described himself as an 'Iron Founder and reed Maker' in his will of 1804, but this was after the formation of his ironfounding partnership with Williams.

[5] *Manchester Mercury*, 26 Dec. 1797. Marshall was apparently a cotton-spinner, who had supplied most of the capital for the business, which was managed by Williams. Personal differences appear to have caused the break-up of the partnership. Williams's address was in Miller's Lane.

'Williams and Knight, iron-founders, 6, Miller's street', appearing in the 1797 directory—but this, too, was short-lived, being followed in that of 1800 by 'Peel, Williams and Co. iron founders, Millar's-street'.[1]

William Ward Williams had already had considerable experience in ironfounding. According to an obituary sketch, he was born on 12 July 1772;[2] his birthplace is not stated, but he was apparently of Welsh extraction. He was said to have raised himself by his industry, integrity, and talent, from 'a situation in life comparatively obscure' to a position of considerable wealth and public esteem. He had been 'brought up from his earliest years a practical iron-founder' and had acquired experience at the Carron ironworks in Scotland and at those of Messrs. Walkers of Rotherham in Yorkshire. 'To an intimate knowledge of the practical detail of his business, Mr. Williams united a vigorous and active mind, strong natural talent, much boldness of character, a constitution unusually robust, and a spirit enterprising in the highest degree.'

It would appear, therefore, that Williams rather than Peel provided most of the technical and business ability of the new partnership, in which one of George Peel's brothers also apparently had an interest. Perhaps the Peels provided additional capital for expansion of the ironfounding business, for very soon the firm established a new works, the 'Phoenix Foundry,' in Swan Street,[3] a continuation of Miller Street, at the top of Shudehill.

There is only scrappy evidence about their early business activities. The directories for the most part describe them simply as ironfounders, but in one they appear as 'iron-founders, and engine makers and roller manufacturers'[4] from which it seems that they were producing steam engines and textile machinery. We are told, indeed, that they made high-pressure engines for Trevithick in 1805.[5] They soon established for themselves a high reputation in mechanical engineering, especially in the manufacture of gear wheels.

Mr. Williams, very early after entering into business, commenced making and arranging a large and extensive series of wheel models, applicable to any

[1] Bancks's *Manchester and Salford Directory*, 1800. George Peel still appears also as 'reed maker, 3 Halliwell street', and William Williams as 'iron founder, house, 2 Ledger street'. Thomas Knight, 'engineer, St. George's road', was probably Williams's former partner. [2] *Manchester Courier*, 20 July 1833.
[3] They first appear there in Dean & Co.'s *Manchester and Salford Directory* (1804). The exact address is given in Dean's 1811 directory as 17 Swan Street. See Musson's article, previously cited, in the *Lancs. & Chesh. Antiq. Soc. Transactions*, for the precise location of this foundry, together with a plan and engraving of it. [4] *Holden's Triennial Directory*, 1805-7.
[5] E. C. Smith, 'Joshua Field's Diary of a Tour in 1821 through the Provinces', Part II, *Newcomen Soc. Trans.*, Vol. XIII (1932-3), p. 44, n. 9. See below, p. 464.

purpose of millwright work, of which a printed list was from time to time published, and gratuitously circulated amongst the public; the importance and utility of this undertaking was universally acknowledged by all mechanical men; favourable mention of it has been made by several eminent writers on mechanics, and it was warmly eulogised by the late Robertson Buchanan, in his 'Essays on the Teeth of Wheels'.[1]

A copy of this printed catalogue has fortunately survived,[2] showing the 'various sizes of Patterns lately made from dry Mahogany [of] Bevil, Spur, and Mitre Geer Upon the most approved Principle, At Peel & Williams' [works]'. It contains a long list of gear-wheels, with their diameter, number of teeth, pitch, and breadth of cog; special reference is made to bevel gear, etc. for corn mills. The catalogue also contains engraved illustrations of shafting, gearing, etc., and of other engineering goods produced by the firm: an 'Iron Loom', a hand-operated 'Water Press', and one of their 'Portable Steam Engines', which were made 'upon this Plan from 4 to 22 Horse power'.

Simon Goodrich, the later-famous railway engineer, noted in his journal on 2 July 1806 having received such a printed catalogue, dated 10 June 1806, from 'Peel, Williams & Co.'s Iron and Brass Foundry, Shudehill Pits, Manchester'.[3] In addition to the above-mentioned gearing patterns, it included 'models for water-wheel castings, mill gearing and engine work, weighing machines for wagons and carts. Factory heating on the most approved plan, and spinning rollers of every sort'. This illustrated catalogue shows that the firm was by this time producing millwork (shafting, etc.) made from wrought iron, as well as cast-iron gearing, and that Fairbairn's later claim to have inaugurated the revolution in millwork, by substituting slender wrought-iron constructions, operating at much higher speeds, in

---

[1] *Manchester Courier*, 20 July 1833. R. Buchanan in *An Essay on the Teeth of Wheels* (revised by P. Nicholson, 1808), p. 161, states: 'Messrs. Peel, Williams and Co. have, after great time, trouble and expense, made and arranged a very great number of patterns of wheels, so as to suit almost every case that can in practice occur. They have published a complete list of them, which they intend inserting also in the "Repertory of Arts." In my opinion, what they have done is a material national benefit; their expense, I am informed, for patterns, has not been less than four thousand pounds.' This statement is repeated in the reprint of the *Essay* in Buchanan's *Practical Essays on Mill Work and other Machinery* (2nd edn., 1823), Vol. I, p. 229. I have discovered no trace of the intended publication in the *Repertory of Arts and Manufactures*.

[2] In Manchester Central Reference Library (Local History Section, MSC. 621). See Musson's article in the *Lancs. & Chesh. Antiq. Soc. Transactions* for photographic copies of the machinery illustrated in this catalogue.

[3] E. A. Forward, 'Simon Goodrich and his work as an engineer', *Newcomen Soc. Trans.*, Vol. XVIII (1937–8), p. 5. I am grateful to Dr. Chaloner for drawing my attention to this reference.

place of the cumbrous old wooden shafts and drums,[1] is open to question. As we have previously noted, this revolution had been commenced earlier by Smeaton and Rennie,[2] though it had probably not gone very far by the early nineteenth century. Peel and Williams were evidently progressive in this field, and also in making 'waterwheel castings', along with their Manchester contemporary, T. C. Hewes.[3]

Their distribution of printed catalogues is interesting early evidence of standardised mass production of engineering goods in anticipation of demand. The Carron Company was issuing similar catalogues of its products from at least as early as the 1770s,[4] and it is possible that Williams may have brought the idea with him from that firm. The Coalbrookdale Company did not prepare any until the early nineteenth century.[5]

James Watt, who visited the works in about 1817, is said to have expressed his admiration of their wheel models or patterns. Buchanan's statement of their capital expenditure on this project indicates that the firm had grown rapidly to a very considerable size. Williams is said to have been warmly attached 'to all improvements in mechanical science', and to have considerably extended the structural uses of iron. 'He very early recommended the adoption of iron roofs (for fire proofing purposes), and in 1813 and 1814, the writer believes, he constructed two very large ones on an improved principle, of cast and malleable iron combined, for the then Marquess of Queensberry and Lord Somers. From that period to the present no improvement has taken place in their construction. . . .'[6]

George Peel died rather suddenly on 27 October 1810.[7] He had married on 4 March 1799 Rebecca, daughter of Richard Barlow, cotton merchant, of Stand, between Manchester and Bury[8], by whom he had two sons, Joseph, born 31 May 1801, and George, born 27 February 1803. Under the terms of his will[9] his interests in the

---

[1] Pole, *op. cit.*, pp. 112–15. See below, pp. 481–2.

[2] See above, pp. 24, 69.

[3] See above, pp. 445–6.

[4] I am indebted for this information to Professor Roy Campbell.

[5] A. Raistrick, *Dynasty of Ironfounders. The Darbys and Coalbrookdale* (1953), p. 14.   [6] *Manchester Courier*, 20 July 1833.

[7] *Manchester Mercury*, 30 Oct. 1810. But the fact that Peel appointed his *father*, already an elderly man, as one of the two executors of his will in 1804 indicates that he had probably been in failing health for some time.

[8] Foster, *op. cit.; Manchester Guardian*, 31 May 1887 (his son George's obituary); Swindells, *op. cit.*, Vol. V, pp. 133–4.

[9] Dated 9 Jan. 1804, proved 13 March 1811, preserved in the Lancs. Record Office.

ironfounding business were left on trust to his father, Joseph Peel, of Fazeley, Staffs., and his 'relation and esteemed friend Robert Peel the younger of Mosley Street in Manchester aforesaid esquire', to be sold; the money obtained therefrom to be invested on behalf of his wife and two sons. Probate was issued on 13 March 1811, the personal estate and effects being valued at between £9,000 and £10,000.

Precisely how George Peel's property in the ironfounding business was disposed of is not evident. The name of Peel was retained by the firm, whose title, however, according to the Manchester directories, was slightly altered from 'Peel, Williams and Co.' to 'Peel and Williams.' It is possible that George Peel's brother still retained his interest in the business. Moreover, at some date, probably not many years after George Peel's death, his widow married his partner, William Williams,[1] and in due course George Peel's sons, Joseph and George, entered the business.

Meanwhile Williams continued very successfully to expand the firm's engineering activities. Just before George Peel's death, the firm acquired a second foundry, the 'Soho Foundry', Ancoats.[2] This foundry had been established some years previously by David Whitehead[3] and had become one of the leading makers of Trevithick's high-pressure steam engines.[4] Peel, Williams & Co. may also have made such engines, for three Manchester foundries were at work on them in 1804–5.[5] David Whitehead's death in 1807 soon gave them an opportunity of expanding their business by acquisition of the Soho Foundry.[6] The advertisement of sale, which appeared in January 1810, provides an

---

[1] Davies, *op. cit.*; Williams's will, proved 28 Jan. 1834, in the Lancs. Record Office. Rebecca Williams died 21 Aug. 1832 (*Manchester Guardian*, 1 Sept. 1832).

[2] Pigot's *Manchester and Salford Directory*, 1811, in which George Peel's name still appears. The address of the Soho Foundry is given as Ashton Street in Pigot's 1813 directory, but this area was only just being developed and the street name was soon altered to Pollard Street, which is off Great Ancoats Street.

[3] Whitehead first appears in Dean & Co's *Manchester and Salford Directory*, 1804, as 'Whitehead: David and Co. Soho Iron-Foundry, near the Ashton Canal Bridge, Ancoats'.

[4] F. Trevithick, *Life of Richard Trevithick* (2 vols., 1872), Vol. I, pp. 104, 173, 324–5; Vol. II, pp. 139–40. H. W. Dickinson and A. Titley, *Richard Trevithick the Engineer and the Man* (Cambridge, 1934), p. 44. Forward, *op. cit.*, p. 5. The latter states that 'it was through a younger Peel that Bennet Woodcroft, in 1860, secured the Trevithick model now in the Science Museum'.

[5] Trevithick, *op. cit.*, Vol. I, pp. 173, 324–5. One of these firms, in addition to Whitehead's, was that of Bateman and Sherratt, who at first refused to pay Trevithick for the patent right (*ibid.*, Vol. II, p. 128)—just as they had earlier pirated Watt's engine (see above, p. 423—but submitted on threat of legal action.

[6] Whitehead's will, proved 16 June 1807, is preserved in the Lancs. Record Office.

interesting description of this 'Extensive Iron Foundry,' which was to be sold or let:[1]

All that capital and extensive Iron Foundry, called '*The Soho Foundry*', in Manchester, heretofore occupied by the late Firm of '*David Whitehead & Co.*' The Foundry is 75 yards long by 25 wide; contains air-furnaces, cupolas, stoves, cranes of extraordinary power, an excellent smithy and finisher's shop, and extraordinary well lighted pattern-makers' and turners' shops, extending 100 yards in length. Adjoining the Foundry, is a most complete boring-mill and turning shop, replete with every apparatus, on the very best principle, for boring and turning every kind of heavy or small iron and brass work.

The boring-mill and turners' shop, are worked by an excellent steam-engine, of 18 horses power, which also works the blasts for the cupolas; and there is additional power, which may be applied to other purposes.

There is an extensive yard, with stable, cart-house, sheds and other conveniences; and also, six Cottage-Houses, for the accommodation of workmen belonging the Foundry.

The premises have likewise belonging to them, a commodious wharf on the bank of the Ashton Canal, which communicates with other Canals, by means whereof, coals and metal are advantageously brought without any expense of land carriage, and goods conveyed to all parts of the kingdom.

After acquiring this foundry, Peel and Williams had sufficient additional capacity considerably to extend the range of their engineering activities. The Manchester directories describe them as 'ironfounders and steam-engine boiler-makers' in 1815 and as 'ironfounders, steam-engine and boiler manufacturers' in 1817.[2] There seems to have been a differentiation of products between their two foundries, as appears from their description in 1819–20 as 'iron founders, roller and spindle makers, Phoenix Foundry, Shudehill; steam-engine manufs. and gas-light erectors, Soho Foundry, Ancoats'.[3] The next two directories appear to indicate that the Phoenix Foundry was mainly confined to iron and brass founding—roller and spindle making being transferred to the Soho Foundry, where 'water-presses' or hydraulic presses were also among the firm's products.[4] That they were among the leading Manchester gas-lighting erectors is shown by the fact that in August 1817 the Police Commissioners gave them the contract to build the first public gas-works, through which, however, they were involved in a scandal concerning the alleged fraudulent influence of Williams, who was a member of the Gas Committee.[5]

[1] *Manchester Mercury*, 23 Jan. 1810.
[2] Pigot and Deans' *Manchester and Salford Directory*, 1815 and 1817.
[3] *Ibid.*, 1819–20.   [4] *Ibid.*, 1821–2 and 1824–5.
[5] *Manchester Observer*, Feb. and March 1819; A. Redford and I. S. Russell, *The History of Local Government in Manchester* (1939), Vol. I, pp. 268–9. The firm

An indication of their relative importance among Manchester engineering firms at this time is provided by examination of the Manchester poor-rate assessments in 1820.[1] The assessment (annual value) of the Soho Foundry was £230, of the Phoenix Foundry £210, a total of £440. Assessments on other leading iron foundries and engineering works were those of Radfords and Waddington (Waterloo Foundry, David St., £200; Workshop, Hanging Ditch, £130); Richard Ormrod (St. George's Foundry, Minshull St., £210); Galloway, Bowman & Co. (Oxford St. foundry, £170); Hewes & Wren (Dale St. works, £130); Ebenezer Smith & Co. (Griffin Foundry, Swan St., £100).[2] These figures indicate that Peel and Williams were at that time easily the biggest ironfounding and engineering firm in Manchester.[3] Their poor-rate assessments, indeed, were bigger than those of most cotton factories except the largest, such as those of McConnel and Kennedy, A. & G. Murray, and Thos. Houldsworth. We can arrive at an approximate capital valuation of the firm from the fact that factories and workshops were assessed at about $5\frac{1}{2}$ to 6 per cent. of such valuation.[4] This gives an approximate valuation of £4,000 for the Soho Foundry and £3,650 for the Phoenix Foundry, a total of £7,650.

They may well have been employing several hundred men at this time, for we know that Hewes and Wren, with an assessment on their one foundry between a third and a quarter of that of Peel and Williams, were employing 140–150 men only four years later.[5] This figure, as Clapham has pointed out, was the largest engineering employment figure quoted before the Select Committee on Artizans and Machinery of 1824,[6] but Peel and Williams were certainly a much bigger firm at

indignantly rejected the 'malevolent and unfounded assertions' (*Manchester Mercury*, 16 March 1819). Details of their work and charges for building the retorts, tar cisterns, washing apparatus, gasometer, piping, etc. are given in the firm's newspaper advertisement and in the *Manchester Observer*, 13 and 20 March 1819.

[1] Manchester Poor Rate Book, 1820 (Manchester Central Reference Library, M9/40/2/85).

[2] It is interesting to note that the assessment upon Fairbairn and Lillie, then just starting in Mather Street was only £8.

[3] It is possible, however, that they may have been exceeded in size by the Salford firm of J & T. Sherratt (formerly Bateman and Sherratt). Unfortunately the Salford poor-rate books for this period have not survived, but the annual value of Sherratts' foundry in Hardman Street is given as £400 in the Salford Church Ley Book, 1834; the firm also had other properties.

[4] Valuation Book per Messrs. Wallis, Bellhouse and Johnson, July 1812 (Manchester Cen. Ref. Lib., M9/40/1/20).

[5] *Select Committee on Artizans and Machinery, Fourth Report, Parl. Papers* (1824), Vol. V, p. 340.

[6] J. H. Clapham, *An Economic History of Modern Britain* (1939), Vol. I, p. 155. Hewes is mis-spelt 'Herves' both in the Committee Report and in Clapham.

this date, and so too probably were several other Manchester and Salford foundries. A rough impression of the size of the labour force employed by Peel and Williams—and of some of their chief products— is provided by the description of their contingent in the Coronation procession in Manchester in 1821.[1] This contingent was sufficiently large to form a distinct group,[2] whereas most other groups, e.g. of masons, joiners, etc., were general bodies of workers in each trade. William Williams and young George Peel headed their men, who followed in separate sections, each with its own banner: the principal engineer, draughtsmen, and head clerks; four engineers bearing a brass working model of a steam engine; the smiths, with their hammers; models of a mitre wheel, a spur wheel, and a pair of bevel wheels, on poles, carried by three millwrights, followed by the general body of millwrights and pattern makers; the engineers, fitters, turners, and borers; the moulders in brass and iron; a 3 h.p. steam engine at work on a cart; and finally the boilermakers, with their rivetting hammers. The workmen, in their separate sections, were walking six abreast and obviously formed a very large contingent. A high degree of specialization among the workers is also evident.

We are fortunate in possessing a description of the Soho Foundry at this time by an observant engineer, Joshua Field, of Maudslay, Sons & Field, the famous London engineering firm, who visited Manchester in 1821, during a tour of the provinces.[3] Field gave a fairly detailed description of the works—the 'engine factory', as he called it—with a rough sketch plan showing the foundry with three cupola furnaces, the boiler-making shop, forges, offices, etc., arranged around the four sides of a rectangular yard, and with a wharf on a way-leave of the Manchester-Ashton-Stockport canal. The Soho Foundry was

a large place and well calculated to do business, the tools and machinery rough, but they seem to have much work in hand. The works are driven by a 24 horse engine newly erected and well fitted up. They have a large & a small boring machine, hollow bar with a rack coming out at the end and moved along by a rachet & lever. Several large rough lathes driven by gear, a horizontal drilling and boring machine, rough screw tackle. Here are a greater

---

[1] *Authentic Particulars of the Processions in Manchester and Salford on Thursday the 19th of July, 1821, in Celebration of the Coronation of His most Gracious Majesty King George the IVth* (Chetham's Library, Manchester). Also *Manchester Guardian*, 21 July 1821.

[2] Another big engineering firm, Ormrod's 'St. George's Foundry', Minshull Street, London Road, also formed its own group. See above, p. 453.

[3] His diary is preserved in the Science Museum, South Kensington. It is reprinted, with introduction and notes by J. W. Hall and Capt. E. C. Smith respectively, in the *Newcomen Soc. Trans.*, Vol. VI (1925), pp. 1–41, and Vol. XIII (1932–3), pp. 15–50.

number of [steam] engines in hand than any place we have yet seen. Large cotton Mill engine, 1 of 60 [H.P.] countermanded, but several of 40 and 30 [H.P.]. There is a half circular pit at the end of the foundry where the Nozzles are put to the Cylinders and the cistern and air pumps fitted . . . Here are also many small portable engines in hand. One painted very fine for France. These small engines are not so good as the large. We were shewn the pattern loft in which the patterns of engine work & wheels were pretty well arranged, rather dirty and cracked by being over the furnaces, and running all round the foundry. The cupolas are worked by a blast cylinder under the beam of the engine and a water regulator. In a room behind the engine is a model of an engine in brass abt. 1 in. Cylinder—they have a slide which is round and packed like a D slide on one side. They are making all their small engines upon this plan, otherwise they use the D slide to the largest cotton mill engines.

The boiler department is not good having very few tools. They make very large waggon Boilers & Punch by a machine worked by hand, very heavy, and [there] is also a pair of Shears having 2 Segments connected by wheels . . . It may be observed that they make their boilers about $\frac{1}{2}$ larger than is generally done in London and in some instances as large again; but they know very little about the consumption of coal. They never pitch and trim their wheels[1] but take care in the casting to have them good.

Field also went to see the Phoenix Foundry, which he said was 'a common foundry for the town', and which had 'an old factory connected with it for making Spinning Cylinders and other common work. They have machines for grooving, rolling, grinding, &c., driven by an 8 H. Engine'.

Here we have what must be one of the most interesting contemporary descriptions of an early engineering works. It reveals a considerable development of steam-powered machine-tools and engineering technique, though Field was rather disdainful of some of the tools and methods employed—coming as he did from Maudslay's, at that time the most technically advanced engineering works in the country. This —and, indeed, considerably earlier evidence[2]—suggests that by 1820 power-driven mechanical engineering had developed more extensively, in Lancashire at any rate, than economic historians such as Clapham have hitherto realized.[3]

We are doubly fortunate in that excellent engravings of the two foundries of Peel and Williams a few years before this date have also survived.[4] They are by the local printer and engraver Thomas Slack

---

[1] To 'pitch and trim' wheels was to bring the teeth to pitch and correct shape by chipping and filing, which was necessary if they were roughly cast.
[2] See above, chaps. XII and XIII. See also the sale advertisement of the Soho Foundry in 1810 (above, p. 465).   [3] See especially Clapham, *op. cit.*, pp. 151–6.
[4] Surviving copies are in the possession of Mr. Arnold Hyde, of Macclesfield and Manchester, and also in the Manchester Central Reference Library (Local

and show in great detail not only the buildings, but also a large assortment of the firm's products spread out in the yards. The Phoenix works consists of a foundry containing three or four furnaces, with an engine-house and other buildings at one end and a three-storied building at the other, while on the opposite side of the yard stands a large four-storied building which looks like a converted cotton mill, doubtless the 'old factory' which Field mentioned. In the yard there are laid many gear wheels of various sorts and sizes, large pans and vats, pipes, railings, etc. The Soho works have a neater, more modern appearance, the predominant feature being the foundry, with a second storey above the furnaces; in the yard are waggon boilers, numerous gear-wheels, pipes, railings, a vat, a garden roller, etc., with a large derrick which was probably used for lifting the firm's own products. From the chimneys of both foundries the smoke of prosperous industrialism is shown proudly rising in thick black clouds.

So impressive was the appearance of the Phoenix Foundry—'the very extensive foundry of Peel, Williams, and Co.'—that James Butterworth waxed lyrical about it in the year 1822:[1]

> Under their hand, the vast capacious cauldrons rise
> Whose heating fires roll their black volumes to the skies;
> Their bowels charged with condens'd vapour prove,
> Of force tremendous! Wheels mechanic move, thereby,
> And millions of twink'ling spindles dance convolvent,
> Through that vast power, within their ample confines pent!

William Williams rose in wealth and social prestige, entering the ranks of the ruling Tory clique in Manchester, becoming a member of the Union Club, the Scramble Club, and John Shaw's Club, and a Police Commissioner.[2] He is said to have been 'an admirer and promoter of all public works, was one of the first projectors of the Botanical Gardens [at Old Trafford] in this town, and with him the idea of establishing a general and cattle market on Shudehill, first originated'. He was also distinguished for many Christian qualities, 'amongst which

---

History Section, MSC. 621). The latter copies have a manuscript note, 'Foundry of Peel & Williams March 25th 1814'. Reduced versions were reproduced at the time upon the firm's beautiful letterheads, of which a specimen is preserved in Chetham's Library, Manchester. These engravings have been reproduced in Musson's article in the *Lancs. & Chesh. Antiq. Soc. Transactions.*

[1] Butterworth, *op. cit.*, p. 185.

[2] At first he lived next-door to the works, but then moved to Piccadilly and Oldham Street, and in the later years of his life resided at Medlock Hall, Holt Town, and finally Broughton Priory.

was a kindly feeling for the poor, a disposition to improve the condition of the working classes, and a desire to assist the unfortunate'.[1]

Under his direction the business steadily progressed, the Soho Foundry becoming 'one of the most noted in Lancashire for its boilers and engines'.[2] When Joseph and George Peel, the sons of his late partner, came of age, they entered into partnership with their step-father and the firm altered its name to 'Peel, Williams, and Peel'.[3] The increasing variety of the firm's products is revealed by the Manchester directories, which describe them at this time as general iron and brass founders, millwrights, and engineers, manufacturing steam engines, marine engines, textile spinning machines, hydraulic presses, gas apparatus, and 'all description of hammered irons'. An unfortunate example of their marine engineering work was referred to by Joshua Field when in Liverpool: one of the Liverpool-Dublin paddle steamers, the *Waterloo*, built in 1819, was 'made by a Millwright at Belfast who had the cylinders &c. made by Peel & Williams of Manchester', but the engine was 'badly contrived and continually out of order' and had to be replaced by another engineer.[4] In the early 'thirties they gave up the Phoenix Foundry and concentrated their production in the Soho Foundry.[5] The latter had recently been about doubled in size by the building of an extension—the 'Soho New Foundry'—on the opposite (north) side of the Ashton canal wayleave.[6]

On 9 July 1833 Williams died, childless, in his 61st year.[7] In his will, Joseph and George Peel were to have the option of continuing for a period of five years to occupy 'the foundries Buildings and places of Business now occupied by us in our said Business of Iron Founders', at the expiration of which period they were to have the option of pur-

---

[1] *Manchester Courier*, 20 July 1833.   [2] *Manchester Guardian*, 31 May 1887.

[3] George Peel is said to have entered the business in 1825. Inst. of Civil Engineers, *Mins. of Proceedings*, Vol. XC (1887), p. 436. His elder brother Joseph probably came in a few years earlier. The title of 'Peel, Williams, and Peel' first appears in E. Baines, *History, Directory, and Gazetteer of the County of Lancaster* (1825), at which time Joseph and George Peel were still living with their step-father in Medlock Hall.

[4] Field's diary, 4 Sept. 1821.

[5] In the I.C.E. *Mins. of Proceedings*, Vol. XC (1887), p. 436, the Phoenix Foundry is said to have been given up in 1825, but the Manchester directories show it under Peel, Williams and Peel down to 1830; it does not appear in 1832. It is shown on Bancks & Co.'s large-scale map of Manchester in 1831. Adshead's map of 1851 shows a silk mill on the site, which is occupied at the present day by the large C.W.S. cold-storage depot for fish, poultry, etc.

[6] This extension first appears on Pigot & Son's 1829 directory map. The works are shown in excellent detail on the maps of Bancks & Co. in 1831 and Adshead in 1851.

[7] *Manchester Guardian* and *Manchester Courier*, 13 July 1833. His will dated 18 May 1833, proved 28 Jan. 1834, is now in the Lancs. Record Office.

chasing the same at valuation; otherwise the property was to be sold by public auction or by private contract. Probate was issued on 28 January 1834, Williams's personal estate and effects being valued at between £10,000 and £12,000.

Joseph and George Peel evidently took up the second option, for the firm carried on for many more years under their partnership but retaining the title of 'Peel, Williams, and Peel'. Throughout their later years the Manchester directories show very little change in the description of the firm: they were now well established in general iron-founding and engineering, and particularly well known for their steam engines, boilers, gasometers, and hydraulic presses. With the development of the railways, they not surprisingly entered the field of locomotive building, their first two engines, the *Soho* and *Manchester*, being produced in 1839.[1] The firm was referred to at that date, along with Nasmyth, Gaskell & Co., of Patricroft, and Sharp, Roberts & Co., of Manchester, as being 'among the first in importance' of locomotive-building and engineering firms in Manchester.[2] But they do not seem to have developed locomotive building to any great extent. They tendered unsuccessfully for the construction of two locomotives for the Manchester and Leeds Railway in June 1840,[3] and in March 1847 we hear that they had tried an engine on the Manchester and Birmingham line.[4] They were apparently still in the locomotive-building field in the 1850s, for a locomotive was included, with steam engines and a hydraulic press, among the special products of the firm illustrated in their bill-heads.[5] But this branch of their business appears to have been comparatively limited.

As hydraulic press and steam-engine makers, they maintained a considerable reputation. Several of their products attracted attention at the International Exhibition of 1862.[6] Some of their presses were 'used for

---

[1] A description of these engines and their trials is given by W. H. Wright and S. H. P. Higgins in the *Journal* of the Stephenson Locomotive Society, Vol. XXXII, no. 376 (Oct. 1956). It is based mainly on reports in the Manchester press, especially the *Manchester Courier*, 5 Oct. and 21 Dec. 1839, and *Herapath's Railway Magazine*, 18 Dec. 1841.

[2] Love and Barton, *Manchester As It Is* (1839), p. 213.

[3] Manchester and Leeds Railway, Minutes, 8 June 1840.

[4] Manchester and Birmingham Railway, Minutes, 4 March 1847.

[5] A few of their invoices of the 1850s, on which some old accounts dated 1835 have been copied, are preserved in the Manchester Cen. Ref. Lib. (Local History Section, MSC. 621). The accounts give details of unimportant miscellaneous charges relating to boiler installations.

[6] *The International Exhibition of 1862. The Illustrated Catalogue of the Industrial Department. British Division*, Vol. I (1862), Class VIII.—Machinery in General, p. 54; *International Exhibition, 1862, Reports by the Juries* (1863), Class VIII.—Machinery in General, p. 10.

expressing the syrup or juice from beetroot, in the sugar manufactories of Southern Russia', and they were also used for pressing or packing cloth, paper, and hay.

Joseph Peel died in 1866, but his brother George remained in command until his death in 1887, at which time Peel, Williams and Peel were described as 'the oldest established firm of machinists in Manchester'.[1] At least two of George's sons, George Chapman and Robert Eldon, had been in the business for some years, and also his brother Joseph's son George down to the early 'seventies.[2] But the firm appears to have closed down immediately after the death of George Peel senior. Its demise may have been due to senile lack of energy and flexibility. The middle 'eighties, of course, were among the darkest years of the 'Great Depression', and falling profits may have contributed to the foundry's closure. But one gets a strong impression that the second and third generations of Peels in this firm had devoted an excessive amount of their time to public and social activities, neglecting their business.[3] They seem to have lived like gentlemen upon the fruits of the capital and industry of the first George Peel and of William Williams, who had made their firm the biggest engineering concern in Manchester. In the latter half of the century it seems to have been gradually running down, finally to disappear in 1887. A considerable part of the Soho Foundry, however, still stands solidly. It has been occupied for many years by the Union Alkali Co. Ltd., but the old firm's title—'PEEL, WILLIAMS & PEEL ENGINEERS', surmounted by a set of gearwheels—still remains embossed on the front of the building in Pollard Street.

[1] *Manchester Guardian*, 31 May 1887. This was not true, however, for W. & J. Galloway & Sons could trace their origins back to the 1790s, and Ormerod, Grierson & Co. to the 1780s, though the Ormerod family connection appears to have been broken in the early 1860s.

[2] According to the Manchester directories.

[3] See the obituaries of Joseph and George Peel: *Manchester Courier*, 16 June 1866; *Gentleman's Magazine*, Aug. 1866; *Manchester Guardian*, 31 May 1887; Inst. of Civil Engineers, *Mins. of Proceedings*, Vol. XC (1887), pp. 435-6.

XV   The Growth of Mass-Production Engineering[1]

Credit is nowadays usually given to the Americans for the pioneering of standardized mass-production and assembly-line manufacture, and there is no doubt that from the mid-nineteenth century onwards they did take the lead in many aspects of mechanical engineering.[2] There is evidence, however, that in some fields they were preceded in the application of such methods by certain early British engineering firms. This is not surprising, in view of the fact that these methods were made possible by the invention of 'self-acting' machine-tools, most of which were brought out in Britain in the late eighteenth and early nineteenth century.[3]

This country, as is well known, led the world in the early years of the

[1] This chapter, by A. E. Musson, utilizes material previously published in his article on 'James Nasmyth and the Early Growth of Mechanical Engineering', *Econ. Hist. Rev.*, 2nd ser., Vol. X, no. 1 (1957). That article, however, has been greatly extended by much additional material on Nasmyth and by new sections on other engineers such as Roberts and Fairbairn. In the first half of the nineteenth century Manchester became the biggest centre of engineering in Britain, stimulated by the rapid growth of the cotton industry, railways, etc., and this chapter surveys some of the more striking developments.

[2] D. L. Burn, 'The Genesis of American Engineering Competition, 1850–1870', *Economic History*, Jan. 1931; H. J. Habakkuk, *American and British Technology in the Nineteenth Century* (1962).

[3] On the general development of nineteenth-century mechanical engineering, see R. Willis, 'Machines and Tools, for Working in Metal, Wood, and other Materials', *Lectures on the Results of the Great Exhibition* (1852), pp. 291–320; S. Smiles, *Lives of the Engineers* (1861–2) and *Industrial Biography* (1863); W. Pole (ed.), *The Life of Sir William Fairbairn . . . Partly written by himself* (1877); S. Smiles (ed.), *James Nasmyth, Engineer. An Autobiography* (1883). For modern works on the subject, see J. W. Roe, *English and American Tool Builders* (1916); A. P. M. Fleming and H. J. S. Brocklehurst, *A History of Engineering* (1925); A. F. Burstall, *A History of Mechanical Engineering* (1963); Institution of Mechanical Engineers, *Engineering Heritage* (1963); the various works on machine-tools by Professor R. S. Woodbury; L. T. C. Rolt, *Tools for the Job: A Short History of Machine Tools* (1965).

Industrial Revolution, in the development of new methods of manufacturing and working iron and in the production of steam engines and machinery for the growing factory system. The increasing demand for iron machinery could not have been met without a revolution in the methods of making machines, without the development, in other words, of mechanical engineering. Most of the early machinery was made of wood and leather, later of cast iron, since the working of wrought iron was a laborious and costly process, requiring much manual labour with hammer, chisel, and file. The early 'engineers', recruited from many trades—millwrights, smiths, carpenters, iron-founders, clock-makers, etc.[1]—were all handicraft workers. It is true, as we have seen, that they had lathes, drills, and gear-cutting engines, and that by the late eighteenth century some machine-makers were working to gauge; some of their machine-tools were even power-operated. But these appear to have been exceptional examples; for the most part, engineering equipment was rudimentary, manually operated and lacking in precision. We have already referred to the well-known difficulties of James Watt in getting parts for his steam engine accurately manufactured; cylinders, pistons and valves required work of hitherto unattainable accuracy. Similar problems had to be faced in the manufacture of machinery for the textile and other industries. There was no precise standardization and parts were not interchangeable; indeed, it was often a most difficult job fitting together the parts of one machine and getting it to work.

This situation persisted into the early decades of the nineteenth century. We have already noticed William Fairbairn's remarks on the general prevalence of hand labour, except for some 'very imperfect lathes, and a few drills', at the time when he first came to Manchester in 1814.[2] James Nasmyth has left similar testimony.[3] 'Up to within the last thirty years', he wrote in 1841, 'nearly every part of a machine had to be made and finished . . . by mere manual labour; that is, on the

---

[1] See above, chap. XIII.

[2] See above, p. 427. Elsewhere he elaborated on these remarks, stating in 1867 that 'fifty years ago, tools and all other descriptions of machinery were chiefly made by hand. Blacksmiths' forges, stocks and dies, and lathes for turning metal and wood, were in existence, but were very imperfect in construction . . . and void of those motions which constitute the slide-rest and self-acting machine. Planing, slotting and paring machines were unknown; and the machine for cutting the teeth of wheels was only just making its appearance.' Section by W. Fairbairn in T. Baines, *Lancashire and Cheshire, Past and Present* (1867), Vol. II, pt. ii, p. cliv.

[3] J. Nasmyth, 'Remarks on the Introduction of the Slide Principle in Tools and Machines employed in the Production of Machinery', in R. Buchanan, *Practical Essays on Mill Work* (3rd edn., revised by G. Rennie, 1841), pp. 393–418.

dexterity of the *hand* of the workman, and the correctness of his *eye*, had we entirely to depend for accuracy and precision in the execution of such machinery as was then required; consequently, the enormous expense [as well as the inaccuracy] ... proved a formidable barrier.' The progress of the Industrial Revolution, Nasmyth points out, was impeded by this 'almost entire dependence upon manual dexterity'.[1]

These problems were tackled by a number of brilliant engineers, led by Bramah, Maudslay, and Clement, all of London, which was at first pre-eminent in mechanical engineering. Maudslay, according to James Nasmyth and others, was the greatest of these pioneers. It was in his workshop that some of the leading figures in the second generation of mechanical engineers received their early training, including Roberts, Nasmyth, and Whitworth. These men, and others like William Fairbairn, established themselves in Manchester, which, with the rapid growth of the cotton industry and of railways, became the most important engineering area in Britain, producing not only textile machinery, but water wheels, steam engines, boilers, railway locomotives, machine tools, and a mass of miscellaneous engineering products.

These London and Manchester engineers gradually solved the problems of mechanizing machine-making, inventing machines to make machines—self-acting machine-tools, such as lathes, planing machines, drilling machines, grooving, slotting, and paring machines, punching and shearing machines—which, as Nasmyth pointed out, made possible 'almost mathematical accuracy and precision' in the manufacture of machinery. The basic principle in all such machine-tools was that of the slide-rest, 'the substitution of a mechanical contrivance in place of the human hand, for *holding, applying,* and *directing* the motions of a cutting tool to the surface of the work'. Such machines could be operated 'with such absolute precision' that they could produce objects of any required shape with 'accuracy, ease, and rapidity'. These machine-tools made possible 'the great era in the history of mechanism':[2] without them the enormous possibilities of steam power and mechanization could never have been developed. They form, in fact, the very basis of the modern machine age, making possible cheap, standardized mass-production.

The first steps in invention are always the most difficult, and the early progress of machine-tool making was comparatively slow.[3] Apart from the technical difficulties and expense, there was a good deal of secrecy maintained by some inventors, while manual craftsmen

---

[1] *Ibid.*, pp. 394–5.   [2] *Ibid.*, pp. 395–402.
[3] As was pointed out by Willis, *op. cit.*, p. 317.

displayed hostility to mechanization.[1] Nasmyth states that, after the invention of the lathe slide-rest, that of the planing machine was of most fundamental importance.

It has done more within the last 10 or 15 years for reducing the cost, and for extending the use of perfect machinery, than had been the case by all the improvements in mechanism for the last century. There is no form which is so frequently required and essential to any piece of mechanism as the plane surface . . . The vast expense attendant on the production of such, by the tedious and unsatisfactory process of chipping and filing, caused every engineer to avoid by all means any arrangements which rendered such forms necessary, however essential they might be to the perfect action of the machine . . . The introduction of the planing machine at once altered the entire system, inasmuch as forms and arrangements became practically possible, which formerly the engineer dared not think of using.

The result was the production not only of 'most strikingly superior' machinery for general manufacture, but also of greatly improved machine-tools at a 'very much reduced cost'. Hence 'in a very short time a most important branch of engineering business, namely, tool-making, arose'.[2]

Professor Willis, writing on machine-tools in 1851, the year of the Great Exhibition, also described the metal-planing machine as 'the greatest boon to constructive mechanism since the invention of the lathe'. Its actual invention, however, is shrouded in obscurity. 'We can only learn that, somewhere about 1820 or 1821 [or a few years earlier], a machine of this kind was made by several engineers', including Fox of Derby, Roberts of Manchester, Clement and Rennie of London, Murray of Leeds, Spring of Aberdeen, and perhaps others.[3] Its use seems to have spread slowly at first, but soon it was improved and manufactured in increasing numbers by such engineers as Nasmyth and Whitworth, with revolutionary results.

These striking developments in machine-tool manufacture were noted by a Parliamentary committee in the early 1840s, which reported that 'tools have introduced a revolution in machinery, and tool-making has become a distinct branch of mechanics, and a very important trade, although twenty years ago it was scarcely known'.[4] In

---

[1] On the other hand, of course, such opposition, and the high cost of skilled labour, were incentives to mechanization in many instances. See below, pp. 485, 491, 505–7.   [2] Nasmyth, *op. cit.*, pp. 403–4.
[3] Willis, *op. cit.*, p. 314. See also Smiles, *Industrial Biography* (1876 edn.), p. 178, and Roe, *op. cit.*, chap. v. One of Roberts's early planing machines, made in 1817, is preserved in the Science Museum in London.
[4] *Select Committee on Exportation of Machinery, Parliamentary Papers* (1841), Vol. VII, *Second Report*, p. vii.

evidence before this committee, William Jenkinson, of Jenkinson and Bow, manufacturers of cotton-spinning machinery in Salford, divided the industry into three sections:[1] 'the manufacture of steam-engines, mill-gearing, hydraulic presses, and such other heavy machinery, I should call one class; the next, and a separate branch, I should say, was tool-making; and the third I should call [textile] machine-making, with its various branches of spindle and fly-making, and roller-making'.

This tendency towards engineering specialization we have previously noted in the late eighteenth and early nineteenth century.[2] Separation between the various branches, however, was never clearcut: throughout the first half of the nineteenth century, and beyond, most firms remained general engineers. Makers of textile machinery were the first to become specialized, on account of the rapid expansion in the market for such products; but many of these continued to manufacture a wide range of other goods. Sharp, Roberts & Co., for example, the most notable textile machine-makers in this period, also produced machine-tools, locomotives, and many other articles. Parr, Curtis, and Madeley, 'the most extensive makers of cotton spinning machinery in Manchester' in 1851, were also 'extensive millwrights and toolmakers'.[3] Nevertheless, one can certainly discern a trend towards increasing specialization.

Tool-making, of course, had its origins, as we have shown, in the old-established trades of instrument-making and clock-making. John Wyke, of Liverpool, for example, who produced not merely a wide variety of hand tools, but also lathes and wheel-cutting engines, was a forerunner of the Lancashire machine-tool makers of the nineteenth century.[4] These tools, however, became bigger, more elaborate, more accurate, and more costly, as a result of developments by Maudslay and other engineers.[5] Jenkinson, in answer to a question before the 1841 Select Committee on Exportation of Machinery on whether tools were used as much in 1824 as then, replied, 'No; I should say that where we had 50£ expended in tools at that time, we have 1,000£ expended now.'[6] And later he expanded on the development of such machine-tools. 'What used to be called tools were simple instruments . . . such as hammers and chissels and files; but those now called tools are in fact machines . . . made at a very great cost, from 100£ up to 2,000£ each; I consider that the tools have wrought a great revolution in machine-making.'[7] He went on to point out that machine-tools were now 'self-

---

[1] *Ibid., First Report*, Q. 1,299.   [2] See above, chap. XIII.
[3] H. G. Duffield, *The Stranger's Guide to Manchester* (1851), pp. 174–5.
[4] See above, p. 437.
[5] See above, p. 63, n. 1, for Holtzappfel's early manufacture of lathes.
[6] *First Report* (1841), Q. 1,312.   [7] *Ibid.*, Q. 1,314.

acting, and go on without the aid of the men', so that 'machinery is made by almost labourers', employed at much lower wages than those of the skilled men formerly required; the machinery was also better made and at lower cost.

The greatest figure in this early development of the Manchester machine-tool industry was undoubtedly Richard Roberts (1789–1864), famous for his invention of the self-actor mule, but also prolific in other engineering achievements.[1] Roberts, after working at one time at Maudslay's in London, settled in Manchester in 1816 and by 1821, though his workshop was still fairly small, he was already establishing a name for himself, as Joshua Field observed on a visit to Manchester at that time.[2] In addition to his metal-planing machine, he also invented improved slide and screw-cutting lathes, a wheel-cutting engine, a slotting machine, etc. An advertisement in the *Manchester Guardian*, of 5 May 1821, referred especially to his new, improved 'cutting engines' for making gear-wheels, and to his improved screw-cutting engine; he would cut gear-wheels and screws of all sizes for cotton-spinners and other manufacturers.

His abilities in the design and manufacture of such machine-tools enabled him to become, in partnership with the Sharp brothers, the leading Manchester manufacturer firstly of cotton-spinning machinery and later of locomotives. In both these fields he pioneered the development of standardized mass-production, by use of gauges and templates.[3] Andrew Ure, in 1835, strongly emphasized the importance of these developments.[4]

Where many counterparts or similar pieces enter into spinning apparatus, they are all made so perfectly identical in form and size, by the self-acting tools, such as the planing and key-groove cutting machines, that any one of them will at once fit into the position of any of its fellows in the general frame.

[1] Smiles, *Industrial Biography* (1863), pp. 265–72; Roe, *op. cit.*, pp. 59–62; H. W. Dickinson, 'Richard Roberts, His Life and Inventions', *Newcomen Soc. Trans.*, Vol. XXV (1945–7), pp. 123–7, and *The Engineer*, 14 Feb. and 17 March 1947; see also *D.N.B.*

[2] Joshua Field's 'Diary of a Tour in 1821 through the Provinces', *Newcomen Soc. Trans.*, Vol. XIII (1932–3), pp. 24–6. Roberts was then employing about 12 to 14 men, and his machine-tools were manually powered.

[3] *The Engineer*, 20 Feb. 1863; Smiles, *Industrial Biography* (1863), p. 271; Institution of Civil Engineers, *Minutes of Proceedings*, Vol. XXIV (1864), p. 537; Roe, *op. cit.*, p. 62.

[4] A. Ure, *The Philosophy of Manufactures* (1835), p. 37. Professor Habakkuk is clearly incorrect in stating (*op. cit.*, p. 120) that the idea of standardized interchangeable parts, after early application by Brunel and Bentham in block-making, 'was not taken up by any other English industry', until it was introduced from America for gun-making in the 1850s. Maudslay and Roberts practised it, and so, as we shall see, did Nasmyth and Whitworth.

For these and other admirable automatic instruments, which have so greatly facilitated the construction and repair of factory machines, and which are to be found at present in all our considerable cotton mills, this country is under the greatest obligations to Messrs. Sharp, Roberts, and Co. of Manchester.

Roberts subsequently applied these and other machine-tools to locomotive manufacture and by the early' fifties the Atlas Works which he and the Sharps had established off Oxford Road was an extraordinarily impressive place, filled with power-driven machine-tools of all kinds and employing nearly a thousand men.[1]

Roberts was not the only machine-tool maker in Manchester in this period. The Swiss engineer, J. G. Bodmer, was an equally prolific mechanical genius,[2] and Joseph Whitworth, who set up his first workshop there about seventeen years after Roberts, was eventually to eclipse all others in reputation (though perhaps undeservedly).[3] And these were only some of the most outstanding among a rapidly growing number of such engineering firms. In Slater's directory of Manchester and Salford of 1845, almost a hundred firms were listed as millwrights, engineers and machinists, machine makers, and ironfounders.[4] Many of these were large concerns, as shown by evidence before the Select Committee on Export of Machinery in 1841. Returns were collected from Manchester, Salford, Bolton, Rochdale, Bury, and other manufacturing towns around Manchester, 'from 115 [engineering] firms employing power to the amount of 1,811 horses, and these were calculated, when in full work, to employ 17,382 hands'; the estimated

---

[1] Duffield, *op. cit.*, pp. 176–87.

[2] See above, p. 62.

[3] The author has written about Whitworth's achievements elsewhere and so will not deal with them here. See A. E. Musson, 'Sir Joseph Whitworth; Toolmaker and Manufacturer', *Chartered Mechanical Engineer*, March 1963, reprinted in *Engineering Heritage* (Institution of Mechanical Engineers, 1963), and 'The Life and Engineering Achievements of Sir Joseph Whitworth, 1803–1887', in the *Whitworth Exhibition* catalogue, July–Aug. 1966 (Inst. of Mech. Eng.). Whitworth undoubtedly owed much to Maudslay, in whose London workshop he was for some time employed. Maudslay, for instance, made important contributions towards achieving true plane metal surfaces, accurate measurement and standardization, using measuring machines, gauges, and 'self-acting' lathes, etc. But Whitworth was largely responsible for bringing these into general engineering practice, through the Institutions of Civil and Mechanical Engineers, the British Association, his own publications, and the magnificent practical demonstrations of his own machine-tools; as a result of his improvements in machine-tool design and construction, by the time of the Great Exhibition (1851) he was pre-eminent in this field. (Subsequently, moreover, he also pioneered improvements in rifles, cannon, and steel production.) In his later life, however, his immense authority tended to exercise a conservative influence on British machine-tool development.

[4] Taking account of the fact that many firms appeared under more than one head.

total capital of these firms was £1,515,000.[1] Thus the average employment figure was 151, average capital £13,174, and average horse-power 16. This evidence again suggests that Sir John Clapham considerably underestimated the growth and size of engineering firms in the first half of the nineteenth century, especially in Lancashire.[2]

Some of the most important of these firms—Bateman and Sherratt; Ebenezer Smith & Co.; Ormrod's; Radfords and Waddington; Hewes and Wren; Peel, Williams & Co.; Galloway, Bowman and Glasgow; Sharp, Roberts & Co.; and Whitworth's—have already been examined in earlier chapters and elsewhere. But one of the most outstanding, that of William (later Sir William) Fairbairn (1789–1874), has tended to be neglected.[3] He was not, as we have seen, one of the 'first generation' of Manchester engineers—in 1820 his firm, then just established, was an infant among a number of older and far larger concerns[4]—but it was a lusty child and grew prodigiously thereafter, so that by the 1840s it was probably the biggest in the city. Fairbairn, like Watt, Rennie, Telford and many other outstanding engineers, was a native of Scotland; he came of humble farming stock, but benefited from the Scottish system of parish education. At the end of his schooling, he was apprenticed in 1804 to John Robinson, millwright, of Percy Main colliery, Northumberland, where he acquired a sound knowledge of steam engines and mining operations. At the same time, he continued his studies every evening, with the aid of the North Shields subscription library, on a self-imposed programme including arithmetic, algebra, geometry, trigonometry, and mensuration, together with literature and history, while he also developed an interest in 'mechanical philosophy' and astronomy.[5] At the end of his apprenticeship, and after millwrighting work at Newcastle and Bedlington, he went to London in 1811, where, despite difficulties with the Millwrights' Society, he managed to get work, including a period at Penn's famous workshop in Greenwich; he also continued to frequent a local library, and was introduced, through

---

[1] S.C. on *Exportation of Machinery*, 1841, First Report, Evidence of W. Jenkinson, Qs. 1,301 and 1,305. In the Leeds, Bradford, Bingley, and Keighley area, engineering was much less developed: the number of firms was not stated, but the total number of 'hands' was 5,000; capital, £407,000; horse-power, 442.

[2] An Economic History of Modern Britain (1950 reprint), Vol. I, pp. 154 and 448–9.

[3] Probably for two reasons: firstly because of the existence of the interesting *Life*, written partly by himself and edited by William Pole; and secondly because the firm's business records have not survived.

[4] See above, p. 466.

[5] It is interesting to note that many leading engineers, such as Smeaton, Maudslay, Fairbairn, and Nasmyth, were interested in astronomy, often making their own instruments, telescopes, orreries, etc. They probably learnt a good deal in this way about precision engineering and mathematical calculation.

a scientifically minded clergyman named Hall, to the Society of Arts and to Alexander Tilloch, editor of the well-known *Philosophical Magazine*. After about two years in the metropolis, he set off to broaden his experience still further in Bath, Bristol, and other parts of the south-west, then travelled to Dublin, where he worked on nail-making machinery, before returning to England and settling in Manchester towards the end of 1813. As we have seen, he worked there first for Adam Parkinson and then for Thomas Hewes, from whom he acquired valuable knowledge of machine-making, millwork, water-wheels, etc.[1] At the same time he continued his studies in natural philosophy, geometry, and algebra, as well as history, etc., with the 'examples of Franklin, Ferguson, and Watt . . . always before me'. Fairbairn, indeed, was yet another outstanding example of 'self-education', and of the considerable intellectual powers displayed by many millwrights of his day, to which he himself attested,[2] and which, as we shall see, were to be outstandingly demonstrated in his career.

At the end of 1817, Fairbairn joined with James Lillie, a fellow-workman, in a millwrighting partnership, which began in a small shed in High Street, with a single lathe, powered by a muscular Irishman. They immediately achieved success with millwork manufactured first for Adam and George Murray and then for McConnel and Kennedy, fellow Scots, who were now among the leading Manchester cotton-spinners.[3] As we have seen, Fairbairn may perhaps have exaggerated the novelty of the light wrought-iron mill-shafting and smaller pulleys which he introduced, in place of the earlier cumbrous wooden shafts and drums;[4] but he and Lillie certainly played a leading role in the 'revolution' in mill construction, and before many years had passed they could justifiably claim to be 'the leading millwrights of the district'.[5] By 1830, moreover, the experimental researches of Fairbairn and Eaton Hodgkinson were also beginning to bear practical fruit in cast-iron construction of factories and also of bridges.[6] Orders for millwork flowed in, not only from manufacturers in Britain, but also from the continent, and by the early 'thirties they were, according to Andrew Ure, 'celebrated over the world'. Before their time, Ure stated, the science of mill architecture was little understood, and mills were built by those who were 'often utterly ignorant of statics or dynamics, or the laws of equilibrium and impulse'. But Fairbairn investigated the

[1] See above, pp. 69–71.　　　　　　　　[2] See above, pp. 73 and 429.
[3] For their earlier careers, see above, p. 440.
[4] See above, pp. 446 and 462–3.
[5] *Life*, pp. 113–14. This statement was supported by A. Ure, *Philosophy of Manufactures* (1835), pp. 32–7. See also *ibid.*, pp. 266 and 273, for their improvements in silk-throwing mills.　　　　[6] See above, pp. 117–18.

materials, sizes, and proportions of the various components of millwork, and was soon able to undertake the entire construction of large textile factories:

> The capitalist has merely to state the extent of his resources, the nature of his manufacture, its intended site, and facilities of position in reference to water or coal, when he will be furnished with designs, estimates, and offers of the most economical terms, consistent with excellence, according to a plan, combining elegance of external aspect, with solidity, convenience, and refinement in the internal structure. As engineer he becomes responsible for the masonry, carpentry, and other work of the building, for the erection of a sufficient power, whether of a steam-engine or water-wheel, to drive every machine it is to contain, and for the mounting of all the shafts and great wheels [gears] by which the power of the first mover is distributed.[1]

Two of the most outstanding examples of such factories were Orrell's mill at Stockport, driven by twin steam engines, and Ashworth's at Egerton, powered by 'a gigantic water-wheel of sixty feet diameter, and one hundred horses' power'.[2] Machine-tools enabled such millwrighting firms to mass-produce shafting and gearing: 'One millwright establishment in Manchester [Fairbairn's?] turns out from three hundred to four hundred yards of shaft-geering every week, finely finished, at a very moderate price, because almost every tool is now more or less automatic, and performs its work more cheaply and with greater precision than the hand could possibly do.'[3]

Fairbairn and Lillie acquired a great reputation for their water-wheels, in the construction of which, as of millwork, Fairbairn continued and improved upon the work of his former employer, Thomas Hewes.[4] Particularly famous were their four great breast-wheels, each 50 ft. in diameter and $10\frac{1}{2}$ ft. wide and each developing 120 horse-power, built in 1824–7 for Buchanan and Finlay's celebrated cotton mills at Catrine in Ayrshire,[5] followed by similar wheels for the same proprietors' Deanston mills. On the continent, similarly, they erected large water-wheels for Hans Caspar Escher's cotton mills at Zürich, which led to numerous other orders for mills in Switzerland, France, and other countries.[6]

[1] Ure, *op. cit.*, p. 33.
[2] *Ibid.*, pp. 33–5 and 109–12. Fairbairn and Lillie also constructed ventilating fans for cotton factories (*ibid.*, p. 220).
[3] *Ibid.*, p. 37.   [4] See above, pp. 69–71, 98, 445–7.
[5] *Mills and Millwork* (2nd edn., 1864–5), pp. 91–2, 129–32; *Life*, pp. 121–3. See also above, p. 70, n. 2.
[6] *Life*, pp. 123–9. For Escher and his relations with Fairbairn, see W. O. Henderson, *Industrial Britain under the Regency* (1968), pp. 2–7.

This rapid growth of business necessitated expansion of their works capacity, so Fairbairn and Lillie moved first (1818) to an old building in Mather Street, and then, in 1821, to Canal (or Cannel) Street, Ancoats,[1] which was later greatly expanded and equipped with steam-powered tools and a foundry. By 1830 they had apparently accumulated a balance of £40,000 and were employing 'upwards of 300 hands'.[2] Up to that time they had concentrated almost entirely on millwork and water-wheels, but were venturing into new fields, including iron shipbuilding.[3] Here again, Fairbairn proceeded in a practical-scientific manner, with trials in 1830–1 of small iron steamboats on the Forth and Clyde Canal and on the rivers Irwell and Mersey, with the results worked out mathematically.[4] His tests on the strength of iron frames ribs, plates, and rivets for shipbuilding were closely related to his work on iron beams and pillars in factory and bridge construction, and just as he played a leading role in revolutionizing millwork, so too he pioneered in the development of iron steamships. The first such ship, built by Aaron Manby, had only been launched in 1821, and at the time when Fairbairn began these experiments there was still great uncertainty as to the suitability of iron for ship construction.

The boats which he first built, in the early 1830s, were made in sections at the Manchester works,[5] but these were all fairly small, flat-bottomed, stern paddle-wheelers, for use on canals and rivers or along the coast, and Manchester was hardly suitable for building larger ships. His partner, Lillie, was dubious about this extension of the firm's activities, and also about a cotton-mill venture at Egerton, so in 1832 the partnership was dissolved, Fairbairn continuing the Canal Street works on his own. In the following years he expanded his activities

---

[1] According to a lease dated 19 Oct. 1821, kindly loaned by Mrs. E. Brickell. Fairbairn states, however, in his *Life* that their business continued to be conducted in Mather Street until the orders for the Catrine water-wheels were received. They were certainly in Canal Street by 1825 (*Manchester Guardian*, 19 Feb. 1825).

[2] *Life*, p. 129. On p. 314, however, a balance of £30,000, is stated. In the following year a large part of the works and its contents were destroyed by fire, damage being estimated at £7,000–8,000; but the premises had only been insured for £4,000 (*Manchester Guardian*, 13 Aug. 1831).

[3] For details, see his *Life*, chaps. IX, X, and XIX, and his *Treatise on Iron Shipbuilding* (1865).

[4] See his *Remarks on Canal Navigation, illustrative of the advantages of the use of Steam as a Moving Power on Canals* (1831).

[5] The *Manchester Guardian* proudly proclaimed Manchester's emergence as a shipbuilding centre, e.g. 26 Feb. and 2 April 1831 (the *Lord Dundas*, built for the Forth–Clyde Canal and tried out on the Irwell); 5 May 1832, quoting from the *Glasgow Chronicle* (referring to the *Cyclops* built twelve months earlier, and a new one, the *Manchester*, both for the Forth–Clyde Canal and coastal trade); 16 March 1833 (*La Reine des Belges*, built for the Ostend–Bruges canal); 4 April 1835 (iron steamboats for the Ouse and Humber trade between Selby and Hull).

prodigiously. In 1835 he established a shipbuilding concern at Millwall, on the Thames, where in the next thirteen years 'upwards of a hundred vessels' were built, both for home and overseas customers, including the Admiralty. By about 1840 at the two works, in Manchester and Mill-wall, he claimed to be employing a total of over 2,000 hands.[1] The Millwall business, however, proved unprofitable and eventually had to be sold, in 1848, with losses totalling over £100,000. There were many pioneering difficulties to cope with, there was fierce competition, and the Thames was a less favourable location for iron shipbuilding than Merseyside, Clydeside, etc., while Fairbairn's personal attention was divided between Manchester and London.

Meanwhile, the Manchester business had continued to prosper, as Fairbairn not only carried on his earlier millwork and water-wheel construction, but also developed a very large manufacture of steam engines and boilers, in which he patented improvements such as the riveting machine (1837) and the 'Lancashire boiler' (1844), and carried out further experiments on strength of materials and safety measures.[2] Steam engines, of course, were now rapidly displacing water-wheels, and they were also required for steamships.[3] A natural development from these engineering activities was locomotive building, which he commenced in about 1838. According to his *Life*, more than 600 loco-motives were built at the Manchester works, for both home and foreign railways.[4] But E. L. Ahrons, who made a detailed investigation of Fairbairn's locomotive building, found this figure much too high: 'by the time the works were closed they had constructed rather more than 400 locomotives', of which about 90 per cent. were built for United Kingdom railways and the remainder for various European, Indian and colonial railways.[5] Fairbairn also continued to construct

[1] *Life*, p. 157. But see below, p. 485.

[2] *Ibid.*, chap. XVI and p. 316. See also *A Century's Progress* (British Engine, Boiler & Electrical Insurance Co. Ltd., 1954). For a view of the inside of Fair-bairn's works in 1861, showing steam-engine beams, cylinders, and boilers, together with a Fairbairn crane and plate-punching machine, see Chaloner and Musson, *Technology and Industry*, plate 123.

[3] The 600-h.p. engines for the naval frigates *Megaera* (2,000 tons) and *Odin*, for example, were made in Manchester (*Records of Manchester*, 1868, under date 17 June 1846, referring to a visit to the works by Ibrahim Pasha, Viceroy of Egypt). [4] *Life*, pp. 316–17.

[5] E. L. Ahrons, 'Short Histories of Famous Firms, No. II, W. Fairbairn and Sons, Manchester', *The Engineer*, 20 Feb. 1920, pp. 184–5. (See also *ibid.*, p. 357, for an additional French locomotive.) Ahrons had actually listed 392 locomotives built by the firm between 1839 and 1862. More recent research has shown that his list was not complete, but P. C. Dewhurst has expressed the view that 'the missing ones cannot well have been more than twenty-five or so' (*The Locomotive*, 15 March 1930, pp. 80–1, in an article referring to a Brazilian locomotive by Fairbairn's).

iron bridges, many of them for railways, his most famous, and most controversial, achievement in this field being the Britannia (Menai) and Conway wrought-iron tubular bridges, built in conjunction with Robert Stephenson; by 1870 he claimed to have built nearly a thousand on this principle, with spans varying from 40 to 300 feet.[1] He also built a large number of cranes, especially those of a wrought-iron tubular construction, which he patented (1850). In all these works he carried out further experimental researches, stemming from his early interest in strengths of materials, based on similarly scientific procedures, and making considerable use of applied mathematics.

Operations on this scale necessitated mechanized mass-production and careful factory organization. These features particularly impressed a visitor to the Manchester works in 1839:[2]

In this establishment the *heaviest* description of machinery is manufactured, including steam engines, water wheels, locomotive engines, and mill geering. There are from 550 to 600 hands employed in the various departments;[3] and a walk through the extensive premises . . . affords a specimen of industry, and an example of practical science, which can scarcely be surpassed. In every direction of the works the utmost *system* prevails, and each mechanic appears to have his peculiar description of work assigned with the utmost economical subdivision of labour . . . Smiths, strikers, moulders, millwrights, mechanics, boiler makers, pattern makers, appear to attend to their respective employments with as much regularity as the working of the machinery they assist to construct.

They were aided by power-operated lathes, planing machines, and riveting machines—the latter a product of a boilermakers' strike, resulting from the introduction of labourers into that department; Fairbairn and his works manager, Robert Smith, had then devised this machine, 'which superseded the labour of 45 out of the 50 of his boiler makers'.[4] With such machine-tools, a great variety of steam engines

---

[1] *Life*, chap. XIII.

[2] Love and Barton, *Manchester As It Is* (1839), pp. 210 ff.

[3] It was also mentioned that in the Millwall establishment 'upwards of 400 hands are employed in the manufacture of steam engines, and in the building of iron steam boats, and other vessels constructed of the same material'. In the *Report on the Sanitary Condition of the Labouring Population of Great Britain* (8vo edn., 1842), p. 250, Fairbairn is said to have employed a total of 'between one and two thousand workpeople' in his 'manufactories of machinery'. (He was giving evidence on drink and sobriety among the working classes.)

[4] See also his *Life*, pp. 163–4. Fairbairn himself, as a workman, had run foul of the Millwrights' Society in London (see above, p. 480), and early in the development of the Canal Street works in Manchester Fairbairn and Lillie experienced a 'turn-out' of millwrights against their taking on an additional apprentice, and therefore advertised for forty to fifty workmen 'unconnected with the "Manchester Millwrights' Club"' (*Manchester Guardian* and *Manchester Gazette*, 19 Feb. 1825). In 1845, however, we find Fairbairn addressing a boiler-

were produced, ranging from 8 to 400 horse-power; one of the latter weighed 200 tons or more and cost £5–6,000. Huge castings were also made in the foundry: 'Castings of 12 tons weight are by no means uncommon: the beam of a 300 horses' power steam engine weighs that amount . . . A fly-wheel, for an engine of 100 horses' power, measures in diameter 26 feet, and weighs about 35 tons.' Fairbairn's works had produced 'some of the largest water-wheels ever manufactured', including one of 62 ft. diameter, and 'the heaviest mill-geering'. The works consumed 60 tons of cast and wrought iron weekly, or 3,120 tons annually. It had an immense stock of wooden patterns 'worth many thousand pounds'. An immense international trade was carried on:

This extensive concern forwards its manufactures to all parts of the world . . . *this* article is for Calcutta, *that* for the West Indies; this for St. Petersburgh, that for New South Wales: and there are, besides, men belonging to it *located* in various parts of Europe, who are employed, under the direction of Mr. Fairbairn, in superintending the erection of work manufactured on these premises.

A speaker at an Anti-Corn-Law meeting in Bolton at that time, referring to the exportation of machinery, even went so far as to say that 'Mr. Fairbairn, of Manchester, did little for [sale in] England; and the men at his works were constantly employed in making machinery to be sent abroad'.[1] This was a great exaggeration, but Fairbairn undoubtedly did develop a large export trade. An outstanding example was his work for the Turkish Government in the early 1840s, including the construction of iron furnaces, forges, and rolling mills, engineering workshops and machine-tools, a large woollen mill, mainly of iron construction, powered by a 100 horse-power water-wheel, together with steam-powered cotton and silk mills.[2] To these we can add a large water-powered factory at Gefle in Sweden, and also engineering work

---

makers' delegate meeting in Manchester, in a spirit of compromise and moderation, though he was opposed to strict apprenticeship regulations and any idea of fixing prices and wages (*Manchester Guardian*, 22 March 1845). In 1851–2 the firm was involved in the great engineers' strike (*Life*, pp. 322–7). Together with other engineering employers (e.g. Nasmyth, see below, pp. 505–7), Fairbairn refused to tolerate the restrictions which skilled craftsmen were trying to enforce at a time when machine-tools were rendering traditional handicraft skills largely obsolete (cf. above, pp. 477–8). He was also concerned to resist any trade-union restrictions on employers' authority, or any 'socialist doctrines'.

[1] *Manchester Guardian*, 27 April 1839.

[2] *Ibid.*, 22 March and 20 Dec. 1843; *Life*, chap. XI; Fairbairn's 'Description of a Woollen Factory erected in Turkey', *Min. Proc. Inst. Civ. Eng.*, Vol. III (1843), pp. 125 ff.

in Russia, all of which countries Fairbairn himself visited.[1] But he also continued his millworking activities in England, of which the huge steam-powered woollen mills built in the early 1850s for Titus Salt at Saltaire, near Bradford, were the supreme example.[2] At about this same period he was also engaged on important Government contracts, including work in the Portsmouth and Plymouth dockyards and the famous Small Arms Factory at Enfield.

The most notable feature of all these works, apart from their immense scale and variety, was Fairbairn's constant effort at combining applied science and technology. As he himself confessed, he was imbued with 'a strong desire to distinguish myself as a man of science',[3] and to remove 'the anomalous separation of theory and practice'.[4] These scientific-technological interests were amply demonstrated in his active membership of various scientific and engineering societies and in the enormous volume of his publications on structural engineering. Soon after he had established himself in Manchester he became a member of the Literary and Philosophical Society, to which he read several papers and of which he eventually became president in 1855–1860.[5] He also spoke frequently at meetings of the British Association, over which he presided in 1861.[6] We have already referred to the outstanding scientific-technological researches which he carried out with Eaton Hodgkinson, F.R.S., on the strength of materials, especially cast-iron beams and pillars, which were published in the Manchester Society's memoirs and in the British Association reports.[7] In 1850 he was elected a Fellow of the Royal Society and subsequently delivered several papers on different subjects of mechanical science; in 1860, therefore, he was awarded one of the Society's gold medals, 'for his various experimental enquiries on the properties of the materials

---

[1] *Life*, pp. 363–7. For more information about the cotton spinning and weaving mill which Fairbairn erected at Gefle, see P. Carlberg, 'Personal Contacts between the Manchester area and Gefle in Sweden a Hundred Years Ago', *Lancs. & Chesh. Antiq. Soc. Trans.*, Vol. LXX (1960), pp. 57–63. This includes a photograph of the huge water-wheel, 43 feet in diameter and 11 feet wide.

[2] *Life*, pp. 327–8; a more detailed description is given in Fairbairn's work *On the Application of Cast and Wrought Iron to Building Purposes* (1854).

[3] *Life*, p. 157.

[4] Presidential address to British Association, Manchester, 1861.

[5] He also enjoyed informal 'philosophical and scientific discussions' at home in the evenings, with friends such as Eaton Hodgkinson, Bennett Woodcroft (scientific adviser to the Patent Office), and James Nasmyth (*Life*, pp. 155–6).

[6] In 1853 and 1862 he was president of the Mechanical Section. In his addresses on all these occasions he particularly emphasized the progress and importance of 'mechanical science'.

[7] See above, pp. 77, 117–18 and 481. See also R. W. Bailey, 'The Contribution of Manchester Researches to Mechanical Science', *Excerpt Minutes of Proceedings of the Meeting of the Institution of Mechanical Engineers in Manchester, 25th June 1929*.

employed in mechanical construction, contained in the "Philosophical Transactions", and in the publications of other scientific societies'. He joined the Institution of Civil Engineers in 1830, and in 1847 was among the founding members of the Institution of Mechanical Engineers, of which he was elected president in 1854-5 and to whose *Transactions* he contributed many papers. His scientific-technological eminence was also recognized by the award of honorary degrees at Edinburgh (1860) and Cambridge (1862), and earlier, in 1852, by his election as a corresponding member of the National Institute of France, not only on account of his practical engineering achievements, but also, as Poncelet the great French engineer put it, for his 'récherches expérimentales entreprises à vue d'éclairer la science de construction'.

Fairbairn's immense interest in 'mechanical science' or 'practical science' is demonstrated not only in the numerous papers which he read to these various learned bodies, but also by his activities at a lower level in the Manchester Mechanics' Institute and the Manchester School of Design.[1] In addition to helping in their foundation and management, he frequently gave lectures both to them and to other groups of intelligent working men. Many of these lectures are gathered in his *Useful Information for Engineers* (1856, 1860, 1866). He also published many other works, among which the following are outstanding: *On the Application of Cast and Wrought Iron to Building Purposes* (1856, 1857, 1864); *Iron: its History, Properties, and Processes of Manufacture* (1861, 1865, 1869); *Treatise on Mills and Millwork* (1861-3, 1864-5); *Treatise on Iron Shipbuilding* (1865); and a host of papers, mostly concerned with the strength of materials (especially cast and wrought iron), and with steam engines and boilers.[2] In all this literature, as in his own engineering constructions, there was a blend of practical engineering and applied science, the necessity for which Fairbairn constantly reiterated. On account of the limitations of his own theoretical knowledge, he frequently secured the collaboration of scientific men, such as Eaton Hodgkinson and later Thomas Tate, mathematics master at Battersea Training College, and W. C. Unwin, subsequently professor of engineering at the Royal East Indian Engineering College, Cooper's

---

[1] He played a leading role in establishing the former, to which he also lectured: see M. Tylecote, *The Mechanics' Institutes of Lancashire and Yorkshire before 1851* (Manchester, 1957), pp. 34 (n. 7), 43, 129-30. He was also one of the founders of the Manchester School of Design: see *Manchester Guardian*, 21 Feb. 1838 and 6 Sept. 1843. And he participated in an attempt to establish an 'Institution of Practical Science': *ibid.*, 27 March 1839. It is worth noting that Fairbairn's interest in establishing these institutions was shared by other Manchester engineers such as Richard Roberts, James Nasmyth, George and Joseph Peel, etc.

[2] See the long list in the Appendix to his *Life*.

Hill. He also enjoyed contacts and friendships with many scientists, both British and foreign, who clearly held him in high esteem.[1]

An equally outstanding engineer—fellow-Scot, contemporary, and friend of Fairbairn—was James Nasmyth, best known, of course, for his invention of the steam hammer in 1839, but also one of the most versatile engineers of his day in the invention and improvement of machine-tools.[2] For such a career, he had a very favourable family background, education, and training, especially by comparison with other engineers, such as Roberts and Fairbairn, who were of humble parentage and 'came up the hard way', with varied practical experience in different ironfounding and engineering workshops, and were mainly self-taught. Nasmyth was born in 1808 of a long line of substantial Edinburgh builders and architects; his father, Alexander Nasmyth, was a reputable Scottish portrait and landscape painter and architect, while his mother, Barbara Foulis, came of a landed aristocratic family.[3] Nasmyth's father was also a keen amateur mechanic, and had assisted Patrick Miller and William Symington in developing their steamboat, which first steamed on Loch Dalswinton in 1787; according to James Nasmyth, he invented the 'bow-and-string' bridge and compression riveting (later patented by Fairbairn and Smith of Deanston). He had a workroom in his house, equipped with lathes and other tools, which he taught his son to handle. It was from his father, too, that young James obtained his ability and early training in draughtsmanship, the

[1] Fairbairn withdrew after 1853 from active management of the business, leaving it mainly to his sons, and concentrated on his writing and consultancy work. The firm was later converted into a limited company, but lost its early pioneering drive and was wound up after Fairbairn's death in 1874, when general trade depression threatened. It may be, as Ahrons suggested, 'that the firm had too many irons in the fire', but it would also seem that Fairbairn was too much concerned in his later life with being a public figure and establishing his status as a scientist and writer, and that his sons were less able men. Many of the works buildings are still standing today, having been occupied for many years by Messrs. Parker (Ancoats) Ltd., a timber firm.

[2] In addition to the general histories of engineering previously cited, see his *Autobiography* (ed. Smiles, 1883); Smiles, *Industrial Biography* (1876 edn.), pp. 275–98, and articles in the *Engineer*, 23 May 1856, 9, 16, and 23 May 1890, 18 and 25 Sept. 1908, 19 March 1920, and 23 May 1941. Nasmyth has a place in the *D.N.B.*, and there are also some brief notes about him in a pamphlet by H. Richardson, *James Nasmyth: A Note on the Life of a Pioneer Engineer* (1929), and in an address delivered by C. A. Gibb to the Watt Club, Edinburgh, 16 Jan. 1943, on *James Nasmyth, Engineer, 1808–90*. Many of Nasmyth's technical drawings of machine-tools, steam-hammers, etc., together with his rough sketchbook, are still preserved in the Library of the Institution of Mechanical Engineers, where I have been kindly allowed to examine them. A considerable collection of Nasmyth records has now been built up in Eccles Public Library. The following account is preparatory to a book on Nasmyth.

[3] Nasmyth's maternal uncle was Sir James Foulis.

importance of which he strongly emphasized in his *Autobiography*. He also enjoyed the conversation of his father's scientific as well as artistic friends, including Sir James Hall,[1] Professors Playfair and Leslie, Dr. Brewster, and others.

Nasmyth's formal education began unhappily in a private school, and was then continued at the Edinburgh High School, which he entered at nine years of age. Here he was bored by the 'rote and cram' system of 'classic learning so called', and left after three years. He acquired much more knowledge from frequent visits to an iron foundry and chemical laboratory owned by the fathers of two school friends, where he gained valuable practical experience. His education being continued privately, he was able to enjoy arithmetic and geometry, whilst he constantly practised drawing or occupied himself in his father's workshop, making his own tools and chemical apparatus. By the time he was seventeen, he was making model steam engines, including one for the Edinburgh School of Arts and another for Professor Leslie, of Edinburgh University, for use in his lectures on natural philosophy, which young Nasmyth was allowed to attend; the Professor also often conversed with him on mechanics, dynamics, etc. Another fruitful friendship was with Robert Bald, the famous Scottish mining engineer, who, in addition to his long practical experience 'had a large acquaintance with literature and science'.

From 1821 to 1826 Nasmyth attended evening classes at the newly founded Edinburgh School of Arts, 'our first technical college', where he studied mechanical philosophy, geometry, mathematics, and chemistry, and had access to a library of scientific books. At the same time, with money from his model engines, he purchased tickets of admission to similar day-time classes in the University. He also continued his workshop activity and, having access to the smithy and iron-foundry of George Douglass, began to make larger steam engines, and even a steam road carriage for the Scottish Society of Arts. He also visited many local works to study and draw steam engines.

Thus Nasmyth combined theory with practice in a remarkable training, which was finally capped with an appointment in 1829 as Henry Maudslay's private assistant, after he had impressed the 'Great Mechanic' with his machine drawings and a model engine. He spent two invaluable years with Maudslay, where his knowledge of machine-tools, steam engines, etc., and of precision engineering with use of true plane surfaces, measuring-machine, and gauges, was enormously increased; whilst there, Nasmyth invented (in 1829) his first machine-

---

[1] It was after his father's old friend, President of the Edinburgh Royal Society, that young Nasmyth was named James Hall.

tool, for cutting square or hexagonal nuts or bolt-heads, by means of a revolving file or cutter, an early example of a milling machine.[1] He also made the acquaintance of the famous scientist, Michael Faraday, thus beginning a friendship which lasted till Faraday's death.

In February 1831, however, Maudslay died and Nasmyth then determined to set up in business for himself. He returned first of all to Edinburgh, where, in a temporary workshop, he made a set of machine-tools—lathe, planing machine, boring and drilling machines, etc.—preparatory to establishing himself in either Liverpool or Manchester. He had visited Lancashire in the previous year, to see Stephenson's 'Rocket', and had also seen some of the large engineering establishments, such as those of Fawcett, Preston & Co., in Liverpool, and Sharp, Roberts & Co. in Manchester; he was particularly impressed by the machine-tools of Roberts, whom he regarded as 'one of the true pioneers of modern mechanical mechanism'. He 'found that the demand for machine-making tools was considerable, and that their production would soon become an important department of business'.[2] The opening of the Liverpool–Manchester Railway and the subsequent construction of other lines had largely increased this demand. At the same time, the scarcity, exorbitant wage demands, irregularity, and carelessness of skilled engineering workers 'gave an increased stimulus to the demand for self-acting machine-tools'.[3]

It was in this branch of engineering, therefore, that Nasmyth decided to establish himself, starting in an old factory flat in Dale Street, Manchester, in 1834, as 'a mechanical engineer and machine-tool maker'. He brought down from Edinburgh his outfit of machine-tools, but otherwise had little capital. In fact he had only £63 in cash, but was proffered a loan by Daniel Grant, the cotton manufacturer of Ramsbotham, and also by Edward Loyd, the Manchester banker, who soon recognized his potential.[4] But he did not need much capital, because the returns were quick: sometimes he got one-third payment in advance and settlement on delivery, so keen was the demand for his self-acting machine-tools: 'planing machines, slide lathes, drilling, boring, slotting machines, and so on'. He also made printing machines for the Cowper brothers, small steam engines, etc.

Orders 'poured in'. It was not long, therefore, before Nasmyth had to seek larger premises. He decided to build a big works at Patricroft, near Manchester, on land leased by himself and his brother George from Thomas Joseph Trafford, Esq., and George Cornwall Legh, Esq., very

---

[1] *Autobiography*, pp. 145–6 and 409; R. S. Woodbury, *History of the Milling Machine* (1960), pp. 24–6. [2] *Ibid.*, p. 182.
[3] *Ibid.*, p. 199–200. [4] *Ibid.*, pp. 184–91.

favourably situated at the junction of the Liverpool–Manchester Railway and the Bridgewater Canal, from which the works derived its name, the Bridgewater Foundry.[1] To secure additional capital for the undertaking, he formed a partnership with Holbrook Gaskell (who assisted him in the administrative side of the business), formerly of the iron-merchanting and nail-making firm of Yates and Cox, in Liverpool, and with Messrs. Birley & Co., the large Manchester cotton-spinning and manufacturing firm, which had spare capital to invest outside the textile trade. The firm was named Nasmyths, Gaskell & Co.[2]

The Birleys withdrew from the partnership in 1838, but not, as Dickinson suggests,[3] because of the burning-down of Macintosh & Co.'s water-proof clothing factory, in which they also had a substantial interest.[4] The fire was on 25 August 1838,[5] while a new partnership was previously formed by Nasmyths, Gaskell & Co. on 30 June of that year.[6] The Birleys' places as sleeping partners were taken by Henry

---

[1] *Autobiography*, chap. XI.

[2] The original deeds concerning these leases and the articles of co-partnership appear not to have survived, but their terms can be deduced from later legal documents now in the Lancashire Record Office, e.g. co-partnership agreements of 30 June 1838 (L.R.O., DDX 260/1), 25 Feb. 1843 (L.R.O., DDX 260/3), and 11 June 1850 (L.R.O., DDX 260/6), and an indenture of 30 June 1850. There is also in Eccles Public Library a schedule of 12 Dec. 1930 listing all the deeds (land leases and co-partnership agreements) then in the possession of Nasmyth, Wilson & Co., dating back to the firm's origin, with brief details of their contents.

From this evidence, it can be established that on 31 Dec. 1835 land covering an acreage of 13,536 sq. yds., bounded by the Liverpool and Manchester Railway, the Bridgewater Canal, and Green Lane, Patricroft, was leased from Thomas Joseph Trafford Esq., by James and George Nasmyth; the land was assigned from 25 March 1836, for a term of 999 years at an annual rent of £112 16s. od. On 30 May 1836 an adjoining area of 5,030 sq. yds, was leased for the same period from George Cornwall Legh, Esq., at a yearly rental of £36 13s. 6½d. And finally, on 31 Oct. 1836, an agreement was made for the lease of two more adjoining plots, containing 46,463 sq. yds., which were not, however, assigned until 25 June 1841, on a 999-year lease, for a yearly rent of £345 14s. od.

A letter dated 11 July 1836 from James Nasmyth to Holbrook Gaskell (still with Yates and Cox in Liverpool) indicates that the articles of co-partnership were then being drawn up by George Humphrys, a Manchester solicitor.

Information regarding these transactions, especially on Holbrook Gaskell's entry into the partnership, is also to be found in R. Dickinson, 'James Nasmyth and the Liverpool Iron Trade', *Transactions of the Historic Society of Lancashire and Cheshire*, Vol. 108 (1956), pp. 83–104.   [3] *Op. cit.*, p. 101.

[4] T. Hancock, *Personal Narrative of the Origin and Progress of the Caoutchouc or India-Rubber Manufacture in England* (1857), pp. vi and 81–2. In Pigot's *Manchester and Salford Directory* of 1838, the addresses of Messrs. Birley & Co. and Charles Macintosh & Co. are the same, 14 Back George Street and Cambridge Street, Chorlton-on-Medlock. Nasmyths, Gaskell & Co. also had a Manchester address at 14 Back George Street, as well as at Patricroft.

[5] *Records of Manchester* (1868), under the year 1838.

[6] Co-partnership agreement, L.R.O., DDX 260/1.

Garnett, a Manchester merchant, and George Humphrys, the Manchester attorney, to whom Nasmyth had been originally introduced by Edward Loyd, the banker, with whom he had consulted regarding the establishment of the Bridgewater Foundry.[1] As to how much capital was invested in the firm, there is no evidence, but the relative proportions are indicated by the clause concerning division of profits in the 1838 co-partnership agreement: after payment of 5 per cent on share capital, the residue of profits was to be divided into 18 equal parts, of which James Nasmyth was to receive 4, George Nasmyth 4, Holbrook Gaskell 4, Henry Garnett 3, and George Humphrys 3.[2]

The capital thus provided enabled the Bridgewater Foundry to be built on an impressive scale. Beginning with just under four acres in 1836, the firm acquired land totalling nearly fourteen acres by 1841. The works, however, did not cover this entire area. According to a report in the *Manchester Guardian* of 18 June 1845: 'The whole of the [factory] premises cover an area of about eight acres, besides a space of about six acres more, on which are erected the cottages of overseers or foremen, and workpeople.'[3] This report included a rough plan of part of the works, including a ball furnace, brass-foundry, boiler and engine houses, grinding shed, and iron-foundry, in a north-south line, with a smithy on the eastern side and dressing shop on the western, all lying between the Bridgewater Canal and Green Lane;[4] in a line to the south of the foundry (but not shown on the plan) were 'the planing and heavy turning shops, and the large and handsome five stories building seen from the [Liverpool–Manchester] railway'. These buildings are

---

[1] *Autobiography*, p. 203.

[2] This agreement also reveals that the two Nasmyths and Gaskell were 'to devote their whole time to the business and . . . not carry on any other', whereas Garnett and Humphrys were clearly sleeping partners.

It may be mentioned here that George Nasmyth withdrew from the business in 1843 (agreement of 25 Feb. 1843, L.R.O., DDX 260/3); Humphrys in 1848 (agreement of 30 June 1848, referred to in schedule of 12 Dec. 1930 in Eccles Public Library); and Gaskell in 1850 (agreement of 11 June 1850, L.R.O., DDX 260/6, and deed of 30 June 1850, referred to in the above schedule). After 1850, therefore, James Nasmyth and Henry Garnett were the only remaining partners, and Nasmyth was in sole active control of the business.

George Nasmyth became a consulting engineer in London, while Holbrook Gaskell switched to the chemical industry, joining with Henry Deacon (who had trained as an engineer at the Bridgewater Foundry) in the firm of Gaskell, Deacon & Co. at Widnes.

[3] See also *Autobiography*, pp. 206 and 216, where Nasmyth refers to the building of houses for the foremen and workmen. In Love and Barton, *op. cit.*, p. 214, it was stated that 'the greater part' of the workmen 'live in cottages which the proprietors have erected for their accommodation'.

[4] A considerable part of these buildings had been blown down by a terrible boiler explosion.

strikingly illustrated in the sketch of the Bridgewater Foundry in Nasmyth's *Autobiography.*

Some very interesting light is thrown on Nasmyth's engineering ideas by surviving letters which he wrote to his partner Holbrook Gaskell while the Bridgewater Foundry was building.[1] These clearly show that Nasmyth was keenly alive to the possibilities of standardized production in advance of orders. 'These are indeed glorious times for the Engineers,' he wrote, for he was overwhelmed with customers.

I never was in such a state of bustle in my life such quantity of people come knocking at my little office door from morning till night . . . The demand for work is realy quite wonderfull and I will do all in my power to bring about what I am certain is the true view of the Business viz. to have such as planing machines and lathes, etc., etc., all ready to supply the parties who come every day asking for them. If we had such a stock ready made we could in every instance sell them at 20 per cent better prices and please the parties much more and avoid all sort of unpleasant work of hurrying on the work. I could fill 20 pages with my views on this head . . . for I am quite up in the clouds about the prospect it opens. We could at once take the lead in the Business if we put it into force. It is *now* as foolish to wait for orders for such machines as to wait for orders for a ton of iron bars and tell the parties when they apply to you that you must have 4 months as the ore has to be got out of the ground before you can supply and that you will write to the mine to smelt their order. Such machines are now become as much articles of current demand as files or anything else of that nature.[2]

In a letter to Gaskell on the following day, Nasmyth again enthused on this subject. 'What a noble business we might become if we only could establish the *ready made* concern. Depend [on it] it's the true view of the business, and we would make twice the returns with not $\frac{1}{2}$ the annoyance arising from being made to promise time for Delivery with the continual risk of Disapointing the parties.'[3]

It seems clear from these letters that the idea of having ready-made machine-tools was a novel one, though other types of machinery, especially textile machines, for which there was a huge demand, had been mass-produced for some years before this date. Nasmyth put his idea into practice in the new Bridgewater Foundry (started in 1836, completed in 1837), and instead of waiting for orders sent out printed catalogues describing the various types, sizes, and prices of machine-tools that were available for sale to customers. Three such catalogues

---

[1] These letters were very kindly loaned by Mr. R. H. Gaskell (grandson of Holbrook Gaskell) of Wheatstone Park, Codsall Wood, near Wolverhampton, who has since deposited them in Eccles Public Library.

[2] Letter dated 11 July 1836.   [3] Letter dated 12 July 1836.

have survived, one of 1839, another of 1849, and a third of the early 'fifties.[1]

This mass-production of machine-tools (and later, as we shall see, of other machinery) necessitated planned factory lay-out, and Nasmyth appears to have been one of the pioneers of assembly-line production. In one of his letters to Holbrook Gaskell while the Bridgewater Foundry was being built, he proposed that the buildings should be *'all in a line* . . . In this way we will be able to keep all in good order'.[2] What Nasmyth intended by this plan is clearly shown in a description of the works which appeared in a little-known booklet, *Manchester As It Is*, published in 1839.[3]

With a view to secure the greatest amount of convenience for the removal of heavy machinery from one department to another, the entire establishment has been laid out with this object in view; and in order to attain it, what may be called the straight line system has been adopted, that is, the various workshops are all in a line, and so placed, that the greater part of the work, as it passes from one end of the foundry to the other, receives in succession, each operation which ought to follow the preceding one, so that little carrying backward and forward, or lifting up and down, is required . . . By means of a railroad, laid through as well as all round the shops, any casting, however ponderous or massy, may be removed with the greatest care, rapidity, and security.

The whole of this establishment is divided into departments, over each of which a foreman, or responsible person, is placed, whose duty is not only to see that the men under his superintendence produce good work, but also to endeavour to keep pace with the productive powers of all the other departments. The departments may be thus specified:—The drawing office, where the designs are made out; and the working drawings produced . . . Then come the pattern-makers . . . next comes the Foundry, and the iron and brass moulders; then the forgers or smiths. The chief part of the produce of the last named pass on to the turners and planers . . . Then comes the fitters and filers . . . in conjunction with this department is a class of men called erectors, that is, men who put together the framework, and the larger part of most machines, so that the last two departments . . . bring together and give the last touches to the objects produced by all the others.

It was stated that 'nearly one uniform width is preserved throughout all the workshops', namely 70 ft., the height of each being 21 ft. 'The total length of shops on the ground floor . . . amounts, in one line, to nearly 400 ft.', while the upper floors of the five-storied building were each 100 ft. long, 60 ft. wide, and 12 ft. high. At that time the ground floor of this building was occupied by the foundry, containing four

[1] Now in Eccles Public Library. See below, p. 499, for their contents.
[2] Letter dated 11 July 1836 (Nasmyth's italics).
[3] Love and Barton, *Manchester As It Is* (1839), pp. 213-19.

cupolas, with a combined capacity of 36 tons; from these the molten iron was carried by rail, in a 'great cauldron, or pot', holding 6 or 7 tons, to any part of the foundry, and lifted off by a massive crane, for making castings.

This extremely interesting description of the Bridgewater Foundry shows that Nasmyth had clearly recognized the advantages of line-production and a smooth flow of work. We are also fortunate in still having a sketch dated 1840, among the surviving Nasmyth drawings,[1] of the inside of what appears to have been the erecting shed, where the various components were being finally assembled to form the finished product, in this case locomotive engines,[2] which are shown pro-gressing in line down the shed. Nasmyth sought by every means to save time and labour. In addition to railroads, he made considerable use of cranes and blocks and pulleys for lifting and moving heavy work.[3]

These methods enabled Nasmyth to turn out a large quantity and variety of machine-tools. It is impossible in the space available here to give a detailed account of these and of the various improvements which Nasmyth introduced.[4] As he himself stated, 'each new tool that I constructed had some feature of novelty about it',[5] but 'I patented very few of my inventions. The others I sowed broadcast over the world of practical mechanics.'[6] Most of the patents which he took out were, in fact, not for machine-tools, but for other mechanical im-provements.[7] But it is as a tool-maker that Nasmyth was pre-eminent.

---

[1] In the Institution of Mechanical Engineers, London.

[2] See below, pp. 503-4, for locomotive manufacture.

[3] Love and Barton, *op. cit.*, pp. 216-17; *The Engineer*, 23 May 1856, pp. 280-2. For an earlier example of similarly 'scientific' factory planning, with 'flow' production, see E. Roll, *An Early Experiment in Industrial Organization* (1930), pp. 169-88, referring to Boulton and Watt's plans for the Soho Foundry about forty years previously.

[4] Nasmyth described many of them in the Appendix to his *Autobiography*. A number of the firm's drawings are preserved in the Institution of Mechanical Engineers. See also Buchanan's *Essays on Millwork* (3rd edn, 1841), Appendix B, p. 411 *et seq.* The late Dr. H. W. Dickinson has given a good description of some of Nasmyth's machine-tools in *The Engineer*, 23 May 1941, pp. 337-9.

[5] *Autobiography*, p. 190.   [6] *Ibid.*, p. 439.

[7] These were related mainly to his interests in ironfounding and forging and in the manufacture of steam engines and railway locomotives, as the following chronological list illustrates: hardened steel (instead of brass) bearings or journals for locomotives, steam engines, etc. (1839, no. 8,023); self-acting railway-carriage brakes (1839, no. 8,299); atmospheric railways, in conjunction with Charles May (1844, no. 10,358); independent driving of factory machines, using small high-pressure engines and piped steam (1849, no. 12,675); improve-ments in the making and adjustment of calico-printing rollers (1850, no. 13,261, in conjunction with John Barton); hydraulic presses for making bricks, tiles, etc., in conjunction with Herbert Minton (1851, no. 13,608, and 1856, no. 948); re-versible rolling-mill (1853, no. 1,680); steam-puddling of iron, foreshadowing

We have previously noticed the nut-shaping or milling machine which he invented while at Maudslay's; another invention of the same period was his spiral-wire drive for small drills, as later used by dentists.[1] In 1836, at the time when the Bridgewater Foundry was building, he made two of his major advances in machine-tool design. He greatly improved on Roberts' key-grooving machine, for cutting 'key grooves' or mortices in the eyes of metal wheels and pulleys, whereby they could be 'keyed' or wedged on driving shafts; since the previous upright type of machine could only take wheels of limited radius in its 'jaws', Nasmyth rearranged its form, fixing the wheel on a flat table, with the cutting tool working from below, instead of above, so that wheels of any size could be cut.[2] Prior to this improvement, key-grooves in larger wheels had still to be cut 'by the laborious and costly process of chipping and filing'. Nasmyth also invented the 'lever planing machine', or shaping machine, using a contrivance which became generally known as 'Nasmyth's steam arm', for planing the smaller detailed parts of machinery, both flat and curved, whether exterior or interior.[3] Previously planing machines had been used mainly 'in the execution of larger parts of machine manufacture', and 'a very considerable proportion of the detail parts still continued to be executed by hand labour', with chisel and file, at great cost yet with considerable inaccuracy, even when highly skilled workmen were employed.

---

the Bessemer process (1854, no. 1,001); hydraulic or steam crane for ironworks (1854, no. 2,744); rolling and planishing tin-plates, in conjunction with James Brown (1856, no. 1,308); hydraulic pumps and presses for packing cotton, etc. (1856, no. 2,577, in conjunction with Robert Wilson).

In addition to these inventions, Nasmyth also introduced many other improvements in these same branches of manufacture, which he did not patent, e.g. a method of propelling canal boats and barges by means of a steam engine and submerged chain (1825); an instrument for measuring expansion of metals (1826); a 'superheater' in steam engines (1827); improvements in the pistons and air-pumps of steam engines (1836); self-adjusting bearings for the shafts of machinery (1838); a safety-ladle for foundry work (1838); a double-faced wedge-shaped sluice-valve for water mains (1839); a universal flexible joint for steam and water pipes (1843); an improvement in blowing fans (1844); a direct-action steam-powered suction fan for ventilating coal mines (1844); an improved method of welding iron (1845); introduction of the V-anvil (1845); an improved safety-valve for boilers (1847); hydraulic punching of holes in thick iron bars and plates, at the request of Faraday (1848); steam-powered drilling of rocks (1854). These inventions, described at the end of Nasmyth's *Autobiography*, were only 'the more prominent' ones; to have described all of his minor 'contrivances' would, he said, 'have required another volume'.

[1] *Autobiography*, pp. 408–9.

[2] *Ibid.*, pp. 409–11; *Engineer*, 23 May 1890, p. 426, and 23 May 1941, p. 338. This was the only machine-tool patented by Nasmyth (in 1838, no. 7,815), apart from the steam-hammer and pile-driver.

[3] *Autobiography*, pp. 415–16; *Engineer*, 23 May 1941, p. 338.

Nasmyth's 'steam arm', controlled merely by 'an intelligent lad', could perform such work far more quickly, accurately, and cheaply.

Another of his improvements, in 1837, was a gear-changing contrivance for reversing the action of slide-lathes, so that the turning tool could cut either towards the head stock or away from it.[1] In 1840, in order to facilitate production of twenty locomotives for the Great Western Railway, Nasmyth devised further improvements in his machine-tools, including another 'contrivance' to enable segmental work on the eccentrics of these engines to be turned on ordinary lathes.[2] 'These tools ... rendered us more independent of mere manual strength and dexterity, while at the same time they increased the accuracy and perfection of the work ... At the same time they had the important effect of diminishing the cost of production ...' They could also be used for manufacturing other types of machinery, and were added, of course, to Nasmyth's machine-tool specialities for sale to other engineering firms.

From 1842 onwards, as we shall see, Nasmyth's energies were concentrated mainly on developing and producing steam-hammers and pile-drivers, which may be regarded as massive machine-tools. In the late 'forties his fertile genius was producing more improvements, notably a slotting or grooving drill, invented in 1847, for cutting cotter slots in piston-rods, etc., and keyways in driving shafts.[3] Again, this transformed a process previously requiring much laborious hand-chiselling and filing to a purely mechanical one by a 'self-acting' machine, two of which could be managed by 'an intelligent lad'. This machine and another, a double or 'ambidextrous' turning lathe, brought out in 1850,[4] were particularly mentioned by *The Engineer* in 1856 as 'two of the most interesting of Mr. Nasmyth's new tools'.[5] Particular reference was also made to his paring or shaping machines, and to various features of the 'very great variety of tools' to be seen in his workshops.[6]

The excellence of Nasmyth's designs and construction is testified by the fact that in the third edition of Buchanan's *Essays on Millwork* (1841) almost as large a number (twenty-one) of his machine-tools were

[1] *Autobiography*, p. 417.   [2] *Ibid.*, pp. 237–8 and 424–5.
[3] *Ibid.*, pp. 432–3; *Engineer*, 23 May 1941, p. 338. This is referred to in drawings and sales catalogues as Nasmyth's 'patent grooving machine', but he actually had no patent for it.
[4] *Autobiography*, p. 436; *Engineer*, 23 May 1941, pp. 338–9. Again the term 'patent' is erroneously applied to this machine in the Nasmyth drawings.
[5] *Engineer*, 23 May 1856, pp. 280–2.
[6] In 1839 there were 'fifty-six turning lathes, of all sizes, at work in this establishment', and planing machines were 'extensively used'. Love and Barton, *op. cit.*, pp. 217–18.

figured as those of all other makers put together (thirty-three). Their variety is also strikingly demonstrated in the machine-tool catalogues issued by the firm in 1839, 1849, and the early 1850s.[1] Since there was no very striking change in the kinds of machine-tools listed between 1839 and 1849, except for the famous steam-hammer and pile-driver,[2] and since the prices were not actually entered in the 1839 list, we will briefly survey the contents of the later one. There were eight ordinary planing machines, of various sizes, capable of taking work-pieces up to 20 ft. long and 4 ft. square, and a few fitted with Nasmyth's 'steam arm', for shaping small detailed parts of machinery; they ranged in price from £75 to £270 and more.[3] Next came the 'nut-cutting and facing machine' (£65), three machines 'for screwing bolts and tapping nuts' (£65–£130), and a 'large wheel dividing apparatus . . . for cutting the teeth of either wood or iron wheels' (£270). A dozen drilling machines were listed, ranging from a small portable hand-drill, costing £10, to a large machine with a 36 ft. bed and three sliding tables for £330. Three types of boring mill were supplied, to bore cylinders of up to 12 ft. diameter; only the price of the smallest (£150) was given. A large number of turning and slide lathes—about a dozen sizes of each, with head-stocks from 6 in. to 30 in.—were listed, together with lathe parts (head-stocks, chucks, beds, and slide-rests), for turning shafts, wheels, etc., cutting screws, and boring. Then came seven key-grooving (or slotting) and paring machines ranging from £55 to £600 (for a 17½-tonner) and more; no price was given for the largest, a great 35-ton machine. A number of boiler-making machines were also manufactured, for bending boiler plates (£160) and for punching and shearing (£55–£350). At the end of the list came a hydraulic press (£150) and miscellaneous items such as boring braces, screw jacks, foundry ladles, furnace fans, and double-faced sluice cocks for water mains.[4]

All these machine-tools were 'self-acting'. Their standardized mass-production—for use in making standardized machinery for the mass-

[1] *Prices of Engineering and Other Tools, Manufactured by Nasmyths, Gaskell & Co., Engineers, Tool Makers, & Manufacturers of Locomotive Steam Engines, Bridgewater Foundry, Patricroft, near Manchester* (1839); *Engineering and Other Tools, Manufactured by Nasmyth, Gaskell & Co., Engineers, Tool Makers, Manufacturers of Locomotive Steam Engines, Hydraulic Presses, Patent Steam Hammers, Steam Pile Drivers, &c., Bridgewater Foundry, Patricroft, near Manchester* (1849). The list of the early 'fifties was exactly the same as that for 1849, with the name of the firm altered in manuscript from 'Nasmyth, Gaskell & Co.' to 'James Nasmyth & Co.'

[2] These will be dealt with separately below, pp. 500–3.

[3] The prices of all machines were not listed.

[4] Another very interesting invention of Nasmyth's, not included in these catalogues, is an early grinding machine. See R. S. Woodbury, *History of the Grinding Machine* (1959), pp. 42–4; Burstall, *op. cit.*, pp. 228–9.

production of standardized 'end products'—obviously heralded the modern machine age. There is little doubt that Britain led the way in the manufacture of such machine-tools. That the U.S.A. were inferior in this department was revealed by the reports of Commissions which visited America in the early 1850s.[1] Whitworth found that 'engine tools' there were 'similar to those in use in England some years ago, being much lighter than those now in use, and turning out less work in consequence'.[2] The Ordnance officials also considered that American machine-tools 'were generally behind those of England'.[3] On the other hand, the Americans led the way in the mechanization of many manufactures, especially of light metal goods.[4] Nasmyth testified to their superiority in the mass-production of small arms, for example, after visiting the factory established by Colt at Pimlico in the early 'fifties, where he found 'perfection and economy such as I have never seen before'. He contrasted the American innovating energy with the 'traditional notions and attachment to old systems' that were widespread in England.[5]

Nasmyth himself, in designing his machine-tools, altered or discarded 'traditional forms and arrangements', emphasizing the need for simplicity, direct-action, and utility.[6] He also paid careful attention to scientific principles. In his 'Remarks' on machine-tools in Buchanan's *Essays on Millwork*, for example, he pointed out that the design and setting of cutting tools had been inadequately studied, without any attempt 'to reduce the subject to ... plain and general principles'. These deficiencies he began to remedy, and as the late Dr. H. W. Dickinson has observed, Nasmyth 'was the first, so far as we are aware, to analyse the angles that a cutting tool must embody, and to design a gauge to enable a tool to be ground to these correct angles'.[7]

His grasp of basic principles and rejection of 'traditional forms' were triumphantly demonstrated in his famous invention of the steam-hammer.[8] Earlier types of steam-powered hammers had merely followed the form of the old water-powered tilt or helve hammers, in

---

[1] *New York Industrial Exhibition: Special Reports of Mr. George Wallis and Mr. Joseph Whitworth (Parliamentary Papers*, 1854, Vol. XXXVI). *Report of the* [Ordnance Department's] *Commission on the Machinery of the United States* (P.P., 1854–5, Vol. L).   [2] Whitworth, *op. cit.*, p. 112.

[3] *Report on Machinery*, p. 578.   [4] See Burn, *op. cit.*

[5] *Report of Select Committee on Small Arms* (P.P., 1854, Vol. XVIII), Q. 1,367. See also Nasmyth's *Autobiography*, pp. 362–3.

[6] *Autobiography*, p. 439.   [7] *Engineer*, 23 May 1941, p. 338.

[8] Here only a brief account will be given of this, Nasmyth's most famous invention. It is a complicated story, with rival claims by Schneider and Bourdon, of Le Creusot Ironworks in France, and also by Robert Wilson, works manager at the Bridgewater Foundry. Here it can only be said that Nasmyth's claims to be

which the tail of the hammer was 'tripped' by cams fitted around a shaft rotated by a water-wheel, whereby the head was raised, to fall by force of gravity; the water-wheel was simply replaced by a steam engine. Nasmyth, however, eliminated all the cumbersome gearing and cams and fixed the hammer head directly to the end of the piston-rod of an inverted steam engine.[1] Thus, while simplifying the whole structure, he added considerably to the speed and force of the hammer blows and, by increasing the gap between hammer head and anvil, permitted much larger forgings to be worked; at the same time, the action of the hammer could be precisely controlled. This invention was motivated by an enquiry from Francis Humphries, engineer of the Great Western Steamship Company, in regard to the difficulty of forging the massive paddle-shaft of Brunel's *Great Britain*. Nasmyth's original sketch of 24 November 1839 is still preserved in the Institution of Mechanical Engineers, London. The idea had to be temporarily abandoned, however, owing firstly to a switch from paddle to screw propulsion for the *Great Britain*, and secondly to the serious trade depression of the late 'thirties and early 'forties, when ironmasters and engineers were unwilling to invest money in new equipment. His sketch, however, was shown to Messrs. Schneider and Bourdon, of Le Creusot Ironworks, when they visited the Bridgewater Foundry,[2] and in April 1842, whilst himself visiting Le Creusot, he was astonished to find there a steam-hammer based on his original idea. Returning to England, he at once took a patent[3] and built his own first hammer which was put to work in his Bridgewater Foundry towards the end of 1842 or early in 1843.[4] Some difficulty was at first experienced in making the hammer 'self-acting', but this was solved by Robert Wilson, the works manager,[5] and the first sales were made to the Low Moor

---

the original inventor are borne out by contemporary evidence, but that Robert Wilson was mainly responsible for making the hammer self-acting. The full story will be told in the forthcoming book on Nasmyth.

[1] This idea was not new, having been propounded in Watt's patent of 28 April 1784 (no. 1,432) and William Deverell's of 6 June 1806 (no. 2,939). But neither of these engineers appears to have actually constructed such a hammer, while Nasmyth was apparently ignorant of their patents.

[2] Nasmyths, Gaskell & Co. gave free access to foreign visitors, believing that sight of their various machine-tools would encourage sales; Nasmyth also showed them ideas in his 'Scheme Book'. Holbrook Gaskell let Schneider and Bourdon examine the steam-hammer sketch while Nasmyth was away from the works. This statement in Nasmyth's *Autobiography*, p. 245, is confirmed by letters from Gaskell and was actually admitted by Schneider.   [3] In June 1842, no. 9,382.

[4] It was ultimately disposed of in 1845 to Muspratt & Co. of the Newton-le-Willows chemical works, where it was used for breaking stones.

[5] Nasmyth, however, took out a patent for this in his own name (1843, no. 9,850).

Ironworks, of Bradford; Rushton and Eckersley, and Benjamin Hick & Son, of Bolton; Hawks, Crawshay & Sons, of Gateshead; John Penn, of Greenwich, and other leading ironfounding and 'engineering firms. Their successful operation in these works, and booming trade in the middle 'forties, rapidly boosted sales to many other establishments, including Government arsenals and dockyards, both at home and abroad.[1] Nasmyth wrote on 22 July 1844 to Daniel Gooch, engineer of the Great Western Railway, who was considering purchasing a steam-hammer for their works at Swindon:[2] 'We are quite throng making them for all quarters. The demand is rising rapidly, already we are at No. 22 having made all sizes from 5 ton in the hammer block down to 1 cwt. . . . it is now quite self-acting and thumps away from 220 blows per minute to any slow rate and intensity.'

The railway 'mania' and the growth of iron shipbuilding greatly increased sales. The hammer proved particularly useful for making large forgings by stamping masses of hot iron into dies or moulds. Nasmyth also applied the steam-hammer principle to the construction of pile-drivers,[3] which came into use mainly for dock construction. Another application was to the hewing, splitting, and dressing of stones.[4] Even after the economic crisis of 1847, orders kept up very well in the subsequent depression, and as trade recovered in the early 'fifties, demand rose even more sharply.

We are like to be smothered with Steam Hammers at the works . . . I never remember such an in rush of orders for Hammers & pile drivers since the enclosed list was printed about 3 months ago.[5] About 40 more [orders for] Hammers have come in and most of them gone out to[o]. We never were so busy with them before and what is most gratifying is that most of them are for folks . . . who have had them from us before, each coming back for more and more and bigger and bigger ones.[6]

As a visitor from *The Engineer* reported in 1856, 'steam-hammers are an article of constant demand at the Bridgewater Foundry, and I

[1] This is evident from the firm's surviving order books and also from an advertising leaflet issued towards the end of 1852, *James Nasmyth and Co.'s Patent Steam Hammers and Pile Drivers. Testimonials to their Efficiency* (in Eccles Public Library), which listed all sales to that date. The date of this pamphlet is confirmed by a letter from Nasmyth to George Loch, 4 Feb. 1853 (Patent Office Library). A complete list of all the steam-hammers made at the Bridgewater Foundry, compiled by Mr. R. Arbuthnott, a former Director of Nasmyth, Wilson & Co., is now in the Eccles Public Library.

[2] J. Nasmyth to D. Gooch, 22 July 1844, British Transport Commission Archives, London, H.R.P. 1/8.

[3] *Autobiography*, chap. XV. He took out a patent for his steam pile-driver in 1843 (no. 9,850).　　　　　　　　　　　[4] Patented in 1843 (no. 10,413).

[5] See above, n. 1, for the list referred to.

[6] J. Nasmyth to G. Loch, 4 Feb. 1853 (Patent Office Library).

observed the frames of a good many in various stages of completion'.[1] They were, in fact, mass-produced and advertised in catalogues together with lathes, planing machines, etc.[2] In the 1849 catalogue, for example, thirteen sizes were listed, with hammer-blocks varying from 3 to 80 cwt., at prices from £150 to £700.[3] By the end of 1856, when Nasmyth retired, a total of about 490 steam-hammers had been sold.[4] Pile-drivers were not in such great demand: by the end of 1852 it appears that only thirteen had been sold, both at home and abroad.[5]

Machine-tools, steam-hammers, and pile-drivers were not the only products of the Bridgewater Foundry. Nasmyth had been greatly impressed by the possibilities opened up by the growth of railways. Whilst still a youth, he had constructed a road steam-carriage,[6] and in 1830 he had travelled from London to Liverpool especially to see Stephenson's 'Rocket', and was present at the opening of the Liverpool–Manchester Railway on 15 September, which was followed by a 'rage' for railway building.[7] Naturally, therefore, Nasmyth soon turned his attention to the manufacture of railway locomotives, for which there was a great demand and for which the Bridgewater Foundry, with its equipment of machine-tools, was 'peculiarly adapted'.[8] It appears that Nasmyth had this manufacture in mind at the time of the foundry's establishment, for it was stated in 1839 that, 'Besides the manufacture of every description of engineers' tools, another branch of business for which this establishment has been erected, is that of locomotive engines. . . .'[9]

The firm's first locomotive, built in 1838-9 and appropriately named 'Bridgewater', was an experimental one, which was tried out on the Liverpool and Manchester Railway in the latter year and eventually disposed of to J. Waring, contractor for the Manchester and Birmingham line, in April 1841. The first orders were from the London and Southampton and the Manchester and Leeds Railways—three for each

---

[1] *Engineer*, 23 May 1856, p. 282.

[2] In the lists of 1849 and the early 'fifties, and probably earlier.

[3] Excluding the cost of anvil and base-plate, and of boiler and boiler mounting, which varied from £30 to £200 and from £80 to £195 respectively, for hammers with blocks from 3 to 60 cwt.; prices for these accessories to hammers with 70- and 80-cwt. blocks were not listed.

[4] According to Mr. R. Arbuthnott's list, based on the firm's order books.

[5] According to the advertising list then issued.

[6] See above, p. 490.

[7] *Autobiography*, pp. 155-7. His sketch of the 'Rocket', dated 13 Sept. 1830, is preserved in Edinburgh Public Library.

[8] *Ibid.*, p. 237. There is space here only for a brief survey of the firm's locomotive building, which will be examined in detail in the forthcoming book on Nasmyth.

[9] Love and Barton, *op. cit.*, p. 218.

line—delivered in 1839.[1] In the early 'forties the firm was very busy manufacturing locomotives for various railway companies—the Midland Counties, the Birmingham and Gloucester, the Manchester and Leeds, and especially the Great Western, which ordered twenty in September 1840; for these, as we have seen, Nasmyth contrived special machine-tools.[2]

After delivery of the last of these engines in October 1842, there was a break in the firm's locomotive building, probably because of the slackening in railway demand. This was the period when manufacture of the steam-hammer began. Towards the end of 1844, however, with the commencement of the great 'railway mania', an order was received from Robert Stephenson & Co. for fifteen locomotives for the Dover or South-Eastern line. The foundry was kept very busy in the middle 'forties on these engines and on others for the London and Birmingham Railway. Demand slackened somewhat after the crisis of 1847, but between then and the middle of 1853 work was continued, though at a more leisurely pace, on substantial orders for the York, Newcastle and Berwick Railway and the Great Northern Railway. After completion of the latter's engines, however, locomotive building at Patricroft ceased until the early 'sixties.

Altogether, between 1838 and 1853 a total of 109 locomotives was built. These, it will be noticed, were almost entirely for home railway companies. In 1841 two were made for the Kaiser Ferdinand Nordbahn Railway in Austria, and two of the engines manufactured for Robert Stephenson & Co. in 1844–6 went to the Paris–Orleans Railway. Not until much later, however, from the late 'seventies onwards, did the Bridgewater Foundry begin to develop a large manufacture of locomotives for foreign railways.

Another speciality of Nasmyth's was his manufacture of small high-pressure steam engines, which he made in considerable numbers for a variety of purposes, but especially for the direct driving of machines, instead of by shafting and gearing from one large engine.[3] He also specialized in the production of small pumping engines for feeding boilers. We even find Boulton, Watt & Co. ordering from Nasmyth 'one of your little steam engines and pumps for feeding boilers, which Mr. Blake was informed by Mr. Nasmyth you had *always*

---

[1] A complete list of Nasmyth locomotives, compiled by Mr. R. Arbuthnott from the order books, is now in Eccles Public Library. It is more accurate than the similar list in the Stephenson Locomotive Society's library. See also S. Rendell, 'The Steam Locomotive Fifty Years Ago and Now', *Transactions of the Manchester Association of Engineers* (1906), pp. 5 and 8.

[2] See above, p. 498.

[3] *Autobiography*, pp. 313–15.

ready'.[1] When supplying such a pumping engine to the Great Western Railway Company in the same year, Nasmyth stated that 'We make a vast number of them for feeding Boilers . . .'[2] It is clear that these engines were being mass-produced by Nasmyth, in advance of orders, like his machine-tools. He similarly turned out a large number of hydraulic presses (worked by his steam pumping engine), for which he took out patents in 1851 and 1856. In a description of the Bridgewater Foundry in the latter year we read of the assembly, ready for sale, of 'a regiment of donkey pumps all in a line . . . marshalled columns of ambi-dextrous lathes and grooving machines; hydraulic presses for making lead pipes; wrought-iron cranes for forges, etc.', in addition to steam-hammers and other machines.[3]

All these manufacturing activities required a large labour force, despite increasing mechanization. Already by 1839, only three years after the establishment of the works, 'about 300 men' were employed.[4] By 1845 there were 'about 420 hands at present employed in the estab-lishment',[5] and by the early 'fifties this figure had been greatly increased. Nasmyth stated before the Royal Commission on Trade Unions in 1868 that the highest number of workmen he ever employed was 1,500.[6] At first he had 'no difficulty in obtaining abundance of skilled workmen in South Lancashire and Cheshire', from the long-established metal-working trades,[7] but he soon experienced difficulties with the engineers' trade society when he began to train local labourers on the machine-tools. He believed in 'free trade in ability' and freedom of individual contract, and refused to put up with the union's apprenticeship and wage restrictions; he raised men according to their aptitudes and skills, and paid wages likewise, as against the union's aim of 'indolent equality'. Neither would he tolerate a trade-union 'closed shop'.[8] The result was a serious strike soon after the establishment of the Bridge-water Foundry.[9] Nasmyth succeeded, however, in breaking the strike

[1] Letter dated 13 April 1846, Boulton and Watt Collection, B.R.L. (Author's italics).

[2] Letter to Daniel Gooch, engineer of the G.W.R., dated 16 May 1846, British Transport Commission Archives, H.R.P. 1/8.   [3] *Engineer*, 23 May 1856.

[4] Love and Barton, *op. cit.*, p. 214.   [5] *Manchester Guardian*, 18 June 1845.

[6] *Royal Commission on Trades Unions, Tenth Report* (*P.P.*, 1868, Vol. XXXIX), Q. 19,137.   [7] *Autobiography*, pp. 214–15.

[8] *Ibid.*, pp. 217–18 and 222–8; *R.C. Report* (1868), Qs. 19,095–19,340.

[9] This 'Turn-Out of Mechanics' at the Bridgewater Foundry was referred to in the *Manchester Guardian*, 16 Nov. 1836, where it was stated to have been caused by the employment of 'two or three men whom Messrs. Nasmyths had brought from Scotland, but who not being members of the mechanics' club, became obnoxious to the other men, who demanded their discharge'.

According to Nasmyth's evidence before the Royal Commission on Trade Unions (1868 *Report*, Q. 19,108), the strike was 'caused by the employment of a

by importing sixty-four men from Scotland, who 'had been bred to various branches of mechanics. Some had been blacksmiths, others carpenters, stone masons, brass or iron founders; but all of them were *handy* men.'[1]

That Nasmyth's labour problems were not thereby completely solved, however, is evident from remarks by Charles Greville, who visited the Bridgewater Foundry in November 1845. He found that the men were getting 'from twenty to thirty-two shillings a week', but that they 'love to change about, and seldom stay very long at one place; some will go away in a week, and some after a day'.[2]

Nasmyth did take on some indentured 'premium' apprentices, but found them troublesome, and preferred 'to employ intelligent well-conducted young lads, the sons of labourers or mechanics, and advance them by degrees according to their merits'. The union restrictions and the high wages demanded by skilled engineering workers led to the more rapid introduction of 'self-acting' machine-tools, which 'displaced hand-dexterity' and could be worked by boys and labourers, at much lower wages; indeed several of these machines could be superintended by a single such worker.[3] But this ultimately led to further labour trouble, in 1851–2, with the newly founded Amalgamated Society of Engineers, which demanded 'that every machine must have a Union man to superintend it, and that he must be paid the full Union regulation wages'. Nasmyth thereupon combined with other engineering employers to crush these demands and a great strike took place, as a result of which he further increased the numbers of his 'self-acting' machines and unskilled workers:[4] 'instead of having the old proportion of one boy to four mechanics, I had four boys to one mechanic nearly'.[5]

Nasmyth was thus an engineer of the type lauded by Andrew Ure

---

man who had not served a regular apprenticeship to the trade', but whom the firm refused to discharge. The strike lasted three months, gravely imperilling the Bridgewater Foundry (Qs. 19,111–12).

[1] *Autobiography*, p. 226; *R.C. Report* (1868), Q. 19,112. Relatives followed them, so that the majority of workers in the Bridgewater Foundry became Scottish (Q. 19,180).

[2] *Greville Memoirs, 1814–60* (1938), Vol. V, p. 239. According to Nasmyth's evidence in 1868, in the early years 'we were paying for skilled mechanics somewhere about 22s. or 24s. a week, [but] when I left off it was more like 36s. a week' (*R.C. Report* (1868), Q. 19,305). A boy of fourteen or fifteen would be started at 3s. a week; a semi-skilled labourer would get 15s. to 21s. a week (Qs. 19,133 and 19,175–6).

[3] *Autobiography*, pp. 307–9; *R.C. Report* (1868), Qs. 19,133–4 and 19,175.

[4] *Autobiography*, pp. 309–11; *R.C. Report* (1868), Qs. 19,133–7. For the general background, see J. B. Jeffereys, *The Story of the Engineers* (1946), chaps. i and ii.   [5] *R.C. Report* (1868), Q. 19,134.

and Samuel Smiles, staunch advocates of mechanization, opponents of craft demarcations, and high priests of individual enterprise and 'self-help'. He believed in encouraging effort by paying wages and bonuses according to ability. He took particular care to choose capable foremen and to give then an interest in increasing output. Robert Willis, for example, was engaged, in 1852, from Kirkcaldy, as general works manager, at an annual salary of £200, rising to £225 in his second year, and £250 in the third, with a rent-free house;[1] a year later, in order 'to encourage and reward you for increased exertion & give you a direct interest in the results of your zeal', it was agreed that he should have a bonus of £1 per £1,000 of output (sales) up to £50,000 and double this rate above that figure.[2]

In these various ways, Nasmyth sought to improve both mechanical and human efficiency. According to his evidence before the Royal Commission on Trade Unions, 'self-acting' machine-tools, worked by 'boys and labourers', enabled him to reduce his labour force by 'fully one-half'.[3] But he was deterred from further expansion by trade-union restrictions. If increased profits had been invested in plant, he could have employed four or five hundred or even a thousand more workers and expanded the business, but he declined to do so because of the trouble he had had with the unions and the additional trouble which expansion might entail. He therefore invested his profits in three per cent. Consols.[4]

Labour troubles appear to have been mainly responsible for Nasmyth's early retirement, despite his victory over the union in 1851–2.

I was so annoyed with walking on the surface of this continually threatening trade union volcano . . . that I was very glad to give it up and retire from the business . . . at least 10 years before the age at which I would otherwise have retired . . . I heard the distant rolling of thunder, and saw that there was a storm coming on, and so I made a compromise between my desire to be very rich and my desire to be very tranquil and comfortable, and resigned the business into the hands of my sleeping partner, and appointed a practical manager to take my place.[5]

[1] J. Nasmyth to R. Willis, 29 April 1852.
[2] Ditto, 1 June 1853. By a letter of 26 May 1854 his basic salary was raised from £250 to £300, with the same bonus arrangement.
[3] *R.C. Report* (1868), Q. 19,137.
[4] *Ibid.*, Q. 19,139.
[5] *Ibid.*, Qs. 19,222 and 19,299. The partnership was dissolved at the end of 1856 and a new one was formed between Henry, Charles, and Robert Garnett and Robert Wilson. The latter had been works manager at the Bridgewater Foundry in 1843–5, before going to the Low Moor Ironworks, Bradford.

Pressure of business also contributed to Nasmyth's early retirement, which he planned several years in advance. Thus we find him writing in 1851 to George Loch,

I have never been so pulled about to all quarters of the compass as I have been this winter on business affairs which have not only taken me so often from Friends but have cut deep into the little time I can devote to Hobbeys but as I trust its all in a good cause and may tend to realize 'the cottage in Kent' and bring 1856 pleasantly visable to the naked eye I do not grumble.[1]

At the end of 1856 he did, in fact, retire, to a house which he named 'Hammerfield', at Penshurst, in Kent. In twenty years he had made a fortune. When he died in 1890, he still had nearly a quarter of a million pounds[2]—testimony to his engineering genius, his business judgement, and the tremendous profits to be made in those pioneering days from machine-tools, steam-hammers, and railway locomotives. This wealth might also be said to reflect his ruthless drive and exploitation of labour. But this would have to be balanced by the employment and trade which his energies created, and by the reduced costs brought about by his machine-tools and new methods of production: lower prices meant that 'both the public and the working man partake of the benefit'.[3]

Appropriately, just before he retired, when *The Engineer* decided to make 'A Tour of the Provinces', its first visit was to the famous Bridgewater Foundry at Patricroft.[4] Some of the machine-tools then seen were still working in the shops of Nasmyth, Wilson & Co. in 1920,[5] and many of the buildings are still standing today as part of a Royal Ordnance factory. While these survivals may be a criticism of later failure to modernize, they are also evidence of the excellence and enduring quality of Nasmyth's designs.

His achievements were based on an early training and education which closely combined theoretical knowledge and practical experience. The latter was not enough: it had to be illuminated by a rational understanding of mechanical principles, as is evident from his 'Remarks' in Rennie's edition of Buchanan's *Practical Essays on Millwork*.[6] That he was no mere practical mechanic is also apparent from his friendship with Faraday, Herschel, and other scientists. He continued to maintain his

---

[1] J. Nasmyth to G. Loch, March 1851 (Patent Office Library). Similarly on 4 Feb. 1853 he wrote of his 'toiling away here single handed' (he was now the sole active partner), and being 'nailed to the [work]shops', but comforted himself with the thought that booming trade would 'yield results that will come in handy for "the cottage in Kent" '.

[2] According to the probate, dated 6 Aug. 1890, in Somerset House.

[3] *R.C. Report* (1868), Q. 19,193.   [4] *Engineer*, 23 May 1856.

[5] *Ibid.*, 19 March 1920.   [6] See above, p. 500.

scientific interests and delivered many papers to the Manchester Literary and Philosophical Society, the British Association, and the Royal Society.[1] Some of these interests—notably astronomy—were outside engineering,[2] and Nasmyth himself emphasized the importance of 'plain common sense' in designing mechanical structures,[3] but he was obviously something more than a first-rate practical mechanic. In the broad sense, he was undoubtedly a 'scientific engineer'. He was also a pioneer of 'scientific management'. In his ideas of factory lay-out, work 'flow', standardization, and mass-production, he was one of the leading creators of the modern machine age. That he was not so successful in coping with the consequent human problems is perhaps not surprising, when we today are still grappling with trade-union 'restrictive practices' and the social effects of 'automation'.

[1] See the list of his papers in the Royal Society's *Catalogue of Scientific Papers, 1800–1900* (19 vols., 1867–1925). They were mainly on astronomy, but included several on metallurgy, geology, etc.

[2] His interest in chemistry, however, may possibly have contributed something to his knowledge of metallurgy. That he retained this early interest is shown by a letter of 11 Jan. 1849 (Patent Office Library), in which he thanked James Loch for the loan of a book by Bunsen, the German chemist.

[3] *Autobiography*, p. 439.

Subject Index

Aberdeen, 53, 256–9, 292, 476
Académie Royale des Sciences, 194, 252–3, 261, 310
Adelphi Society, London, 164
Agriculture, application of science (botany and chemistry) to, 18, 25, 26, 31, 55, 82, 128, 130, 152, 178, 195, 234, 235, 236, 381, 291; machinery developed for, 63, 68, 374, 378–9, 390, 391, 471–2; water-power used in, 39, 68
Albion mills, London, 24, 208
Alchemy, contributions to early development of chemistry, 2, 25, 29
Alkali. See Potash and Soda
Alum, manufacture and uses of, 36, 64, 168, 174, 322, 341
Amalgamated Society of Engineers, 485, n. 4, 506
America (U.S.A.), scientific-technological links and comparisons with, 62, 63, 64, 86, 191, 194, 246, 347, 446, 473, 478, n. 4, 500
Anderson's Institution, Glasgow, 181–2
Anglesey, Mona Vitriol Company of, 170, 246
Apothecaries. See Pharmacy
Architecture and Building, applied science and technology in, 10, 13, 18, 20, 22, 24–5, 27, 38–9, 42, 44–5, 73, 74–9, 100, 107, 110, 111–12, 115–18, 120, 122, 127, 141, 154, 164–5, 172, 174, 183, 372–92 (passim), 445–6, 463, 481–2, 487–8. See also Bridge-building, Cotton (factory construction), and Engineering
Askesian Society, 137
Assaying. See Metallurgy
Astronomy, study of, and the making

of astronomical instruments, 10–23 (passim), 28, 32–4, 40–2, 77, 78, 102, n. 3, 103, 105, 106, n. 7, 108, 110, 114, n. 8, 132, 139, 145, 148, 151–2, 153, 158, 160–1, 164–5, 166, 171, 172, 176, 480, 509
Austria, Birmingham artificers enticed to, 218, 222–4
Aylesbury, 384

Ballistics and gunnery, study of, 11–15 (passim), 23, 103, 161, 164–5, 177. See also Saltpetre and gunpowder
Balloons, ballooning, and pneumatic chemistry, 177–8, 191, 208, 275
Bath, 104, 164, 239, 319, 481; philosophical society at, 89, 194
Bedford, 384
Belper, 70–1
Berlin Academy of Sciences, 220
Beverley, 104
Bilston, 218
Birmingham, science and industry in, 55, 57, 68, 72, 88, 89, 104, 108, 142–7, 153, 163, 169, 181, 190–2, chaps. V and VI (passim), 239, 278, chap X (passim), 377, 382, 389, 430. See also Boulton, Watt, Lunar Society, and Steam Engine
Black-lead, 18
Blacksmiths. See Smiths
Bleaching, applied science (use of alkali, chlorine, lime, vitriol, etc.) in, 7, 61, 78, 79, 82–3, 84, 92–3, 111, 122, 124, 130, 133, 134–5, 137, 153–4, 174, 178, 186–7, 188, 231, 240–4, 246, n. 1, 249, chap. VIII, 342, 343, 345, 347–8, 349, 350; lectures on, 92–3, 111, 130, 174, 242, 329–30,

Name Index